PROCESS CHEMISTRY
FOR WATER AND WASTEWATER
TREATMENT

PROCESS CHEMISTRY
FOR WATER AND WASTEWATER
TREATMENT

LARRY D. BENEFIELD

Department of Civil Engineering
Auburn University
Auburn, Alabama

and

JOSEPH F. JUDKINS, JR.

Department of Civil Engineering
Auburn University
Auburn, Alabama

and

BARRON L. WEAND

Department of Civil Engineering
Virginia Polytechnic Institute
Blacksburg, Virginia

PRENTICE-HALL, INC., *Englewood Cliffs, New Jersey 07632*

Library of Congress Cataloging in Publication Data

Benefield, Larry D.
  Process chemistry for water and wastewater treatment.

  Includes bibliographies and index.
    1. Water—Purification.   2. Sewage—Purification.
3. Sanitary chemistry.   I. Judkins, Joseph F.
II. Weand, Barron L.   III. Title.
TD353.B43   628.1'66   81-8690
ISBN 0-13-722975-5   AACR2

Editorial production supervision
and interior design by: James M. Chege

Cover design by: Mario Piazza

Manufacturing buyer: Joyce Levatino

Printed in the United States of America

10 9 8 7 6 5 4 3 2

PRENTICE-HALL INTERNATIONAL, INC., *London*
PRENTICE-HALL OF AUSTRALIA PTY. LIMITED, *Sydney*
PRENTICE-HALL OF CANADA, LTD., *Toronto*
PRENTICE-HALL OF INDIA PRIVATE LIMITED, *New Delhi*
PRENTICE-HALL OF JAPAN, INC., *Tokyo*
PRENTICE-HALL OF SOUTHEAST ASIA PTE. LTD., *Singapore*
WHITEHALL BOOKS LIMITED, *Wellington, New Zealand*

CONTENTS

CONTENTS

PREFACE

Chemical processes are commonly employed at both water and wastewater treatment plants, and the design and/or operation of a particular process is normally based on a stoichiometric model. Yet, in many instances stoichiometric models greatly oversimplify the actual events which occur during a chemical change. In such cases, it is necessary to apply equilibrium or kinetic models to adequately predict the response of a chemical system to a specified set of environmental conditions.

Because of the complex nature of natural waters, the formulation of an equilibrium or kinetic model is generally a very difficult undertaking, especially for engineers with only limited skills in the area of water chemistry. When confronted with such problems, the engineer normally probes the published literature for guidance but often becomes overwhelmed by the rigorous treatment of the subject presented by most sophisticated water chemists.

A need exists in the environmental engineering area for a textbook which presents the subject of water chemistry in a manner that can be readily followed and understood by engineers who do not have strong backgrounds in chemistry. This book was written in an attempt to satisfy that need. It is a text written by engineers for engineering students to use in undergraduate or firstyear graduate courses in environmental engineering chemistry, or for practicing engineers to use as a reference when a basic understanding of the fundamentals governing the response of common chemical processes used in water and wastewater treatment plants is required. When used as a text, it will provide a firm foundation for engineering students who wish to pursue the theoretical and complex subject of aqueous chemistry, or when used as a reference by practicing engineers, it will provide the background necessary to understand the literature on topics not specifically covered in this book.

The contents of this volume are organized in a manner which requires the reader to first master certain basic chemical principles (Chapters 1–6). The application of these basic principles to the understanding of a number of chemical processes commonly encountered in water and wastewater treatment is then presented in Chapters 7–14. By use of such a format, repetition is minimized. However, it also means that the applications chapters are not complete in themselves but rather build upon the material contained in the fundamentals chapters. Hence, if the book is to be an effective instructional tool, the material contained in Chapters 1–6 must be thoroughly understood before the reader's attention is turned toward Chapters 7–14.

It was difficult to decide the specific ordering of Chapters 7 through 14, and the authors realize that some readers, because of personal preferences, might prefer a different sequence for the chapters than the one used here. However, there is a logical explanation for presenting the chapters in the way they have been. Since coagulation normally precedes all other chemical processes at a treatment plant, it is presented as the first applications chapter, which appropriately follows the chapter on fundamentals of surface and colloidal chemistry. Water softening was considered as an appropriate choice to follow coagulation. However, it is easier to introduce the use of Caldwell-Lawrence diagrams by applying them to water stabilization problems. As a result, water stabilization is discussed in Chapter 8, and a discussion of water softening and conditioning is postponed until Chapter 9. Because ion exchange is a popular process for water softening, the discussion on ion exchange equilibria is inserted as Chapter 10. Column behavior in ion exchange and adsorption processes are very similar, so the material on adsorption is felt to be appropriate for Chapter 11. Fluoride removal (Chapter 12) follows adsorption because adsorption is one of the most commonly employed techniques for removing fluoride ions from water. Iron and manganese removal would have been a logical choice to follow fluoride removal, since both are common to water treatment operations. However, the basic principle of both iron and manganese removal is oxidation to higher insoluble oxidation states. It was therefore decided to insert the chapter on applications of redox chemistry (Chapter 13) ahead of the chapter on iron and manganese removal (Chapter 14). The authors must apologize for the omission of any discussion on phosphorus removal. A Chapter covering this subject had been written but escalating production costs required that the chapter be removed.

A word of appreciation is due to Charlotte Burleson, Dawn Horne, and Gary Steed, who typed the manuscript for publication, and to the many graduate students who were particularly helpful in suggesting improvements to the original draft of the manuscript. Special appreciation is extended to Mary Benefield for editing the final version of the manuscript.

In conclusion, we extend our gratitude to Brown and Caldwell Consulting Engineers for permission to use the Caldwell-Lawrence Water Conditioning Diagrams found in this text.

LARRY D. BENEFIELD
JOSEPH F. JUDKINS
BARRON L. WEAND

PROCESS CHEMISTRY
FOR WATER AND WASTEWATER
TREATMENT

1

BASIC PRINCIPLES

1-1 CHEMICAL EQUATIONS

The purpose of a chemical equation is to express what happens during a chemical change. For example, the addition of sodium hydroxide to a solution of iron(III) chloride results in the precipitation of iron(III) hydroxide. Instead of writing an explanation of what happens during this particular chemical change, it is possible to convey the appropriate meaning through the use of chemical symbols presented in the form of a chemical equation:

$$FeCl_3 + 3NaOH \longrightarrow Fe(OH)_3 + 3NaCl \qquad (1\text{-}1)$$

Equation 1-1 shows that one molecule of iron(III) chloride will combine with three molecules of sodium hydroxide to form one molecule of iron(III) hydroxide and three molecules of sodium chloride.

Before a chemical equation can be written, it is necessary that the chemical changes which occur during the reaction be determined through experimentation. The reader is normally spared the laborious task of obtaining all these equations, however, since they are compiled in textbooks. When writing a chemical equation, it must be understood that atoms do not disappear during a chemical reaction. As a result, there must be the same number of atoms and same weight of material at both the beginning and the end of the reaction. This means that all chemical equations that are written must be balanced; i.e., all the materials that are shown on the left-hand side of an equation must also be shown on the right-hand side of the equation.

UNBALANCED:

$$FeCl_3 + NaOH \longrightarrow Fe(OH)_3 + NaCl \qquad (1\text{-}2)$$

The balanced form of equation 1-2 is given in equation 1-1.

1

Symbols commonly used for writing chemical equations are presented in Table 1-1. General rules for writing and balancing equations are listed in Table 1-2.

TABLE 1-1 Symbols commonly used in chemical equations.

Symbol	Meaning
$\longrightarrow$	Yields, produces (points to products)
$\rightleftharpoons$	Reversible reaction: equilibrium between reactants and products
$\uparrow$	Gas (written immediately after a substance)
$\downarrow$	Solid or precipitate (written immediately after a substance)
(s)	Solid (written after a substance) $CaCO_{3(s)}$
(g)	Gas (written after a substance) $CO_{2(g)}$
(aq)	Aqueous solution (substance dissolved in water) $Ca^{2+}_{(aq)}$
(ℓ)	Liquid (written after substance)

TABLE 1-2 Rules for writing and balancing equations (after Hein, 1970).

1. The correct formulas for the reactants and products must be known or ascertained by use of valence, oxidation numbers, or experimental data.
2. The formulas for the reactants are written to the left of the arrow and those of the products to the right of the arrow.
3. The equation is then ready to be balanced.
   (a) Count and compare the number of atoms of each element on both sides of the equation to determine those that are unbalanced.
   (b) Balance each element, one at a time, by placing small whole numbers in front of the formulas of the substances containing the unbalanced element, so that the number of atoms of each element is the same on both sides of the equation. A number placed in front of a substance multiplies each element in the substance by that number.
   (c) Check all elements after each individual element is balanced to see if, in balancing one element, others have become unbalanced.
   (d) It is usually advantageous to balance all elements other than H and O first; then balance H; and finally balance O.
   (e) Groups of elements such as SO_4^{2-}, which remain unchanged from one side of the equation to the other, may be balanced as a unit.
   (f) The final balanced equation should contain whole number coefficients in the smallest ratio possible. Compare:

$$4HgO \longrightarrow 4Hg + 2O_2 \quad \text{(incorrect form)}$$
$$2HgO \longrightarrow 2Hg + O_2 \quad \text{(correct form)}$$

The following information is provided by a balanced chemical equation:

1. The components of the reaction and its products.

2. Molecular formulas for the reactants and products.

3. The number of moles of each substance given in the reaction.

4. The number of grams of each substance involved in the reaction.

The following equation illustrates this information:

(2) $FeCl_3$ + 3NaOH $\longrightarrow$ $Fe(OH)_3$ + 3NaCl

(1) iron(III)   sodium   iron(III)   sodium
     chloride   hydroxide   hydroxide   chloride

(3)   1   3   1   3

(4) 162.5 $40 \times 3 = 120$ 106.9 $58.4 \times 3 = 175.2$

These types of data are useful when expressing the quantitative relationships that exist between substances in a chemical reaction.

Heat in Chemical Reactions

Energy may be either absorbed or released in a chemical reaction. Chemical reactions may be classified as either *endothermic* or *exothermic*. An endothermic reaction absorbs heat. An exothermic reaction liberates heat, and its products will exist in a lower and more stable energy state than the reactants.

1-2 TYPES OF CHEMICAL REACTIONS

In general, there are two types of chemical reactions: (a) a reaction in which no change in the oxidation state occurs, and (b) a reaction in which changes in the oxidation state do occur.

Metathetical Reactions

Metathetical reactions are reactions in which there are no changes in the oxidation state of any of the elements involved in the reaction. This type of reaction generally occurs because one or more of the products is shifted away from the field of the reaction in such ways as: the liberation of a gas, the formation of an insoluble precipitate, or the formation of a slightly ionized substance (Moeller and O'Conner, 1972). Examples of these types of reactions are:

LIBERATION OF A GAS:

$$CO_{3\,(aq)}^{2-} + 2H_{(aq)}^+ \longrightarrow CO_{2\,(g)} + H_2O_{(\ell)} \tag{1-3}$$

PRECIPITATION:

$$Ca_{(aq)}^{2+} + CO_{3\,(s)}^{2-} \longrightarrow CaCO_{3\,(s)} \tag{1-4}$$

SLIGHTLY IONIZED SUBSTANCE:

$$CN_{(aq)}^- + H_2O_{(\ell)} \longrightarrow OH_{(aq)}^- + HCN_{(aq)} \tag{1-5}$$

Oxidation-Reduction (Redox) Reactions

In a redox reaction some of the atoms or ions undergo a change in oxidation number. To discuss such reactions, it is necessary to understand the terms used. Oxidation is the loss of electrons resulting in an increase in the oxidation number of one or more atoms. Reduction is a gain in electrons resulting in a decrease in the oxidation number of one or more atoms. The two processes must always occur simultaneously. An oxidizing agent is an atom, ion, or molecule that takes up electrons from other substances, whereas a reducing agent is one that gives up electrons to other substances.

A good example of a redox reaction is the chlorine oxidation of manganese(II) and the subsequent removal of manganese by precipitation of manganese(IV) dioxide:

$$\overset{0}{Cl}_{2(g)} + \overset{+2}{Mn}^{2+}_{(aq)} + 2\overset{+1}{H}_2\overset{-2}{O}_{(\ell)} \longrightarrow \overset{+4}{Mn}\overset{-2}{O}_{2(s)} + 2\overset{-1}{Cl}^{-}_{(aq)} + 4\overset{+1}{H}^{+}_{(aq)} \qquad (1\text{-}6)$$

Increase in oxidation no.
Loss of 2e⁻/ Mn atom

Decrease in oxidation no.
Gain of 1 e⁻ / Cl atom

Manganese loses 2 e^- per Mn atom when changing from Mn^{2+} to MnO_2. This corresponds to an increase in oxidation number. Thus, Mn^{2+} is oxidized. Chlorine gains one electron per chlorine atom when changing from molecular chlorine to the chloride form. In this case chlorine is reduced. It is important to remember that in a balanced equation the number of electrons lost in the oxidation process must equal the number gained in the reduction process.

If oxidation-reduction reactions are to be dealt with quantitatively, they must be balanced. Many redox reactions are difficult to balance by inspection, and as a result several methods have been developed to aid in balancing equations of this type. One such method, called the *half-reaction method*, is illustrated in example problem 1-1.

EXAMPLE PROBLEM 1-1: Balance the following chemical equation, where the proposed reaction is assumed to occur in aqueous solution:

$$NO_{3\,(aq)}^{-} + SO_{3\,(aq)}^{2-} \longrightarrow N_{2(g)} + SO_{4\,(aq)}^{2-}$$

Solution:

1. Assign oxidation numbers to all elements in the reaction:

$$\overset{+5}{N}\overset{-2}{O}_3^{-} + \overset{+4}{S}\overset{-2}{O}_3^{2-} \longrightarrow \overset{0}{N}_2 + \overset{+6}{S}\overset{-2}{O}_4^{2-}$$

2. Separate the basic equation into two half-reactions: one half-reaction should illustrate the oxidation step and the second half-reaction should illustrate the reduction step.

Oxidation:

$$SO_{3(aq)}^{2-} \longrightarrow SO_{4(aq)}^{2-}$$

Reduction:

$$NO_{3(aq)}^{-} \longrightarrow N_{2(g)}$$

3. Balance all atoms in the two half-reactions with the exception of oxygen and hydrogen:

$$SO_{3(aq)}^{2-} \longrightarrow SO_{4(aq)}^{2-}$$
$$2NO_{3(aq)}^{-} \longrightarrow N_{2(g)}$$

4. To each half-reaction add the number of electrons involved in the electron transfer process:

$$SO_{3(aq)}^{2-} \longrightarrow SO_{4(aq)}^{2-} + 2e^{-}$$
$$10\,e^{-} + 2NO_{3(aq)}^{-} \longrightarrow N_{2(g)}$$

5. Balance the charge of each half-reaction, using either H^{+} or OH^{-} ions:

$$SO_{3(aq)}^{2-} + 2H_{(aq)}^{+} \longrightarrow SO_{4(aq)}^{2-} + 2\,e^{-} + 4H_{(aq)}^{+}$$
$$10\,e^{-} + 2NO_{3(aq)}^{-} + 12H_{(aq)}^{+} \longrightarrow N_{2(g)}$$

6. Balance all the O and H atoms in the half-reactions, using H_2O:

$$SO_{3(aq)}^{2-} + 2H_{(aq)}^{+} + H_2O_{(\ell)} \longrightarrow SO_{4(aq)}^{2-} + 4H_{(aq)}^{+} + 2\,e^{-}$$
$$10\,e^{-} + 2NO_{3(aq)}^{-} + 12H_{(aq)}^{+} \longrightarrow N_{2(g)} + 6H_2O_{(\ell)}$$

7. Balance the half-reactions to ensure that equal numbers of electrons are involved in each half-reaction. For this example this means that the oxidation half-reaction must be multiplied by 5.

$$5SO_{3(aq)}^{2-} + 10H_{(aq)}^{+} + 5H_2O_{(\ell)} \longrightarrow 5SO_{4(aq)}^{2-} + 20H_{(aq)}^{+} + 10\,e^{-}$$
$$10\,e^{-} + 2NO_{3(aq)}^{-} + 12H_{(aq)}^{+} \longrightarrow N_{2(g)} + 6H_2O_{(\ell)}$$

8. Add the two half-reactions to obtain the overall reaction:

$$2NO_{3(aq)}^{-} + 5SO_{3(aq)}^{2-} + 2H_{(aq)}^{+} \longrightarrow N_{2(g)} + 5SO_{4(aq)}^{2-} + H_2O_{(\ell)}$$

1-3 CALCULATIONS FROM CHEMICAL EQUATIONS

Chemical equations express a definite weight relationship among the reactants and products in a reaction. This is the basis of stoichiometry, the science that deals with the measurement of relative proportions of elements and compounds in chemical reactions.

Mole Ratio Calculations

When working with chemical reactions, it is often desirable to compute the quantity of material which is produced from, or required to react with, a given quantity of another substance. The method of molar ratios is convenient for this purpose. To solve problems by this method the following procedure should be employed:

1. Construct a balanced chemical equation for the chemical reaction of interest.
2. Compute the number of moles of each substance of known quantity present in the chemical equation.

$$\text{No. of moles known substance} = \frac{\text{weight of substance present}}{\text{molecular weight of substance}} \qquad \textbf{(1-7)}$$

3. Using the molar coefficients given in the balanced equation, compute the molar ratio between the substance of known quantity and the desired substance.

$$\text{molar ratio} = \frac{\text{molar coefficient of desired substance}}{\text{molar coefficient of known substance}} \qquad \textbf{(1-8)}$$

4. Compute the number of moles of the desired substance.

$$\text{No. of moles desired substance} = \text{No. of moles known substance} \times \text{molar ratio} \qquad \textbf{(1-9)}$$

5. Convert the number of moles of desired substance into the required units (e.g., from moles to pounds).

The method of molar ratios is illustrated in example problem 1-2.

EXAMPLE PROBLEM 1-2: Wastewater containing significant amounts of lead are generated during the production of television picture tubes and automobile batteries. Lead is normally removed from such wastewater by precipitating it in the form of a slightly soluble salt. One method which is commonly used is the addition of trisodium phosphate to form lead phosphate. Lead phosphate is very insoluble under acid conditions, and at pH 3.5 the solubility of lead is approximately 0.15 mg/ℓ. The chemical reaction which describes this process is

$$3Pb^{2+} + 2Na_3PO_4 \cdot 12H_2O \rightleftharpoons Pb_3(PO_4)_2 + 6Na^+ + 24H_2O$$

Because of the extremely low solubility of lead phosphate it can be assumed that the forward reaction goes to completion. For this situation, how many pounds (dry weight) of lead phosphate sludge are formed per million gallons of wastewater treated, if excess trisodium phosphate is added and the wastewater initially contains 20 mg/ℓ of Pb^{2+}?

Solution:

1. An examination of the chemical equation shows that it is balanced.

2. Compute the number of moles/ℓ represented by 20 mg/ℓ of Pb^{2+}.

$$\begin{matrix} \text{No. of moles} \\ \text{of } Pb^{2+} \end{matrix} = \frac{20 \times 10^{-3}}{207.2} = 9.65 \times 10^{-5} \text{ mole}/\ell$$

3. Calculate the molar ratio of lead phosphate to lead:

$$\text{molar ratio} = \tfrac{1}{3}$$

4. Determine the number of moles of lead phosphate produced:

$$\begin{matrix} \text{No. of moles}/\ell \\ \text{of } Pb_3(PO_4)_2 \end{matrix} = (9.65 \times 10^{-5})(\tfrac{1}{3})$$
$$= 3.22 \times 10^{-5} \text{ mole}/\ell$$

5. Compute the pounds of lead phosphate formed per million gallons of wastewater treated. Use the value of 8.34 as a conversion factor to convert mg/ℓ to lb/MG.

$$\frac{lb}{MG} = 3.22 \times 10^{-5}\frac{mole}{\ell} \times 811.6 \times 10^3 \frac{mg}{mole} \times 8.34 \frac{lb/MG}{mg/\ell}$$
$$= 217.95 \frac{lb}{MG}$$

1-4 SOLUTIONS

In chemical terminology the word "solution" indicates a system in which one or more substances are uniformly and homogeneously dissolved or blended into another substance. There are two components in a solution: the solute (the substance which is dissolved) and the solvent (the substance which does the dissolving and which is present in the greatest quantity).

There are three states of matter: solid, liquid, and gas. From these three states it is possible to have nine different types of solutions. However, the one of most interest in water process chemistry is the solids dissolved in liquid solution.

Concentration of Solutions

The amount of one substance that will dissolve in another substance is called the *solubility* of the substance which is dissolved. For example, 0.015 g of calcium carbonate will dissolve in 1000 g of water at a temperature of 0°C. Thus, the solubility of $CaCO_3$ is 0.015 g per 1000 g of water at 0°C. (Note: Other factors such as pH affect $CaCO_3$ solubility. These are discussed later.)

A saturated solution contains a dissolved solute in equilibrium with undissolved solute. This equilibrium may be expressed as

$$\text{solute}_{(undissolved)} \rightleftharpoons \text{solute}_{(dissolved)}$$

In a saturated solution two processes are occurring simultaneously: crystallization and dissolution; i.e., the solute is dissolving into and crystallizing out of the solvent at the same time. A state of equilibrium is established when the rate of dissolution is equal to the rate of crystallization. At this point the amount of solute in solution will remain a constant value.

Solution temperature is very important when considering the position of equilibrium in a saturated solution. The solubility of a substance is temperature dependent, and as a result the amount of dissolved solute in solution at the equilibrium position of a saturated solution will vary with temperature.

A solution is said to be undersaturated when it contains less solute per unit volume than does its corresponding saturated solution. In other words, more solute can be dissolved into an undersaturated solution.

The concentration of a solution expresses the amount of solute which is dissolved in a particular quantity of solvent. Either chemical or physical units can be used to express solution concentrations. Some common physical and chemical methods of expressing concentration are the following:

1. PHYSICAL:

(a) Weight — Percent Solution

The solute concentration may be expressed as a percentage of the total weight of the solution. Thus, a reagent bottle that is labeled "Sodium Hydroxide, NaOH, 10%" means that every 100 g of the solution contains 10 g of NaOH and 90 g of water.

$$\text{weight} - \text{percent} = \frac{\text{g-solute}}{\text{g-solute} + \text{g-solvent}} \times 100 \qquad \textbf{(1-10)}$$

However, it is often impractical to prepare a solution by weight alone. To circumvent this problem, volume measurements are often employed, using a density conversion factor to convert to weight. For example, the weight of sulfuric acid (H_2SO_4) present in 1000 mℓ of a solution containing 13% H_2SO_4 by weight can be obtained as follows:

$$\text{Wt. } H_2SO_4 = 1000 \text{ m}\ell \times 1.52 \frac{\text{g}}{\text{m}\ell} \times \frac{13}{100} = 197.6 \text{ g}$$

The value of 1.52 represents the density of concentrated sulfuric acid.

2. CHEMICAL:

(a) Molarity

The number of moles of the solute in solution is not apparent when solution concentration is given in terms of weight — percent. It is much more informative to express concentration in units which identify the number of moles of solute present per unit volume of solution. The molar method of expressing concentration is one way of accomplishing this.

A 1 molar solution contains 1 mole or 1 g-molecular weight per liter of solution. It is important to understand that the volume of solute and the volume of solvent together add up to 1 ℓ. Hence, the molarity of a solution is the number of moles of solute per liter of solution. A capital M is used to denote molarity.

$$M = \frac{\text{g-solute}}{(\text{g-mol. wt. solute})(\text{liters of solution})} \qquad \textbf{(1-11)}$$

To prepare a solution of a given molarity, dissolve the appropriate quantity of solute in something less than the required volume of solvent and then dilute with solvent to the final volume.

EXAMPLE PROBLEM 1-3: What is the molarity of 200 mℓ of a solution containing 2.0 g of sodium chloride (NaCl)?

Solution:

1. Compute the g-molecular weight of NaCl:

$$\text{g-molecular wt.} = 23 + 35.5 = 58.5$$

2. Determine the solution molarity from equation 1-11:

$$M = \frac{2.0}{(58.5)(0.2)} = 0.17$$

(b) Normality

The number of g-equivalent weights of solute present in 1 ℓ of solution is expressed as the normality or normal concentration, N, of a solution. Thus, a 1 N solution contains 1 g-equivalent weight of solute per liter of solution. The objective of using equivalent weights is to equate different weights of substances that have the same reacting capacity.

$$N = \frac{\text{g-solute}}{(\text{g-eq. wt. of solute})(\text{liters of solution})} \qquad \textbf{(1-12)}$$

It must be remembered that the number of chemical equivalents in a given amount of substance is defined only in terms of specific reactions involving that substance. The term *equivalent weight* is defined as the number of grams of substance that will provide Avogadro's number (A_n) of units of reaction (e.g., A_n number of protons transferred, A_n number of electrons transferred).

For the general case, g-equivalent weight can be defined mathematically as

$$\text{g-eq. wt.} = \frac{\text{g-molecular wt.}}{C_c} \qquad \textbf{(1-13)}$$

where C_c represents the combining capacity of the solute. The value of C_c depends on the reaction for which it is to be used. To determine a value for C_c, the following assumptions are generally made:

acids: C_c = number of available protons, where n represents this number in the chemical formula $H_n A$

bases: C_c = number of available hydroxyls, where n represents this number of hydroxyls in the chemical formula $B(OH)_n$

salts: C_c = total number of positive or negative charges per formula

oxidizing
or reducing
agents: C_c = number of electrons transferred per formula.

EXAMPLE PROBLEM 1-4: Compute the equivalent weight of potassium dichromate as used in the following reactions:

1. $Cr_2O_7^{2-} + 2Pb^{2+} + H_2O \longrightarrow 2PbCrO_4 + 2H^+$
2. $Cr_2O_7^{2-} + 14H^+ + 6e^- \longrightarrow 2Cr^{3+} + 7H_2O$

Solution:

1. This step should always be performed to determine what type of reaction is occurring, i.e., whether it is a metathetical reaction or a redox reaction.

   In this problem reaction 1 is a metathetical reaction, whereas reaction 2 is a redox reaction in which Cr changes from the +6 oxidation state to the +3 oxidation state. Thus, three electrons are transferred in reaction 2.

2. Compute the g-equivalent weight from equation 1-13.

   REACTION 1: The salt $K_2Cr_2O_7$ has a total of two positive charges; hence,

$$\text{g-eq. wt.} = \frac{294.2}{2} = 147.1$$

   REACTION 2: There are three electrons transferred in the redox reaction; hence,

$$\text{g-eq. wt.} = \frac{294.2}{3} = 98.1$$

This example illustrates the danger of assigning a normality value to stock solutions of many chemical species which are capable of competing in more than one type of reaction.

The common convention in environmental engineering is to use the $CaCO_3$ equivalent form to express chemical concentration. The following relationship can be used to convert from concentration in terms of mg/ℓ to its equivalent as $CaCO_3$:

$$\begin{matrix} \text{mg/}\ell \text{ as} \\ CaCO_3 \end{matrix} = \begin{bmatrix} \text{mg/}\ell \text{ of} \\ \text{substance} \end{bmatrix} \times \begin{bmatrix} \text{equivalent weight of } CaCO_3 \\ \overline{\text{equivalent weight of substance}} \end{bmatrix} \qquad \text{(1-14)}$$

It is also worth keeping in mind that the relationship between normality and molarity is given by the expression

$$N = C_c M \qquad \text{(1-15)}$$

The Dilution Equation

It is often necessary to prepare a solution of one concentration from another more concentrated one. If a solution is diluted by adding solvent, the amount of solute does not change. The number of moles of solute in a sample is given by the product of the molarity and the sample volume. Since the number of moles of solute does not change on dilution, the following relationship is valid:

$$M_1V_1 = M_2V_2 = \begin{bmatrix} \text{number of moles} \\ \text{of solute} \end{bmatrix} \qquad \textbf{(1-16)}$$

The same reasoning can be applied when concentration is expressed as normality, so that a more general form of equation 1-16 is

$$(\text{conc.})_1 \, V_1 = (\text{conc.})_2 \, V_2 \qquad \textbf{(1-17)}$$

In equation 1-17 concentration can be in any units of weight and volume, but the units must be the same for both solutions. The most common units are normalities and molarities.

EXAMPLE PROBLEM 1-5: Find the volume of concentrated H_2SO_4 solution (36 N) which must be diluted to 1000 mℓ to make 1000 mℓ of 0.1 N solution.

Solution:

1. Ensure that the concentration and volume units are the same for the initial and final conditions and then substitute the required values into equation 1-17.

$$(36 \ N) \ V_1 = (0.1 \ N)(1000 \ \text{m}\ell) = 2.78 \ \text{m}\ell$$

Acids and Bases

Several theories have been proposed to answer the question: "What are acids and bases?" Two such theories are the Arrhenius theory and the Brönsted-Lowry theory.

1. In 1887 Arrhenius suggested that acids were substances which separated (ionized) in solution to produce hydrogen ions (H^+, or free protons) and that bases were substances which ionized to produce hydroxide ions (OH^-).

2. In 1923 the Brönsted proton transfer theory was introduced. This theory states that an acid is any substance that dissociates in solution to produce a proton and that a base is a substance that combines with or accepts a proton. Thus, according to the Brönsted theory, a base does not necessarily produce hydroxide ions, but is any substance that can accept a hydrogen ion in a reaction. Since an H^+ ion is a proton, such reactions are called *proton transfer reactions*, or *protolysis reactions* (note that H^+ ions do not exist in solution but rather combine with a water molecule to form a hydronium ion, H_3O^+. However, for simplicity the H^+ notation is normally used in this text).

In general, any acid-base reaction can be described by the reaction

$$\text{acid} + \text{base} \longrightarrow \text{conjugate base} + \text{conjugate acid} \qquad \textbf{(1-18)}$$

The conjugate base of an acid is the remainder of the acid after the proton has been released by the acid. The conjugate acid of a base is formed when the base accepts a proton from the acid. Table 1-3 contains a list of some anions and their conjugate acids.

TABLE 1-3 Anions and their conjugate acids (after Smoot, Price, and Barret, 1971).

Anion	Name	Conjugate Acid
HSO_3^-	Hydrogen sulfite	H_2SO_3
NO_2^-	Nitrite	HNO_2
CH_3COO^-	Acetate	CH_3COOH
SO_3^{2-}	Sulfite	HSO_3^-
ClO^-	Hypochlorite	$HClO$
CN^-	Cyanide	HCN
CO_3^{2-}	Carbonate	HCO_3^-
S^{2-}	Sulfide	HS^-

Salts

An acid may react with a base to form a salt and water. Such a reaction can be understood by noting that an acid is composed of one or more hydrogen atoms combined with a nonmetallic atom or molecular group, whereas one type of base is composed of one or more hydroxyl groups combined with a metallic atom. In solution the hydrogen ions from the acid unite with the hydroxyl group from the base to form water. If the water is evaporated, the residue left behind will be a new compound composed of the negative nonmetallic ions from the acid and the positive metallic ions from the base. This new ionic compound is called a *salt*. The formation of a salt is illustrated in equation 1-19.

$$\underset{\text{acid}}{HCl} + \underset{\text{base}}{KOH} \longrightarrow \underset{\text{salt}}{K^+ : Cl^-} + \underset{\text{water}}{H_2O} \tag{1-19}$$

Salts are formed from neutralization reactions, i.e., the reaction between an acid and a base. However, when salts are dissolved with water, the resulting solution may be altered because of a reaction between the salt and water. Such a reaction is called a *protolysis* or *hydrolysis reaction* and is discussed in detail later.

Electrolytes and Nonelectrolytes

Nonelectrolytes are substances whose aqueous solutions do not conduct electricity, whereas electrolytes are substances whose aqueous solutions conduct electricity. This difference in electrical conductivity is due to the fact that electrolytes exist as ions or are capable of producing ions in solution. Nonelectrolytes do not have this property. Table 1-4 lists some common electrolytes and nonelectrolytes. The compounds listed in this table suggest that it is acids, bases, and salts which, in general, are found to be electrolytes.

At this point, it is appropriate to distinguish between the terms *dissociation* and *ionization*. Although it is customary to use these terms interchangeably, the two words actually have different meanings. In the ionization process, ions are produced by the

TABLE 1-4 *Examples of electrolytes and nonelectrolytes.*

Electrolytes	Nonelectrolytes
H_2SO_4	$C_6H_{12}O_6$ (glucose)
H_3PO_4	C_2H_5OH (ethanol)
HCl	$CO(NH_2)_2$ (urea)
NaOH	O_2
NH_4OH	
NaCl	

reaction of a compound with water (recall that in HCl the bond is covalent, not ionic), whereas dissociation describes the process in which substances that are ionic in the pure state dissolve in water to give solutions that contain ions (recall that in salts such as KCl the bond is ionic, not covalent).

Depending on the degree of dissociation or ionization, electrolytes are classified as either strong or weak. Strong electrolytes are close to 100% ionized, whereas weak electrolytes are considerably less than 100% ionized. Double arrows ($\rightleftharpoons$) are used to indicate the equilibrium state in chemical reactions involving a weak electrolyte. A single arrow ($\rightarrow$) is used to denote the reaction of a strong electrolyte.

$$HCl_{(aq)} \longrightarrow H^+_{(aq)} + Cl^-_{(aq)} \qquad \textbf{(1-20)}$$

$$HF_{(aq)} \rightleftharpoons H^+_{(aq)} + F^-_{(aq)} \qquad \textbf{(1-21)}$$

A list of some strong and weak electrolytes is presented in Table 1-5.

TABLE 1-5 *Examples of strong and weak electrolytes*

Strong Electrolytes	Weak Electrolytes
H_2SO_4	H_2CO_3
HCl	H_2S
HNO_3	HF
KOH	HNO_2
NaOH	HOCl
$Ca(OH)_2$	NH_4OH
Most salts	

When the molar ratio method is used to determine the amounts of material consumed and produced by chemical reactions, it is assumed that all reactants change to products without considering that products may react to form the original reactants. In a reaction which progresses to completion, all the reactants are converted to products according to the molar coefficients given in the balanced chemical equation. However, many chemical reactions do not go to completion because the products which are formed react to produce the starting reactants. A reaction of this type is called a *reversible reaction*. Both processes, the forward reaction (the reaction to the right) and the reverse reaction (the reaction to the left), occur simultaneously. To indicate that the reaction is reversible, a double arrow is used in the chemical equation.

Equilibrium

As an introduction to the principles of chemical equilibrium a general chemical reaction will be analyzed qualitatively to show what an equilibrium state actually is. Consider the following chemical reaction, which is assumed to occur in aqueous solution:

$$A + B \;\rightleftharpoons\; C + D \tag{1-22}$$

The double arrows indicate that both the forward reaction (the reaction of species A and B to form species C and D) and the reverse reaction (the reaction of species C and D to form species A and B) may occur.

Assume that species A and B are mixed together in aqueous solution. To follow the process as reaction 1-22 proceeds in the forward direction, the concentration of either A, B, C, or D can be measured as a function of time. In this case assume that the concentration of C is measured as a function of time and follows the relationship shown in Figure 1-1(a). This figure indicates that after a short period of time the concentration of C becomes constant and no longer varies with time. Since a state of equilibrium is considered to exist when the rate of the forward reaction equals the rate of the reverse reaction, it might be a temptation to assume that equilibrium has been established when the concentration of C ceases to change with time. However, this may not be true. In fact, equilibrium should only be assumed when the same concentration of C is approached in the reverse direction for the same total solution concentration of C. Therefore, to verify that equilibrium is established in the system, mix C and D together in aqueous solution and measure the concentration of C as a function of time. If a response similar to that shown in Figure 1-1(b) is obtained, it can be assumed that a state of equilibrium has been reached.

Figures 1-1(a) and 1-1(b) can be superimposed to produce Figure 1-1(c). This figure shows that, although the final equilibrium concentration of C is the same regardless of whether equilibrium is approached from the forward or reverse direction, the *rate of approach* to equilibrium depends on the initial concentrations of the various chemical species involved in the reaction. In general, the greater the initial concentrations of the reactants the greater will be the rate at which equilibrium is approached.

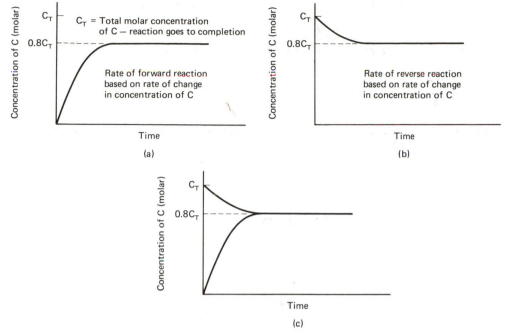

FIGURE 1-1 *Variation in concentration of species C with time.*

With this brief introduction to the concept of chemical equilibrium a more quantitative approach is now taken to derive the *law of chemical equilibrium*. This law is derived from the fundamentals of reaction kinetics.

Kinetic Approach to Equilibrium

Every chemical reaction proceeds at a characteristic speed or rate which depends on: (a) the concentration of the reacting species, (b) the solution temperature, and (c) the presence of catalytic agents. Consider again the hypothetical reaction

$$A + B \; \rightleftharpoons \; C + D \tag{1-22}$$

where

$$A + B \; \longrightarrow \; C + D$$

is the forward reaction, and

$$C + D \; \longrightarrow \; A + B$$

is the reverse reaction. If the *reaction mechanism* is such that a collision between a single particle of A and a single particle of B is necessary for the forward reaction to occur and a collision between a single particle of C and a single particle of D is necessary for the reverse reaction to occur, then initially the rate of the forward reaction will be greater than the rate of the reverse reaction because the initial concentrations of A and B are greater than C and D, and therefore the collision frequency is greater. Furthermore, if the initial concentration (number of particles per unit volume) of either A or B were doubled, the number of collisions per unit time between A and B should also double resulting in a two-fold increase in the rate of the forward reaction. On the other hand, if the initial concentration of both A and B were doubled the number of collisions per unit time between A and B should quadruple, resulting in a four-fold increase in the rate of the forward reaction. This line of reasoning suggests that, for the reaction represented by equation 1-22, the rates of the forward reaction and the reverse reaction may be expessed mathematically as

$$r_f = K_1[A][B] \tag{1-23}$$

$$r_r = K_2[C][D] \tag{1-24}$$

where r_f = rate of forward reaction

        r_r = rate of reverse reaction

   K_1, K_2 = proportionality constants (rate constants) which account for all factors that affect the speed of the reaction except concentration.

In equations 1-23 and 1-24, [   ] represents chemical species concentration in moles per liter.

The variation in the rate of the forward and reverse reactions with time are illustrated in Figure 1-2. In this figure the point at which the rate of the forward reac-

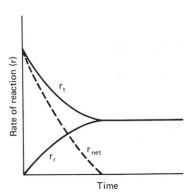

FIGURE 1-2 *Variation in reaction rate with time.*

tion, r_f, is equal to the rate of the reverse reaction, r_r (with a net rate of reaction, r_{net}, of zero) is the point at which equilibrium is established in the reaction. Hence, chemical equilibrium represents a dynamic state where two or more opposing chemical reactions are taking place at the same time and at the same rate. Thus, at equilibrium the following relationship is valid for the chemical reaction represented by equations 1-22:

$$r_f = r_r$$

or

$$K_1[A][B] = K_2[C][D] \tag{1-25}$$

collecting all variable terms on one side and all constant terms on the other gives

$$\frac{K_1}{K_2} = \frac{[C][D]}{[A][B]} \qquad (1\text{-}26)$$

Since the ratio of two constant terms is a constant, the ratio K_1/K_2 can be replaced by a single constant term. In this case the ratio K_1/K_2 is represented by the term $(K_c)_{eq.}$ so that equation 1-26 can be written as

$$(K_c)_{eq.} = \frac{[C][D]}{[A][B]} \qquad (1\text{-}27)$$

where $(K_c)_{eq.}$ is referred to as the *concentration equilibrium constant* or sometimes as the *apparent equilibrium constant* for the reaction under consideration.

For the reaction represented by equation 1-22 it was assumed that the reaction mechanism was such that a collision between a single particle of A and a single particle of B was necessary for the forward reaction to occur, and a collision between a single particle of C and a single particle of D was necessary for the reverse reaction to occur. Consider now a reaction mechanism such that a simultaneous collision between two particles of A and a single particle of B is necessary for the forward reaction to occur, and a simultaneous collision between two particles of C and a single particle of D is necessary for the reverse reaction to occur. Such a reaction may be represented by equation 1-28.

$$2A + B \;\rightleftharpoons\; 2C + D \qquad (1\text{-}28)$$

Following the same line of reasoning used for the development of equations 1-23 and 1-24, the rates of the forward reaction and the reverse reaction may be expressed mathematically as

$$r_f = K_1[A]^2[B] \qquad (1\text{-}29)$$

$$r_r = K_2[C]^2[D] \qquad (1\text{-}30)$$

Then, at equilibrium

$$r_f = r_r$$

or

$$K_1[A]^2[B] = K_2[C]^2[D] \qquad (1\text{-}31)$$

The relationship for the concentration equilibrium constant is therefore

$$(K_c)_{eq.} = \frac{[C]^2[D]}{[A]^2[B]} \qquad (1\text{-}32)$$

On the basis of the preceding discussion, for the general reaction

$$aA + bB \;\rightleftharpoons\; cC + dD \qquad (1\text{-}33)$$

where a, b, c and d represent the stoichiometric number of particles of A, B, C, and D, respectively, involved in the completely balanced reaction, the concentration equilibrium constant may be expressed by the relationship

$$(K_c)_{eq.} = \frac{[C]^c[D]^d}{[A]^a[B]^b} \qquad (1\text{-}34)$$

Equation 1-34 is a mathematical representation of the *law of chemical equilibrium*, and in its general form can be applied to many kinds of reactions such as those of dissociation, ionization, precipitation (solubility) protolysis, and complex ion formation. Note that the actual reaction mechanism may differ considerably from the stoichiometry of the general reaction. However, the equilibrium properties of a system do not depend on the pathway by which equilibrium is reached, and as a result the general form of equation 1-34 is valid for any reaction at equilibrium. It should also be noted that the use of activities or effective concentrations in the development of equation 1-34 would have been more rigorously correct than the use of actual analytical concentrations (molar concentrations). The concept of activity is discussed in some detail in a later section.

There are three important points related to equation 1-34 which should be understood:

1. The concentration of A, B, C, and D in this equation exist only at equilibrium. However, there is an infinite number of sets of concentrations which correspond to the equilibrium state.

2. Standard convention is that the numerical value of the equilibrium constant reflects the value of the ratio of the product of the product concentrations to the product of the reactant concentrations. Thus, a small value of $(K_c)_{eq}$ means that equilibrium is reached after only a small fraction of the reactant material has been converted to product. On the other hand, a large $(K_c)_{eq.}$ value implies that a large fraction of the reactant material has been converted to product when equilibrium is established.

3. The initial concentration of reactant and product is important in determining the direction in which a reaction will move toward establishing equilibrium. For example, even though $(K_c)_{eq.}$ may be quite large, a reaction may move from right to left if the initial concentration of the products is large and the initial concentration of the reactants is small.

It is important to note that the concept of chemical equilibrium provides no information on the rate at which equilibrium is attained in a chemical system, nor does it state the direction from which equilibrium is approached.

1-6 CHEMICAL THERMODYNAMICS

It is a fundamental law of nature that any chemical system will tend to undergo an irreversible change from some initial, nonequilibrium state to a final, equilibrium state. Once equilibrium has been attained, no further change will occur unless the system is stressed in some way. To expand on this basic concept, consider the hypothetical chemical reaction

$$aA + bB \rightleftharpoons cC + dD \qquad (1\text{-}33)$$

To determine if the reaction represented by equation 1-33 is really possible, substances A and B are mixed together and the solution monitored for the disappearance of either A or B or for the appearance of either C or D. If the position of equilibrium lies to the right, then the chemical reaction described by equation 1-33 is said to be spontaneous (some writers refer to such a reaction as being *feasible*).

At this point, it is appropriate to ask the question: "What makes a reaction occur spontaneously?" To answer this question, it is necessary to understand that chemical change and energy are always related. When the reactant material contains large amounts of energy relative to the products, the chemical reaction occurs spontaneously. Thus, a spontaneous chemical reaction is a process in which matter moves toward a more stable state.

When describing equilibrium in thermodynamic terms it is necessary to define a property of a chemical system relating the equilibrium state to the concentration of the various species involved in the equilibrium state. To identify this parameter consider that a change in heat content during a chemical reaction is referred to as an enthalpy change (ΔH). When a chemical reaction releases large amounts of heat (has a large negative enthalpy change), it is said to be strongly exothermic. On the other hand, when large amounts of heat are absorbed during a chemical reaction (a large positive enthalpy change), the reaction is said to be strongly endothermic. Strongly exothermic reactions occur spontaneously, whereas strongly endothermic reactions do not occur spontaneously. However, some weakly endothermic reactions occur spontaneously, whereas some weakly exothermic reactions are not spontaneous in nature. This implies that enthalpy change alone cannot be used to judge the spontaneity of a chemical reaction. It has been found that a change in the degree of randomness or disorder (a change in entropy) is also involved in causing reactions to occur. As was the case for enthalpy, a definite amount of entropy is associated with each chemical species. When a given chemical species is in the gaseous state (where the molecules are highly disordered with respect to each other), the entropy of this species is much greater than when it is in the solid state (where the molecules are highly ordered with respect to each other). For any given state, the entropy of a particular chemical species depends on such factors as the translational, rotational, and vibrational motion of the species and the attractive forces between individual atoms or molecules.

The actual driving force for a chemical reaction is a combination of enthalpy change and entropy change. This driving force is called the *Gibbs free energy change* (ΔG) of the reaction. An expression for ΔG which includes both enthalpy and entropy terms can be derived from the second law of thermodynamics. This equation has the form

$$\begin{bmatrix} \text{change in} \\ \text{free energy} \end{bmatrix} = \begin{bmatrix} \text{change in} \\ \text{enthalpy} \end{bmatrix} - [\text{temperature}] \times \begin{bmatrix} \text{change in} \\ \text{entropy} \end{bmatrix}$$

or

$$\Delta G = \Delta H - T \Delta S \qquad (1\text{-}35)$$

Experimental observations have shown that a decrease in free energy is associated with reactions which occur spontaneously, whereas a reaction with a positive free energy change will occur only if energy is supplied to the system to drive the reaction. Reactions having a negative ΔG are termed *exergonic;* those that have a positive ΔG are called *endergonic.* If the ΔG for a reaction equals zero, the reaction is at equilibrium.

A negative free energy change (where ΔG is negative) indicates that a reaction has a tendency to occur. However, the rate of certain spontaneous reactions is so slow that it may take many years to detect any change in the reactant concentration. As a result, to predict whether a given spontaneous reaction will be useful, it is necessary to know both the point at which equilibrium is established as well as the rate at which the reaction proceeds.

Free Energy of Formation

To provide a uniform basis for the calculation of the free energy change associated with a particular chemical reaction, standard state conditions for the chemical system are assigned. Other standard state conditions are defined later, but for now consider a pressure of 1 atmosphere (atm) and a temperature of 298 degrees Kelvin (K) (25°C) as standard state conditions. A free energy change under standard state conditions is denoted by $\Delta G°$.

Every chemical *element* is assigned a free energy of formation value of zero at standard state. For *compounds,* standard molar free energies of formation (the free energy change accompanying the formation of 1 mole of compound) are employed. This means that every compound will have a $\Delta G°_{formation}$ value for 1 mole, 1 atm, and 298°K.

The free energy change for a chemical reaction is defined as the difference between the free energies of the final and initial states:

$$\Delta G = G_{final} - G_{initial} \tag{1-36}$$

Integrating equation 1-36 with the concept of free energy of formation, the following equation, which describes the free energy change of a reaction under standard state conditions, can be developed:

$$\Delta G°_{reaction} = \Sigma \, \Delta G°_{\substack{formation \\ of \ products}} - \Sigma \, \Delta G°_{\substack{formation \\ of \ reactants}} \tag{1-37}$$

If $\Delta G°_{reaction}$ is negative, then reactants in their standard state should be converted to products in their standard state, whereas a positive $\Delta G°_{reaction}$ indicates that reactants in their standard state should not be converted to products in their standard state. If $\Delta G°_{reaction}$ is computed to be 0, then equilibrium exists between reactants and products in their standard state. It should be noted that $\Delta G°_{reaction}$ is expressed in terms of KCal, since $\Delta G°_{formation}$ is normally given as KCal/mole and these values are multiplied by the molar coefficients in the balanced chemical equation to get $\Delta G°_{reaction}$.

Standard state free energy of formation data for many chemical species of interest in process chemistry are presented in Appendix I.

EXAMPLE PROBLEM 1-6: Iron(II), Fe^{2+}, is a soluble chemical species that may be present in ground waters which exist under anaerobic conditions. When exposed to air, iron(II) may slowly oxidize to iron(III), which forms an insoluble hydroxide precipitate that, if not removed from the water, will interfere with certain water uses. The common method for removing iron from a water supply is to chemically oxidize iron(II) to iron(III) and then remove the iron in the form of its insoluble hydroxide precipitate in a sedimentation phase. With this in mind, determine the spontaneity (feasibility) of the following reaction under standard state conditions:

$$Fe^{2+}_{(aq)} + 2HCO^-_{3(aq)} + Cl_{2(g)} + H_2O_{(\ell)}$$
$$\rightleftharpoons Fe(OH)_{3(s)} + 2CO_{2(g)} + 2Cl^-_{(aq)} + H^+_{(aq)}$$

Solution:

1. From Appendix I determine the free energy of formation data for chemical species in the balanced equation.

Chemical Species	Free Energy of Formation (KCal/mole)
$Fe^{2+}_{(aq)}$	−20.30
$HCO^-_{3(aq)}$	−140.31
$Cl_{2(g)}$	0
$H_2O_{(1)}$	−56.69
$Fe(OH)_{3(s)}$	−166.0
$CO_{2(g)}$	−94.25
$Cl^-_{(aq)}$	−31.35
$H^+_{(aq)}$	0 (by convention)

2. Using equation 1-37, compute $\Delta G^\circ_{reaction}$.

$$\Delta G^\circ_{reaction} = [-166.0 + 2(-94.25) + 2(-31.35)]$$
$$- [-20.30 + 2(-140.31) + (-56.69)]$$
$$= -59.59 \text{ KCal}$$

Since the $\Delta G^\circ_{reaction}$ is negative, the left-to-right reaction is spontaneous (feasible) under standard state conditions. During the preliminary development stage of a chemical treatment system (particularly when dealing with an industrial waste problem), it is advisable that proposed chemical reactions be checked for their feasibility.

Enthalpy of Formation

Exactly the same convention used for the calculation of $\Delta G^\circ_{reaction}$ is used to calculate $\Delta H^\circ_{reaction}$. This means that chemical elements in the standard state ($P = 1$ atm, $T = 298°K$) are assigned a value of zero enthalpy of formation. All compounds are assigned a standard molar enthalpy of formation, $\Delta H^\circ_{formation}$ (the enthalpy of formation accompanying the formation of 1 mole of compound).

Values of $\Delta H^{\circ}_{\text{reaction}}$ can be computed from equation 1-38, which is similar to equation 1-37 (the equation used to calculate $\Delta G^{\circ}_{\text{reaction}}$).

$$\Delta H^{\circ}_{\text{reaction}} = \Sigma \, \Delta H^{\circ}_{\substack{\text{formation of} \\ \text{product}}} - \Sigma \, \Delta H^{\circ}_{\substack{\text{formation of} \\ \text{reactant}}} \tag{1-38}$$

Standard state enthalpy of formation data are presented in Appendix I.

Nonideality Corrections

All equilibrium expressions presented up to this point have assumed an ideal solution; i.e., it has been assumed that each ion in solution behaves independently of any other ion. However, such an assumption is valid only for very dilute ionic solutions.

Even though ions in an electrolyte solution are separated by layers of solvent molecules, they will still interact with each other to some degree. The more concentrated the solution, the stronger will be the forces of interaction. The interaction between ions in solution affects the colligative properties of the solution (e.g., freezing point, vapor pressure, boiling point) much more than the presence of an equal number of molecules of a nonelectrolyte. Such a system response is described as a deviation from ideality and is accounted for on the basis of the effects of solute-solute interactions and solute-solvent interactions on the behavior of the solute ions. In order to describe such a system thermodynamically, a new concentration term called *activity* is introduced. *Activity* is the effective concentration required to retain the general thermodynamic treatment of an ideal solution. In other words, by working with activities it is possible to consider nonideal solutions as being ideal solutions.

To relate activities to analytical concentrations, a proportionality constant called the *activity coefficient* is introduced. The equation for the relationship is

$$a = \gamma C \tag{1-39}$$

where a = activity (moles/ℓ)

γ = activity coefficient

C = analytical concentration (moles/ℓ). It should be noted that a more rigorous development would express C as molal concentration and not molar concentration. Molar concentrations approach molal concentrations for dilute solutions but vary significantly for systems involving brines and sea water. Hence, since the primary concern in this text will be with dilute solutions, the development of activity coefficient relationships will be based on molar concentrations.

For the low solution concentrations which are typically encountered in process chemistry, negative deviations from ideal behavior are experienced and the activity coefficient has a value less than unity. However, the value of the activity coefficient approaches unity as the solution approaches infinite dilution because the magnitude of the chemical and physical interactions which occur due to the presence of the ionic species becomes insignificant.

Ionic Strength Considerations

The nonideal behavior exhibited by electrolyte solutions is due to solute-solute and solute-solvent interactions. The magnitude of these interactions is dependent on several factors, two of which are solute concentration and the electrical charge associated with each ionic species in solution. The ionic strength of a solution, I, is a value obtained by integrating the effects of these two parameters into a single expression suitable for use in the theoretical equations which have been derived for the calculation of activity coefficients.

Ionic strength is determined by the following relationship:

$$I = \frac{1}{2} \sum_{i=1}^{i=i} C_i Z_i^2 \qquad \text{(1-40)}$$

where I = ionic strength

C_i = analytical concentration of the ith species (moles/ℓ)

Z_i = oxidation number of the ith species.

Note that the normal convention is to report ionic strength as a dimensionless quantity. However, to be strictly correct, it should be reported as having concentration units.

EXAMPLE PROBLEM 1-7: Compute the ionic strength of a solution containing 0.01 M CaCl$_2$ and 0.001 M Na$_2$SO$_4$.

Solution:
  1. Determine the oxidation number for each ionic species in solution.
     (a) CaCl$_2$
         (i) Ca^{2+}
         (ii) Cl$^-$

     (b) Na$_2$SO$_4$
         (i) Na$^+$
         (ii) SO$_4^{2-}$
  2. Compute the ionic strength of the solution from equation 1-40.

$$I = \tfrac{1}{2}[(M_{Ca^{2+}})(2)^2 + (M_{Cl^-})(1)^2 + (M_{Na^+})(1)^2 + (M_{SO_4^{2-}})(2)^2]$$
$$= \tfrac{1}{2}[(0.01)(4) + (2)(0.01) + (2)(0.001)(1) + (0.001)(4)]$$
$$= 0.033$$

Jurinak (1976) suggests that in most cases the ionic strength of natural waters (including raw water to be processed for potable and industrial use but not wastewater) can be estimated with a reasonable degree of accuracy by including the concentration of only the major chemical species in the calculation of I. The chemical species concentrations to be measured and used in the calculation of I are

1. Sodium
2. Calcium
3. Magnesium
4. Chloride
5. Sulfate
6. Bicarbonate
7. Carbonate

It is logical to assume that ionic strength is related to the total dissolved solids concentration (TDS) of a solution. Following this line of reasoning, Langelier (1936) proposed that ionic strength could be estimated from the equation

$$I = (2.5 \times 10^{-5})(\text{TDS}) \qquad \text{(1-41)}$$

where   TDS = total dissolved solids concentration of the water (mg/ℓ)—This equation applies when the TDS concentration is less than 1000 mg/ℓ.

To account for the presence of nonionic silica, which also contributes to the TDS value, Kemp (1971) has proposed the following relationship:

$$I = (2.5 \times 10^{-5})(\text{TDS} - 20) \qquad \text{(1-42)}$$

Note that neither equation 1-41 nor equation 1-42 accounts for the presence of dissolved organic material (nonionic), which may make a major contribution to the TDS concentration, especially in wastewater. However, Kemp (1971) indicates that the activity coefficient, γ, is relatively insensitive to small changes in ionic strength, and as a result these equations generally provide a reasonable approximation of the ionic strength of the solution.

The ionic strength of a natural water (where the TDS concentration is due mainly to mineral salts) can also be predicted from the measurement of its electrical conductance, EC. Kemp (1971) has proposed the following relationship between EC and ionic strength:

$$I = (2.5 \times 10^{-5})(\text{EC})(g) \qquad \text{(1-43)}$$

where   EC = electrical conductance at 20°C (μmho cm^{-1})

   g = proportionality factor, which generally has a value within the range of 0.55 to 0.70. (A value of 0.67 is commonly accepted for g.)

Care should be exercised when applying equation 1-43 because there are times when g may fall outside the range quoted for this factor. Note that the value of the product of EC and g is equivalent to the TDS value in equation 1-41. Thus, a comparison between the measured TDS concentration and the product of the measured EC value and g provides the required information for judging the applicability of equation 1-43.

EXAMPLE PROBLEM 1-8: In the summer, water in Boulder Creek has an electrical conductivity of 0.0005 mho/cm at 20°C. What is the ionic strength of this water?

Solution:

1. Convert the electrical conductivity measurement in mho/cm to μmho/cm:

$$0.0005 \frac{\text{mho}}{\text{cm}} \times 10^6 \frac{\mu\text{mho}}{\text{mho}} = 500 \ \mu\text{mho/cm}$$

2. Assuming that $g = 0.67$, compute I from equation 1-43:

$$I = (2.5 \times 10^{-5})(500)(0.67) = 0.0084 \ M$$

Jurinak (1976) stated that the ionic strength of water sample from most streams and lakes can be expected to have a value near 0.01 M. However, he also notes that the ionic strength value shows a seasonal dependence and is usually highest in the summer and fall, when stream flows are at a minimum.

Theoretical Equations for the Activity Coefficient

The activity coefficient for ionic chemical species can be estimated from one of the following relationships:

1. The Debye-Hückel limiting law for solutions whose ionic strengths do not exceed 0.005 M:

$$\log \gamma = -AZ^2 \sqrt{I} \qquad \textbf{(1-44)}$$

where γ = activity coefficient, with the subscripts M, D, and T representing the absolute value of the oxidation number of the chemical species in question (e.g., γ_M represents the activity coefficient for monovalent ions, γ_D represents the activity coefficient for divalent ions, and γ_T the activity coefficient for trivalent ions)

Z = oxidation number of the chemical species of interest

$A = 1.82 \times 10^6 (DT)^{-3/2}$

T = temperature of the solution (°K)

D = dielectric constant for water, which is generally taken to have a value of 78.3

I = ionic strength of the solution with respect to all ionized solutes (moles/ℓ).

2. The extended Debye-Hückel relationship for solutions whose ionic strengths do not exceed 0.1 M:

$$\log \gamma = -AZ^2 \left(\frac{\sqrt{I}}{1 + Bb\sqrt{I}} \right) \qquad \textbf{(1-45)}$$

where $B = 50.3(DT)^{-1/2}$

b = adjustable parameter corresponding to the size of the ion, in angstroms (angstrom units where $1 \text{ Å} = 10^{-8}$ cm). Value of b for different ionic species are presented in Table 1-6.

TABLE 1-6 Values of the parameter b for 130 selected ions in Angstrom units (after Butler, 1964).

b	Charge 1
9	H^+
8	$(C_6H_5)_2CHCOO^-$, $(C_3H_7)_4N^+$
7	$OC_6H_2(NO_3)_3^-$, $(C_3H_7)_3NH^+$, $CH_3OC_6H_4COO^-$
6	Li^+, $C_6H_5COO^-$, $C_6H_4OHCOO^-$, $C_6H_4ClCOO^-$, $C_6H_5CH_2COO^-$, $CH_2^=CHCH_2COO^-$, $(CH_3)_2CCHCOO^-$, $(C_2H_5)_4N^+$, $(C_3H_7)_2NH_2^+$
5	$CHCl_2COO^-$, CCl_3COO^-, $(C_2H_5)_3NH^+$, $(C_3H_7)NH_3^+$
4	Na^+, $CdCl^+$, ClO_2^-, IO_3^-, HCO_3^-, $H_2PO_4^-$, HSO_3^-, $H_2AsO_4^-$, $Co(NH_3)_4(NO_2)_2^+$, CH_3COO^-, CH_2ClCOO^-, $(CH_3)_4N^+$, $(C_2H_5)_2NH_2^+$, $NH_2CH_2COO^-$, $^+NH_3CH_2COOH$, $(CH_3)_3NH^+$, $C_2H_5NH_3^+$
3	OH^-, F^-, CNS^-, CNO^-, HS^-, ClO_3^-, ClO_4^-, BrO_3^-, IO_4^-, MnO_4^-, K^+, Cl^-, Br^-, I^-, CN^-, NO_2^-, NO_3^-, Rb^+, Cs^+, NH_4^+, Tl^+, Ag^+, $HCOO^-$, H_2 (citrate)$^-$, $CH_3NH_3^+$, $(CH_3)_2NH_2^+$

b	Charge 2
8	Mg^{2+}, Be^{2+}
7	$(CH_2)_5(COO)^{2-}$, $(CH_2)_6(COO)^{2-}$, (congo red)$^{2-}$
6	Ca^{2+}, Cu^{2+}, Zn^{2+}, Sn^{2+}, Mn^{2+}, Fe^{2+}, Ni^{2+}, Co^{2+}, $C_6H_4(COO)^{2-}$, $H_2C(CH_2COO)^{2-}$, $CH_2CH_2(COO)^{2-}$
5	Sr^{2+}, Ba^{2+}, Ra^{2+}, Cd^{2+}, Hg^{2+}, S^{2-}, $S_2O_4^{2-}$, WO_4^{2-}, Pb^{2+}, CO_3^{2-}, SO_3^{2-}, MoO_4^{2-}, $Co(NH_3)_5Cl^{2+}$, $Fe(CN)_5NO^{2-}$, $H_2C(COO)_2^{2-}$, $(CH_2COO)_2^{2-}$, $(CHOHCOO)_2^{2-}$, $(COO)_2^{2-}$, H (citrate)$_2^{2-}$
4	Hg_2^{2+}, SO_4^{2-}, $S_2O_3^{2-}$, $S_2O_8^{2-}$, SeO_4^{2-}, CrO_4^{2-}, HPO_4^{2-}, $S_2O_6^{2-}$

b	Charge 3
9	Al^{3+}, Fe^{3+}, Cr^{3+}, Sc^{3+}, Y^{3+}, La^{3+}, In^{3+}, Ce^{3+}, Pr^{3+}, Nd^{3+}, Sm^{3+}
6	Co (ethylenediamine)$_3^{3+}$
5	Citrate^{3-}
4	PO_4^{3-}, $Fe(CN)_6^{3-}$, $Cr(NH_3)_6^{3+}$, $Co(NH_3)_6^{3+}$, $Co(NH_3)_5H_2O^{3+}$

b	Charge 4
11	Th^{4+}, Zn^{4+}, Ce^{4+}, Sn^{4+}
6	$Co(S_2O_3)(CN)_5^{4-}$
5	$Fe(CN)_6^{4-}$

b	Charge 5
9	$Co(S_2O_3)_2(CN)_4^{5-}$

It is difficult to apply this relationship in process chemistry because the system of interest is generally composed of a mixture of different ionic species.

3. The Güntelberg relationship for solutions whose ionic strengths do not exceed 0.1 M:

$$\log \gamma = -AZ^2 \left(\frac{\sqrt{I}}{1 + \sqrt{I}} \right) \tag{1-46}$$

4. The Davies relationship for solutions whose ionic strengths do not exceed 0.5 M:

$$\log \gamma = -AZ^2 \left(\frac{\sqrt{I}}{1 + \sqrt{I}} - 0.2I \right) \tag{1-47}$$

Stumm and Morgan (1970) indicate that Davies has later suggested the use of 0.3 rather than 0.2 in the last term of the equation.

In process chemistry both the Güntelberg and Davies relationships are commonly used to calculate activity coefficients.

EXAMPLE PROBLEM 1-9: Analysis of a water reveals a TDS concentration of 200 mg/ℓ. If a saturated equilibrium condition is established at 25°C between the solid and aqueous phases of $CaCO_3$, what would the analytical Ca^{2+} concentration be if the equilibrium carbonate concentration were 20 mg/ℓ expressed as $CaCO_3$? The thermodynamic equilibrium constant (based on activities) for $CaCO_3$ is $10^{-8.32}$ at 25°C.

Solution:

1. Convert CO_3^{2-} concentration from mg/ℓ as $CaCO_3$ to moles/ℓ.

    (a) Apply equation 1-14 and convert to concentration as CO_3^{2-}:

    $$CO_3^{2-}(mg/\ell) = 20 \left[\frac{30}{50} \right] = 12 \text{ mg/}\ell$$

    (b) Apply equation 1-11 and convert mg/ℓ to moles/ℓ:

    $$CO_3^{2-}(moles/\ell) = \frac{12 \times 10^{-3}}{60} = 0.2 \times 10^{-3} \text{ mole/}\ell$$

2. Estimate the ionic strength of the solution from equation 1-41:

    $$I = (2.5 \times 10^{-5})(200) = 0.005 \ M$$

3. Set up the solubility equilibrium expression for $CaCO_3$:

    $$CaCO_{3(s)} \rightleftharpoons Ca^{2+}_{(aq)} + CO^{2-}_{3(aq)}$$

    $$(K_a)_{eq.} = (Ca^{2+})(CO_3^{2-})$$

where the ( ) denotes activities.

4. State the equilibrium expression in terms of analytical concentrations:

$$(K_a)_{eq.} = \gamma_D[Ca^{2+}]\gamma_D[CO_3^{2-}]$$

5. Using equation 1-46, compute values for the activity coefficients which are present in the concentration equilibrium expression:

$$\log \gamma_D = -(1.82 \times 10^6)[(78.3)(298)]^{-3/2}(2)^2 \left[\frac{\sqrt{0.005}}{1 + \sqrt{0.005}}\right] = -0.134$$

$$\gamma_D = 10^{-0.134} = 0.73$$

6. Substitute known values into the concentration equilibrium expression and determine $[Ca^{2+}]$:

$$[Ca^{2+}] = \frac{10^{-8.32}}{(0.73)^2(0.2 \times 10^{-3})} = 9.38 \times 10^{-5.32} \text{ mole}/\ell$$

7. Convert Ca^{2+} concentration from moles/ℓ to mg/ℓ as $CaCO_3$ by applying equations 1-11 and 1-14:

$$Ca^{2+} (\text{mg}/\ell \text{ as } CaCO_3) = [(9.38 \times 10^{-5.32})(40)(10^3)]\left[\frac{50}{20}\right] = 4.49$$

The Effect of Ionic Strength on the Value of the Equilibrium Constant

All waters of interest in process chemistry contain varying amounts of dissolved ionic material. Thus, it is necessary to correct the analytical concentration of each ionic species to the thermodynamically correct activity term.

For the general reaction

$$aA + bB \quad \rightleftharpoons \quad cC + dD \tag{1-33}$$

it is possible to write an equilibrium expression of the form

$$(K_a)_{eq.} = \frac{(C)^c(D)^d}{(A)^a(B)^b} \tag{1-48}$$

where $(K_a)_{eq.}$ = activity equilibrium constant, which is also known as the thermodynamic equilibrium constant

  ( ) = activity (moles/ℓ).

Since

$$a = \gamma C \tag{1-39}$$

or

$$( ) = \gamma [\ ]$$

for any real (nonideal) system, it is possible to write

$$(K_a)_{eq.} = \frac{[\gamma C]^c [\gamma D]^d}{[\gamma A]^a [\gamma B]^b} \tag{1-49}$$

or

$$(K_a)_{eq.} = \frac{[C]^c [D]^d}{[A]^a [B]^b} \left[\frac{\gamma_C^c \gamma_D^d}{\gamma_A^a \gamma_B^b} \right] \tag{1-50}$$

Substituting for the first term on the right-hand side of equation 1-50 from equation 1-34 gives

$$(K_a)_{eq.} = (K_c)_{eq.} \left[\frac{\gamma_C^c \gamma_D^d}{\gamma_A^a \gamma_B^b} \right] \tag{1-51}$$

or

$$(K_c)_{eq.} = \frac{(K_a)_{eq.}}{\left[\dfrac{\gamma_C^c \gamma_D^d}{\gamma_A^a \gamma_B^b} \right]} \tag{1-52}$$

In general, the equilibrium constants cited in the literature are thermodynamic equilibrium constants. Equation 1-52 illustrates how such constants should be modified when calculations are to be carried out using analytical concentrations rather than activities.

The law of chemical equilibrium may also be derived from thermodynamic considerations (the previous derivation was based on reaction kinetics) and the relationship between species activity and free energy change for equation 1-33 expressed as

$$\Delta G_{reaction} = \Delta G^\circ_{reaction} + RT \ln \left[\frac{(C)^c (D)^d}{(A)^a (B)^b} \right] \tag{1-53}$$

where R = gas constant (1.987 calories/mole-°K)

T = temperature (°K).

Equation 1-53 shows that the free energy change $\Delta G_{reaction}$ for a particular chemical reaction is a function of the activities of the reactants and products as well as the standard free energy change $\Delta G^\circ_{reaction}$ of the reaction.

It is possible to develop the relationship between $\Delta G^\circ_{reaction}$ and $(K_a)_{eq.}$ by considering that when the reaction has reached a state of equilibrium:

1. There is no net conversion of reactant to product, and hence $\Delta G_{reaction} = 0$.

2. $(K_a)_{eq.} = \dfrac{(C)^c (D)^d}{(A)^a (B)^b}$.

Thus, at equilibrium equation 1-53 reduces to the form

$$0 = \Delta G^\circ_{reaction} + RT \ln (K_a)_{eq.}$$

or

$$\Delta G^\circ_{reaction} = -RT \ln (K_a)_{eq.} \tag{1-54}$$

Since $\Delta G°$ denotes the standard state free energy change which implies a reaction temperature of 25°C, equation 1-54 may be further reduced to

$$\Delta G°_{\text{reaction}} = -(1.98)(298)(2.303) \log (K_a)_{\text{eq}}.$$

or

$$\Delta G°_{\text{reaction}} = -1363 \log (K_a)_{\text{eq}}. \qquad (1\text{-}55)$$

The relationship between $(K_a)_{\text{eq}}$ and $\Delta G°_{\text{reaction}}$ is presented in Table 1-7.

The data presented in Table 1-7 should be interpreted as follows: if $\Delta G°_{\text{reaction}}$ is negative, the position of equilibrium will lie to the right (the equilibrium concentration

TABLE 1-7 Relationship between $\Delta G°_{\text{reaction}}$ and $(K_a)_{\text{eq}}$. (after Conn and Stumpf, 1972).

$(K_a)_{\text{eq}}$	$\log (K_a)_{\text{eq}}$	$\Delta G°_{\text{reaction}}$
0.001	−3	4089
0.01	−2	2726
0.1	−1	1363
1	0	0
10	1	−1363
100	2	−2726
1000	3	−4089

of products will be large relative to the equilibrium concentration of reactants). However, a positive $\Delta G°_{\text{reaction}}$ indicates that the position of equilibrium will lie to the left (the equilibrium concentration of products will be small relative to the equilibrium concentration of reactants). If $\Delta G°_{\text{reaction}}$ is 0, then reactants and products are in equilibrium under standard state conditions, i.e., there is no net conversion of reactants to products under standard state conditions.

Equation 1-55 is important because it provides a means of computing values for many thermodynamic equilibrium constants from the vast quantity of thermodynamic data ($\Delta G°_{\text{formation}}$ data) which are available. Note that $\Delta G°_{\text{reaction}}$ is defined as the change in free energy when reactants and products are present in their standard states. The standard concentration state for solutes in solution is unit activity: for gases, 1 atm; and for solvents such as water, unit activity. If a hydrogen ion is produced or utilized in a reaction, its activity in the standard state is taken as 1 M or pH = 0.

EXAMPLE PROBLEM 1-10: Chlorine gas is commonly used to disinfect water for potable use. However, it is hypochlorous acid, HOCl, which is formed when chlorine gas is passed through water that is the primary disinfecting agent in most water treatment situations. Determine the equilibrium constant at standard state for the following reaction:

$$Cl_{2(g)} + H_2O_{(\ell)} \rightleftharpoons HOCl_{(aq)} + H^+_{(aq)} + Cl^-_{(aq)}$$

Solution:

1. From appendix I obtain $\Delta G^\circ_{formation}$ for each chemical species in the balanced chemical equation.

Species	$\Delta G_{formation}$ (KCal/mole)
$Cl_{2(g)}$	0.00
$H_2O_{(\ell)}$	−56.69
$HOCl_{(aq)}$	−19.11
$H^+_{(aq)}$	0.00
$Cl^-_{(aq)}$	−31.35

2. Compute the $\Delta G^\circ_{reaction}$ value from equation 1-37:

$$\Delta G^\circ_{reaction} = [(-31.35) + (0.00) + (-19.11)] - [(0.00) + (-56.69)]$$
$$= 6.23 \text{ KCal}$$

3. Calculate $(K_a)_{eq.}$ from equation 1-55:

$$\log (K_a)_{eq.} = \frac{6.23}{-1.364}$$
$$= -4.57$$

or

$$(K_a)_{eq.} = 10^{-4.57}$$

Hence, the position of equilibrium lies to the left; i.e., the reaction occurs to only a small extent under standard state conditions and is referred to as not being feasible.

A chemical reaction tends to proceed spontaneously from a nonequilibrium state to an equilibrium state because of the change in free energy between reactants and products. It was shown earlier that the free energy change for a chemical reaction is related to the free energy change of the reaction at standard state and to the activities of the products and reactants by the relationship

$$\Delta G_{reaction} = \Delta G^\circ_{reaction} + RT \ln \frac{(C)^c(D)^d}{(A)^a(B)^b} \qquad (1\text{-}53)$$

If the reaction quotient, Q, is defined as the product to reactant ratio of nonequilibrium activities, then

$$Q = \frac{(C)^c(D)^d}{(A)^a(B)^b} \qquad (1\text{-}56)$$

At equilibrium, Q has a value equal to $(K_a)_{eq.}$. Substituting Q into equation 1-53 gives

$$\Delta G_{reaction} = \Delta G^\circ_{reaction} + RT \ln Q \qquad (1\text{-}57)$$

A further substitution from equation 1-54 for $\Delta G^\circ_{\text{reaction}}$ in equation 1-57 gives

$$\Delta G_{\text{reaction}} = -RT \ln (K_a)_{\text{eq}} + RT \ln Q \qquad \textbf{(1-58)}$$

Equation 1-58 can be rearranged into the form

$$\Delta G_{\text{reaction}} = RT \ln \frac{Q}{(K_a)_{\text{eq.}}} \qquad \textbf{(1-59)}$$

ΔG Determinations at Other Than Standard State

Few if any reactions of interest in process chemistry occur at pH = 0 (standard state for H^+). Generally, reactants and products of the various reactions of interest exist at something other than standard state. Thus, it is desirable to be able to compute the free energy change for the actual reaction conditions. The required procedure for such a determination is illustrated in example problem 1-11.

EXAMPLE PROBLEM 1-11: The $\Delta G^\circ_{\text{reaction}}$ for the reaction presented in example problem 1-10 was found to be 6.23 KCal.

$$Cl_{2(g)} + H_2O_{(\ell)} \rightleftharpoons HOCl_{(aq)} + H^+_{(aq)} + Cl^-_{(aq)}$$

(a) Compute $\Delta G_{\text{reaction}}$, assuming that all reactants and products are in their standard state except H^+. The pH of the solution is 7.

(b) Compute $\Delta G_{\text{reaction}}$, assuming that only water is in its standard state, whereas all solutes are at an activity of 0.01 M, the pH = 7, and Cl_2 is at 2 atm pressure.

Solution:

Part (a):

1. Compute $\Delta G_{\text{reaction}}$ from equation 1-57:

$$\Delta G_{\text{reaction}} = \Delta G^\circ_{\text{reaction}} + 2.303\ RT \log Q$$

where

$$Q = \frac{(HOCl)(H^+)(Cl^-)}{(Cl_2)(H_2O)}$$

Since only the activity of the hydrogen ion is not at standard state, equation 1-57 reduces to

$$\Delta G_{\text{reaction}} = 6.23 + (298)(1.98)(10^{-3})(2.303) \log 10^{-7}$$
$$= 6.23 + (-9.51)$$
$$= -3.3 \text{ KCal}$$

Note that by changing the pH of the solution, the equilibrium position for the reaction has shifted from the left to the right.

Part (b):

1. Compute $\Delta G_{reaction}$ from equation 1-57:

$$\Delta G_{reaction} = 6.23 + (298)(1.98)(10^{-3})(2.303) \log \frac{(0.01)(10^{-7})(0.01)}{(2)}$$
$$= 6.23 + (-15.30)$$
$$= -9.1 \text{ KCal}$$

Simultaneous Reactions

In process chemistry the general case is to have not one but several reactions occurring simultaneously. To analyze such a system, the engineer must group all the reactions occurring in the system into one of two different classes:

1. Those reactions which have no reactants or products which are common to any other reaction occurring in the system. Reactions which fall into this class are treated individually. The position of equilibrium for such reactions is indicated by the $\Delta G_{reaction}$ value for any specified reaction; i.e., the $\Delta G_{reaction}$ value is computed by assuming that each reaction occurs alone in the system.

2. Those reactions which have reactants and/or products which are common to other reactions occurring in the system. Reactions which fall into this class cannot be treated individually, but, rather, all reactions which have chemical species in common must be considered together. A unique $\Delta G_{reaction}$ value is determined for the entire group of reactions from the algebraic sum of the individual $\Delta G_{reaction}$ values for the reactions in the group. The overall $\Delta G_{reaction}$ value determines the position of equilibrium for the system of reactions.

The required procedure for computing the overall $\Delta G_{reaction}$ value for a system of reactions is illustrated in example problem 1-12.

EXAMPLE PROBLEM 1-12: Precipitation in the form of calcium carbonate is a common method for removing the cation calcium which causes hardness from water. In this method, the pH of the water is raised by the addition of lime, $Ca(OH)_2$. The solubility of calcium carbonate is reduced with increasing pH; thus, an increase in pH causes $CaCO_{3(s)}$ to precipitate.

Any simultaneous chemical reaction which influences the residual soluble calcium concentration must be considered if the softening process is to be accurately modeled. One such reaction is the formation of the $CaCO_3^\circ$ ion-pair.

Will the following reactions occur simultaneously?

1. $CaCO_{3(s)} \rightleftharpoons Ca^{2+}_{(aq)} + CO^{2-}_{3(aq)}$: $(K_a)_{eq} = 10^{-8.34}$

2. $Ca^{2+}_{(aq)} + CO^{2-}_{3(aq)} \rightleftharpoons CaCO^\circ_{3(aq)}$: $(K_a)_{eq} = 10^{3.2}$

where $(CO_3^{2-})_{(aq)} = 10^{-5} \text{ mole}/\ell$

$(Ca^{2+})_{(aq)} = 10^{-2.5} \text{ mole}/\ell$

$(CaCO_3^\circ)_{(aq)} = 10^{-5.5} \text{ mole}/\ell$

Temperature $= 25°C$
are initial conditions.

Solution:

1. Compute the $\Delta G_{\text{reaction}}$ value for reaction 1 from equation 1-59:

$$\Delta G_{\text{reaction}} = RT \ln \left[\frac{Q}{(K_a)_{\text{eq.}}} \right]$$

(a) Calculate the value of the reaction quotient, Q:

$$Q = \frac{(Ca^{2+})_{(aq)}(CO_3^{2-})_{(aq)}}{(CaCO_3)_{(s)}}$$

The activity of the solid phase is taken to be unity; therefore

$$Q = (Ca^{2+})(CO_3^{2-}) = (10^{-2.5})(10^{-5}) = (10^{-7.5})$$

(b) Compute $\Delta G_{\text{reaction}}$:

$$\Delta G_{\text{reaction}} = 1.36 \log \left[\frac{10^{-7.5}}{10^{-8.34}} \right] = 1.14 \text{ KCal}$$

2. Compute the $\Delta G_{\text{reaction}}$ value for reaction 2 from equation 1-59:

(a) Calculate the value of the reaction quotient, Q:

$$Q = \frac{(CaCO_3)_{(aq)}}{(Ca^{2+})_{(aq)}(CO_3^{2-})_{(aq)}} = \frac{(10^{-5.5})}{(10^{-2.5})(10^{-5})} = 10^2$$

(b) Compute $\Delta G_{\text{reaction}}$:

$$\Delta G_{\text{reaction}} = 1.36 \log \left[\frac{10^2}{10^{3.2}} \right] = -1.63 \text{ KCal}$$

3. Sum the individual reaction free energy changes together to obtain the value of the driving force for the system of reactions:

$$\Delta G_{\text{system}} = \sum_{i=1}^{i=i} \Delta G_i = (1.14) + (-1.63) = -0.49 \text{ KCal}$$

Therefore, both reactions proceed as written. The effect of ion-pairing is to increase the residual soluble calcium concentration.

1-7 FACTORS AFFECTING CHEMICAL EQUILIBRIUM

Le Châtelier's principle states that if stress is applied to a system in equilibrium, the system will act to relieve the stress and restore equilibrium, but under a new set of equilibrium conditions. Consider the hypothetical equilibrium represented by equation 1-33:

$$aA + bB \rightleftharpoons cC + dD \qquad\qquad (1\text{-}33)$$

According to Le Châtelier's principle, if the concentration of either A or B is increased, the equilibrium shifts to the right (the rate of the reaction to the right increases), whereas if the concentration of either C or D is increased, the equilibrium shifts to the left (the rate of the reaction to the left increases). The equilibrium always shifts in the direction that reduces the concentration of the chemical species which was added. In other words, when the concentration of one or more of the chemical species existing in the system at equilibrium is changed, the concentration of all the species will change and a new equilibrium composition will be established.

The position of equilibrium is also affected by the presence of catalysts or by changes in pressures or temperature. In this text the discussion is limited mainly to the effects of changes in concentration and temperature.

Concentration Effects

In the derivation of the law of chemical equilibrium using the kinetic approach, it was shown that (a) the rates of the forward and reverse reactions are dependent on the concentrations of the reacting chemical species, and (b) when equilibrium has been attained the rates of the forward and reverse reactions are equal. The rate equations for the general reaction

$$aA + bB \rightleftharpoons cC + dD \qquad (1\text{-}33)$$

are

$$r_f = K_1[A]^a[B]^b \qquad (1\text{-}60)$$

$$r_r = K_2[C]^c[D]^d \qquad (1\text{-}61)$$

at equilibrium

$$r_f = r_r$$

and the concentrations for A, B, C, and D indicated in equations 1-60 and 1-61 are equilibrium concentrations. After equilibrium has been attained, if the concentration of one of the reacting species is changed, the rate of the appropriate reaction will change according to its rate equation and the system will no longer be at equilibrium. For example, if the concentration of A is increased, the rate of the forward reaction will increase. The system will shakedown to a new equilibrium state where the equilibrium concentrations of the reacting species are entirely different from the equilibrium concentrations of the first equilibrium state. However, the ratio of the concentration given by equation 1-34; i.e.,

$$(K_c)_{eq.} = \frac{[C]^c[D]^d}{[A]^a[B]^b} \qquad (1\text{-}34)$$

will remain unchanged. Since the equilibrium concentrations of the reacting species are different for the new equilibrium state, the rates of the forward and reverse reactions will be different from their values at the previous equilibrium state but will still be equal.

Temperature Effects

In most cases the reactions of interest in process chemistry are conducted at temperatures other than 25°C. Therefore, when computing the equilibrium position for a reaction or system of reactions, it is necessary to adjust equilibrium constants for nonstandard state temperatures. Van't Hoff proposed the following relationship for the variation of the equilibrium constant with absolute temperature:

$$\frac{d \ln (K_a)_{eq.}}{dT} = \frac{\Delta H^{\circ}_{reaction}}{R(T)^2} \qquad \textbf{(1-62)}$$

Integrating equation 1-62 gives

$$\ln (K_a)_{eq.} = \text{constant} - \frac{\Delta H^{\circ}_{reaction}}{RT} \qquad \textbf{(1-63)}$$

If the forward reaction (left-to-right) for a chemical system at equilibrium is endothermic, then the backward reaction (right-to-left) is exothermic. Le Châtelier's principle states that the equilibrium position of a chemical system will shift in such a manner as to relieve any stress imposed on it. This means that if heat is added to a chemical system at equilibrium, the equilibrium position will shift so that a portion of the heat energy added to the system can be utilized. Thus, if heat is added to the aforementioned chemical system, the position of equilibrium is shifted to the right (the endothermic reaction is favored), which means that the value of the equilibrium constant is increased.

The effect of temperature (as described by equation 1-63) on the magnitude of the equilibrium constant is illustrated in Figure 1-3. The slope term in equation 1-63 is

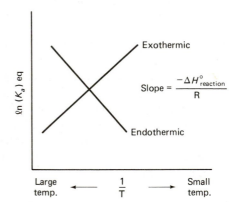

FIGURE 1-3 Relationship between equilibrium constant and temperature.

given by $-\Delta H^{\circ}_{reaction}/R$. The trace which is derived from this equation can have either a positive or negative slope, depending on the sign of $\Delta H^{\circ}_{reaction}$. If the reaction of interest is exothermic, $\Delta H^{\circ}_{reaction}$ will have a negative value and the slope of the trace will be positive and will indicate a decrease in the magnitude of $(K_a)_{eq.}$ with an increase in temperature. The opposite is true when the reaction of interest is endothermic.

Assuming that $\Delta H^{\circ}_{reaction}$ remains constant within the temperature range of T_1 to T_2, integration of equation 1-62 between the limits of T_1 to T_2 gives an expression of the form

$$\ln \left\{ \frac{[(K_a)_{eq.}]_2}{[(K_a)_{eq.}]_1} \right\} = \frac{-\Delta H^{\circ}_{reaction}}{R} \left[\frac{1}{T_2} - \frac{1}{T_1} \right]$$

or

$$\ln \left\{ \frac{[(K_a)_{\text{eq.}}]_2}{[(K_a)_{\text{eq.}}]_1} \right\} = \frac{\Delta H^\circ_{\text{reaction}}}{R} \left[\frac{T_2 - T_1}{T_1 T_2} \right] \qquad \text{(1-64)}$$

If $[(K_a)_{\text{eq.}}]_1$ is known for T_1, then by using equation 1-64 it is possible to compute $[(K_a)_{\text{eq.}}]_2$ for the temperature change from T_1 to T_2 if $\Delta H^\circ_{\text{reaction}}$ is known. This equation has been used to estimate the effect of temperature over a limited range for a number of chemical reactions. However, it should always be remembered that inherent in the use of this equation is the assumption that $\Delta H^\circ_{\text{reaction}}$ remains constant. In certain reactions of interest in process chemistry, this is not the case, and in such situations care should be taken when adjusting the equilibrium constant for temperature variations. In later discussions empirical relationships will be presented which can be used to adjust the equilibrium constant for temperature variations for those reactions which do not follow the van't Hoff relationship.

EXAMPLE PROBLEM 1-13: In example problem 1-12 it was noted that calcium carbonate precipitation was the mechanism for calcium removal during lime softening.

$$CaCO_{3(s)} \rightleftharpoons Ca^{2+}_{(aq)} + CO^{2-}_{3(aq)}$$

At 25°C the equilibrium constant, $(K_a)_{\text{eq.}}$, for this reaction is $10^{-8.34}$. However, softening at most water treatment plants is carried out at temperatures somewhat lower than 25°C. With this in mind, determine what effect softening at 10°C will have on calcium removal.

Solution:

1. Compute $\Delta H^\circ_{\text{reaction}}$ from equation 1-38:

   (a) Determine $\Delta H^\circ_{\text{formation}}$ for each compound in the reaction of interest from Appendix I.

Compound	$\Delta H^\circ_{\text{formation}}$ (KCal/mole)
$CaCO_{3(s)}$	-288.45
$Ca^{2+}_{(aq)}$	-129.77
$CO^{2-}_{3(aq)}$	-161.63

   (b) Calculate $\Delta H^\circ_{\text{reaction}}$:

   $$\Delta H^\circ_{\text{reaction}} = [(-129.77) + (-161.63)] - [-288.45]$$
   $$= -2.95 \text{ KCal}$$

2. Determine $[(K_a)_{\text{eq.}}]_{10°C}$ from equation 1-64, which has been rearranged into the form

   $$\log [(K_a)_{\text{eq.}}]_{10°C} = \left[\frac{\Delta H^\circ_{\text{reaction}}}{2.303 R} \left(\frac{T_2 - T_1}{T_1 T_2} \right) \right] + \log [(K_a)_{\text{eq.}}]_{25°C}$$

Therefore,

$$\log [(K_a)_{eq.}]_{10°C} = \left[\frac{-2950 \text{ calories}}{(2.303)(1.98)}\left(\frac{283 - 298}{(283)(298)}\right)\right] + \log 10^{-8.34}$$

$$= 0.116 - 8.34$$

or

$$[(K_a)_{eq.}] = 10^{-8.22}$$

Thus, a reduction in temperature will increase the magnitude of the equilibrium constant slightly. This means that the position of equilibrium will be shifted to the right, which will increase the equilibrium soluble calcium concentration.

PROBLEMS

1-1. Determine the oxidation number of each element in the following compounds:

(a) $HOCl$ (g) $Mg(OH)_2$

(b) $NaOCl$ (h) $CaCO_3$

(c) $FeCl_3$ (i) $Ca(HCO_3)_2$

(d) $Al_2(SO_4)_3$ (j) NH_4NO_3

(e) H_3PO_4 (k) HNO_2

(f) $Ca(OH)_2$ (l) $K_2Cr_2O_7$

1-2. How many moles of water can be obtained from 1000 g of each of the hydrates?

(a) $Al_2(SO_4)_3 \cdot 18H_2O$

(b) $CuSO_4 \cdot 5H_2O$

1-3. When a water treatment plant operator buys $Al_2(SO_4)_3 \cdot 18H_2O$ to be used in the coagulation process, what percent aluminum is he buying?

1-4. Name the compounds listed in Problem 1-1.

1-5. Both nitrogen and phosphorus are considered to be the key nutrients in the eutrophication process. When measuring the concentration of the various chemical species containing these elements, it is common practice to express the concentration value in the most elemental form. To illustrate this concept, assume that the ammonia concentration of a wastewater effluent is found to be 12.15 mg/ℓ. To express the nitrogen concentration in its elemental form, the ammonia concentration is multiplied by the weight fraction of nitrogen in the ammonia molecule; i.e., $12.15(\frac{14}{17}) = 10$ mg/ℓ. To denote that concentration is expressed in the elemental form, the concentration is followed by a bracket containing the chemical species measured, a hyphen, and then the chemical symbol for the nutrient, e.g., 10 mg/ℓ (NH_3-N).

Express the following concentrations in their elemental form:

(a) 25 mg/ℓ of NH_3

(b) 15 mg/ℓ of NO_2^-

(c)   18 mg/ℓ of NO_3^-

(d)   29 mg/ℓ of PO_4^{3-}

1-6. The following unbalanced equations do not involve oxidation and reduction. Convert them to balanced equations.

(a) $FeSO_4 \cdot 7H_2O + Ca(HCO_3)_2 \longrightarrow Fe(HCO_3)_2 + CaSO_4 + H_2O$

(b) $Fe(HCO_3)_2 + Ca(OH)_2 \longrightarrow Fe(OH)_2 + CaCO_3 + H_2O$

(c) $Fe_2(SO_4)_3 + Ca(HCO_3)_2 \longrightarrow Fe(OH)_3 + CaSO_4 + CO_2$

(d) $Fe_2(SO_4)_3 + Ca(OH)_2 \longrightarrow Fe(OH)_3 + CaSO_4$

(e) $H_2SO_3 + Ca(HCO_3)_2 \longrightarrow Ca(HSO_3)_2 + CO_2 + H_2O$

(f) $Ca(OH)_2 + CO_2 \longrightarrow CaCO_3 + H_2O$

(g) $FeCl_3 + H_2O \longrightarrow Fe(OH)_3 + H^+ + Cl^-$

(h) $FeCl_3 + Ca(OH)_2 \longrightarrow Fe(OH)_3 + CaCl_2$

1-7. The following unbalanced equations involve oxidation and reduction. Convert them to balanced equations.

(a) $H_2S + Cl_2 \longrightarrow SO_2 + Cl^-$

(b) $Fe(HCO_3)_2 + Cl_2 \longrightarrow Fe(OH)_3 + CO_2 + Cl^-$

(c) $Cr_2O_7^{2-} + Fe^{2+} \longrightarrow Cr^{3+} + Fe^{3+}$

1-8. When alum is added to water containing natural alkalinity, the following reaction occurs:

$$Al_2(SO_4)_3 \cdot 14H_2O + 3Ca(HCO_3)_2 \longrightarrow 2Al(OH)_3 + 3CaSO_4 + 6CO_2 + 18H_2O$$

If alkalinity is expressed as mg/ℓ of $CaCO_3$, compute the alkalinity concentration required for each mg/ℓ of alum added to the water.

1-9. If insufficient natural alkalinity is available to complete the alum reaction, lime is added to make up the alkalinity deficit.

$$Al_2(SO_4)_3 \cdot 18H_2O + 3Ca(OH)_2 \longrightarrow 2Al(OH)_3 + 3CaSO_4 + 18H_2O$$

Compute the lime concentration required for each mg/ℓ of alum added to the water.

1-10. Calculate in pounds per million gallons the theoretical weight of alum, $Al_2(SO_4)_3 \cdot 18H_2O$, required to give a dose of 10 mg/ℓ as $Al_2(SO_4)_3$. The water contains 2.5 mg/ℓ of natural alkalinity, expressed as $CaCO_3$. How much lime as $Ca(OH)_2$ must be added to complete the reaction with alum?

1-11. How many grams of K_2SO_4 are contained in 100 mℓ of 0.40 N solution?

1-12. How many mℓ of 2.5 M H_2SO_4 would be neutralized by 20 mℓ of 3.0 M NaOH?

1-13. How many mℓ of 0.05 N $HC_2H_3O_2$ (acetic acid) and how many mℓ of 0.05 N H_2SO_4 would be neutralized by 10 mℓ of 0.5 N NaOH?

1-14. A sodium hydroxide solution of unknown concentration was standardized against a perchloric acid ($HClO_4$) solution labeled 0.1 N. If 100 mℓ of NaOH solution were neutralized by 70 mℓ of the acid, what is the normality of the NaOH? If, in fact, the perchloric acid solution had been standardized for a redox reaction in which the chlorine atom was reduced to the chloride ion, what would be the actual normality of the NaOH solution?

1-15. Calculate the ionic strength of a solution which was analyzed and found to have the following chemical composition:

Ions Present	Conc. (mg/ℓ)
Ca^{2+}	58.0
Mg^{2+}	85
Na^+	10.1
Fe^{2+}	0.33
CO_3^{2-}	0.41 as $CaCO_3$
HCO_3^-	183.0
SO_4^{2-}	23.59 as $CaCO_3$
Cl^-	15.0
NO_3^-	4.0
F^-	1.6

1-16. When chlorine gas is added to pure water the following reaction occurs:

$$Cl_{2(g)} + H_2O_{(\ell)} \rightleftharpoons HOCl_{(aq)} + H^+_{(aq)} + Cl^-_{(aq)}$$

At hydrogen ion activities less than about 10^{-4} M, this reaction can be considered to achieve completion. Hypochlorous acid (HOCl) is a weak electrolyte and ionizes according to the reaction

$$HOCl_{(aq)} \rightleftharpoons H^+_{(aq)} + OCl^-_{(aq)}$$

If $(K_a)_{eq.}$ for this reaction is 2.5×10^{-8} at 25°C and $(H^+) = 1$ M, where will the position of equilibrium be if the temperature is 10°C and the $(H^+) = 10^{-6}$ M? Assume all other chemical species to be at unit activity. What happens to the position of equilibrium if the hydrogen ion activity is decreased to 10^{-8} M?

REFERENCES

BALLINGER, J.T., and WOLF, L.J., *Chemical Science and Technology*, I, St. Louis Community College at Florissant Valley, MO (1976).

BUTLER, J.N., *Ionic Equilibrium—A Mathematical Approach*, Addison-Wesley Publishing Co., Inc., Reading, MA (1964).

CONN, E.E., and STUMPF, P.K., *Outlines of Biochemistry*, 3rd ed., John Wiley & Sons, Inc., New York (1972).

GILL, S.J., and NORMAN, A.D., *An Advanced Introduction To Chemistry*, The C. V. Mosby Company, St. Louis, MO (1975).

HEIN, M., *Foundations of College Chemistry*, 2nd ed., Dickenson Publishing Company, Inc., Encino, CA (1970).

JURINAK, J.J., "Chemistry of Aquatic Systems," Lecture Notes, Utah State University (1976).

KEMP, P.H., "Chemistry of Natural Waters—I," *Water Research*, **5,** 297 (1971).

LANGELIER, W.F., "Effect of Temperature on the pH of Natural Waters," *J. Am. Water Works Assoc.*, **28,** 1500 (1936).

MOELLER, T., and O'CONNOR, R., *Ions in Aqueous Systems*, McGraw-Hill Book Company, New York (1972).

SMOOT, R.C., PRICE, J., and BARRETT, R.L., *Chemistry—A Modern Course*, Charles E. Merrill Publishing Company, Columbus, OH (1971).

STUMM, W., and MORGAN, J.J., *Aquatic Chemistry*, Wiley-Interscience, New York (1970).

TIMM, J.A., *General Chemistry*, 4th ed., McGraw-Hill Book Company, New York (1966).

2

ACID-BASE EQUILIBRIA

2-1 FUNDAMENTALS

The ionization of water is described by equation 2-1:

$$2H_2O \rightleftharpoons H_3O^+ + OH^- \tag{2-1}$$

According to this equation, water functions as both a Brönsted acid and a Brönsted base. Since the concentration of H_3O^+ and OH^- ions in pure water is very small, i.e., water exhibits ideal behavior with respect to the concentration of H_3O^+ and OH^- ions, the equilibrium constant expression for equation 2-1 has the following form [using the simplified reaction $(H_2O \rightleftharpoons H^+ + OH^-)$]:

$$(K_a)_{eq.} = \frac{(H^+)_{(aq)}(OH^-)_{(aq)}}{(H_2O)_{(\ell)}} \tag{2-2}$$

The value of $(K_a)_{eq.}$ has been found to be 1.8×10^{-16} mole/ℓ at 25°C.

The concentration (or activity) of H_2O in pure water may be computed to be $(1000/18)$ or 55.5 moles/ℓ. In dilute aqueous solutions such as those encountered in process chemistry, the concentration of H_2O can be considered as a constant and is normally incorporated into the value of $(K_a)_{eq.}$ so that the equilibrium constant expression for the ionization of water takes the form

$$K_w = (55.5)(1.8 \times 10^{-16}) = (H^+)_{(aq)}(OH^-)_{(aq)} \tag{2-3}$$

where K_w is termed the *ion product of water* and has an approximate value of 10^{-14} at 25°C. Equation 2-3 shows that for any aqueous solution, if the concentration of either H^+ or OH^- ion is known, the concentration of the other species can always be calculated.

42

The "p" Notation

Because the values for the concentration of H^+ and OH^- ions and for the equilibrium constants are usually very small numbers, it is convenient to handle them in logarithmic terms, which have the following forms:

$$pH = -\log (H^+) \tag{2-4}$$

$$pOH = -\log (OH^-) \tag{2-5}$$

$$p(K_a)_{eq} = -\log (K_a)_{eq.} \tag{2-6}$$

A useful term relating pH and pOH to the ion product of water can be developed by taking the logarithm of both sides of equation 2-3:

$$\log 10^{-14} = \log (H^+) + \log (OH^-) \tag{2-7}$$

or

$$-14 = \log (H^+) + \log (OH^-) \tag{2-8}$$

Multiplying both sides of equation 2-8 by -1 gives

$$14 = -\log (H^+) - \log (OH^-) \tag{2-9}$$

By substituting from equations 2-4 and 2-5 for $-\log (H^+)$ and $-\log (OH^-)$, equation 2-9 reduces to

$$14 = pH + pOH \tag{2-10}$$

The pH and pOH of neutral solutions are both equal to 7, whereas the pH of acidic solutions is something less than 7 and the pH of alkaline solutions is greater than 7.

It is important to understand that pH is defined as the negative log of the hydrogen ion activity and that a pH meter measures activity and not analytical concentration.

Equilibrium Calculations for Acids and Bases

A weak acid or base is a weak electrolyte, which means they are only partially ionized in an aqueous solution, and as a result equilibrium is established between the ionized and unionized chemical forms. The position of equilibrium will depend on the degree of ionization that occurs.

According to the Brönsted theory, an acid is a proton donor, whereas a base is a proton acceptor. This implies that a chemical equation representing the ionization of a weak monoprotic (containing one hydrogen atom per acid molecule) acid may be written as

$$HA + H_2O \rightleftharpoons H_3O^+ + A^-$$

or, in a more simplified form, as

$$HA \rightleftharpoons H^+ + A^- \tag{2-11}$$

The equilibrium constant expression for equation 2-11 has the form

$$K_a = \frac{(H^+)(A^-)}{(HA)} \tag{2-12}$$

where K_a represents the thermodynamic equilibrium constant for the acid reaction. Solving equation 2-12 for (H^+) gives

$$(H^+) = \frac{K_a(HA)}{(A^-)} \tag{2-13}$$

Taking the logarithm of both sides of this expression yields

$$\log(H^+) = \log K_a + \log(HA) - \log(A^-) \tag{2-14}$$

Multiplying both sides of equation 2-14 by -1 gives

$$-\log(H^+) = -\log K_a - \log(HA) + \log(A^-)$$

which can be expressed as

$$pH = pK_a + \log\left[\frac{(A^-)}{(HA)}\right] \tag{2-15}$$

or

$$pH = pK_a + \log\left[\frac{(\text{proton acceptor})}{(\text{proton donor})}\right] \tag{2-16}$$

Equation 2-16 is known as the *Henderson-Hasselbalch equation* and is very useful in calculating the pH of a solution containing a mixture of a weak acid and its salt.

Certain acids are capable of releasing more than one hydrogen ion on ionization of an acid molecule. For example, a diprotic acid will release two hydrogen ions when the acid molecule is completely ionized. The ionization of a diprotic acid is considered to be a two-step process:

STEP 1:

$$H_2A \rightleftharpoons HA^- + H^+ \tag{2-17}$$

$$K_{a1} = \frac{(HA^-)(H^+)}{(H_2A)} \tag{2-18}$$

where K_{a1} represents the thermodynamic equilibrium constant for the first ionization or loss of the first proton.

STEP 2:

$$HA^- \rightleftharpoons A^{2-} + H^+ \tag{2-19}$$

$$K_{a2} = \frac{(A^{2-})(H^+)}{(HA^-)} \tag{2-20}$$

where K_{a2} represents the thermodynamic equilibrium constant for the second ionization or loss of the second proton.

An acid capable of liberating three hydrogen ions on ionization of an acid molecule is called a *triprotic acid*, and equilibrium is represented by three different reactions. When acids have more than one ionization constant, the constants generally differ in value by several orders of magnitude. The Henderson-Hasselbalch equation can be applied to each individual step in the ionization of a polyprotic acid.

The chemical equation representing the reaction of a weak base in water may be written as

$$A^- + H_3O^+ \rightleftharpoons HA + H_2O$$

or

$$A^- + H^+ \rightleftharpoons HA \tag{2-21}$$

The equilibrium constant expression for this reaction is

$$K_b = \frac{(HA)}{(A^-)(H^+)} \tag{2-22}$$

where K_b represents the thermodynamic equilibrium constant for the base reaction.

Strong acids and bases can be handled in a manner similar to weak acids and bases. The important thing to remember here is that these chemical species are strong electrolytes, which means that the position of equilibrium for their ionization will be far to the right; i.e., ionization is almost complete.

It is important to understand the effect which the addition of an acid or base has on the ionization of water. If an acid is added to water, hydrogen ions are released according to equation 2-11:

$$HA \rightleftharpoons H^+ + A^- \tag{2-11}$$

When a base is added to water, hydrogen ions are removed according to equation 2-21:

$$A^- + H^+ \rightleftharpoons HA \tag{2-21}$$

In any dilute aqueous solution the equilibrium constant expression for the ionization of water must be satisfied:

$$K_w = (H^+)(OH^-) \tag{2-3}$$

Thus, for K_w to remain constant after the addition of an acid or base, the system must shakedown to a new equilibrium position (Le Châtelier's principle) where the product of the hydrogen ion activity and the hydroxyl ion activity equals K_w. If an acid increases the hydrogen ion activity of a dilute aqueous solution, then the hydroxyl ion activity must necessarily decrease if K_w is to remain constant. The reverse is true for the addition of a base to an aqueous solution.

A general procedure for solving most types of chemical equilibrium problems is outlined in the following steps:

1. Write a balanced chemical equation for each reaction involved in the equilibrium state of the chemical system of interest. One of these reactions will always be the ionization of water for reactions occurring in dilute aqueous solution.

2. Develop an equilibrium constant expression for each reaction described in step 1.

3. Develop any additional expressions required to give an equal number of equations to unknown values which exist in the chemical system at equilibrium. Normally, additional equations come from two sources:
   (a) Mass balance relationships for the chemical species of interest:

$$C_T = \sum_{i=1}^{i=n} C_i \qquad (2\text{-}23)$$

   where C_T = total molar concentration of the chemical species of interest (moles/ℓ)

   C_i = concentration of individual chemical forms which contain the species of interest (moles/ℓ).

   (b) An electron charge balance expression based on the fact that the sum of negative charges must equal the sum of positive charges in aqueous solution:

$$\sum_{i=1}^{i=n} [P_i(C_i^{P_i})] = \sum_{i=1}^{i=n} [N_i(A_i^{N_i})] \qquad (2\text{-}24)$$

   where (C^P) = molar concentration of cationic species C with positive charge P (moles/ℓ)

   (A^N) = molar concentration of anionic species A with negative charge N.

4. Manipulate all the equations until values are determined for all unknown quantities. In many cases these mathematical manipulations may become complex and time-consuming unless certain simplifying assumptions are made. However, the individual who is making such assumptions should always be sure they are valid for the situation in which they are being used.

The four-step approach for solving equilibria problems will now be illustrated through the use of two example problems dealing with acid-base equilibria. In these problems the equations necessary for an exact solution will be developed. Yet, many times the equations necessary for an exact solution are complex and difficult to solve. For this reason, the assumptions necessary to simplify the solution to the example problems will also be outlined and subsequent mathematical operations presented.

EXAMPLE PROBLEM 2-1: Determine the pH of a 10^{-6} molar solution of hydrochloric acid, HCl, assuming that the solution exhibits ideal behavior and the temperature is $25°C$. HCl is a strong acid.

Solution:

A. *EXACT SOLUTION PROCEDURE*

1. Write balanced chemical equations for all reactions involved in the equilibrium state of the chemical system:

   (a) $HCl \rightleftharpoons H^+ + Cl^-$

   (b) $H_2O \rightleftharpoons H^+ + OH^-$

2. Write equilibrium constant expressions for all reactions:

   (a) $K_a = \dfrac{[H^+][Cl^-]}{[HCl]}$

   (b) $K_w = [H^+][OH^-]$

3. Write a mass balance relationship which accounts for the total mass of chlorine in the system by applying equation 2-23:

$$C_T = [HCl] + [Cl^-]$$

4. Write an electron charge balance expression by applying equation 2-24:

$$(1)\,[H^+] = (1)\,[Cl^-] + (1)\,[OH^-]$$

   or

$$[H^+] = [Cl^-] + [OH^-]$$

5. Solving the electron charge balance expression for $[Cl^-]$ and substituting from this equation for $[Cl^-]$ in the mass balance expression gives

$$[HCl] = C_T - [H^+] + [OH^-]$$

6. Substituting from the expression in step 5 for $[HCl]$ in the equilibrium constant expression for hydrochloric acid gives

$$K_a = \frac{[H^+]([H^+] - [OH^-])}{C_T - ([H^+] - [OH^-])}$$

7. Substituting for $[OH^-]$ from the ionization constant expression for water, the expression in step 6 becomes

$$K_a = \frac{[H^+]([H^+] - (K_w/[H^+]))}{C_T - ([H^+] - (K_w/[H^+]))}$$

   This expression expands into an equation of the form

$$[H^+]^3 + (K_a - K_w)[H^+]^2 - K_a C_T[H^+] + K_a K_w = 0$$

   which must be solved by trial and error for $[H^+]$.

B. *APPROXIMATE SOLUTION PROCEDURE:* It is possible to solve for $[H^+]$ by a simpler method than solving a cubic equation. Such a method requires that a number of assumptions be made to reduce the complexity of the problem. The assumptions which are usually made for a solution containing a strong acid or base are as follows:

(a) The water equilibrium reaction is insignificant; i.e., the concentration of H^+ ions contributed by the ionization of H_2O is insignificant compared to the concentration of H^+ ions from the ionization of the acid, and the concentration of OH^- ions contributed by the ionization of water is insignificant compared to the concentration of OH^- ions produced by the base reaction.

(b) Strong acids and bases are considered to be totally ionized in aqueous solution. Thus, equilibrium is not established during the ionization process, but rather the reactions go to completion. For such reactions the molar ratio method can be used to determine the amount of material produced or consumed during the course of the reaction.

1. Write the balanced chemical equation for the ionization of HCl:

$$HCl \longrightarrow H^+ + Cl^-$$

2. Compute the molar concentration of the H^+ ion:
   (a) Calculate the H^+: HCl molar ratio

$$H^+ : HCl = \frac{1}{1} = 1$$

   (b) Calculate the H^+ ion concentration:

$$[H^+] = (1)(10^{-6}) = 1 \times 10^{-6} \text{ mole}/\ell$$

3. Determine the pH of the solution from equation 2-4:

$$pH = -\log[10^{-6}] = 6$$

EXAMPLE PROBLEM 2-2: Calculate the H^+ ion concentration for a 10^{-3} molar solution of nitrous acid, assuming that the solution exhibits ideal behavior and the temperature is 25°C. HNO_2 is a weak acid.

Solution:

A. EXACT SOLUTION PROCEDURE: The exact solution to a monoprotic acid ionization problem requires the solution of the cubic equation presented in step 7 of example problem 2-1, regardless of whether the acid is strong or weak.

B. APPROXIMATE SOLUTION PROCEDURE: To reduce the complexity of the problem, the assumption that the water equilibrium reaction is insignificant is usually made for a solution containing a weak acid or base.

1. Develop the balanced chemical equation for the ionization of nitrous acid:

$$HNO_2 \rightleftharpoons H^+ + NO_2^-$$

2. Write the equilibrium constant expression for the acid equilibrium:

$$K_a = \frac{[H^+][NO_2^-]}{[HNO_2]}$$

3. Let X represent the fraction of the initial nitrous acid concentration which is ionized when the system reaches equilibrium. Then,

$$[HNO_2] = (0.001 - X)$$
$$[H^+] = X$$
$$[NO_2^-] = X$$

The equilibrium constant expression can then be written in terms of X.

$$K_a = \frac{X^2}{0.001 - X}$$

4. Compute a value for K_a from basic thermodynamic data.
   (a) Compute $\Delta G_{reaction}^\circ$ for the reaction in step 1:

$$\Delta G_{reaction}^\circ = [(-8.9) + (0)] - [(-13.30)] = 4.4 \text{ KCal}$$

   (b) Compute K_a by rearranging equation 1-55 into the form

$$K_a = 10^{-\Delta G^\circ_{reaction}/1.363} = 10^{-(4.4/1.363)} = 5.9 \times 10^{-4}$$

5. Substitute for K_a in the equilibrium constant expression and solve for X:

$$5.9 \times 10^{-4} = \frac{X^2}{10^{-3} - X}$$

or

$$X^2 + (5.9 \times 10^{-4})X - 5.9 \times 10^{-7} = 0$$
$$X = \frac{(-5.9 \times 10^{-4}) + \sqrt{(5.9 \times 10^{-4})^2 - 4(1)(-5.9 \times 10^{-7})}}{2(1)}$$
$$= \frac{(-5.9 \times 10^{-4}) + (16.5 \times 10^{-4})}{2}$$
$$= 5.3 \times 10^{-4}$$

When K_a has a value less than 10^{-5}, it is often assumed that X is very small and can be neglected in the denominator term.

6. Determine the pH of the solution from equation 2-4:

$$pH = -[\log 5.3 + \log 10^{-4}] = 3.3$$

Example problems 2-1 and 2-2 illustrate the difficulty of solving even a simple equilibrium problem through basic mathematical operations unless assumptions are made to simplify the manipulations involved. In more complex equilibrium problems it is sometimes difficult to know what chemical species can be neglected, i.e., what assumptions are valid under the conditions of equilibrium established in the system. Therefore, to aid the engineer in solving equilibrium problems, a graphical technique is presented, which can be used to obtain a rapid and accurate solution to many types of chemical systems where equilibrium is established.

2-2 EQUILIBRIUM DIAGRAMS

The equilibrium diagram for an acid-base equilibrium problem is constructed by plotting the logarithm of the concentration of each individual species versus pH. In other words, the concentration of each chemical species is expressed as a function of pH. The general procedure for developing the plotting equations and constructing the log (species) vs. pH diagram for a strong monoprotic acid, a weak monoprotic acid, and a weak diprotic acid is outlined in the following discussion. For convenience, an ideal solution will be assumed so that activity will be equal to analytical concentration.

I. STRONG MONOPROTIC ACID:
   A. Write balanced chemical equations for all reactions involved in the chemical system.
      1. When using the graphical technique, it is always assumed that the ionization reaction of all strong acids or bases goes to completion.
         (a) $HA \longrightarrow H^+ + A^-$
         (b) $H_2O \rightleftharpoons H^+ + OH^-$
   B. Write equilibrium constant expressions for all reactions in which an equilibrium state is established:

$$K_w = [H^+][OH^-]$$

   C. Write a mass balance relationship which accounts for the total mass of the nonmetal group of the acid, i.e., the chemical species denoted by A:

$$C_T = [A^-]$$

   This form of the mass balance expression results from the assumption that complete ionization of the acid occurs.
   D. Define the chemical species which exist in the system at equilibrium. For a strong monoprotic acid solution these are

$$A^-, H^+, OH^-$$

   E. Develop the plotting equation for each chemical species defined in step D.
      1. The plotting equation for $[A^-]$ is obtained by taking the logarithm of both sides of the mass balance expression:

$$\log [A^-] = \log C_T \qquad \textbf{(2-25)}$$

This equation implies that $[A^-]$ is independent of pH and has a value equal to the molar concentration of the acid initially present.

2. The plotting equation for $[H^+]$ is given by rearranging equation 2-4:

$$\log [H^+] = - \, pH \qquad \qquad \textbf{(2-26)}$$

3. The plotting equation for $[OH^-]$ is obtained by rearranging equation 2-10:

$$pOH = pK_w - pH \qquad \qquad \textbf{(2-10)}$$

Since pOH is defined as

$$pOH = -\log [OH^-] \qquad \qquad \textbf{(2-5)}$$

the plotting equation will have the form

$$\log [OH^-] = pH - pK_w \qquad \qquad \textbf{(2-27)}$$

F. Each plotting equation has the same form as the equation of a straight line. To determine the slope of the trace given by each equation, differentiate each plotting equation with respect to pH.

1. The slope of the line defined by equation 2-25 is

$$\frac{d \log [A]^-}{d\mathrm{pH}} = 0$$

Thus, a plot of $\log [A^-]$ vs. pH will give a horizontal trace that will intersect the ordinate at the value given by $\log C_T$.

2. The slope of the line defined by equation 2-26 is

$$\frac{d \log [H^+]}{\mathrm{pH}} = -1$$

A plot of $\log [H^+]$ vs. pH will give a linear trace of slope -1.

3. The slope of the line defined by equation 2-27 is

$$\frac{d \log [OH^-]}{d\mathrm{pH}} = +1$$

A plot of $\log [OH^-]$ vs. pH will give a linear trace of slope $+1$.

G. Construct the $\log$ [species] vs. pH diagram which describes how the logarithm of the concentration of each chemical specie listed in step D varies with pH.

1. A generalized $\log$ [species] vs. pH diagram for a strong monoprotic acid is presented in Figure 2-1. To simplify construction of such a

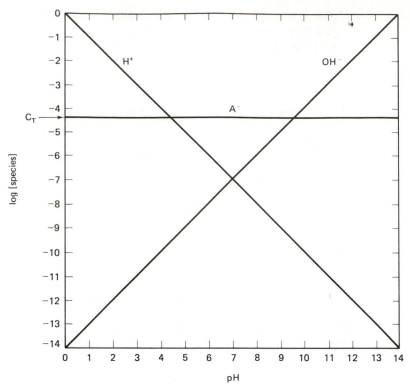

FIGURE 2-1 *Log (species) vs* pH *diagram for a strong monoprotic acid.*

diagram, it should be remembered that only two points are required to define the shape of a linear trace.

II. WEAK MONOPROTIC ACID:

A. Write balanced chemical equations for all reactions involved in the equilibrium state:

    (1) $HA \rightleftharpoons H^+ + A^-$

    (2) $H_2O \rightleftharpoons H^+ + OH^-$

B. Write equilibrium constant expressions for all reactions in which an equilibrium state is established:

    (1) $K_a = \dfrac{[H^+][A^-]}{[HA]}$

    (2) $K_w = [H^+][OH^-]$

C. Write a mass balance expression which accounts for the total mass of the nonmetal group of the acid:

$$C_T = [HA] + [A^-]$$

D. Define the chemical species which exist in the system at equilibrium. For a weak monoprotic acid solution, these are

$$[HA], \quad [A^-], \quad [H^+], \quad [OH^-]$$

E. Develop the plotting equation(s) for each chemical species defined in step D.

1. The plotting equations for the water system, i.e., for $\log [H^+]$ and $\log [OH^-]$, are common to all aqueous solutions. Thus, equations 2-26 and 2-27 will apply to all chemical system involving equilibria in aqueous solutions.

2. The plotting equations for [HA] are developed as follows:

   (a) Rearrange the equilibrium constant expression for the weak monoprotic acid, HA, into the form

   $$[A^-] = \frac{K_a[HA]}{[H^+]}$$

   Substituting from this relationship for $[A^-]$ in the mass balance equation gives

   $$[HA] + \frac{K_a[HA]}{[H^+]} = C_T$$

   which can be further rearranged into the form

   $$[HA] = \frac{C_T}{1 + \frac{K_a}{[H^+]}}$$

   Multiplying the right-hand side of this equation by $[H^+]/[H^+]$ and then taking the logarithm of both sides of the equation gives

   $$\log [HA] = \log [H^+] + \log C_T - \log ([H^+] + K_a)$$

   or

   $$\log [HA] = \log C_T - pH - \log ([H^+] + K_a) \qquad \textbf{(2-28)}$$

   Equation 2-28 represents a curve rather than a straight line. However, certain simplifying approximations can be made which will facilitate the use of this equation when manual construction of an equilibrium diagram is required.

   (b) There are two limiting cases when plotting $\log [HA]$ vs. pH, using equation 2-28:

   Case 1:

   $$[H^+] \gg K_a \quad \text{or} \quad pK_a > pH$$

   For this case, K_a can be neglected so that equation 2-28 reduces to

   $$\log [HA] = \log C_T - pH - \log [H^+]$$

   or

   $$\log [HA] = \log C_T \qquad \textbf{(2-29)}$$

   Equation 2-29 will be applicable when [pH] is at least one unit less than the pK_a. The slope of the line described by equation 2-29 is

$$\frac{d \log [\text{HA}]}{d\text{pH}} = 0$$

Thus, a plot of log [HA] vs. pH in the region where the pH value is at least one unit less than the pK_a will give a horizontal trace which will intersect the ordinate at the value given by log C_T.

Case 2:

$$[\text{H}^+] \ll K_a \quad \text{or} \quad pK_a < \text{pH}$$

For this case, $[\text{H}^+]$ can be neglected in the $([\text{H}^+] + K_a)$ sum term in equation 2-28 so that the equation reduces to

$$\log [\text{HA}] = \log C_T - \text{pH} - \log K_a$$

or

$$\log [\text{HA}] = \log C_T - \text{pH} + pK_a \qquad \textbf{(2-30)}$$

Equation 2-30 will be applicable when $[\text{H}^+]$ is at least 10 times smaller than K_a; i.e., when the pH is at least one unit larger than the pK_a. The slope of the line described by equation 2-30 is

$$\frac{d \log [\text{HA}]}{d\text{pH}} = -1$$

A plot of log [HA] vs. pH in the region where the pH value is at least one unit larger than the pK_a will give a linear trace of slope -1.

3. The plotting equations for $[\text{A}^-]$ are developed as follows:
   (a) Rearrange the equilibrium constant expression for the weak monoprotic acid, HA, into the form

$$[\text{HA}] = \frac{[\text{H}^+][\text{A}^-]}{[K_a]}$$

Substituting from this relationship for [HA] in the mass balance equation gives

$$\frac{[\text{H}^+][\text{A}^-]}{K_a} + [\text{A}^-] = C_T$$

or

$$\left[\frac{[\text{H}^+]}{K_a} + 1\right][\text{A}^-] = C_T$$

which can be further rearranged into the form

$$[\text{A}^-] = \frac{K_a C_T}{[\text{H}^+] + K_a}$$

Taking the logarithm of both sides of this equation gives

$$\log [A^-] = \log K_a + \log C_T - \log ([H^+] + K_a)$$

or

$$\log [A^-] = \log C_T - pK_a - \log ([H^+] + K_a) \qquad (2\text{-}31)$$

(b) There are two limiting cases when plotting $\log [A^-]$ vs. pH using equation 2-31:

Case 1:

$$[H^+] \gg K_a \quad \text{or} \quad pK_a > pH$$

For this case, K_a can be neglected in the sum ($[H^+] + K_a$) so that equation 2-31 reduces to

$$\log [A^-] = \log C_T - pK_a - \log [H^+]$$

or

$$\log [A^-] = \log C_T - pK_a + pH \qquad (2\text{-}32)$$

Equation 2-32 will be applicable when the pH is at least one unit less than the pK_a. The slope of the line described by equation 2-32 is

$$\frac{d \log [A^-]}{d pH} = +1$$

Thus, a plot of $\log [A^-]$ vs. pH in the region where the pH value is at least one unit less than the pK_a will give a linear trace of slope $+1$.

Case 2:

$$[H^+] \ll K_a \quad \text{or} \quad pH > pK_a$$

For this case, $[H^+]$ can be neglected in the ($[H^+] + K_a$) sum term in equation 2-31 so that the equation reduces to

$$\log [A^-] = \log C_T - pK_a - \log K_a$$

or

$$\log [A^-] = \log C_T \qquad (2\text{-}33)$$

Equation 2-33 will be applicable when the pH is at least one unit larger than the pK_a. The slope of the line described by equation 2-33 is

$$\frac{d \log [A^-]}{d pH} = 0$$

A plot of $\log [A^-]$ vs. pH in the region where the pH value is at least one unit greater than the pK_a will give a horizontal trace having an ordinate value equal to $\log C_T$.

F. Construct the equilibrium diagram which describes how the logarithm of the concentration of each chemical species listed in step D varies with pH.
    1. Figure 2-2 illustrates a general equilibrium diagram for a weak

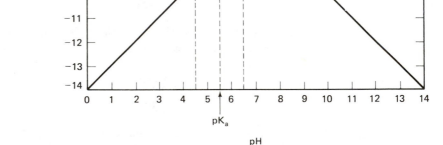

— — — — — — : Construction lines

FIGURE 2-2 *Incomplete equilibrium diagram for a weak monoprotic acid.*

monoprotic acid which could be developed from equations 2-26, 2-27, 2-29, 2-30, 2-32, and 2-33. In this figure, discontinuities which extend one pH unit to either side of the pK_a value are shown for the [HA] and [A$^-$] curves. Within this region the relationship between log [species] and pH is no longer linear. However, in this curved region a relationship exists between [HA] and [A$^-$] which greatly simplifies curve construction: when the pH has a value equal to pK_a, the concentration of unionized acid is one-half its initial value, or

$$[\text{HA}] = [\text{A}^-]$$

when

$$\text{pH} = pK_a$$

Under these conditions

$$[HA] = \tfrac{1}{2}C_T \qquad\qquad\text{(2-34)}$$

$$[A^-] = \tfrac{1}{2}C_T \qquad\qquad\text{(2-35)}$$

Taking the logarithm of both sides of equation 2-34 and equation 2-35 gives

$$\log[HA] = \log C_T - \log 2$$

$$\log[A^-] = \log C_T - \log 2$$

or

$$\log[HA] = \log C_T - 0.3 \qquad\qquad\text{(2-36)}$$

$$\log[A^-] = \log C_T - 0.3 \qquad\qquad\text{(2-37)}$$

Thus, when $pH = pK_a$ both curves [HA] and [A$^-$] pass through a common point, which is located on the pK_a vertical 0.3 units below a horizontal line passing through the ordinate at $\log C_T$. Using this point as a reference, the linear portions of curves [HA] and [A$^-$] may be joined as shown in Figure 2-3.

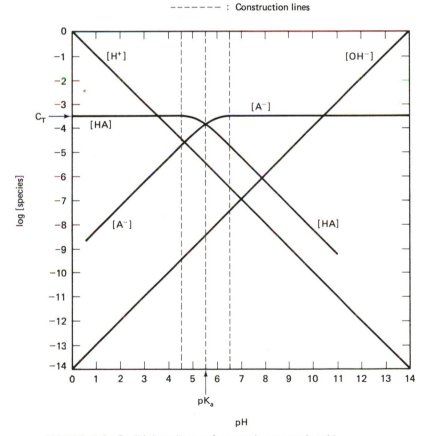

FIGURE 2-3 Equilibrium diagram for a weak monoprotic acid.

III. WEAK DIPROTIC ACID:

A. Write balanced chemical equations for all reactions involved in the equilibrium state:

(1) $H_2A \rightleftharpoons H^+ + HA^-$

(2) $HA^- \rightleftharpoons H^+ + A^{2-}$

(3) $H_2O \rightleftharpoons H^+ + OH^-$

B. Write equilibrium constant expressions for all reactions in which an equilibrium state is established:

(1) $K_{a1} = \dfrac{[H^+][HA^-]}{[H_2A]}$

(2) $K_{a2} = \dfrac{[H^+][A^{2-}]}{[HA^-]}$

(3) $K_w = [H^+][OH^-]$

C. Write a mass balance expression which accounts for the total mass of the nonmetal group of the acid:

$$C_T = [H_2A] + [HA^-] + [A^{2-}]$$

D. Define the chemical species which exist in the system at equilibrium. For a weak diprotic acid solution these are

$$[H_2A], \quad [HA^-], \quad [A^{2-}], \quad [H^+], \quad [OH^-]$$

E. Develop the plotting equation(s) for each chemical species defined in step D.

1. The plotting equations for $[H_2A]$ are developed as follows:

   (a) Rearrange the equilibrium constant expression for the first ionization into the form

$$[HA^-] = \frac{K_{a1}[H_2A]}{[H^+]} \qquad \textbf{(2-38)}$$

Rearrange the equilibrium constant expression for the second acid ionization into the form

$$[A^{2-}] = \frac{K_{a2}[HA^-]}{[H^+]} \qquad \textbf{(2-39)}$$

Substitute for $[HA^-]$ in equation 2-39 from equation 2-38:

$$[A^{2-}] = \frac{K_{a1}K_{a2}[H_2A]}{[H^+]^2} \qquad \textbf{(2-40)}$$

Substitute for $[HA^-]$ and $[A^{2-}]$ in the mass balance expression from equations 2-38 and 2-40:

$$[H_2A]\left[1 + \frac{K_{a1}}{[H^+]} + \frac{K_{a1}K_{a2}}{[H^+]^2}\right] = C_T$$

or

$$[H_2A] = \frac{C_T[H^+]^2}{[H^+]^2 + [H^+]K_{a1} + K_{a1}K_{a2}} \qquad (2\text{-}41)$$

Taking the logarithm of both sides of equation 2-41 gives

$$\log[H_2A] = \log C_T + 2\log[H^+]$$
$$- \log([H^+]^2 + [H^+]K_{a1} + K_{a1}K_{a2})$$

or

$$\log[H_2A] = \log C_T - 2\text{pH}$$
$$- \log([H^+]^2 + [H^+]K_{a1} + K_{a1}K_{a2}) \qquad (2\text{-}42)$$

(b) There are five different cases to consider when plotting $\log[H_2A]$ vs. pH, using equation 2-42.

Case 1:

$$K_{a2} \ll K_{a1} \ll [H^+] \quad \text{or} \quad pK_{a2} > pK_{a1} > \text{pH}$$

For this case any term containing K_{a1} or K_{a2} in the sum $([H^+]^2 + [H^+]K_{a1} + K_{a1}K_{a2})$ can be neglected. (Recall that the second ionization constant for a diprotic acid is usually several orders of magnitude smaller than the first.) Thus, equation 2-42 reduces to

$$\log[H_2A] = \log C_T - 2\text{pH} - \log[H^+]^2$$

or

$$\log[H_2A] = \log C_T \qquad (2\text{-}43)$$

Equation 2-43 will be applicable when the pH is at least one unit less than the pK_{a1}. The slope of the line described by equation 2-43 is

$$\frac{d\log[H_2A]}{d\text{pH}} = 0$$

A plot of $\log[H_2A]$ vs. pH, in the region where the pH value is at least one unit less than the pK_{a1}, will give a horizontal trace which will intersect the ordinate at the value given by $\log C_T$.

Case 2:

$$K_{a2} \ll [\mathrm{H^+}] \quad \text{or} \quad \mathrm{p}K_{a2} > \mathrm{pH}$$

For this case, any term containing K_{a2} in the sum $([\mathrm{H^+}]^2 + [\mathrm{H^+}]K_{a1} + K_{a1}K_{a2})$ can be neglected so that equation 2-42 reduces to

$$\log [\mathrm{H_2A}] = \log C_T - 2\mathrm{pH} - \log ([\mathrm{H^+}]^2 + [\mathrm{H^+}]K_{a1}) \quad \textbf{(2-44)}$$

Equation 2-44 represents a curve rather than a straight line and is applicable when the pH is within the interval one unit either side of $\mathrm{p}K_{a1}$.

Case 3:

$$K_{a2} \ll [\mathrm{H^+}] \ll K_{a1} \quad \text{or} \quad \mathrm{p}K_{a2} > \mathrm{pH} > \mathrm{p}K_{a1}$$

In this case, the terms $[\mathrm{H^+}]^2$ and $K_{a1}K_{a2}$ can be neglected in the sum $([\mathrm{H^+}]^2 + [\mathrm{H^+}]K_{a1} + K_{a1}K_{a2})$ so that equation 2-42 reduces to

$$\log [\mathrm{H_2A}] = \log C_T - 2\mathrm{pH} - (\log [\mathrm{H^+}] + \log K_{a1})$$

or

$$\log [\mathrm{H_2A}] = \log C_T - \mathrm{pH} + \mathrm{p}K_{a1} \quad \textbf{(2-45)}$$

Equation 2-45 will be applicable when the pH is at least one unit less than $\mathrm{p}K_{a2}$ but at least one unit greater than $\mathrm{p}K_{a1}$. The slope of the line described by equation 2-45 is

$$\frac{d \log [\mathrm{H_2A}]}{d\mathrm{pH}} = -1$$

Case 4:

$$[\mathrm{H^+}] \ll K_{a1} \quad \text{or} \quad \mathrm{pH} > \mathrm{p}K_{a1}$$

Here the term $[\mathrm{H^+}]^2$ can be neglected in the sum $([\mathrm{H^+}]^2 + [\mathrm{H^+}]K_{a1} + K_{a1}K_{a2})$ so that equation 2-42 reduces to

$$\log [\mathrm{H_2A}] = \log C_T - 2\mathrm{pH} - \log ([\mathrm{H^+}]K_{a1} + K_{a1}K_{a2}) \quad \textbf{(2-46)}$$

Equation 2-46 represents a curve rather than a straight line and is applicable when the pH is within the interval one unit on either side of $\mathrm{p}K_{a2}$.

Case 5:

$$[\mathrm{H^+}] \ll K_{a2} \quad \text{or} \quad \mathrm{pH} > \mathrm{p}K_{a2}$$

Here the terms $[\mathrm{H^+}]K_{a1}$, and $[\mathrm{H^+}]^2$ can be neglected in the sum

$([H^+]^2 + [H^+]K_{a1} + K_{a1}K_{a2})$ so that equation 2-42 reduces to

$$\log [H_2A] = \log C_T - 2pH - (\log K_{a1} + \log K_{a2})$$

or

$$\log [H_2A] = \log C_T - 2pH + pK_{a1} + pK_{a2} \qquad \textbf{(2-47)}$$

Equation 2-47 will be applicable when the pH is at least one unit greater than pK_{a2}. The slope of the line described by equation 2-47 is

$$\frac{d \log [H_2A]}{dpH} = -2$$

2. The plotting equations for $[HA^-]$ are developed as follows:
    (a) Rearrange the equilibrium constant expression for the first acid ionization into the form

$$[H_2A] = \frac{[H^+][HA^-]}{K_{a1}} \qquad \textbf{(2-48)}$$

Substitute into the mass balance equation for $[H_2A]$ from equation 2-48 and for $[A^{2-}]$ from equation 2-39:

$$[HA^-]\left[1 + \frac{[H^+]}{K_{a1}} + \frac{K_{a2}}{[H^+]}\right] = C_T$$

or

$$[HA^-] = \frac{K_{a1}[H^+]C_T}{[H^+]^2 + [H^+]K_{a1} + K_{a1}K_{a2}} \qquad \textbf{(2-49)}$$

Taking the logarithm of both sides of equation 2-49 gives

$$\log [HA^-] = \log C_T + \log K_{a1} + \log [H^+]$$
$$- \log ([H^+]^2 + [H^+]K_{a1} + K_{a1}K_{a2})$$

or

$$\log [HA^-] = \log C_T - pK_{a1} - pH$$
$$- \log ([H^+]^2 + [H^+]K_{a1} + K_{a1}K_{a2}) \qquad \textbf{(2-50)}$$

(b) Applying the same reasoning used to develop plotting equations for $[H_2A]$, the following equations and slopes may be obtained for $[HA^-]$:

Case 1: pH is at least one unit less than pK_{a1}.

$$\log [HA^-] = \log C_T - pK_{a1} + pH \qquad \textbf{(2-51)}$$

$$\frac{d \log [HA^-]}{dpH} = +1$$

Case 2: pH is within the interval one unit to either side of pK_{a1}.

$$\log [HA^-] = \log C_T - pK_{a1} - pH \\ - \log ([H^+]^2 + [H^+]K_{a1}) \qquad \textbf{(2-52)}$$

Case 3: pH is at least one unit less than pK_{a2} but at least one unit greater than pK_{a1}.

$$\log [HA^-] = \log C_T \qquad \textbf{(2-53)}$$

$$\frac{d \log [HA^-]}{d\text{pH}} = 0$$

Case 4: pH is within the interval one unit to either side of pK_{a2}.

$$\log [HA^-] = \log C_T - pK_{a1} - pH \\ - \log (K_{a1}[H^+] + K_{a1}K_{a2}) \qquad \textbf{(2-54)}$$

Case 5: pH is at least one unit greater than pK_{a2}.

$$\log [HA^-] = \log C_T + pK_{a2} - pH \qquad \textbf{(2-55)}$$

$$\frac{d \log [HA^-]}{d\text{pH}} = -1$$

3. The plotting equations for $[A^{2-}]$ are developed as follows:
   (a) Rearrange the equilibrium constant expression for the second acid ionization into the form

$$[HA^-] = \frac{[H^+][A^{2-}]}{K_{a2}} \qquad \textbf{(2-56)}$$

Substitute from equation 2-56 for $[HA^-]$ in equation 2-48:

$$[H_2A] = \frac{[H^+]^2[A^{2-}]}{K_{a1}K_{a2}} \qquad \textbf{(2-57)}$$

Substitute into the mass balance equation for $[H_2A]$ from equation 2-57 and for $[HA^-]$ from equation 2-56:

$$[A^{2-}]\left[1 + \frac{[H^+]}{K_{a1}K_{a2}} + \frac{[H^+]}{K_{a2}}\right] = C_T$$

or

$$[A^{2-}] = \frac{K_{a1}K_{a2}C_T}{[H^+]^2 + K_{a1}[H^+] + K_{a1}K_{a2}} \qquad \textbf{(2-58)}$$

Taking the logarithm of both sides of equation 2-58 gives

$$\log [A^{2-}] = \log K_{a1} + \log K_{a2} + \log C_T$$
$$- \log ([H^+]^2 + K_{a1}[H^+] + K_{a1}K_{a2})$$

or

$$\log [A^{2-}] = \log C_T - pK_{a1} - pK_{a2}$$
$$- \log ([H^+]^2 + K_{a1}[H^+] + K_{a1}K_{a2}) \qquad (2\text{-}59)$$

(b) The following equations and slopes may be obtained for $[A^{2-}]$:

Case 1: pH is at least one unit less than pK_{a1}.

$$\log [A^{2-}] = \log C_T - pK_{a1} - pK_{a2} + 2pH \qquad (2\text{-}60)$$
$$\frac{d \log [A^{2-}]}{dpH} = +2$$

Case 2: pH is within the interval one unit to either side of pK_{a1}.

$$\log [A^{2-}] = \log C_T - pK_{a1} - pK_{a2}$$
$$- \log ([H^+]^2 + [H^+]K_{a1}) \qquad (2\text{-}61)$$

Case 3: pH is at least one unit less than pK_{a2} but at least one unit greater than pK_{a1}.

$$\log [A^{2-}] = \log C_T - pK_{a2} + pH \qquad (2\text{-}62)$$
$$\frac{d \log [A^{2-}]}{dpH} = +1$$

Case 4: pH is within the interval one unit to either side of pK_{a2}.

$$\log [A^{2-}] = \log C_T - pK_{a1} - pK_{a2}$$
$$- \log (K_{a1}[H^+] + K_{a1}K_{a2}) \qquad (2\text{-}63)$$

Case 5: pH is at least one unit greater than pK_{a2}.

$$\log [A^{2-}] = \log C_T \qquad (2\text{-}64)$$
$$\frac{d \log [A^{2-}]}{dpH} = 0$$

F. Construct the equilibrium diagram which describes how the logarithm of the concentration of each chemical species listed in step D varies with pH.

1. Figure 2-4 illustrates a general equilibrium diagram for a weak diprotic acid which could be developed from equations 2-26, 2-27, 2-43, 2-44, 2-45, 2-46, 2-47, 2-51, 2-52, 2-53, 2-54, 2-55, 2-60, 2-61, 2-62, 2-63, and 2-64. Construction points can be used to aid in curve construction in the region which extends one pH unit to either side of pK_{a1} and pK_{a2}.

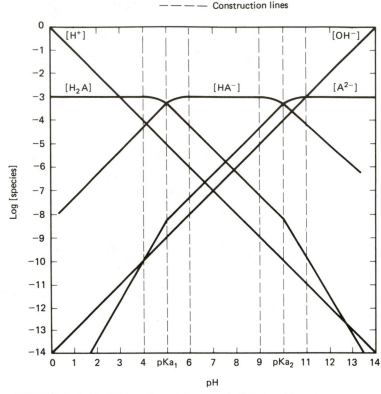

FIGURE 2-4 *Equilibrium diagram for a weak diprotic acid.*

Construction point 1: When the pH has a value equal to pK_{a1}, the following relationships are assumed:

$$[H_2A] = [HA^-]$$
$$[A^{2-}] = 0$$

Under these conditions,

$$[H_2A] = \tfrac{1}{2}C_T$$
$$[HA^-] = \tfrac{1}{2}C_T$$

ACID-BASE EQUILIBRIA CHAPTER 2

Taking the logarithm of both sides of these equations gives

$$\log [H_2A] = \log C_T - 0.3 \qquad (2\text{-}65)$$

$$\log [HA^-] = \log C_T - 0.3 \qquad (2\text{-}66)$$

This implies that when pH = pK_{a1}, both curves [H_2A] and [HA^-] pass through a common point which is located on the pK_{a1} vertical 0.3 units below a horizontal line passing through the ordinate at $\log C_T$.

Construction point 2: When the pH has a value equal to pK_{a2}, the following relationships are assumed:

$$[HA^-] = [A^{2-}]$$

$$[H_2A] = 0$$

Manipulating these equations gives

$$\log [HA^-] = \log C_T - 0.3 \qquad (2\text{-}67)$$

$$\log [A^{2-}] = \log C_T - 0.3 \qquad (2\text{-}68)$$

Construction point 1 can be used to locate the curved sections of the [H_2A] and [HA^-] plots in the region of pK_{a1}. Construction point 2 can be used to locate the curved sections of the [HA^-] and [A^{2-}] plots in the region of pK_{a2}. Although [A^{2-}] is actually curved in the region of pK_{a2}, little error is incurred by plotting them as linear because of their very small concentration values in these regions.

Approximate Method for Constructing Equilibrium Diagrams

Using the equations previously developed as a basis, a rapid approximate method of constructing log [species] vs. pH diagrams is now presented for both weak monoprotic and polyprotic acids (Sillen, 1959, and Loewenthal and Marais, 1976).

A. *MONOPROTIC ACIDS:*

1. Establish the scale for the graph by setting the ordinate to cover the range of 0 to -14 and the abscissa to cover the range of 0 to 14
2. Draw a horizontal dashed line through the value of C_T on the log [species] axis
3. Draw a vertical dashed line through the value of pK_{a1} on the pH axis. The intersection of the horizontal and vertical dashed lines defines a system point.
4. Draw dashed lines at 45° which pass through the system point. One of these lines will have a slope of $+1$; the other line will have a slope of -1. Both lines will be located below the horizontal C_T line.

5. Plot a construction point which lies on the vertical dashed line through the pK_a and is located 0.3 log units below the system point.

6. Sketch two curves through the construction point which is located 0.3 log units below the system point. Each of these curves should intersect a 45° line one pH unit to one side of the pK_a vertical and should also intersect the horizontal C_T line one pH unit on the opposite side of the pK_a vertical.

7. Draw the $[H^+]$ line to pass through the coordinates $(0, 0)$ and $(14, -14)$, and draw the $[OH^-]$ line to pass through the coordinates $(0, -14)$ and $(14, 0)$. The completed equilibrium diagram is shown in Figure 2-5.

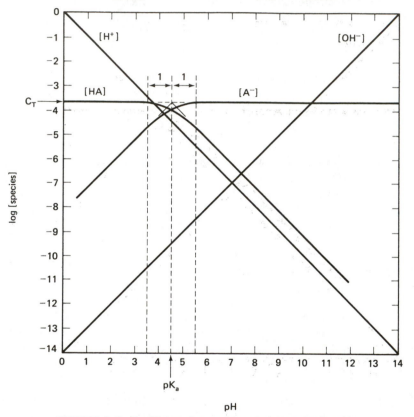

FIGURE 2-5 *Equilibrium diagram for a weak monoprotic acid.*

B. DIPROTIC ACIDS: The log [species] vs. pH diagram for a diprotic acid can be constructed by using the same approximation procedure outlined for monoprotic acids. The procedure for a diprotic acid is based on the premise that the ionization of a diprotic acid can be represented by the ionization of two different monoprotic acids. The C_T value for each monoprotic acid is assumed to be equal to the C_T value for the diprotic acid. One of the monoprotic acids is assumed to have a pK_a equal to pK_{a1}, whereas the other monoprotic acid is assumed to have a pK_a equal to pK_{a2}. After the curves for each acid have been plotted, the curves of species common to each ionization are then joined.

The log [species] vs. pH diagram for a diprotic acid constructed as if it represented the ionization of two different monoprotic acids is shown in Figure 2-6. Little error is

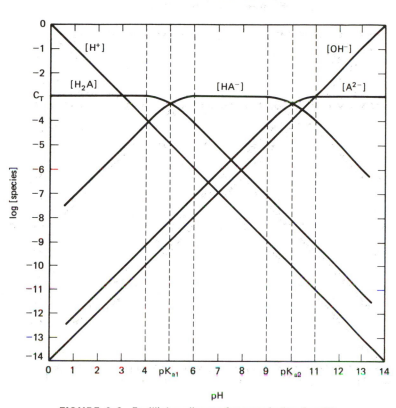

FIGURE 2-6 *Equilibrium diagram for a weak diprotic acid.*

incurred by plotting the lower legs of the $[A^{2-}]$ and $[H_2A]$ lines with slopes of $+1$ and -1 rather than $+2$ and -2 as well as not reflecting the intermediate curved sections joining the sections of different slopes. This is due to the low concentration range relative to C_T that these sections are located in.

Effect of C_T on the Equilibrium Diagram

The value of C_T does not affect the general shape of the equilibrium diagram. If C_T is varied, the whole diagram moves up or down (up if C_T is increased; down if C_T is decreased) along the vertical construction line which passes through the pK_a value on the X axis, but the shape of the diagram remains the same. The effect of varying C_T on the location of the diagram is illustrated in Figure 2-7.

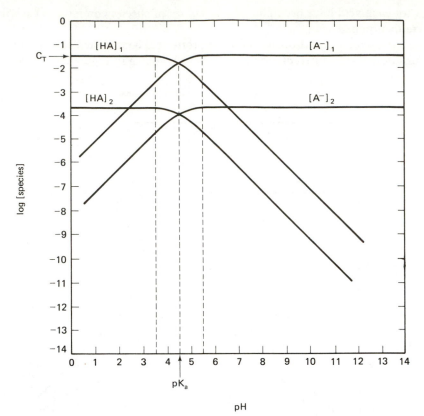

FIGURE 2-7 *Effect of C_T on the equilibrium diagram.*

Solving Acid-Base Equilibria Problems with Equilibrium Diagrams

Before it is possible to apply equilibrium diagrams to solve weak acid-base equilibria problems, it is necessary to introduce the concept of proton levels and outline the procedure for developing a proton balance equation.

The development of a proton balance equation is based on the premise that the sum total of the concentrations of all the chemical species existing at equilibrium which contain more protons than the initial state chemical species must equal the sum total of the concentrations of all the chemical species existing at equilibrium which contain less protons than the initial state chemical species. For example, if HA is added to H_2O, the species HA and H_2O are called the *initial state species*. In aqueous solution HA will ionize to form H^+ or (H_3O^+) and A^-, whereas water undergoes autoionization to form H^+ or (H_3O^+) and OH^-. The species H^+, A^-, and OH^- are referred to as *equilibrium state species*, where H^+ or (H_3O^+) is formed by H_2O gaining one proton, A^- is formed by HA losing one proton, and OH^- is formed by H_2O losing one proton. Thus, the proton balance equation for this system can be written as

$$[H^+] = [A^-] + [OH^-] \qquad (2\text{-}69)$$

In developing the proton balance equation for weak acids or bases and their salts, any chemical species which does not directly affect the equilibrium state is neglected. These species may, however, be included in ionic strength computations. To illustrate this point, consider the case where the salt of a weak acid, NaA, is added to water. Sodium is neglected when developing the proton balance expression, and initial state chemical species are assumed to be A^- and H_2O. In aqueous solution A^- will accept a proton to form HA, whereas water undergoes autoionization to form H^+ and OH^-. Thus, HA and H^+ are formed by initial state species gaining one proton, whereas OH^- is formed by an initial state species losing one proton. The proton balance equation for this system has the form

$$[HA] + [H^+] = [OH^-] \qquad (2\text{-}70)$$

A graphical representation of the proton balance equation for both the HA-H_2O and NaA-H_2O systems is presented in Figure 2-8.

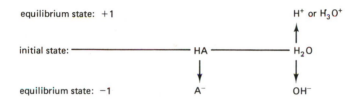

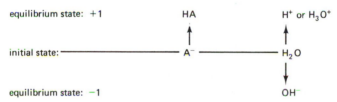

FIGURE 2-8   Graphical representation of proton balance equation.

The following example problems will serve to illustrate the use of equilibrium diagrams in solving weak acid-base equilibria problems.

EXAMPLE PROBLEM 2-3:   Calculate the equilibrium pH of a 0.1 M nitrous acid, HNO_2, solution at 25°C. $pK_a = 3.29$.

Solution:

1. Develop the proton balance equation for the chemical system.

    equilibrium state: $+1$ H^+
                                                                      ↑
    initial state: ————————HNO_2————————H_2O
                                            ↓   ↓
    equilibrium state: -1 NO_2^- OH^-

    Therefore,

    $$[H^+] = [NO_2^-] + [OH^-]$$

2. Construct the equilibrium diagram for the system following the approximation procedure previously outlined. The equilibrium diagram for this problem is presented in Figure 2-9.

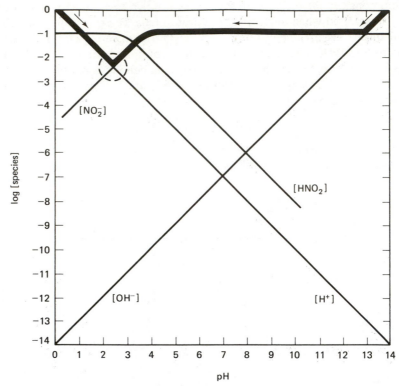

FIGURE 2-9 *Equilibrium diagram for a 0.1 M HNO₂ solution.*

3. The equilibrium state for the system is located at the point on the equilibrium diagram where the proton balance equation is satisfied. This point is located as follows:

(a) Begin following the [OH⁻] curve at the top right-hand corner of the diagram. When the pH is greater than 13, the [OH⁻] curve is above the [NO₂⁻] curve. This means that [HO⁻] ≫ [NO₂⁻] in this region, and as a result [NO₂⁻] can be neglected in the sum [NO₂⁻] + [OH⁻]. However, below pH 13 the [OH⁻] curve drops below the [NO₂⁻] curve, which means that at pH < 13 the [OH⁻] term can be neglected in the sum [NO₂⁻] + [OH⁻]. Here, when the [OH⁻] curve drops below the [NO₂⁻] curve, begin following the [NO₂⁻] curve until it intersects the [H⁺] curve. The equilibrium state is located at that point.

(b) Figure 2-9 shows that the [H⁺] curve and the [NO₂⁻] curve intersect at pH = 2.5. Thus, the equilibrium pH is 2.5. The route to the problem solution is outlined by the heavy line in Figure 2-9.

EXAMPLE PROBLEM 2-4: Calculate the equilibrium pH of a 0.1 M solution of sodium nitrite, $NaNO_2$, at 25°C.

Solution:

1. Develop the proton balance equation for the chemical system.

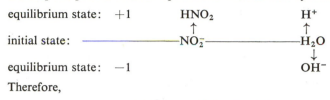

Therefore,

$$[HNO_2] + [H^+] = [OH^-]$$

2. Construct the equilibrium diagram for the nitrous acid system. The equilibrium diagram for this problem is identical to that presented in Figure 2-9.

3. The equilibrium state for the system is located at the point on the equilibrium diagram where the proton balance equation is satisfied. The route to the problem solution is outlined by the heavy line in Figure 2-10. The intersection of the $[HNO_2]$ curve and the $[OH^-]$ curve gives the equilibrium pH as 8.

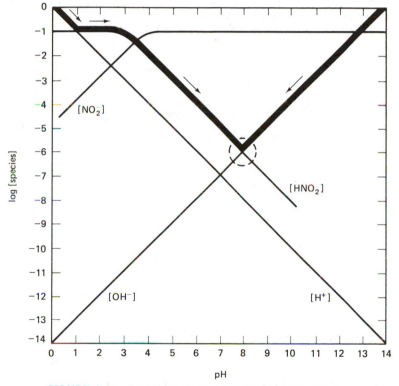

FIGURE 2-10 *Equilibrium diagram for a 0.1 M NaNO₂ solution.*

EXAMPLE PROBLEM 2-5: Calculate the equilibrium pH of a 0.2 M solution of carbonic acid, H_2CO_3, at 25°C. $pK_{a1} = 6.3$, $pK_{a2} = 10.3$.

Solution:

1. Develop the proton balance equation for the chemical system.

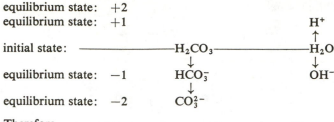

equilibrium state: $+2$
equilibrium state: $+1$ H^+
                                         ↑
initial state: ————————H_2CO_3————————H_2O
                              ↓   ↓
equilibrium state: -1 HCO_3^- OH^-
                              ↓
equilibrium state: -2 CO_3^{2-}

Therefore,

$$[H^+] = [OH^-] + [HCO_3^-] + 2[CO_3^{2-}]$$

Note: Each carbonate molecule is formed by the loss of two protons. Thus, the molar proton loss concentration from this specie is two times the molar concentration of the carbonate ion.

2. Construct the equilibrium diagram for this system following the approximation procedure outlined earlier. The equilibrium diagram for this problem is presented in Figure 2-11.

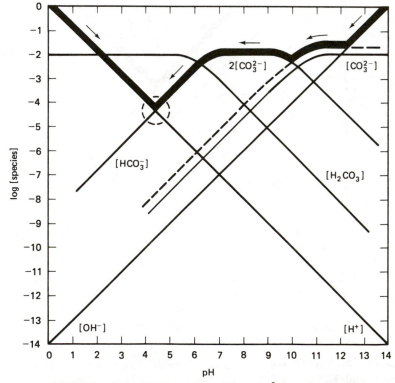

FIGURE 2-11 Equilibrium diagram for a 10^{-2} M H_2CO_3 solution.

         ACID-BASE EQUILIBRIA   CHAPTER 2

Note: The $2[CO_3^{2-}]$ curve is represented by a dashed line parallel to but 0.3 units above the $[CO_3^{2-}]$ curve because

$$\log 2[CO_3^{2-}] = \log [CO_3^{2-}] + \log 2$$

or

$$\log 2[CO_3^{2-}] = \log [CO_3^{2-}] + 0.3$$

3. The equilibrium state for the system is located at the point on the equilibrium diagram where the proton balance equation is satisfied.

   The route to the problem solution is outlined by the heavy line in Figure 2-11. The intersection of the $[H^+]$ curve and the $[HCO_3^-]$ curve gives an equilibrium pH of 4.2.

EXAMPLE PROBLEM 2-6: Calculate the equilibrium pH of a 10^{-2} M solution of $NaHCO_3$ at 25°C.

Solution:

1. Develop the proton balance equation for the chemical system.

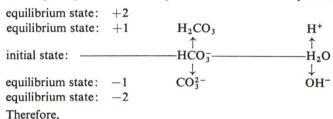

equilibrium state: +2
equilibrium state: +1 H_2CO_3 H^+
                                  ↑   ↑
initial state: ——————————HCO_3^-——————————H_2O
                                  ↓   ↓
equilibrium state: −1 CO_3^{2-} OH^-
equilibrium state: −2

Therefore,

$$[H_2CO_3] + [H^+] = [CO_3^{2-}] + [OH^-]$$

2. Construct the equilibrium diagram for a 10^{-2} M solution of H_2CO_3 (see Figure 2-12).

3. The equilibrium state for the system is located at the point on the equilibrium diagram where the proton balance equation is satisfied.

   The route to the problem solution is outlined by the heavy line in Figure 2-12. The intersection of the $[H_2CO_3]$ curve and the $[CO_3^{2-}]$ curve gives the equilibrium pH as 8.2.

EXAMPLE PROBLEM 2-7: Calculate the equilibrium pH of a 10^{-2} M solution of Na_2CO_3 at 25°C.

Solution:

1. Develop the proton balance equation for the chemical system.

equilibrium state: +2 H_2CO_3
                                  ↑
equilibrium state: +1 HCO_3^- H^+
                                  ↑   ↑
initial state: ——————————CO_3^{2-}——————————H_2O
                                                                      ↓
equilibrium state: −1 OH^-
equilibrium state: −2

Therefore,

$$2[H_2CO_3] + [HCO_3^-] + [H^+] = [OH^-]$$

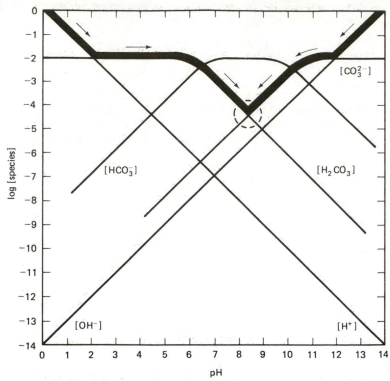

FIGURE 2-12 *Equilibrium diagram for a 10^{-2} M $NaHCO_3$ solution.*

2. Construct an equilibrium diagram for a 10^{-2} M solution of H_2CO_3 which also includes a $2[H_2CO_3]$ curve (see Figure 2-13).

3. The equilibrium state for the system is located at the point on the equilibrium diagram where the proton balance equation is satisfied.

   The route to the problem solution is outlined by the heavy line in Figure 2-13. The intersection of the $[HCO_3^-]$ curve and the $[OH^-]$ curve gives the equilibrium pH as 10.9.

The equivalence point of a solution is the equilibrium pH established when weak acids or bases and/or their salts are added to water. For example, example problem 2-3 shows that the equivalence point of a 0.1 M HNO_2 solution is at a pH of approximately 2.5, whereas example problem 2-4 shows that the equivalence point of a 0.1 M solution of the sodium salt of HNO_2 is at a pH of approximately 8. Thus, for a monoprotic acid system there are two equivalence points: one for the acid and one for its salt. Since a diprotic acid may have two salts, a diprotic acid system will have three equivalence points: one for the acid and one for each of its salts.

It should be understood that the equivalence point is a variable parameter in that it depends on the solution concentration. This point is illustrated in Figure 2-7. The value of the concept of equivalence points will become evident when alkalinity and

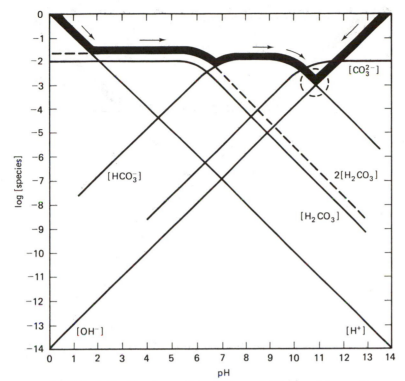

FIGURE 2-13 *Equilibrium diagram for a 10^{-2} M Na_2CO_3 solution.*

acidity are discussed. In an earlier discussion it was noted that ionization is essentially complete when a strong acid or strong base is added to water. As a result, the final state pH can be computed by using the molar ratio method. The final state pH for such systems can also be determined from a log [species] vs. pH diagram. This is illustrated in the following two example problems:

EXAMPLE PROBLEM 2-8: Calculate the solution pH which results when hydrochloric acid, HCl, is diluted with distilled water to give a 10^{-4} *M* HCl solution.

Solution:

1. Develop the proton balance equation for the chemical system.

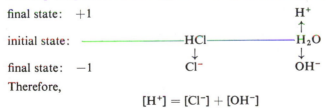

Therefore,

$$[H^+] = [Cl^-] + [OH^-]$$

2. Construct a log [species] vs. pH diagram for the chemical system (see Figure 2-14).

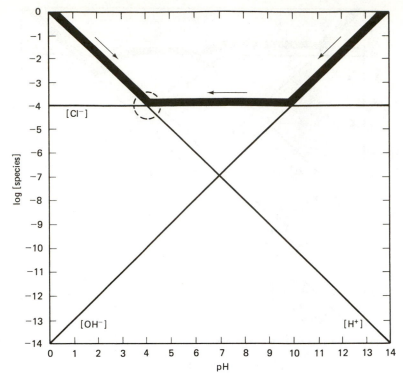

FIGURE 2-14 *Log (species) vs* pH *diagram for a 10^{-4} M HCℓ solution.*

3. The final state for the system is located at the point on the log [species] vs. pH diagram where the proton balance equation is satisfied.

   The route to the problem solution is outlined by the heavy line in Figure 2-14. The intersection of the [H^+] curve and the [Cl^-] curve gives a final pH of 4.

EXAMPLE PROBLEM 2-9: Calculate the solution pH which results when potassium hydroxide, KOH, is diluted with distilled water to give a 10^{-4} M KOH solution.

Solution:

1. Develop the proton balance equation for the chemical system.

$$KOH + H_3O^+ \rightleftharpoons K^+ + 2H_2O$$

final state:	+1	K^+		H^+
		↑		↑
initial state:	———————	—KOH—	——————	—H_2O
				↓
final state:	−1			OH^-

Therefore,

$$[K^+] + [H^+] = [OH^-]$$

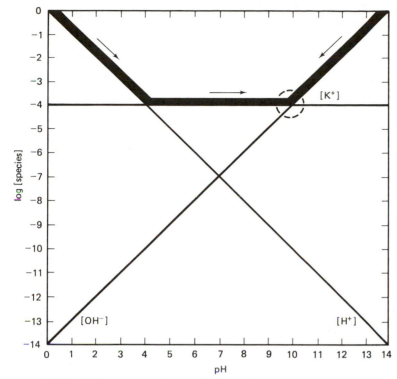

FIGURE 2-15 *Log (species) vs pH diagram for a 10^{-4} M KOH solution.*

2. Construct a log [species] vs. pH diagram for the chemical system (see Figure 2-15).

3. The final state for the system is located at the point on the log [species] vs. pH diagram where the proton balance equation is satisfied.

   The route to the problem solution is outlined by the heavy line in Figure 2-15. The intersection of the $[K^+]$ curve and the $[OH^-]$ curve gives the final pH as 10.

2-3 ALKALINITY AND ACIDITY

In a general sense, alkalinity may be defined as the capacity of a water to neutralize acids, whereas acidity may be defined as the capacity of water to neutralize bases. However, these definitions do very little toward elucidating the concepts upon which these parameters are based. Such concepts are best explained by considering acidity in terms of a solution containing both a strong acid and a weak acid or salt of a weak acid and by considering alkalinity in terms of a solution containing both a strong base and a weak acid or salt of a weak acid. For instance, in example problem 2-4 it is shown that the equivalence point for a 0.1 M solution of $NaNO_2$ is at pH = 8. If a solution of 0.1 M $NaNO_2$ and 10^{-4} M HCl were made up, the proton balance equation would have the form

equilibrium state: $+1$ HNO_2 H^+

initial state: ——————————HCl————NO_2^-————H_2O

equilibrium state: -1 Cl^- OH^-

or

$$[HNO_2] + [H^+] = [Cl^-] + [OH^-]$$

Figure 2-16 shows that the equilibrium pH for this system is 6. A measure of the acidity of this solution would be the equivalents of base required to elevate the solution pH to the NO_2^- equivalence point or to pH $= 8$. Thus, the acidity of a sample might also be thought of as the amount of strong acid present in a weak acid or weak acid salt solution.

Alkalinity may be illustrated in a manner similar to that for acidity, except that for alkalinity consider a solution made up of 0.1 M $NaNO_2$ and 10^{-4} M KOH. The

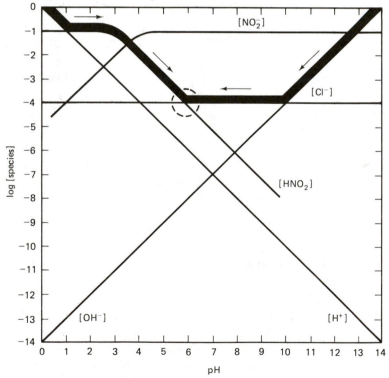

FIGURE 2-16 Log (species) vs pH diagram for a solution containing 0.1 moles/ℓ $NaNO_2$ and 10^{-4} moles/ℓ HCℓ.

proton balance equation for this solution would have the form

equilibrium state: $+1$ $\qquad$ K^+ $\qquad$ HNO_2 $\qquad$ H^+

initial state: $\qquad$ —KOH—$\qquad$—NO_2^-—$\qquad$—H_2O

equilibrium state: -1 $\qquad$ OH^-

or

$$[K^+] + [HNO_2] + [H^+] = [OH^-]$$

Figure 2-17 shows that the equilibrium pH for this system will be 10. A measure of the alkalinity of this solution would be the equivalents of acid required to lower the solution pH to the NO_2^- equivalence point of 8. Alkalinity, therefore, may be thought of as the amount of strong base present in a weak acid or weak acid salt solution.

Alkalinity and acidity of aqueous solutions as related to water and wastewater treatment are normally based on the carbonic acid (H_2CO_3) system. Carbonic acid is a weak diprotic acid, which means that the carbonic acid system has three equivalence points associated with it. As a result, it is possible to obtain three different alkalinity

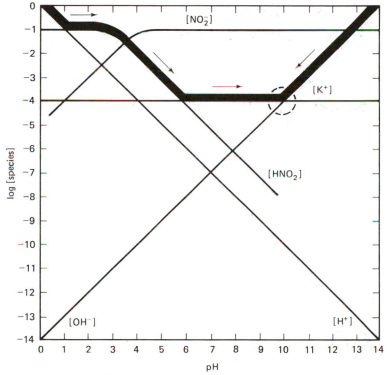

FIGURE 2-17 Log (species) vs pH diagram for a solution containing 0.1 moles/ℓ NaNO$_2$ and 10^{-4} moles/ℓ KOH.

and three different acidity measurements. These three different forms of alkalinity and acidity are explained in the following discussion:

A. *Alkalinity:*

1. The solution pH lies above the CO_3^{2-} equivalence point. The equivalents of acid required to lower the pH to the CO_3^{2-} equivalence point are a measure of the "caustic alkalinity."

2. The solution pH lies above the HCO_3^- equivalence point. The equivalents of acid required to lower the pH to the HCO_3^- equivalence point are a measure of the "phenolphthalein alkalinity."

3. The solution pH lies above the H_2CO_3 equivalence point. The equivalents of acid required to lower the pH to the H_2CO_3 equivalence point are a measure of the "total alkalinity," which is usually referred to as simply "alkalinity." Assuming that the difference in pH between the solution and the equivalence point is due to the presence of a strong base, the equation describing total alkalinity may be obtained from a proton balance equation as follows:

equilibrium state: $+2$
equilibrium state: $+1$ $\quad\quad\quad\quad\quad$ B^+ $\quad\quad\quad\quad\quad\quad\quad\quad\quad$ H^+
$\quad\quad\quad\quad\quad\quad\quad\quad\quad\quad\quad\quad\quad\quad\quad\uparrow\quad\quad\quad\quad\quad\quad\quad\quad\quad\quad\quad\quad\quad\uparrow$
initial state: $\quad$ ————————BOH————H_2CO_3————H_2O
$\quad\quad\quad\quad\quad\quad\quad\quad\quad\quad\quad\quad\quad\quad\quad\quad\quad\quad\quad\quad\quad\quad\downarrow\quad\quad\quad\quad\quad\downarrow$
equilibrium state: -1 $\quad\quad\quad\quad\quad\quad\quad\quad\quad\quad\quad\quad\quad$ HCO_3^- $\quad\quad$ OH^-
$\quad\quad\quad\quad\quad\quad\quad\quad\quad\quad\quad\quad\quad\quad\quad\quad\quad\quad\quad\quad\quad\quad\downarrow$
equilibrium state: -2 $\quad\quad\quad\quad\quad\quad\quad\quad\quad\quad\quad\quad\quad$ CO_3^{2-}

$$[B^+] + [H^+] = 2[CO_3^{2-}] + [HCO_3^-] + [OH^-] \quad\quad (2\text{-}71)$$

where BOH represents a strong base. Since alkalinity may be defined as the amount of strong base in the solution of interest, equation 2-71 may be written as

$$[B^+] = 2[CO_3^{2-}] + [HCO_3^-] + [OH^-] - [H^+] \quad\quad (2\text{-}72)$$

$$[Alk] = 2[CO_3^{2-}] + [HCO_3^-] + [OH^-] - [H^+] \quad\quad (2\text{-}73)$$

Writing equation 2-73 in terms of equivalent concentrations gives

$$[Alk]_e = [CO_3^{2-}]_e + [HCO_3^-]_e + [OH^-]_e - [H^+]_e \quad\quad (2\text{-}74)$$

where [ ]$_e$ represents concentration in eq/ℓ.

A schematic representation of the three forms of alkalinity for the carbonic acid system is presented in Figure 2-18. Note that if the initial pH of the solution lies between the HCO_3^- and CO_3^{2-} equivalence points, it is possible to measure total

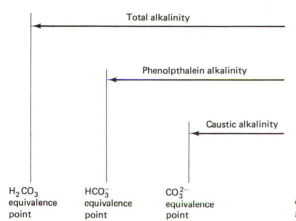

FIGURE 2-18 Forms of alkalinity for the carbonic acid system.

alkalinity and phenolphthalein alkalinity. On the other hand, if the initial pH of the solution lies between the H_2CO_3 and HCO_3^- equivalence points, only total alkalinity can be measured. If the initial pH of the solution is below the H_2CO_3 equivalence point, no alkalinity is present in the solution.

B. *Acidity*

1. The solution pH lies below the H_2CO_3 equivalence point. The equivalents of base required to raise the pH to the H_2CO_3 equivalence point are a measure of the "mineral acidity."

2. The solution pH lies below the HCO_3^- equivalence point. The equivalents of base required to raise the pH to the HCO_3^- equivalence point are a measure of the "CO_2 acidity."

3. The solution pH lies below the CO_3^{2-} equivalence point. The equivalents of base required to raise the pH to the CO_3^{2-} equivalence point are a measure of the "total acidity," which is usually referred to as simply "acidity." If the difference in pH between the solution and the equivalence point is due to the presence of a strong acid, the equation describing total acidity may be obtained from a proton balance equation as follows:

equilibrium state: $+2$		H_2CO_3	
equilibrium state: $+1$		$\uparrow$ HCO_3^-	H^+
initial state:	——————HA——————	$\uparrow$ CO_3^{2-}——————	$\uparrow$ H_2O
equilibrium state: -1	$\uparrow$ A^-		$\uparrow$ OH^-
equilibrium state: -2			

or

$$2[H_2CO_3] + [HCO_3^-] + [H^+] = [A^-] + [OH^-] \qquad \textbf{(2-75)}$$

where HA represents a strong acid. Since acidity may be defined as the amount of strong acid in the solution, equation 2-75 may be written as

$$[A^-] = 2[H_2CO_3] + [HCO_3^-] + [H^+] - [OH^-] \qquad \textbf{(2-76)}$$

or

$$[Acd] = 2[H_2CO_3] + [HCO_3^-] + [H^+] - [OH^-] \qquad \textbf{(2-77)}$$

Writing equation 2-77 in terms of equivalent concentrations gives

$$[Acd]_e = [H_2CO_3]_e + [HCO_3^-]_e + [OH^-]_e - [H^+]_e \qquad \textbf{(2-78)}$$

A schematic representation of the three forms of acidity for the carbonic acid system is presented in Figure 2-19. If the initial pH of the solution lies between the

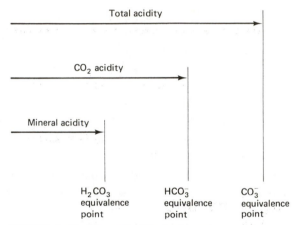

FIGURE 2-19 *Forms of acidity for the carbonic acid system.*

H_2CO_3 and HCO_3^- equivalence points, it is possible to measure CO_2 acidity and total acidity. On the other hand, an initial pH between the HCO_3^- and CO_3^{2-} equivalence points means that only total acidity can be measured. If the initial pH of the solution is above the CO_3^{2-} equivalence point, no acidity is present in the solution.

A mass balance expression for the carbonic acid system has the form

$$C_T = [H_2CO_3] + [HCO_3^-] + [CO_3^{2-}] \qquad \textbf{(2-79)}$$

If equations 2-73 and 2-77 are added together, the resulting equation is

$$2[H_2CO_3] + 2[HCO_3^-] + 2[CO_3^{2-}] = [Alk] + [Acd] \qquad \textbf{(2-80)}$$

Since this expression has a value equal to $2\,C_T$, it follows that the relationship between alkalinity, acidity, and total carbonic specie concentration is

$$C_T = \frac{[Alk] + [Acd]}{2} \qquad \textbf{(2-81)}$$

2-4 THE CARBONIC ACID SYSTEM

It is generally assumed that the pH of most natural waters is controlled by the carbonic acid system. The applicable equilibrium reactions are

$$CO_{2(g)} + H_2O_{(l)} \rightleftharpoons (H_2CO_3)_{(aq)} \rightleftharpoons H^+_{(aq)} + HCO^-_{3(aq)} \qquad \text{(2-82)}$$

$$HCO^-_{3(aq)} \rightleftharpoons H^+_{(aq)} + CO^{2-}_{3(aq)} \qquad \text{(2-83)}$$

$$H_2O_{(l)} \rightleftharpoons H^+_{(aq)} + OH^-_{(aq)} \qquad \text{(2-84)}$$

The development of subsequent equations in this section is based on the assumption that no CO_2 exchange occurs between the atmosphere and water (i.e., a closed system is assumed), which is generally a valid condition for most chemical treatment processes, since the rate of CO_2 exchange is usually much slower than the rate at which the chemical reaction of interest occurs. Since only a small fraction of the total CO_2 dissolved in water is hydrolyzed to H_2CO_3, it is convenient to sum the concentrations of dissolved CO_2 and H_2CO_3 to define a new concentration term, $H_2CO_3^*$. Equilibrium constant expressions may now be written for the carbonic acid system which have the form

$$K_1 = \frac{[H^+][HCO_3^-]}{[H_2CO_3^*]} \qquad \text{(2-85)}$$

$$K_2 = \frac{[H^+][CO_3^{2-}]}{[HCO_3^-]} \qquad \text{(2-86)}$$

$$K_w = [H^+][OH^-] \qquad \text{(2-87)}$$

The total carbonic species concentration in solution is usually represented by C_T and defined in terms of a mass balance expression:

$$C_T = [H_2CO_3^*] + [HCO_3^-] + [CO_3^{2-}] \qquad \text{(2-88)}$$

Equations 2-85, 2-86 and 2-87 are based on the assumption that the solution containing the carbonic species exhibits ideal behavior, so that the equilibrium constants K_1, K_2, and K_w represent thermodynamic equilibrium constants. Most solutions encountered in process chemistry, however, will exhibit nonideal behavior, the magnitude of which is determined by the ionic environment of each charged species. Equations 2-85, 2-86 and 2-87 must, therefore, be corrected by using activities in place of analytical concentrations.

Using the activity correction illustrated in either equation 1-46 or 1-47, equations 2-85, 2-86 and 2-87 can be correctly expressed as follows:

$$K_1 = \frac{\gamma_m[H^+]\gamma_m[HCO_3^-]}{[H_2CO_3^*]}$$

or

$$K_1' = \frac{K_1}{(\gamma_m)^2} = \frac{[H^+][HCO_3^-]}{[H_2CO_3^*]} \qquad \text{(2-89)}$$

$$K_2 = \frac{\gamma_m[H^+]\gamma_D[CO_3^{2-}]}{\gamma_m[HCO_3^-]}$$

or

$$K_2' = \frac{K_2}{\gamma_D} = \frac{[\text{H}^+][\text{CO}_3^{2-}]}{[\text{HCO}_3^-]} \tag{2-90}$$

$$K_w = \gamma_m[\text{H}^+]\gamma_m[\text{OH}^-]$$

or

$$K_w' = \frac{K_w}{(\gamma_m)^2} = [\text{H}^+][\text{OH}^-] \tag{2-91}$$

The prime indicates thermodynamic constants which have been corrected for temperature and ionic strength. It is important to remember that pH is a measure of hydrogen ion activity and is defined as

$$\text{pH} = -\log\left(\gamma_m[\text{H}^+]\right) \tag{2-92}$$

Thus,

$$[\text{H}^+] = \frac{10^{-\text{pH}}}{\gamma_m} \tag{2-93}$$

The thermodynamic equilibrium constant values are for standard state conditions, which implies 25°C. To develop an apparent equilibrium constant, the 25°C constant is first adjusted for temperature, and then the ionic strength corrections are applied. The equilibrium constants K_1, K_2, and K_w may be adjusted for temperature by applying the following relationships (Loewenthal and Marais, 1976):

$$\text{p}K_1 = \frac{17{,}052}{T} + 215.21(\log T) - 0.12675(T) - 545.56 \tag{2-94}$$

$$\text{p}K_2 = \frac{2902.39}{T} + 0.02379(T) - 6.498 \tag{2-95}$$

$$\text{p}K_w = \frac{4787.3}{T} + 7.1321(\log T) + 0.010365(T) - 22.801 \tag{2-96}$$

where T represents the solution temperature in degrees Kelvin.

Since carbonic acid is a weak acid, the distribution of the various carbonic species (H_2CO_3^*, HCO_3^-, and CO_3^{2-}) depends on the pH of the solution. A convenient method for determining the distribution of the various species involves defining a set of ionization fractions, α, which represent the fraction of the total carbonic species concentration, C_T, present as a given species. To develop expressions which describe the various ionization fractions, the following terms are defined:

$$\alpha_0 = \frac{[\text{H}_2\text{CO}_3^*]}{C_T} \tag{2-97}$$

$$\alpha_1 = \frac{[\text{HCO}_3^-]}{C_T} \tag{2-98}$$

$$\alpha_2 = \frac{[\text{CO}_3^{2-}]}{C_T} \tag{2-99}$$

where α_0 represents the fraction of the total carbonic species which is in the H_2CO_3^*

form, α_1 the fraction in the HCO_3^- form, and α_2 the fraction in the CO_3^{2-} form. Solving equation 2-89 for $[HCO_3^-]$ and equation 2-90 for $[CO_3^{2-}]$ yields

$$[HCO_3^-] = \frac{K_1'[H_2CO_3^*]}{[H^+]} \tag{2-100}$$

$$[CO_3^{2-}] = \frac{K_2'[HCO_3^-]}{[H^+]} \tag{2-101}$$

Substituting for $[HCO_3^-]$ in equation 2-101 from equation 2-100 gives

$$[CO_3^{2-}] = \frac{K_1'K_2'[H_2CO_3^*]}{[H^+]^2} \tag{2-102}$$

Substituting for $[HCO_3^-]$ and $[CO_3^{2-}]$ in equation 2-88 from equations 2-100 and 102 gives

$$C_T = [H_2CO_3^*] + \frac{K_1'[H_2CO_3^*]}{[H^+]} + \frac{K_1'K_2'[H_2CO_3^*]}{[H^+]^2}$$

or

$$C_T = [H_2CO_3^*]\left[1 + \frac{K_1'}{[H^+]} + \frac{K_1'K_2'}{[H^+]^2}\right] \tag{2-103}$$

A further substitution for C_T in equation 2-97 from equation 2-103 shows that

$$\alpha_0 = \frac{1}{1 + \dfrac{K_1'}{[H^+]} + \dfrac{K_1'K_2'}{[H^+]^2}} \tag{2-104}$$

Following a similar development for α_1 and α_2, the expressions derived for α_1 and α_2 are

$$\alpha_1 = \frac{1}{\dfrac{[H^+]}{K_1'} + 1 + \dfrac{K_2'}{[H^+]}} \tag{2-105}$$

$$\alpha_2 = \frac{1}{\dfrac{[H^+]^2}{K_1'K_2'} + \dfrac{[H^+]}{K_2'} + 1} \tag{2-106}$$

Calculations involving the carbonic system in aqueous solution requires knowledge of the concentration of five species: $H_2CO_3^*$, HCO_3^-, CO_3^{2-}, OH^-, and H^+. Thus, five independent equations are needed to arrive at a solution. Three of the required equations have already been written: equations 2-89, 2-90, and 2-91. A fourth equation, (2-93), is available from a pH measurement, and the remaining equation must be developed from some measurable quantity defined in terms of the chemical species of interest. The measurable quantity most often used in this development is alkalinity (Loewenthal and Marais, 1976).

Total alkalinity has been defined as the number of moles of hydrogen ions required to convert one liter of a solution into an equivalent carbonic acid solution (Loewenthal and Marais, 1976). On the assumption that alkalinity is due to the presence of a strong base in a carbonic acid solution, equation 2-74 applies:

$$[Alk]_e = [CO_3^{2-}]_e + [HCO_3^-]_e + [OH^-]_e - [H^+]_e \qquad \text{(2-74)}$$

To reformulate this in terms of moles per liter, recall that univalent species are numerically equivalent in terms of moles or equivalents, whereas both alkalinity (expressed in terms of equivalents of $CaCO_3$) and carbonate show a 2:1 ratio of equivalents to moles (i.e., 1 mole of CO_3^{2-} represents two equivalents). Therefore, equation 2-74 can be modified as follows:

$$2[Alk] = 2[CO_3^{2-}] + [HCO_3^-] + [OH^-] - [H^+] \qquad \text{(2-107)}$$

or

$$[Alk] = [CO_3^{2-}] + \tfrac{1}{2}[HCO_3^-] + \tfrac{1}{2}[OH^-] - \tfrac{1}{2}[H^+] \qquad \text{(2-108)}$$

where all concentrations are in moles per liter.

With the formulation of equation 2-108, measurement of pH and alkalinity enables the determination of $H_2CO_3^*$, HCO_3^-, and CO_3^{2-} concentrations. The specific equations are equation 2-101 and a rearrangement of equation 2-89.

$$[CO_3^{2-}] = \frac{K_2'[HCO_3^-]}{[H^+]} \qquad \text{(2-101)}$$

$$[H_2CO_3^*] = \frac{[H^+][HCO_3^-]}{K_1'} \qquad \text{(2-109)}$$

Substituting equation 2-101 into equation 2-108 and solving for bicarbonate gives

$$[HCO_3^-] = \frac{2[Alk] - [OH^-] + [H^+]}{\left[\dfrac{2K_2'}{[H^+]} + 1\right]} \qquad \text{(2-110)}$$

Similar equations can be developed for CO_3^{2-} and $H_2CO_3^*$ by substituting equation 2-110 into equations 2-101 and 2-109.

$$[CO_3^{2-}] = \frac{2[Alk] - [OH^-] + [H^+]}{\left[2 + \dfrac{[H^+]}{K_2'}\right]} \qquad \text{(2-111)}$$

$$[H_2CO_3^*] = \frac{2[Alk] - [OH^-] + [H^+]}{\left[\dfrac{2K_1'K_2'}{[H^+]^2} + \dfrac{K_1'}{[H^+]}\right]} \qquad \text{(2-112)}$$

Substituting for $[H_2CO_3^*]$, $[HCO_3^-]$, and $[CO_3^{2-}]$ in equation 2-88 from equations 2-110, 2-111 and 2-112, an expression for C_T can be derived which has the form

$$C_T = \frac{1}{\alpha_0}\left[\frac{2[Alk] - \dfrac{K_w'}{[H^+]} + [H^+]}{\dfrac{2K_1'K_2'}{[H^+]^2} + \dfrac{K_1'}{[H^+]}}\right] \qquad \text{(2-113)}$$

From equation 2-113 the following relationship can be developed for total alkalinity:

$$[\text{Alk}] = \frac{1}{2}\left\{ C_T\alpha_1\left[\frac{2K_2'}{[\text{H}^+]} + 1\right] + \frac{K_w'}{[\text{H}^+]} - [\text{H}^+]\right\} \qquad (2\text{-}114)$$

2-5 BUFFERING IN WATER SYSTEMS

A buffer is defined as a solution which will experience only small changes in pH when relatively large amounts of either acids or bases are added to it. In most cases a buffer solution is composed of a mixture of a weak Brönsted acid and its salt, although a weak Brönsted base and its salt will also provide a buffer solution. The mechanism by which a buffer solution resists large pH changes can be illustrated by considering a solution composed of a mixture of acetic acid and its sodium salt. For example, assume that a solution is 0.05 M acetic acid and 0.05 M sodium acetate (a 0.1 M acetate buffer), and a small amount of sodium hydroxide is added to the solution. The following reactions occur:

$$
\begin{array}{ccccc}
CH_3COOH & \rightleftharpoons & H^+ & + & CH_3COO^- \\
 & & + & & \\
NaOH & \longrightarrow & OH^- & + & Na^+ \\
 & & \updownarrow & & \\
 & & H_2O & &
\end{array}
$$

The hydrogen ions produced by the ionization of acetic acid react with the hydroxyl ions from sodium hydroxide to form water. The removal of hydrogen ions in the water reaction causes more acetic acid to ionize. Thus, as long as there are unionized acetic acid molecules in solution to furnish hydrogen ions, the pH change due to the addition of hydroxyl ions will be small. A similar explanation can be presented to explain the resistance to change in pH when an acid such as HCl is added to the acetate buffer. In this case the following reactions occur:

$$
\begin{array}{ccccc}
CH_3COO:Na & \longrightarrow & CH_3COO^- & + & Na^+ \\
 & & + & & \\
HCl & \longrightarrow & H^+ & + & Cl^- \\
 & & \updownarrow & & \\
 & & CH_3COOH & &
\end{array}
$$

The hydrogen ions produced by the ionization of HCl react with the acetate ions from sodium acetate to form acetic acid, and as long as there are acetate ions in solution, the pH change due to the addition of hydrogen ions will be small.

The point at which a buffer displays its greatest ability to resist pH change is where the pH of the solution is equal to the pK_a of the acid making up the buffer solution. This can be explained by examining the titration curve for the buffer acid.

Titration Curves

A titration curve may be obtained in the laboratory by adding small amounts of a strong base to a weak acid solution and measuring the pH after each addition. The titration curve for a 0.1 M acetic acid solution is presented in Figure 2-20. The curve

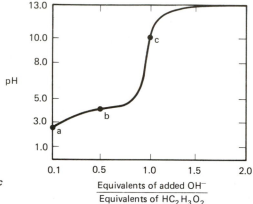

FIGURE 2-20 *Titration curve for a 0.1 M acetic acid solution (after Gray and Haight, 1967).*

has an S shape, which suggests that the pH does not change at a constant rate with the addition of strong base. The chemical events responsible for the shape of the titration curve can be elucidated by considering points *a*, *b*, and *c* located on the plot presented in Figure 2-20.

Point a: This point represents the equilibrium pH established in a 0.1 M acetic acid solution. At this point no base has been added. Furthermore, the concentration of the ionized acid form is much smaller than the unionized form, so that the unionized acid concentration can be considered equal to the initial concentration.

Point b: This point represents the pH established when the concentration of unionized acid equals the concentration of ionized acid; i.e.,

$$[CH_3COOH] = [CH_3COO^-] \qquad \text{(2-115)}$$

The validity of equation 2-115 can be substantiated by considering the Henderson-Hasselbalch equation (equation 2-15) for the acetate buffer.

$$pH = pK_a + \log\left[\frac{[CH_3COO^-]}{[CH_3COOH]}\right] \qquad \text{(2-15)}$$

When the measured solution pH is equal to the acid pK_a value, equation 2-15 reduces to

$$\log\left[\frac{[CH_3COO^-]}{[CH_3COOH]}\right] = 0$$

which suggests that

$$\frac{[CH_3COO^-]}{[CH_3COOH]} = 1$$

or

$$[CH_3COO^-] = [CH_3COOH]$$

An examination of the titration curve in the vicinity of point b also shows that, around this point, the solution pH will show the smallest change per unit of strong base added. Because of this effect most workers consider that the useful range of a buffer is 1.5 pH units on either side of its pK_a value.

Point c: This point represents the pH established when the concentration of ionized acid equals the initial acid concentration.

$$[CH_3COO^-] = 0.1 \ M$$

In other words, the pH at this point is the equivalence point for a 0.1 M acetate salt solution. Further addition of a strong base past point c will result in a continued increase in pH. The limiting pH is set by the pH of the titrant. In Figure 2-20 it is assumed that a 0.1 M solution of NaOH is used as the titrant. Thus, when the effects of dilution become insignificant, the pH will approach 13.

Figure 2-21 illustrates the titration curve for a 0.01 M sodium carbonate solution. Again, the chemical events responsible for the shape of the curve will be explained by discussing various points along the curve length.

Point a: This point represents the equivalence point of a 0.01 M Na$_2$CO$_3$ solution. At this point it is assumed that

$$[CO_3^{2-}] = 0.01 \ M$$

Point b: This point is located where the measured solution pH is equal to pK_2 (the pK_a for HCO$_3^-$). At point b,

$$[CO_3^{2-}] = [HCO_3^-] = \frac{0.01 \ M}{2}$$

Around this point the solution exhibits a strong resistance to pH change; i.e., the pH will change very little per unit of strong acid added.

Point c: This point is located where the measured solution pH is equal to the equivalence point pH of $NaHCO_3$. At this point it is assumed that

$$[HCO_3^-] = 0.01 \ M$$

Point d: This point is located where the measured solution pH is equal to pK_1 (the pK_a for $H_2CO_3^*$). At point *d*,

$$[HCO_3^-] = [H_2CO_3^*] = \frac{0.01 \ M}{2}$$

Around this point the solution also exhibits a strong resistance to pH change.

Point e: This point is located where the measured solution pH is equal to the equivalence point pH of $H_2CO_3^*$. At this point is it assumed that

$$[H_2CO_3^*] = 0.01 \ M$$

Figure 2-21 shows that in region I the concentrations of CO_3^{2-} and HCO_3^- are relatively high and the solution exhibits a strong resistance to pH change around pH 10.3, whereas the concentrations of HCO_3^- and $H_2CO_3^*$ are relatively high in region II, where the solution exhibits a strong resistance to pH change around pH 6.3.

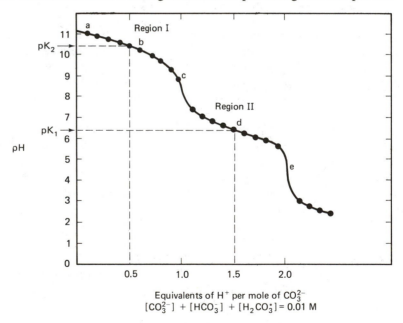

FIGURE 2-21   *Titration curve for a 0.01 M Na_2CO_3 solution (after Gray and Haight, 1967).*

EXAMPLE PROBLEM 2-10:   A solution of ammonia in water is neutralized with dilute HCl to a pH of 8.5. What are the relative concentrations of free base and conjugate acid? $pK_a = 9.2$.

Solution:

1. Substitute the appropriate pH and pK_a values into equation 2-15 and solve for the $[NH_3]/[NH_4^+]$ ratio.

$$8.5 = 9.25 + \log \frac{[NH_3]}{[NH_4^+]}$$

$$\log \frac{[NH_3]}{[NH_4^+]} = -0.75$$

$$\frac{[NH_3]}{[NH_4^+]} = 0.18$$

EXAMPLE PROBLEM 2-11: The pH of a 0.1 M solution of sodium bicarbonate is adjusted to 7.0 with a strong acid. A 100 mℓ portion of this solution is obtained, and to this volume 100 mℓ of 0.08 N HCl is added. What are the final $H_2CO_3^*$ and HCO_3^- concentrations and resulting pH?

Solution:

1. Determine the initial distribution of $[H_2CO_3^*]$ and $[HCO_3^-]$ species:

$$7.0 = 6.35 + \log \frac{[HCO_3^-]}{[H_2CO_3^*]}$$

$$\frac{[HCO_3^-]}{[H_2CO_3^*]} = 4.47$$

Since

$$[HCO_3^-] + [H_2CO_3^*] = 0.1$$

The concentrations of the $[H_2CO_3^*]$ species is given by

$$4.47[H_2CO_3^*] + [H_2CO_3^*] = 0.1$$
$$[H_2CO_3^*] = 0.018 \ M$$

It is then possible to determine the $[HCO_3^-]$ concentration:

$$[HCO_3^-] + 0.018 = 0.1$$
$$[HCO_3^-] = 0.082 \ M$$

Since only a one-proton shift is involved, the number of milliequivalents (meq.) per 100 mℓ is given by

$$\text{meq. of } H_2CO_3^* \text{ per 100 m}\ell = 100 \times 0.018 = 1.8 \text{ meq.}$$
$$\text{meq. of } HCO_3^- \text{ per 100 m}\ell = 100 \times 0.082 = 8.2 \text{ meq.}$$

2. Compute the meq. of hydrogen ions added to the original volume:

$$\text{meq. of } [H^+] \text{ added} = 100 \times 0.08 = 8.0 \text{ meq.}$$

3. Determine the new distribution of $[H_2CO_3^*]$ and $[HCO_3^-]$ species:

$$HCO_3^- + H^+ \rightleftharpoons H_2CO_3^*$$

The bicarbonate ions are reduced by the amount of hydrogen ions added:

$$\text{meq. of } H_2CO_3^* \text{ per 200 m}\ell = 1.8 + 8.0 = 9.8 \text{ meq.}$$
$$\text{meq. of } HCO_3^- \text{ per 200 m}\ell = 8.2 - 8.0 = 0.2 \text{ meq.}$$

The new concentration of each species is computed as follows:

$$[H_2CO_3^*] = \frac{9.8}{200} = 0.049 \; M$$

$$[HCO_3^-] = \frac{0.2}{200} = 0.001 \; M$$

4. Compute the final pH, using the Henderson-Hasselbalch equation (equation 2-15):

$$pH = 6.35 + \log\frac{0.001}{0.049} = 4.66$$

Buffer Characteristics

For pH buffering, a buffer may be characterized as having a neutralizing capacity and an intensity.

1. Neutralizing Capacity: From equation 2-16 it can be seen that only the molar ratio of the proton acceptor to proton donor determines the pH of the buffer solution. For example, suppose that an acetate buffer was 0.05 M in acetic acid and 0.05 M in sodium acetate; the solution pH would be

$$pH = 4.76 + \log\left[\frac{0.05}{0.05}\right]$$

or

$$pH = 4.76$$

However, if the salt to acid ratio was 6:2, the resulting pH would be

$$pH = 4.76 + \log\left[\frac{6}{2}\right]$$

or

$$pH = 5.23$$

Not only does the effectiveness of a buffer solution depend on the molar ratio of the proton acceptor to proton donor (which establishes location on the titration curve), but it also depends on the initial amounts of the acid and its salt (known as the *limiting buffer capacity*), since one or the other is consumed by reaction with the strong acid or strong base added to the solution. In other words, the actual amounts of the buffer components control the amount of hydrogen ions or hydroxyl ions that may be added without consuming all the buffering component. This principle is illustrated in

ACID-BASE EQUILIBRIA   CHAPTER 2

example problem 2-12. Note that the rate of pH change per unit addition of strong acid or strong base increases steadily as the limiting buffer capacity is approached (i.e., as the buffer is exhausted). This response can be observed in Figure 2-20.

EXAMPLE PROBLEM 2-12: Determine the pH of 100 ml of a 0.1 M acetate buffer (0.05 M in acetic acid and 0.05 M sodium acetate) and the pH of 100 ml of a 0.01 M acetate buffer (0.005 M acetic acid and 0.005 M sodium acetate) after 100 ml of 0.001 M HCl has been added to each solution.

Solution:

1. Compute the initial pH of each buffer solution from equation 2-16:
   (a) For the 0.1 M buffer solution

   $$pH = 4.76 + \log \left[\frac{0.05}{0.05} \right] = 4.76$$

   (b) For the 0.01 M buffer solution,

   $$pH = 4.76 + \log \left[\frac{0.005}{0.005} \right] = 4.76$$

2. Determine the initial distribution of the buffer components in terms of milliequivalents. Recall that

   $$\text{number of meq.} = (\text{volume in } ml) \times (\text{normality})$$

   (a) For the 0.1 M buffer solution,

   $$\text{meq. of } CH_3COOH = 100 \times 0.05 = 5 \text{ meq.}$$
   $$\text{meq. of } CH_3COO^- = 100 \times 0.05 = 5 \text{ meq.}$$

   (b) For the 0.01 M buffer solution,

   $$\text{meq. of } CH_3COOH = 100 \times 0.005 = 0.5 \text{ meq.}$$
   $$\text{meq. of } CH_3COO^- = 100 \times 0.005 = 0.5 \text{ meq.}$$

3. Compute the milliequivalents of hydrogen ions added to the original volume:

   $$\text{meq. of } H^+ \text{ ions added} = 100 \times 0.001 = 0.1 \text{ meq.}$$

4. Determine the distribution of the CH_3COOH and CH_3COO^- chemical species after acid addition:

   $$CH_3COO^- + H^+ \rightleftharpoons CH_3COOH$$

   The acetate ion concentration is reduced by an amount equal to the hydrogen ions added.
   (a) For the 0.1 M buffer solution,

   $$\text{meq. of } CH_3COOH = 0.1 + 5.0 = 5.1 \text{ meq.}$$
   $$\text{meq. of } CH_3COO^- = 5.0 - 0.1 = 4.9 \text{ meq.}$$

Therefore, the new concentration of each species is

$$[CH_3COOH] = \frac{5.1}{200} = 0.025 \ M$$

$$[CH_3COO^-] = \frac{4.9}{200} = 0.024 \ M$$

(b) For the 0.01 M buffer solution,

$$\text{meq. of } CH_3COOH = 0.5 + 0.1 = 0.6 \text{ meq.}$$

Therefore, the new concentration of each species is

$$[CH_3COOH] = \frac{0.6}{200} = 0.003 \ M$$

$$[CH_3COO^-] = \frac{0.4}{200} = 0.002 \ M$$

5. Compute the final pH, using equation 2-15.

   (a) For the 0.1 M buffer solution,

$$pH = 4.76 + \log\left[\frac{0.024}{0.025}\right] = 4.74$$

   (b) For the 0.01 M buffer solution,

$$pH = 4.76 + \log\left[\frac{0.002}{0.003}\right] = 4.58$$

Thus, the pH change is significantly greater in the system with the smaller limiting buffer capacity value.

2. Buffer Intensity: Buffer intensity is defined as the number of moles of strong acid or strong base required to change the pH of 1 liter of solution by one pH unit. However, considering the titration curve presented in Figure 2-20, it is evident that buffer intensity varies with solution composition (as shown by pH, which indicates the molar ratio between proton acceptor and proton donor). This is demonstrated by the fact that more base is required to effect a one unit pH change near the pK_a (midpoint), or point *b*, of the titration curve than is required near the equivalence point for the acid salt (point *c* on the curve). Beyond point *c* buffering action is a result of the water equilibrium reaction. Because of the variation in buffer intensity along the titration curve, it is best to represent buffer intensity by a differential such that

$$\beta = \frac{dC}{dpH} \qquad (2\text{-}116)$$

where β = buffer intensity

dC = differential quantity of strong acid or base added to the solution

dpH = differential change in pH due to the addition of a dC amount of strong acid or strong base.

The generalized titration curve for a weak monoprotic acid is shown in Figure 2-22. Since the slope at any point along the titration curve is given by $d\text{pH}/dC_B$, a graphical representation of β as a function of pH may be obtained by plotting the inverse of the slope of the titration curve versus pH. On the other hand, a mathematical representation of β may be obtained by differentiating the equation defining the titration curve with respect to pH (Weber and Stumm, 1963).

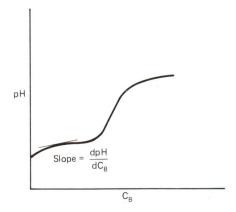

FIGURE 2-22 *Generalized titration curve for a weak monoprotic acid.*

Buffer Intensity of a Homogeneous Aqueous Carbonic Acid Solution with Respect to Strong Acids and Strong Bases

Since the carbonic acid system is the major buffering system of interest in process chemistry, a mathematical representation of β will be developed for this system. Recall that for the carbonic acid system total alkalinity is defined as the amount of strong base present in a carbonic acid solution and is represented mathematically by the proton balance equation

$$C_B = [\text{Alk}] = 2[\text{CO}_3^{2-}] + [\text{HCO}_3^-] + [\text{OH}^-] - [\text{H}^+] \qquad \textbf{(2-73)}$$

Equation 2-73 describes the titration curve for the carbonic acid system based on the amount of strong base remaining after each incremental addition of strong acid. Differentiation of equation 2-73 with respect to pH gives

$$\frac{dC_B}{d\text{pH}} = \beta$$

or

$$\beta = \frac{dC_B}{d\text{pH}} = \frac{d[\text{HCO}_3^-]}{d\text{pH}} + \frac{2d[\text{CO}_3^{2-}]}{d\text{pH}} - \frac{d[\text{H}^+]}{d\text{pH}} + \frac{d[\text{OH}^-]}{d\text{pH}} \qquad \textbf{(2-117)}$$

According to the chain rule from calculus, equation 2-117 may be represented as

$$\frac{dC_B}{d\text{pH}} = \frac{dC_B}{d[\text{H}^+]} \cdot \frac{d[\text{H}^+]}{d\text{pH}} \qquad \textbf{(2-118)}$$

The $d[H^+]/dpH$ term may be determined by recalling that, for an ideal solution,

$$pH = -\log[H^+]$$

or

$$pH = -\left(\frac{1}{2.3}\right) \ln [H^+] \tag{2-119}$$

Therefore,

$$\frac{dpH}{d[H^+]} = \frac{d[-(1/2.3) \ln [H^+]]}{d[H^+]}$$

where

$$\frac{d[-(1/2.3) \ln [H^+]]}{d[H^+]} = -\frac{1}{2.3[H^+]} \tag{2-120}$$

The inverse of equation 2-120 is

$$\frac{d[H^+]}{dpH} = -2.3[H^+] \tag{2-121}$$

Equation 2-117 may, therefore, be expressed as

$$\beta = -2.3[H^+]\left[\frac{dC_B}{d[H^+]}\right]$$

or

$$\beta = -2.3[H^+]\left[\frac{d[HCO_3^-]}{d[H^+]} + \frac{2d[CO_3^{2-}]}{d[H^+]} - \frac{d[H^+]}{d[H^+]} + \frac{d[OH^-]}{d[H^+]}\right] \tag{2-122}$$

The $d[OH^-]/d[H^+]$ term may be evaluated by substituting $K_w/[H^+]$ for $[OH^-]$. Thus,

$$\frac{d(K_w/[H^+])}{d[H^+]} = -\frac{K_w}{[H^+]^2} \tag{2-123}$$

Also, it is noted that the $-d[H^+]/d[H^+]$ term has a value of -1. Hence, equation 2-122 reduces to

$$\beta = -2.3[H^+]\left[\frac{d[HCO_3^-]}{d[H^+]} + \frac{2d[CO_3^{2-}]}{d[H^+]}\right] + 2.3[H^+] + 2.3\frac{K_w}{[H^+]}$$

or

$$\beta = -2.3[H^+]\left[\frac{d[HCO_3^-]}{d[H^+]} + \frac{2d[CO_3^{2-}]}{d[H^+]}\right] + 2.3[H^+] + 2.3[OH^-] \tag{2-124}$$

Recalling that

$$[HCO_3^-] = C_T\alpha_1 \tag{2-125}$$

$$[CO_3^{2-}] = C_T\alpha_2 \tag{2-126}$$

and substituting for the appropriate terms in equation 2-124, the following equation is obtained:

$$\beta = -2.3[H^+]\left[\frac{d(C_T\alpha_1)}{d[H^+]} + \frac{2d(C_T\alpha_2)}{d[H^+]}\right] + 2.3[H^+] + 2.3[OH^-] \quad \textbf{(2-127)}$$

Weber and Stumm (1963) showed that equation 2-127 reduces to

$$\beta = 2.3\left[C_T\frac{\alpha_1^2}{K_1}\left([H^+] + \frac{K_1K_2}{[H^+]} + 4K_2\right) + [H^+] + [OH^-]\right] \quad \textbf{(2-128)}$$

Equation 2-128 may also be expressed in terms of alkalinity as (Weber and Stumm, 1963):

$$\beta = 2.3\left[\frac{\alpha_1([Alk]_e - [OH^-] + [H^+])\left([H^+] + \frac{K_1K_2}{[H^+]} + 4K_2\right)}{K_1\left(1 + \frac{2K_2}{[H^+]}\right)} + [H^+] + [OH^-]\right] \quad \textbf{(2-129)}$$

where $\quad \alpha_1 = \dfrac{K_1}{[H^+] + K_1 + \dfrac{K_1K_2}{[H^+]}}$ $\qquad\qquad$ **(2-130)**

β = buffer intensity (equivalents required per unit pH change per liter of solution)

$[OH^-]$ = hydroxyl ion concentration (moles/ℓ)

$[H^+]$ = hydrogen ion concentration (moles/ℓ)

$[Alk]_e$ = total alkalinity (eq/ℓ)

Figure 2-23 illustrates the variation in β with respect to pH for a 10^{-3} M carbonic acid system as predicted by equation 2-128. In Figure 2-24 the log [species] vs. pH diagram, titration curve, and buffer intensity vs. pH diagram are shown for this same system. Figure 2-24 illustrates two interesting points:

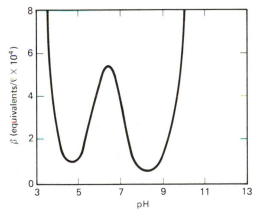

FIGURE 2-23 *Variation of buffer intensity with pH for a $C_T = 10^{-3}$ moles/ℓ homogeneous carbonic acid system (after Lowenthal and Marais, 1976).*

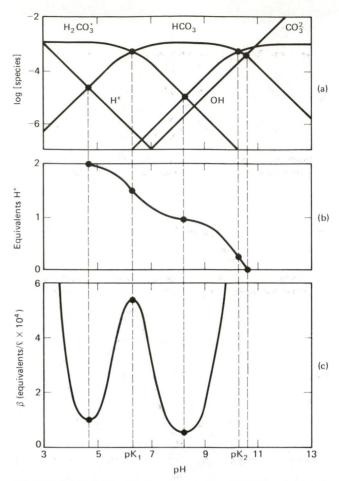

FIGURE 2-24 *(a) Equilibrium diagram for a 10^{-3} M carbonic acid system; (b) Titration curve for a sodium carbonate solution; (c) Buffer intensity diagram for a 10^{-3} M carbonic acid system (after Lowenthal and Marais, 1976).*

1. Within the pH range of most natural waters (6–9) the maximum buffer intensity occurs around the point where the solution pH is equal to pK_1 (around 6.3), whereas the minimum buffer intensity occurs around the point where the solution pH is equal to the equivalence point of a pure sodium bicarbonate system (around 8.3).

2. Points of minimum buffer intensity occur at the H_2CO_3 and HCO_3^- equivalence points. However, a well-defined minimum does not exist at the CO_3^{2-} equivalence point. This happens because at the high pH associated with the CO_3^{2-} equivalence point the hydroxyl ion concentration becomes the primary term in equation 2-128. Because of this, the CO_3^{2-} end point is not easily located during a titration.

Equations 2-128 and 129 can be used to estimate the expected pH change when a known concentration of strong acid or strong base is added to a water buffered by the carbonic acid system. The calculations involved in such a problem are illustrated in example problem 2-13. It should be understood that these types of calculations are valid only as long as equilibrium is attained but $CaCO_{3(s)}$ is not precipitated, and CO_2 is not evolved during the process change.

EXAMPLE PROBLEM 2-13: The wastewater from an industrial plant is discharged into the outfall sewer to a municipal treatment facility, at the rate of 0.2 MGD. The average flow in the sewer is 2 MGD, the wastewater alkalinity is 200 mg/ℓ as $CaCO_3$, and the pH is 7.5. What will the pH of the wastewater be after mixing with the industrial waste if the industrial flow has a pH of 3.5? The temperatures of the municipal and industrial wastewaters are 10°C and 30°C, respectively. Assume that buffering in the wastewater is due to the carbonate system and that ionic strength corrections can be neglected.

Solution:

1. Estimate the temperature of the wastewater after mixing:

$$T = \frac{(10°C)(2.0) + (30°C)(0.2)}{2.0 + 0.2} = 12°C$$

2. Determine the values of K_1 and K_2 from equations 2-94 and 2-95:

$$pK_1 = \frac{17,052}{285} + 215.21 \log (285) - 0.12675(285) - 545.56$$

$$= 6.46 \Rightarrow K_1 \approx 10^{-6.5}$$

$$pK_2 = \frac{2902.39}{285} + 0.02379(285) - 6.498$$

$$= 10.46 \Rightarrow K_2 \approx 10^{-10.5}$$

3. Convert alkalinity concentration to equivalents per liter:

$$[Alk.] = \frac{0.2 \text{ g}/\ell}{\text{eq. wt. of } CaCO_3} = \frac{0.2}{50} = 4 \times 10^{-3} \text{ eq.}/\ell$$

4. Calculate the hydrogen ion concentration:

$$[H^+] = 10^{-pH} = 10^{-7.5}$$

5. Compute the hydroxyl ion concentration:

$$[OH^-] = \frac{K_w}{[H^+]} = \frac{10^{-14}}{10^{-7.5}} = 10^{-6.5}$$

6. Determine the value of alpha, using equation 2-130:

$$\alpha_1 = \frac{10^{-6.5}}{10^{-6.5} + 10^{-7.5} + (10^{-6.5} \times 10^{-10.5}/10^{-7.5})} = 0.91$$

7. Estimate the buffer intensity of the municipal wastewater, using equation 2-129:

$$\beta = 2.3\left[\frac{(0.91)(4\times10^{-3}-10^{-6.5}+10^{-7.5})\{10^{-7.5}+(10^{-6.5}\times10^{-10.5}/10^{-7.5})+4\times^{-10.5}\}}{10^{-6.5}[1+(2\times10^{-10.5}/10^{-7.5})]}\right.$$
$$\left.+10^{-7.5}+10^{-6.5}\right]$$

$$\beta = 8.37\times10^{-4}\ eq./\ell$$

or

$$\beta = 8.37\times10^{-4}\frac{eq.}{\ell}\times3.78\frac{\ell}{gal}\times2\times10^{6}\frac{gal}{day} = 6.3\times10^{3}\ eq./day$$

8. Approximate the addition of hydrogen ions resulting from the industrial waste flow:

$$\Delta[H^+] = 3.2\times10^{-4}\frac{eq.}{\ell}\times3.78\frac{\ell}{gal}\times0.2\times10^{6}\frac{gal}{day}$$
$$= 2.4\times10^{2}\ eq./day$$

9. Compute the expected pH change, using equation 2-116:

$$\beta = \frac{\Delta C}{\Delta pH} = \frac{\Delta[H^+]}{\Delta pH}$$

The above expression can be solved for ΔpH to give

$$\Delta pH = \frac{2.4\times10^{2}\ eq./day}{6.3\times10^{3}\ eq./day} = 0.038$$

Thus, the final pH of the mixture is

$$pH = 7.5 - 0.038 = 7.46$$

Buffer Intensity of a Homogeneous Aqueous Carbonic Acid System with Respect to CO₂

If CO_2 is added to an aqueous solution, the effect is the same as titrating with a weak acid, whereas if CO_2 is removed from an aqueous solution, the effect is the same as titrating with a weak base. To understand how the addition or removal of CO_2 affects the carbonic acid system, it is necessary to consider the equations which describe C_T, alkalinity, and acidity:

$$C_T = [H_2CO_3^*] + [HCO_3^-] + [CO_3^{2-}] \tag{2-88}$$
$$[Acd] = 2[H_2CO_3^*] + [HCO_3^-] + [H^+] - [OH^-] \tag{2-77}$$
$$[Alk] = 2[CO_3^{2-}] + [HCO_3^-] + [OH^-] - [H^+] \tag{2-73}$$

Equations 2-88 and 2-77 show that addition or removal of CO_2 will increase or decrease both C_T and [Acd], whereas equation 2-73 shows that [Alk] is unaffected by CO_2 addition or removal. The addition of CO_2 will not affect alkalinity as long as $CaCO_{3(s)}$

does not precipitate because, for each ion of HCO_3^- or CO_3^{2-} formed from H_2CO_3 ionization, one or two H^+ ions are generated. The reverse is true when CO_2 is removed. Thus, the net change in alkalinity is zero.

Weber and Stumm (1963) have shown that the buffer intensity of a homogenous aqueous carbonic acid system with respect to CO_2 is equal to the rate of change of C_T with respect to pH:

$$\beta = \frac{dC_T}{d\text{pH}} \qquad (2\text{-}131)$$

or equal to one-half the rate of change in acidity with respect to pH:

$$\beta = \frac{1}{2}\left[\frac{d[\text{Acd}]}{d\text{pH}}\right] \qquad (2\text{-}132)$$

From these relationships these workers derived the following equation for β:

$$\beta_{CO_2} = -2.3\left[\frac{\left([\text{Alk}]_e - [\text{OH}^-] + [\text{H}^+]\right)\left([\text{H}^+] + \frac{K_1 K_2}{[\text{H}^+]} + 4K_2\right)}{K_1\left(1 + \frac{2K_2}{[\text{H}^+]}\right)^2} + \frac{[\text{H}^+] + [\text{OH}^-]}{\alpha_1\left(1 + \frac{2K_2}{[\text{H}^+]}\right)}\right]$$

$$(2\text{-}133)$$

Equation 2-133 can be used to estimate the expected pH change when a known concentration of CO_2 is added to or removed from a water buffered by the carbonic acid system. For the results to be valid, however, $CaCO_{3(s)}$ must not precipitate during the process change and equilibrium must be established in the system.

EXAMPLE PROBLEM 2-14: Assume that after water softening the final saturated state of the water was found to contain alkalinity of 70 mg/ℓ, calcium of 15 mg/ℓ (both as $CaCO_3$), and pH of 11.0. Determine the amount of CO_2 required (lb/MG) to lower the pH of the water to 8.0. Assume an ideal solution with a temperature of 25°C so that $K_1 = 10^{-6.3}$ and $K_2 = 10^{-10.3}$.

Solution:

1. Convert alkalinity concentration to equivalents per liter:

$$[\text{Alk}]_e = \frac{0.070 \text{ g}/\ell}{\text{eq. wt. of } CaCO_3} = \frac{0.070}{50} = 1.4 \times 10^{-3} \text{ eq.}/\ell$$

2. Calculate the initial hydrogen ion concentration:

$$[\text{H}^+] = 10^{-\text{pH}} = 10^{-11}$$

3. Compute the initial hydroxyl ion concentration:

$$[\text{OH}^-] = \frac{K_w}{[\text{H}^+]} = \frac{10^{-14}}{10^{-11}} = 10^{-3}$$

4. Determine the value of α_1, using equation 2-130:

$$\alpha_1 = \frac{10^{-6.3}}{10^{-11} + 10^{-6.3} + \left(\frac{10^{-6.3} \times 10^{-10.3}}{10^{-11}}\right)} = 0.17$$

5. Estimate the buffer intensity of the water using equation 2-133:

$$\beta = -2.3 \left[\frac{(1.4 \times 10^{-3} - 10^{-3} + 10^{-11})\left(10^{-11} + \dfrac{10^{-6.3} \times 10^{-10.3}}{10^{-11}} + 4 \times 10^{-10.3}\right)}{10^{-6.3}\left(1 + \dfrac{2 \times 10^{-10.3}}{10^{-11}}\right)^2} \right.$$

$$\left. + \frac{10^{-11} + 10^{-3}}{0.17\left(1 + \dfrac{2 \times 10^{-10.3}}{10^{-11}}\right)} \right]$$

$$= -12.4 \times 10^{-3} \text{ eq.}/\ell$$

The negative sign has no significance when determining chemical dose. Therefore, the absolute value of β is used in subsequent calculations.

6. Compute the ΔpH for the neutralization:

$$\Delta pH = 11 - 8 = 3$$

7. Calculate the CO_2 required per liter of solution from equation 2-131:

$$\Delta C_T = (CO_2)_{req} = \beta \Delta pH$$

$$(CO_2)_{req} = (1.24 \times 10^{-3})(3) = 3.72 \times 10^{-3} \text{ eq.}/\ell$$

8. Calculate the CO_2 requirement on the basis of lb/MG:

$$(CO_2)_{req} = 3.72 \times 10^{-3} \frac{\text{eq.}}{\ell} \times 22 \frac{\text{g}}{\text{eq.}} \times 10^3 \frac{\text{mg}}{\text{g}} \times 8.34$$

$$= 682.5 \text{ lb/MG}$$

Because β will change with pH, a better answer may be obtained by summing the CO_2 dose required for incremental pH changes until the desired pH of 8 is reached.

Use of Titration Curves for Computing Chemical Dose Requirements

In many treatment situations it is often necessary to neutralize wastewater or to make a water more acidic or basic (e.g., chromium reduction with sulfur dioxide is most efficient when conducted at a pH near 3.0). For such cases the engineer is required to compute the acid or alkali dose necessary to achieve the desired treatment pH. However, it is extremely difficult to directly calculate chemical dosages for complex water and wastewater systems because the exact composition of the waste is usually not known. The best method for determining acid or alkali dosages in these complex systems is to titrate a sample of the water or wastewater to be treated and select the required chemical dosage from the titration curve. To construct such a curve small amounts of acid or base are incrementally added to a sample of the water to be treated and the resulting pH is measured. A plot of pH vs. chemical added will give the titration curve.

EXAMPLE PROBLEM 2-15: The titration curve shown in Figure 2-25 was constructed by titrating a 1 ℓ sample of wastewater with 0.1 N H_2SO_4. Determine the acid feed in liters per million gallons of wastewater treated if the pH is to be adjusted to 3, using 12 N HCl.

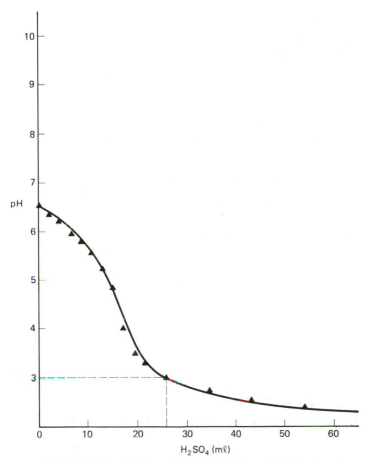

FIGURE 2-25 *Titration curve for acidification of chromium waste.*

Solution:

1. Referring to Figure 2-25, it can be seen that 26 mℓ of 0.1 N H_2SO_4 was required to acidify 1 ℓ of wastewater to pH 3.0. This is equivalent to an acid dose of

$$\frac{0.1 \frac{eq.}{\ell} \times \frac{1\,\ell}{1000\ m\ell} \times 26\ m\ell}{1\ \ell\ of\ wastewater} = 0.0026 \frac{eq.}{\ell}$$

2. Determine the required acid feed rate.

    (a) Compute the acid dose in terms of eq./MG:

$$0.0026 \frac{eq.}{\ell} \times 3.78 \frac{\ell}{gal} \times 1,000,000 \frac{gal}{MG} = 9828 \frac{eq.}{MG}$$

    (b) Convert the acid dose to liters of 12 N HCl per million gallons of water treated.

$$\frac{9828 \ eq./MG}{12 \ eq./\ell} = 819 \ \ell/MG$$

2-6 MEASURING ALKALINITY

As noted earlier, if a carbonic acid system is to be described quantitatively, the concentration of $H_2CO_3^*$, HCO_3^-, CO_3^{2-}, OH^-, and H^+ must be known. This means that five independent equations must be used. Three of these equations come from the equilibrium constant expressions for the ionizations of $H_2CO_3^*$, HCO_3^-, and H_2O, and a fourth equation is obtained from a pH measurement on the solution. The fifth equation must come from some measurable quantity defined in terms of the chemical species of interest. This measurable quantity may be acidity, total carbonic species concentration, or alkalinity. However, the measurable quantity most often used is total alkalinity (Loewenthal and Marais, 1976).

    Total alkalinity is generally determined volumetrically by titrating with 0.02 N H_2SO_4 and is reported in terms of mg/ℓ as $CaCO_3$. The $H_2CO_3^*$ equivalence point is usually located by adding methyl-orange indicator to the solution prior to the titration. When the solution's color changes from yellow to orange during the titration, it is assumed that the $H_2CO_3^*$ equivalence point has been reached. Total alkalinity is then given by this relationship:

$$\frac{\text{total alkalinity}}{\text{(mg/}\ell \text{ as CaCO}_3)} = \frac{(\text{m}\ell \text{ of titrant to end point})(N \text{ of titrant})(50 \times 10^3)}{(\text{m}\ell \text{ of sample titrated})} \quad \textbf{(2-134)}$$

    Although the location of the $H_2CO_3^*$ equivalence point is not significantly affected by ionic strength and temperature, it is affected by the total carbonic species concentration (see Figure 2-7). Therefore, titrating to a predetermined end point (which is the case when methyl-orange is used) may give incorrect results because the equivalence point is dependent on C_T, which is initially unknown. Furthermore, as the acidimetric titration progresses, the pH of the solution continually decreases. This means that the concentration of the $H_2CO_3^*$ species, and hence the concentration of dissolved CO_2, increases as the titration progresses (see Figure 2-11 and equation 2-82), which may result in an evolution of CO_2. Although the loss of CO_2 does not affect alkalinity (see equation 2-73), it does result in a decrease in C_T and therefore will change the location of the $H_2CO_3^*$ equivalence point. Thus, to obtain an accurate alkalinity measurement, the end point should not be predetermined but should always be located for a particular titration.

The end point for the alkalinity titration is difficult to determine by visual observation but may be obtained graphically by applying the following procedure:

1. Arrange the titration data as shown in the first two columns of Table 2-1.

TABLE 2-1 *Titration data for end-point determination.*

Vol. of Titrant (ml)	pH	$\dfrac{\Delta pH}{\Delta V}$	$\dfrac{\Delta^2 pH}{\Delta V^2}$
5.0	7.50		
15.0	7.27	0.02	
20.0	7.05	0.04	
22.0	6.89	0.08	
23.0	6.74	0.15	
23.5	6.66	0.16	
23.8	6.51	0.50	
24.0	6.38	0.65	
24.1	6.29	0.90	0.2
24.2	6.18	1.10	2.8
24.3	5.79	3.90	4.4
24.4	4.96	8.30	−5.9
24.5	4.72	2.40	−1.3
24.6	4.61	1.10	−0.4
24.7	4.54	0.70	
25.5	4.27	0.50	
25.5	4.27	0.24	
26.0	4.16	0.22	
28.0	3.86	0.15	

2. Compute $\Delta pH/\Delta V$ by dividing the difference between adjacent pH measurements by the difference between adjacent volume measurements.

3. Compute $\Delta^2 pH/\Delta V^2$ by taking the difference between adjacent $\Delta pH/\Delta V$ values.

4. Plot $\Delta^2 pH/\Delta V^2$ vs. the corresponding titrant volume. The end point (point of inflection) is located where the second derivative of the function is zero, i.e., where the trace intersects the X axis. If the titration curve and the $\Delta^2 pH/\Delta V^2$ vs. titrant volume diagram are developed together as shown in Figure 2-26, the end-point pH may be established by constructing a vertical line from the $(\Delta^2 pH/\Delta V^2) = 0$ point to the titration curve. The pH corresponding to the point of intersection on the titration curve is the end-point pH.

Once the equivalence point has been established, total alkalinity may be computed from equation 2-134 by using the total milliliters of titrant required to reach the equivalence point.

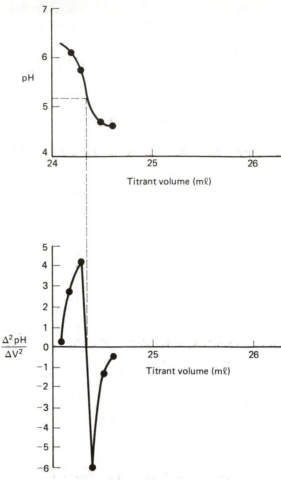

FIGURE 2-26 *Curve construction for end-point determination.*

PROBLEMS

2-1. Compute the equilibrium pH of a 0.001 M acetic acid solution. Assume that the solution exhibits ideal behavior and that the temperature is 25°C.

2-2. Compute the equilibrium pH of a 0.1 M ammonium chloride solution. Assume that the solution exhibits ideal behavior and that the temperature is 25°C. The pK_a for the NH_4^+-NH_3 system is 9.3.

2-3. Calculate the pH of a solution prepared by diluting 0.05 mole of ammonium carbonate ($(NH_4^+)_2CO_3$) to 1 ℓ with water. Assume that the solution exhibits ideal behavior and that the temperature is 25°C.

2-4. Given a liter of 0.2 M acetic acid solution at 25°C, how many grams of sodium acetate would have to be added to raise the pH to 4.0? Assume no volume change occurs.

2-5. Determine the equilibrium pH of a 0.01 M solution of sodium bicarbonate, assuming that the solution exhibits ideal behavior and that the temperature is 25°C. Compare this pH value to the value obtained when the equilibrium constants are adjusted for a temperature of 5°C and an ionic strength of 0.01 M.

2-6. A buffer solution contains 0.1 M of acetic acid and 0.1 M of sodium acetate (a 0.2 M acetate buffer). What is the pH after 4 mℓ of 0.05 M HCl are added to 10 mℓ of the buffer?

2-7. Each week an industry produces 2.0 MG of a strong acid waste (pH = 1.5) during a 48-hr production cycle. The plant engineer would like to discharge the waste into a river that passes near the plant. The river has a 7-day, 10-year low flow of 25 MG/day, an alkalinity of 150 mg/ℓ of $CaCO_3$, and a pH of 7.7. If the temperature of the mixture is 25°C, will it be possible to discharge all the waste into the river within a 7-day period if the river pH must be maintained above 7.0?

2-8. Groundwater containing 60 mg/ℓ of CO_2 at 10°C is pumped through a spray nozzle aerator, and the CO_2 concentration is reduced to 5.6 mg/ℓ. If the alkalinity of the water is 200 mg/ℓ as $CaCO_3$ and the initial pH is 5.8, what is the pH of the water after CO_2 removal? Assume the water temperature remains at 10°C and that ionic strength corrections can be neglected.

REFERENCES

GRAY, H.B., and HAIGHT, G.P., Jr., *Basic Principles of Chemistry*, W.A. Benjamin, Inc., New York (1967).

LOEWENTHAL, R.E., and MARAIS, G.V.R., *Carbonate Chemistry of Aquatic Systems: Theory and Application*, Ann Arbor Science Publishers, Ann Arbor, MI (1976).

WEBER, W.J., and STUMM, W., "Mechanism of Hydrogen Ion Buffering in Natural Waters," *J. Am. Water Works Assoc.*, **55**, 1553 (1963).

3

SOLUBILITY EQUILIBRIA

When the ions of a sparingly soluble salt are brought together in solution it is observed that if the concentration of these ions is sufficiently large, a solid is formed that will settle from the solution. For example, consider the reaction between calcium and carbonate ions:

$$Ca^{2+}_{(aq)} + CO^{2-}_{3(aq)} \rightleftharpoons CaCO_{3(s)}$$

If the concentration of the calcium and carbonate ions is not excessively high, the formation of calcium carbonate may be observed to be time dependent. In this case, when solutions containing Ca^{2+} and CO_3^{2-} ions are mixed, no immediate formation of $CaCO_{3(s)}$ will be observed. However, after a short period of time the presence of very small particles of $CaCO_{3(s)}$ will be noted. As time progresses, the size and number of these calcium carbonate particles will increase. This has led researchers to conclude that precipitation is a two stage process:

1. Nucleation—condensation of ions to very small particles.
2. Particle growth—growth of these particles as a result of diffusion of ions from the solution.

In order for nucleation to occur the solute ions must be attracted strongly enough to displace solvent molecules which separate these ions in solution. However, the existence of a strong force of attraction between the ions is not enough to guarantee that precipitation will occur. The concentration of the ions in solution must also be high enough to ensure a high collision frequency between the solute ions. Furthermore, the force of attraction must be great enough to cause the ions to bind when the collisions occur. In fact, it appears that in every instance a certain degree of supersaturation, with respect to the compound to be precipitated, must be achieved before precipitation will occur. Hence, in the subsequent discussion of solubility equilibria, the reader should not get the impression that supersaturated solutions do not exist. It is possible that precipitation will not occur instantaneously for solutions where the degree of super-saturation is low.

3-1 SOLUBILITY EQUILIBRIUM FOR SLIGHTLY SOLUBLE SALTS

In 1899 Nernst showed that the equilibrium between a solid ionic salt and its solution in water was governed by the solubility equilibrium expression. For a general, slightly soluble compound C_aA_b, the dissolution equation is

$$C_aA_{b(s)} \rightleftharpoons aC_{(aq)} + bA_{(aq)} \qquad \text{(3-1)}$$

and the solubility equilibrium constant expression for an ideal solution is

$$(K_a)_{eq.} = [C]^a[A]^b \qquad \text{(3-2)}$$

where $(K_a)_{eq.}$ is the thermodynamic solubility equilibrium constant for the salt of interest, C represents the cationic component, and A represents the anionic component. Note that the solubility equilibrium constant expression, unlike most other equilibrium constant expressions, has no denominator term because it is assumed that the solid is pure and, hence, has unit activity.

The thermodynamic equilibrium constant is called a *solubility product constant*, K_s, when it refers to the equilibrium situation represented by a saturated solution as described by equation 3-1. In essence, the solubility product constant represents the maximum value the product of the ion concentrations can have for a given set of conditions. No precipitate will form if the product $[C]^a[A]^b$ has a value less than K_s. In such a case, the solution is undersaturated. However, in the case where the product $[C]^a[A]^b$ is greater than K_s, the solution is oversaturated, and C_aA_b will ultimately precipitate until the ion product just equals K_s.

Solubility product constants for a number of slightly soluble salts are presented in Appendix IV. Note that solubility product constants which are determined for freshly formed precipitates will generally be larger than the values obtained for precipitates which have been formed for some period of time. Such an observation is referred to as the *aging phenomenon*. This phenomenon occurs because freshly formed precipitates have crystal lattices which are highly disordered; however, upon aging, larger and more ordered crystals form. The latter inactive crystal form is less soluble than the initial active crystal form.

Common Ion Effect

The solubility behavior of a precipitate in the presence of an electrolyte which contains a common chemical species is important in process chemistry. As an example of such a system, consider the general equilibrium reaction for a slightly soluble salt described by equation 3-1:

$$C_aA_{b(s)} \rightleftharpoons aC_{(aq)} + bA_{(aq)} \qquad \text{(3-1)}$$

According to Le Châtelier's principle, the addition of an electrolyte that contains either C or A ions to the solution will force the position of equilibrium to the left (toward the solid phase). Such a response is referred to as the *common ion effect*, and it results in a decrease in the solubility of the precipitate in question.

The observed decrease in solubility of the precipitate because of the addition of a common ion can also be explained by considering the solubility equilibrium constant expression for equation 3-1:

$$K_s = [C]^a[A]^b \qquad (3\text{-}3)$$

Assume that an electrolyte which contains the chemical species, C, is added to the solution. Immediately after addition of the electrolyte, the value of the ion product $[C]^a[A]^b$ will be greater than K_s, and C_aA_b will precipitate. When a new equilibrium is finally established, the activity (concentration for an ideal solution) of C will have increased, and the activity of A will have decreased from the initial equilibrium values. The decrease in the activity of A is a result of C_aA_b precipitating from solution so that the ion product $[C]^a[A]^b$ will again equal K_s. This means that the solubility of C_aA_b is decreased by the addition of a common ion to the solution.

Indifferent Electrolyte Effect

In process chemistry the formation or dissolution of a precipitate always occurs in the presence of indifferent electrolytes. Although ions from such species do not participate directly in the solubility equilibrium reaction, they do affect the solubility behavior of the precipitate. To understand how the presence of indifferent electrolytes may affect the solubility behavior of a precipitate, consider the simple equilibrium expression

$$CA_{(s)} \rightleftharpoons C_{(aq)} + A_{(aq)} \qquad (3\text{-}4)$$

The equilibrium constant expression for this reaction is given by

$$(K_a)_{eq.} = (C)(A) \qquad (3\text{-}5)$$

or

$$(K_a)_{eq.} = \gamma_m[C]\gamma_m[A] \qquad (3\text{-}6)$$

Equation 3-6 can be written as

$$(K_c)_{eq.} = \frac{(K_a)_{eq.}}{(\gamma_m)^2} \qquad (3\text{-}7)$$

The greater the concentration of indifferent electrolytes, the greater will be the ionic strength of the solution and the smaller the value of the activity coefficient. In process chemistry the value of γ is normally less than 1.0. Hence, the smaller the value of γ, the larger will be the value of $(K_c)_{eq.}$, indicating that the solubility of the solid phase will increase. This means that the solubility of a precipitate will increase with an increase in the concentration of indifferent electrolytes in the solution of interest.

Solubility Calculations

Since a saturated solution represents an equilibrium situation for the reaction of a salt with water, basic equilibrium principles may be used to calculate the solubility of a salt and the equilibrium concentration of ions in solution. Such calculations are presented in example problems 3-1, 3-2, and 3-3.

EXAMPLE PROBLEM 3-1: The solubility product, K_s, for calcium sulfate in water at 25°C is 1.96×10^{-4}. Determine the equilibrium Ca^{2+} concentration for a saturated calcium sulfate solution if ideal behavior is assumed.

Solution:

1. Write the appropriate dissolution reaction.

$$CaSO_{4(s)} \rightleftharpoons Ca^{2+}_{(aq)} + SO^{2-}_{4(aq)}$$

2. Write the appropriate equilibrium constant expression:

$$K_s = [Ca^{2+}][SO_4^{2-}]$$

or

$$1.96 \times 10^{-4} = [Ca^{2+}][SO_4^{2-}]$$

3. Calculate the equilibrium Ca^{2+} concentration.
   Since $CaSO_4$ is a 1:1 salt, the equilibrium molar concentrations of calcium and sulfate ions are identical. Thus, if M represents the molar equilibrium concentration of each species, the equilibrium constant expression may be expressed as

$$1.96 \times 10^{-4} = M^2$$

or

$$[Ca^{2+}] = M = 1.4 \times 10^{-2} \text{ mole}/\ell$$

4. Convert the molar concentration to mg/ℓ.

$$\text{mg } Ca^{2+}/\ell = (1.4 \times 10^{-2})(1000)(40.08) = 561$$

EXAMPLE PROBLEM 3-2: Compute the solubility (mg/ℓ) of cadmium hydroxide in water at 25°C. Assume that the solution exhibits ideal behavior and that the solubility product, K_s, is 5.9×10^{-15}.

Solution:

1. Write the appropriate dissolution reaction:

$$Cd(OH)_{2(s)} \rightleftharpoons Cd^{2+}_{(aq)} + 2OH^-_{(aq)}$$

2. Write the appropriate equilibrium constant expression:

$$K_s = [Cd^{2+}][OH^-]^2$$

or

$$5.9 \times 10^{-15} = [Cd^{2+}][OH^-]^2$$

3. Write the electron charge balance expression, neglecting the presences of H^+ and OH^- ions due to the autoionization of water:

$$\sum_{i=1}^{i=i} [P_i(C_i^{P_i})] = \sum_{i=1}^{i=i} [N_i(A_i^{N_i})]$$

or

$$2[Cd^{2+}] = 1[OH^-]$$

4. Compute the equilibrium molar Cd^{2+} concentration.

   Substitute for $[OH^-]$ in the equilibrium constant expression from the relationship between Cd^{2+} and OH^- given in step 3:

$$5.9 \times 10^{-15} = [Cd^{2+}](2[Cd^{2+}])^2$$

or

$$5.9 \times 10^{-15} = 4[Cd^{2+}]^3$$

Thus,

$$[Cd^{2+}] = 1.14 \times 10^{-5} \text{ mole}/\ell$$

5. Determine the molar solubility of $Cd(OH)_{2(s)}$.

   Since 1 mole of Cd^{2+} ions are produced for each mole of cadmium hydroxide which dissolves, the molar solubility of $Cd(OH)_{2(s)}$ is equal to the equilibrium molar concentration of Cd^{2+}, or 1.14×10^{-5} mole/ℓ.

6. Convert the molar solubility to mg/ℓ:

$$\text{solubility of } Cd(OH)_2 = (1.14 \times 10^{-5})(1000)(146.4)$$
$$\text{mg}/\ell = 1.67$$

EXAMPLE PROBLEM 3-3: Compute the $Ca^{2+}_{(aq)}$ concentration in a saturated $CaCO_{3(s)}$ solution at 25°C, and compare this value to the $Ca^{2+}_{(aq)}$ concentration in a saturated $CaCO_{3(s)}$ solution to which 25 g/ℓ of Na_2CO_3 (strong electrolyte) has been added. $K_s = 8.9 \times 10^{-9}$. Neglect ionic strength corrections.

Solution:

1. Write the appropriate dissolution reaction:

$$CaCO_{3(s)} \rightleftharpoons Ca^{2+}_{(aq)} + CO^{2-}_{3(aq)}$$

2. Write the appropriate equilibrium constant expression:

$$K_s = [Ca^{2+}][CO_3^{2-}]$$

or

$$8.9 \times 10^{-9} = [Ca^{2+}][CO_3^{2-}]$$

3. Calculate the equilibrium Ca^{2+} concentration,

$$8.9 \times 10^{-9} = M^2$$

where $M = $ molar equilibrium concentration of Ca^{2+} or CO_3^{2-}.

Therefore,

$$[Ca^{2+}] = M = 0.94 \times 10^{-4} \text{ mole}/\ell$$

4. Convert the molar concentration to mg/ℓ:

$$\text{mg Ca}^{2+}/\ell = (0.94 \times 10^{-4})(1000)(40.08) = 3.76$$

5. Write the electron charge balance expression, neglecting the presences of H^+ and OH^- ions due to the autoionization of water:

$$[Na^+] + 2[Ca^{2+}] = 2[CO_3^{2-}]$$

6. Compute the molar concentration of Na_2CO_3:

$$[Na_2CO_3] = \frac{25}{106} = 0.236$$

7. Determine the molar concentration of sodium ions after the sodium carbonate has totally dissociated.

Since two sodium ions are produced for each molecule of sodium carbonate which dissociates, the molar concentration of sodium is

$$[Na^+] = 2(0.236) = 0.47 \text{ mole}/\ell$$

8. Solve the equation presented in step 5 for the carbonate ion concentration:

$$[CO_3^{2-}] = \frac{[Na^+] + 2[Ca^{2+}]}{2}$$

9. Substitute from step 8 for $[CO_3^{2-}]$ in step 2 and for $[Na^+]$ from step 7, and solve for the equilibrium calcium concentration,

$$8.9 \times 10^{-9} = \frac{[Ca^{2+}][Na^+] + 2[Ca^{2+}]^2}{2}$$

or

$$[Ca^{2+}]^2 + 0.236[Ca^{2+}] - 4.45 \times 10^{-9} = 0$$

$$[Ca^{2+}] = \frac{(-0.236) \pm [(0.236)^2 - (4)(1)(-4.45 \times 10^{-9})]^{1/2}}{2(1)} \simeq 0$$

In this case the equilibrium calcium concentration is so small that, for all practical purposes, it can be taken as zero.

At this point, it should be noted that the solubility behavior of most slightly soluble salts is much more complicated than is implied in the calculations presented in example problems 3-1, 3-2, and 3-3. In general, the position of equilibrium in a saturated solution is affected by the formation of anionic, cationic and/or nonionic complexes. Such complexes may be formed between the components of the precipitate, with ions produced by the solvent (H^+ or OH^- ions in the case of aqueous solutions), or with other ions present in the solution. These factors, which complicate solubility equilibrium calculations, are discussed in detail in the following section.

3-2 EFFECT OF OTHER SOLUTES ON SALT SOLUBILITIES

Consider the equilibrium reaction described by equation 3-1:

$$C_aA_{b(s)} \rightleftharpoons aC_{(aq)} + bA_{(aq)} \qquad (3\text{-}1)$$

According to Le Châtelier's principle, any ion in solution that can react with the ionic species C or A to decrease their concentration as free hydrated ions will increase the solubility of $C_aA_{b(s)}$. In other words, if a certain concentration of the C or A ions formed from $C_aA_{b(s)}$ dissolution are incorporated into other soluble chemical species, more C_aA_b must be dissolved if saturated equilibrium is to be maintained (i.e., if the ion product $[C]^a[A]^b$ is to equal the solubility product constant, K_s).

Competing Acid-Base Equilibria

If the anion in a compound whose solubility is being considered can form a weak acid with hydrogen ions, the solubility of the salt may be considerably greater than that computed from the saturated equilibrium constant expression. The most important salts encountered in process chemistry which involve both solubility and acid-base equilibria are carbonates (CO_3^{2-}), phosphates (PO_4^{3-}), and sulfides (S^{2-}). Examples of how protolysis reactions (proton transfer reactions) affect the solubility of each of these types of salts are given in the following equations:

1) $$CaCO_{3(s)} \rightleftharpoons Ca^{2+}_{(aq)} + CO^{2-}_{3(aq)}$$
$$+$$
$$H^+_{(aq)}$$
$$\updownarrow$$
$$HCO^-_{3(aq)} \qquad (3\text{-}8)$$
$$+$$
$$H^+_{(aq)}$$
$$\updownarrow$$
$$H_2CO_{3(aq)}$$
carbonic acid

2) $$PbS_{(s)} \rightleftharpoons Pb^{2+}_{(aq)} + S^{2-}_{(aq)}$$
$$+$$
$$H^+_{(aq)}$$
$$\updownarrow$$
$$HS^-_{(aq)} \qquad (3\text{-}9)$$
$$+$$
$$H^+_{(aq)}$$
$$\updownarrow$$
$$H_2S_{(g)}$$
hydrogen sulfide

3)

$$AlPO_{4(s)} \rightleftharpoons Al^{3+}_{(aq)} + PO^{3-}_{4(aq)}$$

$$+$$

$$H^+_{(aq)}$$

$$\Updownarrow$$

$$HPO^{2-}_{4(aq)}$$

$$+$$

$$H^+_{(aq)} \hspace{8cm} \textbf{(3-10)}$$

$$\Updownarrow$$

$$H_2PO^-_{4(aq)}$$

$$+$$

$$H^+_{(aq)}$$

$$\Updownarrow$$

$$H_3PO_{4(aq)}$$
orthophosphoric acid

Thus, it can be seen from these equilibrium expressions that the result of establishing acid-base equilibria is an increase in the solubility of the salt under consideration.

Effects of Complex Ion Formation on Solubility

A chemical species in which a cation is covalently bonded to one or more coordinating groups is referred to as a *complex ion*. The coordinating group is termed the *ligand*. Water is the most common ligand encountered in aqueous solutions. For simplicity, cations in aqueous solution have been written without the coordinated water molecules. However, a more accurate formulation would be to include such groups. For example, Al^{3+} should be written as $Al(H_2O)_6^{3+}$.

The number of ligands present in a given complex ion is difficult to determine accurately. However, Bard (1966) states that "A convenient rule of thumb is that the maximum number of ligands that can associate with a metal ion (the so-called maximum coordination number) is very often twice the ionic charge on the metal ion. Furthermore, maximum coordination numbers larger than 6 are very rare, and metals which are quadrivalent, such as Pt^{4+} and Sn^{4+}, generally have maximum coordination numbers of 6."

To apply the concepts of chemical equilibria to complex ion formation reactions, consider the reaction between metal ion (Me) and ligand (L) in which the charges are omitted:

$$Me + L \rightleftharpoons MeL \hspace{4cm} \textbf{(3-11)}$$

where

$$k_1 = \frac{[MeL]}{[Me][L]} \hspace{4cm} \textbf{(3-12)}$$

Note that in a complex ion formation reaction, the equilibrium constant is given as a formation constant rather than as a dissociation constant which was used to describe solubility equilibria.

A stepwise formation constant is associated with the addition of each ligand to a cation.

$$MeL + L \rightleftharpoons Me(L)_2 \qquad (3\text{-}13)$$

where

$$k_2 = \frac{[Me(L)_2]}{[MeL][L]} \qquad (3\text{-}14)$$

$$\vdots$$

$$Me(L)_{n-1} + L \rightleftharpoons Me(L)_n \qquad (3\text{-}15)$$

where

$$k_n = \frac{[Me(L)_n]}{[Me(L)_{n-1}][L]} \qquad (3\text{-}16)$$

It is also possible to show complex ion formation (when more than one ligand is involved) as a cumulative process.

$$Me + L \rightleftharpoons MeL \qquad (3\text{-}17)$$

where

$$K_1 = \frac{[MeL]}{[Me][L]} \qquad (3\text{-}18)$$

$$Me + 2L \rightleftharpoons Me(L)_2 \qquad (3\text{-}19)$$

where

$$K_2 = \frac{[Me(L)_2]}{[Me][L]^2} \qquad (3\text{-}20)$$

$$\vdots$$

$$Me + nL \rightleftharpoons Me(L)_n \qquad (3\text{-}21)$$

where

$$K_n = \frac{[Me(L)_n]}{[Me][L]^n} \qquad (3\text{-}22)$$

The relationship between the two types of formation constant representation is shown in Table 3-1.

TABLE 3-1 *Relationship between cumulative formation constants and stepwise formation constants.*

Reaction	Cumulative Formation Constant	Stepwise Formation Constant
$Me + L \rightleftharpoons MeL$	K_1	k_1
$Me + 2L \rightleftharpoons Me(L)_2$	K_2	$k_1 k_2$
....	...	...
$Me + nL \rightleftharpoons Me(L)_n$	K_n	$k_1 k_2 \ldots k_n$

As an example of the formation of a complex ion, consider the interaction between the hydrated zinc ion and ammonia.

1) $$\begin{bmatrix} H_2O & OH_2 \\ & Zn & \\ H_2O & OH_2 \end{bmatrix}^{2+} + NH_3 \rightleftharpoons \begin{bmatrix} H_2O & NH_3 \\ & Zn & \\ H_2O & OH_2 \end{bmatrix}^{2+} + H_2O$$

$$Zn(H_2O)_4^{2+} + NH_3 \rightleftharpoons Zn(NH_3)(H_2O)_3^{2+} + H_2O \qquad \textbf{(3-23)}$$

2) $$\begin{bmatrix} H_2O & NH_3 \\ & Zn & \\ H_2O & OH_2 \end{bmatrix}^{2+} + NH_3 \rightleftharpoons \begin{bmatrix} H_3N & NH_3 \\ & Zn & \\ H_2O & OH_2 \end{bmatrix}^{2+} + H_2O$$

$$Zn(NH_3)(H_2O)_3^{2+} + NH_3 \rightleftharpoons Zn(NH_3)_2(H_2O)_2^{2+} + H_2O \qquad \textbf{(3-24)}$$

3) $$\begin{bmatrix} H_3N & NH_3 \\ & Zn & \\ H_2O & OH_2 \end{bmatrix}^{2+} + NH_3 \rightleftharpoons \begin{bmatrix} H_3N & NH_3 \\ & Zn & \\ H_3N & OH_2 \end{bmatrix}^{2+} + H_2O$$

$$Zn(NH_3)_2(H_2O)_2^{2+} + NH_3 \rightleftharpoons Zn(NH_3)_3(H_2O)^{2+} + H_2O \qquad \textbf{(3-25)}$$

4) $$\begin{bmatrix} H_3N & NH_3 \\ & Zn & \\ H_3N & OH_2 \end{bmatrix}^{2+} + NH_3 \rightleftharpoons \begin{bmatrix} H_3N & NH_3 \\ & Zn & \\ H_3N & NH_3 \end{bmatrix}^{2+} + H_2O$$

$$Zn(NH_3)_3(H_2O)^{2+} + NH_3 \rightleftharpoons Zn(NH_3)_4^{2+} + H_2O \qquad \textbf{(3-26)}$$

Equations 3-23, 3-24, 3-25, and 3-26 represent the stepwise formation of the various ammonia complexes of zinc. An aqueous solution which contains zinc ions and ammonia will usually contain all these complexes in equilibrium with each other.

As an example of the effect of complex ion formation on solubility equilibria, consider a situation where zinc is to be removed from a wastewater containing ammonia by precipitating the zinc as its insoluble carbonate:

$$ZnCO_{3(s)} \rightleftharpoons Zn^{2+}_{(aq)} + CO^{2-}_{3(aq)}$$

$$+$$

$$NH_{3(g)}$$

$$\updownarrow$$

$$Zn(NH_3)^{2+}_{(aq)}$$

$$+$$

$$NH_{3(g)}$$

$$\updownarrow$$

$$Zn(NH_3)^{2+}_{2(aq)}$$

$$+ \qquad\qquad (3\text{-}27)$$

$$NH_{3(g)}$$

$$\updownarrow$$

$$Zn(NH_3)^{2+}_{3(aq)}$$

$$+$$

$$NH_{3(g)}$$

$$\updownarrow$$

$$Zn(NH_3)^{2+}_{4(aq)}$$

Equation 3-27 shows that the formation of the ammonia complexes of zinc will increase the solubility of zinc carbonate. This may create a problem in a situation where it is desirable to reduce the soluble zinc concentration to some specified level before the wastewater is discharged. To aid the engineer in analyzing systems in which complex ion formation may significantly affect salt solubility, a comprehensive table of cumulative constants for metal ion complexes is presented in Appendix V.

Complex ions are formed because of the tendency of positive ions to attract regions of high electron density, which occur in anions or neutral molecules having unshared pairs of valence electrons. Many workers refer to the complex ions formed between anions and cations as *ion-pairs*. For example, consider the $CaCO_{3(s)}$ equilibrium reaction in a saturated solution:

$$CaCO_{3(s)} \rightleftharpoons Ca^{2+}_{(aq)} + CO^{2-}_{3(aq)} \qquad\qquad (3\text{-}28)$$

Because of short range coulombic forces between Ca^{2+} and CO_3^{2-} ions, a soluble $CaCO^0_{3(aq)}$ species may form (the superscript 0 denotes an electrically neutral species). This soluble complex is often referred to as the *calcium carbonate ion-pair*, and its formation will increase the solubility of $CaCO_{3(s)}$ by the amount of Ca^{2+} and CO_3^{2-} ions tied up in the complex.

$$CaCO_{3(s)} \rightleftharpoons Ca^{2+}_{(aq)} + CO^{2-}_{3(aq)} \qquad \text{(3-29)}$$

$$\Updownarrow$$

$$CaCO^0_{3(aq)}$$

To avoid confusion, in this text no distinction will be made between complexes formed between cationic and anionic species and between cationic and neutral species. All such complexes will be referred to as *complex ions*, even though some species may have no charge. However, there is one type of complex ion formation which deserves special attention because of its importance in process chemistry. This special case is *hydrolysis* (the reaction of an ion with water).

Effect of Hydrolysis on Solubility

As noted earlier, metal ions exist in aqueous solution as hydrated ions. For example, the Al^{3+} ion is actually surrounded by six molecules of water $[Al(H_2O)_6^{3+}]$. In an aqueous solution complex-forming reactions between the hydrated metal ion and hydroxyl ions occur. Such reactions lead to the formation of *hydroxocomplexes*, which will contain one or more metal ions (mono- or polynuclear complexes). The following generalized equations may be used to describe the hydroxocomplex formation reactions for a trivalent metal ion when only mononuclear complexes are formed:

1) $\quad Me^{3+}_{(aq)} + OH^-_{(aq)} \rightleftharpoons MeOH^{2+}_{(aq)}$ $\qquad$ (3-30)

2) $\quad Me^{3+}_{(aq)} + 2OH^-_{(aq)} \rightleftharpoons Me(OH)^+_{2(aq)}$ $\qquad$ (3-31)

3) $\quad Me^{3+}_{(aq)} + 3OH^-_{(aq)} \rightleftharpoons Me(OH)^0_{3(aq)}$ $\qquad$ (3-32)

4) $\quad Me^{3+}_{(aq)} + 4OH^-_{(aq)} \rightleftharpoons Me(OH)^-_{4(aq)}$ $\qquad$ (3-33)

Since hydrated metal cations are weak acids called *cation acids*, the formation of metal hydroxocomplexes may also be considered as the ionization of these cation acids. To demonstrate this principle, consider the following equations, which describe the ionization reactions for a trivalent metal ion that forms only mononuclear complexes:

1) $\quad Me(H_2O)^{3+}_{6(aq)} \rightleftharpoons Me(H_2O)_5(OH)^{2+}_{(aq)} + H^+_{(aq)}$ $\qquad$ (3-34)

or

$\quad Me^{3+}_{(aq)} + H_2O \rightleftharpoons Me(OH)^{2+}_{(aq)} + H^+_{(aq)}$ $\qquad$ (3-35)

2) $\quad Me^{3+}_{(aq)} + 2H_2O \rightleftharpoons Me(OH)^+_{2(aq)} + 2H^+_{(aq)}$ $\qquad$ (3-36)

3) $\quad Me^{3+}_{(aq)} + 3H_2O \rightleftharpoons Me(OH)^0_{3(aq)} + 3H^+_{(aq)}$ $\qquad$ (3-37)

4) $\quad Me^{3+}_{(aq)} + 4H_2O \rightleftharpoons Me(OH)^-_{4(aq)} + 4H^+_{(aq)}$ $\qquad$ (3-38)

These types of ionization reactions occur whenever salts containing metal ions dissolve in water and, since hydrogen ions are produced, the resulting solution will be acidic. Such reactions significantly increase the solubility of slightly soluble salts and should be considered in solubility calculations. This effect is shown in the following zinc carbonate system:

$$ZnCO_{3(s)} \rightleftharpoons Zn^{2+}_{(aq)} + CO^{2-}_{3(aq)}$$
$$\updownarrow$$
$$ZnOH^{+}_{(aq)}$$
$$\updownarrow$$
$$Zn(OH)^{0}_{2(aq)} \qquad \qquad \text{(3-39)}$$
$$\updownarrow$$
$$Zn(OH)^{-}_{3(aq)}$$
$$\updownarrow$$
$$Zn(OH)^{2-}_{4(aq)}$$

Hydroxocomplex formation constants for aluminum, iron(III), and iron(II) are listed in Table 3-2. The reactions presented in the table show that hydrogen ions are liberated when metal hydroxocomplexes are formed. (Metal hydroxocomplex-forming reactions may also be viewed as reactions which remove hydroxyl ions from solution.) Thus, a log [species] vs. pH diagram may be used to describe the effect of metal hydroxo-complex formation on the solubility of a slightly soluble metal salt. To illustrate the procedure for constructing these diagrams, plotting equations will be developed for the $Al(OH)_{3(s)}$-H_2O system and the $Fe(OH)_{3(s)}$ system, using the constants shown in Tables 3-2 and 3-3. Subsequently, the appropriate diagrams will be constructed by using

TABLE 3-2 *Hydroxocomplex formation constants for aluminum, iron(III), and iron(II) at 25°C and zero ionic strength.*

	Reaction			log K
1)	Al^{3+}	$\rightleftharpoons$	$Al(OH)^{2+} + H^+$	−5.02
2)	$2Al^{3+}$	$\rightleftharpoons$	$Al_2(OH)^{4+}_2 + 2H^+$	−6.27
3)	$6Al^{3+}$	$\rightleftharpoons$	$Al_6(OH)^{3+}_{15} + 15H^+$	−47.00
4)	$8Al^{3+}$	$\rightleftharpoons$	$Al_8(OH)^{4+}_{20} + 20H^+$	−68.7
5)	$13Al^{3+}$	$\rightleftharpoons$	$Al_{13}(OH)^{5+}_{34} + 34H^+$	−97.39
6)	Al^{3+}	$\rightleftharpoons$	$Al(OH)^{-}_4 + 4H^+$	−23.57
7)	Fe^{3+}	$\rightleftharpoons$	$FeOH^{2+} + H^+$	−3.0*
8)	Fe^{3+}	$\rightleftharpoons$	$Fe(OH)^{+}_2 + 2H^+$	−6.4*
9)	$2Fe^{3+}$	$\rightleftharpoons$	$Fe_2(OH)^{4+}_2 + 2H^+$	−3.1*
10)	Fe^{3+}	$\rightleftharpoons$	$Fe(OH)^{0}_3 + 3H^+$	−13.5*
11)	Fe^{3+}	$\rightleftharpoons$	$Fe(OH)^{-}_4 + 4H^+$	−23.5*
12)	Fe^{2+}	$\rightleftharpoons$	$FeOH^+ + H^+$	−8.3
13)	Fe^{2+}	$\rightleftharpoons$	$Fe(OH)^{0}_2 + 2H^+$	−17.2
14)	Fe^{2+}	$\rightleftharpoons$	$Fe(OH)^{-}_3 + 3H^+$	−32.0
15)	Fe^{2+}	$\rightleftharpoons$	$Fe(OH)^{2-}_4 + 4H^+$	−46.4

* Values do not correspond to 25°C or zero ionic strength.

Sources: Black and Chen (1965), Bolt and Bruggenwert (1976), Ringbom (1963), Rouse (1976), Rubin and Kovac (1974), Sillen and Martell (1964).

TABLE 3-3 *Solubility product constants for aluminum hydroxide, iron(III), hydroxide, and iron(II) hydroxide at 25°C.*

	Reaction	$\log K_s$
1)	$Al(OH)_{3(s)} \rightleftharpoons Al^{3+}_{(aq)} + 3OH^-_{(aq)}$	-32.34
2)	$Fe(OH)_{3(s)} \rightleftharpoons Fe^{3+}_{(aq)} + 3OH^-_{(aq)}$	-37.5
3)	$Fe(OH)_{2(s)} \rightleftharpoons Fe^{2+}_{(aq)} + 2OH^-_{(aq)}$	-15.1

Source: Rubin and Kovac (1974).

the plotting equations.

(A)   *$Al(OH)_{3(s)}$-H_2O System:*

1) $Al(OH)_{3(s)} \rightleftharpoons Al^{3+}_{(aq)} + 3OH^-_{(aq)}$ **(3-40)**

where

$$K_{sp} = [Al^{3+}][OH^-]^3 \qquad \textbf{(3-41)}$$

Taking the logarithm of both sides of this equation gives

$$K_{sp} = \log[Al^{3+}] + 3\log[OH^-] \qquad \textbf{(3-42)}$$

at 25°C K_{sp} has a value of $10^{-32.34}$. Equation 3-43 is obtained by substituting this value into Equation 3-42.

$$-32.34 = \log[Al^{3+}] + 3\log[OH^-] \qquad \textbf{(3-43)}$$

From the ionization of water it is known that

$$\log[OH^-] = pH - 14 \qquad \textbf{(3-44)}$$

Substituting for $\log[OH^-]$ in equation 3-42 from equation 3-44 gives

$$-32.4 = \log[Al^{3+}] + 3pH - 42 \qquad \textbf{(3-45)}$$

or

$$\log[Al^{3+}] = 9.66 - 3pH \qquad \textbf{(3-46)}$$

2) $Al^{3+}_{(aq)} \rightleftharpoons Al(OH)^{2+}_{(aq)} + H^+_{(aq)}$ **(3-47)**

where

$$10^{-5.02} = \frac{[Al(OH)^{2+}][H^+]}{[Al^{3+}]} \qquad \textbf{(3-48)}$$

Taking the logarithm of both sides of this equation produces an expression of the form

$$-5.02 = \log[Al(OH)^{2+}] + \log[H^+] - \log[Al^{3+}] \qquad \textbf{(3-49)}$$

or

$$-5.02 = \log\,[\text{Al(OH)}^{2+}] - \text{pH} - \log\,[\text{Al}^{3+}] \qquad \textbf{(3-50)}$$

Substituting for $\log\,[\text{Al}^{3+}]$ in equation 3-50 from equation 3-46 gives equation 3-51:

$$\log\,[\text{Al(OH)}^{2+}] = 4.44 - 2\text{pH} \qquad \textbf{(3-51)}$$

3) $\qquad 2\text{Al}^{3+}_{(aq)} \;\rightleftharpoons\; \text{Al}_2(\text{OH})^{4+}_{2(aq)} + 2\text{H}^+_{(aq)} \qquad \textbf{(3-52)}$

where

$$10^{-6.27} = \frac{[\text{Al}_2(\text{OH})^{4+}_2][\text{H}^+]^2}{[\text{Al}^{3+}]^2} \qquad \textbf{(3-53)}$$

Taking the logarithm of both sides of this equation gives

$$-6.27 = \log\,[\text{Al}_2(\text{OH})^{4+}_2] + 2\log\,[\text{H}^+] - 2\log\,[\text{Al}^{3+}] \qquad \textbf{(3-54)}$$

Substituting for $\log\,[\text{Al}^{3+}]$ in equation 3-54 from equation 3-46, it is possible to write

$$\log\,[\text{Al}_2(\text{OH})^{4+}_2] = 13.05 - 4\text{pH} \qquad \textbf{(3-55)}$$

4) $\qquad 6\text{Al}^{3+} \;\rightleftharpoons\; \text{Al}_6(\text{OH})^{3+}_{15} + 15\text{H}^+ \qquad \textbf{(3-56)}$

where

$$10^{-47.00} = \frac{[\text{Al}_6(\text{OH})^{3+}_{15}][\text{H}^+]^{15}}{[\text{Al}^{3+}]^6} \qquad \textbf{(3-57)}$$

If the logarithm of both sides of this equation is taken, then

$$-47.00 = \log\,[\text{Al}_6(\text{OH})^{3+}_{15}] + 15\log\,[\text{H}^+] - 6\log\,[\text{Al}^{3+}] \qquad \textbf{(3-58)}$$

Substituting from equation 3-46 for $\log\,[\text{Al}^{3+}]$ results in the following plotting equation:

$$\log\,[\text{Al}_6(\text{OH})^{3+}_{15}] = 10.96 - 3\text{pH} \qquad \textbf{(3-59)}$$

5) $\qquad 8\text{Al}^{3+} \;\rightleftharpoons\; \text{Al}_8(\text{OH})^{4+}_{20} + 20\text{H}^+ \qquad \textbf{(3-60)}$

where

$$10^{-68.7} = \frac{[\text{Al}_8(\text{OH})^{4+}_{20}][\text{H}^+]^{20}}{[\text{Al}^{3+}]^8} \qquad \textbf{(3-61)}$$

Equation 3-62 is developed by taking the logarithm of both sides of equation 3-61:

$$-68.7 = \log\,[\text{Al}_8(\text{OH})^{4+}_{20}] + 20\log\,[\text{H}^+] - 8\log\,[\text{Al}^{3+}] \qquad \textbf{(3-62)}$$

This expression reduces to the desired plotting equation when a substitution is made for the $\log\,[\text{Al}^{3+}]$ from equation 3-46.

$$\log\,[\text{Al}_8(\text{OH})^{4+}_{20}] = 8.58 - 4\text{pH} \qquad \textbf{(3-63)}$$

6) $\quad 13Al^{3+} \rightleftharpoons Al_{13}(OH)_{34}^{5+} + 34H^+$ $\hfill$ **(3-64)**

where

$$10^{-97.39} = \frac{[Al_{13}(OH)_{34}^{5+}][H^+]^{34}}{[Al^{3+}]^{13}} \hfill \text{(3-65)}$$

Taking the logarithm of both sides of this equation gives

$$-97.39 = \log[Al_{13}(OH)_{34}^{5+}] + 34\log[H+] - 13\log[Al^{3+}] \hfill \text{(3-66)}$$

Substituting for $\log[Al^{3+}]$ in equation 3-66 from equation 3-46 it is possible to write

$$\log[Al_{13}(OH)_{34}^{5+}] = 28.19 - 5pH \hfill \text{(3-67)}$$

7) $\quad Al^{3+} \rightleftharpoons Al(OH)_4^- + 4H^+$ $\hfill$ **(3-68)**

where

$$10^{-23.57} = \frac{[Al(OH)_4^-][H^+]^4}{[Al^{3+}]} \hfill \text{(3-69)}$$

Equation 3-70 is derived by taking the logarithm of both sides of equation 3-69:

$$-23.57 = \log[Al(OH)_4^-] + 4\log[H+] - \log[Al^{3+}] \hfill \text{(3-70)}$$

Making the appropriate substitution for $\log[Al^{3+}]$ gives

$$\log[Al(OH)_4^-] = pH - 13.91 \hfill \text{(3-71)}$$

(B) $Fe(OH)_{3(s)}$-H_2O System:

Using the same procedure outlined for the $Al(OH)_{3(s)}$-H_2O system, the following plotting equations can be developed for the $Fe(OH)_{3(s)}$-H_2O system:

1) $\quad \log[Fe^{3+}] = 4.5 - 3pH$ $\hfill$ **(3-72)**

2) $\quad \log[FeOH^{2+}] = 1.5 - 2pH$ $\hfill$ **(3-73)**

3) $\quad \log[Fe(OH)_2^+] = -1.9 - pH$ $\hfill$ **(3-74)**

4) $\quad \log[Fe_2(OH)_2^{4+}] = 5.9 - 4pH$ $\hfill$ **(3-75)**

5) $\quad \log[Fe(OH)_3^0] = -9.0$ $\hfill$ **(3-76)**

6) $\quad \log[Fe(OH)_4^-] = pH - 19$ $\hfill$ **(3-77)**

After developing the plotting equations for the system of interest, it is possible to construct a log [species] vs. pH diagram for that system. Such diagrams will show the variation in salt solubility with pH and will also indicate the pH where the minimum salt solubility is to be expected. Log [species] vs. pH diagrams for the $Al(OH)_{3(s)}$-H_2O and $Fe(OH)_{3(s)}$-H_2O systems are presented in Figures 3-1 and 3-2, respectively. In these diagrams the soluble species which shows the maximum solubility at a specified pH always controls the solubility of the salt of that pH. This is analogous to the rate-limiting step for a series of sequential reactions, and it means that the uppermost line at any specified pH in a log [species] vs. pH solubility plot will indicate the boundary of the solid-aqueous phase.

A chemical species which is capable of either accepting or donating a proton is said to be *amphoteric*. Figures 3-1 and 3-2 illustrate the general amphoteric nature of

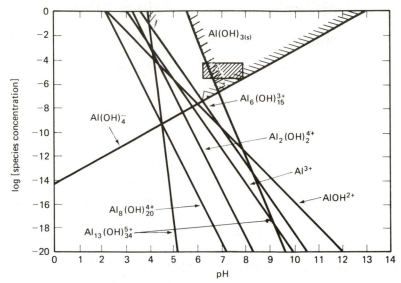

FIGURE 3-1 Solubility diagram for aluminum hydroxide.

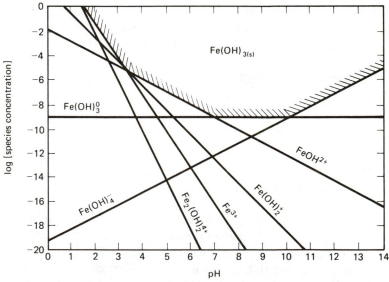

FIGURE 3-2 Solubility diagram for iron (III) hydroxide.

SOLUBILITY EQUILIBRIA CHAPTER 3

metal hydroxides. These figures show that the solubility of $Al(OH)_{3(s)}$ and $Fe(OH)_{3(s)}$ is increased under both acidic and basic conditions. Under acid conditions both form cationic species which increase the solubility of the solid phase, whereas under alkaline conditions both form anionic species which increase the solubility of the solid phase.

Computing Total Soluble Species Concentrations

The presence of all the different soluble chemical species which affect the soluble concentration of any particular chemical component must be accounted for when computing the total soluble analytical concentration of that component. To illustrate this concept, the total soluble calcium concentration in a saturated calcium carbonate solution is considered. In this case, it is assumed that all the soluble chemical species which affect the solubility of calcium carbonate are shown in Figure 3-3. The individual

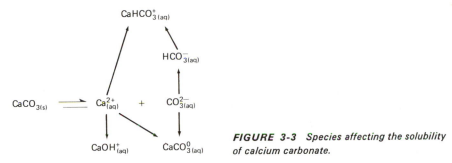

FIGURE 3-3 *Species affecting the solubility of calcium carbonate.*

equilibrium reactions presented in this figure (assuming an ideal solution) are as follows:

1) $\quad CaCO_{3(s)} \rightleftharpoons Ca^{2+}_{(aq)} + CO^{2-}_{3(aq)}$ (3-28)

$\quad K_1 = (K_s)_{CaCO_3} = 8.7 \times 10^{-9} = [Ca^{2+}][CO_3^{2-}]$ (3-78)

2) $\quad Ca^{2+}_{(aq)} + OH^-_{(aq)} \rightleftharpoons CaOH^+_{(aq)}$ (3-79)

$\quad K_2 = (K)_{CaOH^+} = 10^{1.3} = \dfrac{[CaOH^+]}{[Ca^{2+}][OH^-]}$ (3-80)

3) $\quad Ca^{2+}_{(aq)} + HCO^-_{3(aq)} \rightleftharpoons CaHCO^+_{3(aq)}$ (3-81)

$\quad K_3 = (K)_{CaHCO_3^+} = 10^{1.01} = \dfrac{[CaHCO_3^+]}{[Ca^{2+}][HCO_3^-]}$ (3-82)

4) $\quad Ca^{2+}_{(aq)} + CO^{2-}_{3(aq)} \rightleftharpoons CaCO^0_{3(aq)}$ (3-83)

$\quad K_4 = (K)_{CaCO_3^0} = 10^{3.15} = \dfrac{[CaCO_3^0]}{[Ca^{2+}][CO_3^{2-}]}$ (3-84)

For the equilibrium system shown in Figure 3-3, the total soluble analytical calcium concentration, $[Ca]_T$, is given by the following mass balance equation:

$$[Ca]_T = [Ca^{2+}] + [CaOH^+] + [CaHCO_3^+] + [CaCO_3^0]$$ (3-85)

Solving equations 3-80, 3-82, and 3-84 for the appropriate complex ion and then substituting the results into equation 3-85 gives

$$[Ca]_T = [Ca^{2+}]\{1 + K_2[OH^-] + K_3[HCO_3^-] + K_4[CO_3^{2-}]\} \qquad \textbf{(3-86)}$$

At equilibrium the following relationships exist:

$$[Ca^{2+}] = \frac{K_1}{[CO_3^{2-}]} \qquad \text{(rearrangement of 3-78)}$$

$$[OH^-] = \frac{K_w}{[H^+]} \qquad \textbf{(3-87)}$$

$$[HCO_3^-] = \alpha_1 C_T \qquad \text{(rearrangement of 2-98)}$$

$$[CO_3^{2-}] = \alpha_2 C_T \qquad \text{(rearrangement of 2-99)}$$

Substitution of these values into equation 3-86 yields the following:

$$[Ca]_T = \frac{K_1}{\alpha_2 C_T}\left\{1 + \frac{(K_w)K_2}{[H^+]} + K_3\alpha_1 C_T + K_4\alpha_2 C_T\right\} \qquad \textbf{(3-88)}$$

In this equation C_T represents the equilibrium total carbonic species concentration, and $[H^+]$ the equilibrium hydrogen ion concentration.

Figure 3-4 illustrates the relationship between total soluble calcium concentration and pH for an equilibrium C_T value of 10^{-5} mole/ℓ and a solution temperature of 25°C. For convenience, an ideal solution is assumed.

In process chemistry a common mechanism for removing many chemical species is precipitating the species as a slightly soluble salt. When this type of treatment is employed, it is necessary that some type of chemical compound be added to the water to induce precipitate formation. Thus, the engineer must be able to compute the required chemical dose to be used in the treatment process. This type of calculation is illustrated in example problems 3-4, 3-5, and 3-6.

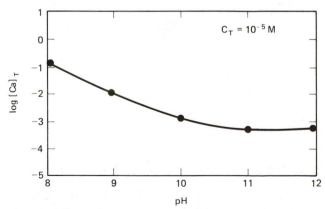

FIGURE 3-4 *Relationship between total soluble calcium concentration and pH at an equilibrium C_T value of 10^{-5} moles/ℓ.*

EXAMPLE PROBLEM 3-4: A manufacturing process generates 100,000 gal/day of wastewater containing 10 mg/ℓ of soluble lead.

(a) Determine the pH required to achieve the minimum soluble lead concentration if lead is to be precipitated as lead sulfide.

(b) Compute the minimum soluble lead concentration which can be achieved if the wastewater pH is maintained at the pH value computed in step (a) and sodium sulfide is added to precipitate PbS.

(c) Calculate the sodium sulfide dose required to attain the soluble lead concentration determined in step (b).

Neglect the ionic strength corrections, and assume that the process temperature is 25°C.

Solution:

1. Establish the appropriate equilibrium constant expressions for the equilibria reactions which affect the solubility of the PbS-H_2O system. For this problem assume that the following equilibrium constant expressions apply:

   (a) $(K_s)_{PbS} = 10^{-28.16} = [Pb^{2+}][S^{2-}]$

   (b) $10^{4.9} = \dfrac{[PbOH^+]}{[H]^+}$

   (c) $10^{-4.5} = Pb(OH)_2^0$

   (d) $10^{-15.4} = [Pb(OH)_3^-][H^+]$

   (e) $10^{-7} = \dfrac{[H^+][HS^-]}{[H_2S]}$

   (f) $10^{-12.9} = \dfrac{[H^+][S^{2-}]}{[HS^-]}$

2. Develop the ionization fraction equations for the H_2S system. For any weak diprotic acid the ionization fraction equations are

   (a) $\alpha_0 = \dfrac{1}{1 + \dfrac{K_1}{[H^+]} + \dfrac{K_1 K_2}{[H^+]^2}} = \dfrac{[H_2S]}{C_T}$

   (b) $\alpha_1 = \dfrac{1}{\dfrac{[H^+]}{K_1} + 1 + \dfrac{K_2}{[H^+]}} = \dfrac{[HS^-]}{C_T}$

   (c) $\alpha_2 = \dfrac{1}{\dfrac{[H^+]^2}{K_1 K_2} + \dfrac{[H^+]}{K_2} + 1} = \dfrac{[S^{2-}]}{C_T}$

   For hydrogen sulfide, $K_1 = 10^{-7}$ and $K_2 = 10^{-12.9}$. In this problem C_T represents the total equilibrium sulfide species concentration.

3. Develop the mass balance equation which describes the total soluble lead concentration:

   $$[Pb]_T = [Pb^{2+}] + [PbOH^+] + [Pb(OH)_2^0] + [Pb(OH)_3^-]$$

   Substituting from the appropriate equilibrium constant expressions, it is possible to write

   $$[Pb]_T = \frac{10^{-28.16}}{[S^{2-}]} + (10^{4.9})[H^+] + 10^{-4.5} + \frac{10^{-15.4}}{[H^+]}$$

Substituting for $[S^{2-}]$ from the α_2 equation gives

$$[Pb]_T = \frac{10^{-28.16}}{\alpha_2 C_T} + (10^{4.9})[H^+] + 10^{-4.5} + \frac{10^{-15.4}}{[H^+]}$$

4. Develop a log $[Pb]_T$ vs. pH diagram for a specified C_T value. Such a plot will indicate the pH at which PbS has its minimum solubility (see Figure 3-5).

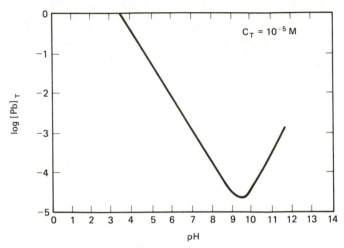

FIGURE 3-5 *Solubility diagram for the PbS-H$_2$O system at an equilibrium C_T value of 10^{-5} moles/ℓ.*

5. Figure 3-5 shows that at pH $= 9.5$ the minimum soluble lead concentration is $10^{-4.6}$ mole/ℓ, or 5.2 mg/ℓ.

6. By varying the equilibrium C_T value, it can be seen that for C_T values greater than 10^{-17} mole/ℓ the total soluble lead concentration is, for all practical purposes, independent of C_T. Thus, the required sodium sulfide dose can be computed from the following mass balance equation:

$$[S]_{dose} = ([Pb]_{initial} - [Pb]_{final}) + [HS^-] + [H_2S]$$

or

$$[S]_{dose} = ([Pb]_{initial} - [Pb]_{final}) + (\alpha_1 C_T) + (\alpha_0 C_T)$$

However, since the required equilibrium C_T value is so small, this equation reduces to

$$[S]_{dose} = ([Pb]_{initial} - [Pb]_{final})$$

Thus,

$$[S]_{dose} = (4.85 \times 10^{-5} M) - (2.5 \times 10^{-5} M)$$

$$= 2.3 \times 10^{-5} \text{ molar}$$

or

$$\text{sodium sulfide dose} = (2.3 \times 10^{-5})(78)(10^3)$$

$$= 1.8 \text{ mg/}\ell$$

Therefore,

$$\text{sodium sulfide dose} = (1.8)(8.34)$$

$$= 15 \text{ lb/MG}$$

EXAMPLE PROBLEM 3-5: A wastewater has a soluble lead(II) concentration of 100 mg/ℓ, a pH of 6, and an alkalinity concentration of 200 mg/ℓ as $CaCO_3$ (assume that alkalinity is due to the carbonic acid system only). Determine the theoretical lime dose required to adjust the pH to the point of minimum $Pb(OH)_2$ solubility if only $Pb(OH)_{2(s)}$ is assumed to precipitate. Neglect ionic strength corrections, and assume that the treatment temperature is 25°C.

Solution:

1. Establish the appropriate equilibrium constant expressions for the equilibrium reactions which affect the solubility of the $Pb(OH)_2$ system. Assume that the following equilibrium constant expressions apply:

   (a) $(K_s)_{Pb(OH)_2} = 10^{-14.9} = [Pb^{2+}][OH^-]^2$

   (b) $10^{4.9} = \dfrac{[PbOH^+]}{[H^+]}$

   (c) $10^{-4.5} = Pb(OH)_2^0$

   (d) $10^{-15.4} = [Pb(OH)_3^-][H^+]$

2. Develop the equation which describes the total soluble lead concentration:

   $$[Pb]_T = [Pb^{2+}] + [PbOH^+] + [Pb(OH)_2^0] + [Pb(OH)_3^-]$$

   Substituting from the appropriate equilibrium constant expressions gives

   $$[Pb]_T = \frac{10^{-14.9}}{[OH^-]^2} + (10^{4.9})[H^+] + 10^{-4.5} + \frac{10^{-15.4}}{[H^+]}$$

3. Develop a log $[Pb]_T$ vs. pH diagram for the $Pb(OH)_2$-H_2O system. Such a plot will indicate the pH at which $Pb(OH)_{2(s)}$ has its minimum solubility (see Figure 3-6).

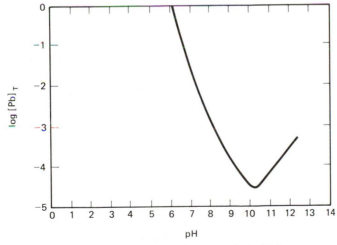

FIGURE 3-6 Solubility diagram for the $Pb(OH)_2$-H_2O system.

4. Figure 3-6 shows that at a pH of 10 the minimum soluble lead concentration is $10^{-4.5}$ mole/ℓ, or 6.5 mg/ℓ.

5. Compute the total carbonic species concentration present in the untreated water.

    (a) Convert alkalinity to moles/ℓ of $CaCO_3$:

    $$[Alk] = \frac{(200)(10^{-3})}{100} = 2 \times 10^{-3} \text{ mole}/\ell$$

    (b) Compute the total carbonic species concentration in the untreated water from equation 2-113:

    $$C_T = \frac{1}{(0.692)} \times \left[\frac{2(2 \times 10^{-3}) - (10^{-14}/10^{-6}) + 10^{-6}}{\frac{2(4 \times 10^{-7})(5 \times 10^{-11})}{(10^{-6})^2} + \frac{(4 \times 10^{-7})}{10^{-6}}} \right]$$

    $$= 1.44 \times 10^{-2} \text{ mole}/\ell$$

6. Compute the alkalinity at the desired pH. The addition of a strong base to increase the pH of a water will increase the OH^- concentration of the solution, and thus the alkalinity by an equivalent amount (see equation 2-74), as long as a carbonate or hydroxyl species does not precipitate. Thus, the alkalinity at the pH to be achieved by lime addition can be computed from equation 2-114 if lead hydroxide is assumed to be infinitely soluble.

    $$[Alk] = \frac{1}{2}\left\{ (1.44 \times 10^{-2})(0.681) \left[\frac{(2)(5 \times 10^{-11})}{10^{-10}} + 1 \right] + \frac{10^{-14}}{10^{-10}} - 10^{-10} \right\}$$

    $$= 9.8 \times 10^{-3}$$

    The C_T value (in moles/ℓ) used here is that obtained from the initial characterization of the untreated water. The hydrogen ion concentration and α_1 are computed for the pH to which the water must be adjusted for $Pb(OH)_{2(s)}$ precipitation.

7. Compute the lime dose required for $Pb(OH)_{2(s)}$ precipitation at a pH of 10.

    (a) Since the change in alkalinity between the initial and final pH levels is directly related to the lime dose required to bring about the pH change, the latter may be calculated as follows (assuming $Pb(OH)_2$ does not precipitate):

    $$\begin{array}{l} \text{lime dose} = [Alk]_{final} - [Alk]_{initial} = \Delta[Alk] \\ (\text{moles}/\ell) \end{array}$$

    $$= (9.8 \times 10^{-3}) - (2 \times 10^{-3})$$

    $$= 7.8 \times 10^{-3} \text{ mole}/\ell$$

    If NaOH is used for pH adjustment, only one-half mole of alkalinity expressed as $CaCO_3$ is generated per mole of NaOH (see equation 2-

108), so that the required NaOH dose would be twice that calculated for lime.

(b) If $Pb(OH)_2$ is allowed to precipitate, a mass balance equation shows that the amount of $Pb(OH)_2$ precipitated will be the same as the change in soluble lead concentration between the initial and final solutions:

$$[Pb]_{precipitated} = [Pb]_{initial} - [Pb]_{final} = \Delta[Pb]$$

Since the loss of OH^- ions from solution during $Pb(OH)_2$ precipitation would cause both a decrease in alkalinity and pH, additional lime (or NaOH) is added to prevent this. The OH^- precipitated as $Pb(OH)_2$ is twice the $\Delta[Pb]$, but lime supplies two OH^- ions per molecule, so that the additional lime required is equal to $\Delta[Pb]$. Thus, the total lime requirement is calculated as follows:

$$\text{total lime dose} = \Delta[AlK] + \Delta[Pb]$$
$$(\text{moles}/\ell)$$

$$= 7.8 \times 10^{-3} + \frac{(100 - 6.5)(10^{-3})}{207.2}$$

$$= 8.2 \times 10^{-3} \text{ mole}/\ell$$

or

$$\text{total lime dose} = (8.2 \times 10^{-3})(10^3)(74)(8.34)$$
$$\text{lb(MG)}$$

$$= 5061 \text{ lb/MG}$$

As before, if NaOH is used the chemical requirement is twice as great as that computed for lime.

EXAMPLE PROBLEM 3-6: A wastewater contains a high soluble copper(II) concentration. Determine the pH where the minimum solubility of $Cu(OH)_2$ occurs, assuming a constant equilibrium, total ammonia nitrogen concentration (NH_4^+—NH_3—N) of 100 mg/ℓ. At this pH, what is the total soluble copper concentration? Neglect ionic strength corrections, and assume that the solution temperature is 25°C.

Solution:

1. Establish the appropriate equilibrium constant expressions for the equilibrium reactions which affect the solubility of the $Cu(OH)_2$ system. Assume that the following equilibrium constant expressions apply:
   (a) $(K_s)_{Cu(OH)_2} = 10^{-18.8} = [Cu^{2+}][OH^-]^2$
   (b) $10^{-20.6} = [Cu_2(OH)_2^{2+}][OH^-]^2$
   (c) $10^{-3.6} = \dfrac{[Cu(OH)_3^-]}{[OH^-]}$
   (d) $10^{-2.7} = \dfrac{[Cu(OH)_4^{2-}]}{[OH^-]^2}$

(e) $\quad 10^{4.31} = \dfrac{[\mathrm{CuNH_3^{2+}}]}{[\mathrm{Cu^{2+}}][\mathrm{NH_3}]}$

(f) $\quad 10^{7.98} = \dfrac{[\mathrm{Cu(NH_3)_2^{2+}}]}{[\mathrm{Cu^+}][\mathrm{NH_3}]^2}$

(g) $\quad 10^{-9.26} = \dfrac{[\mathrm{NH_3}][\mathrm{H^+}]}{[\mathrm{NH_4^+}]}$

2. Develop the ionization fraction equations for the $\mathrm{NH_4^+}$ system. For any weak monoprotic acid the ionization fraction equations are

(a) $\quad \alpha_0 = \dfrac{1}{\dfrac{K_1}{[\mathrm{H^+}]} + 1} = \dfrac{[\mathrm{NH_4^+}]}{C_T}$

(b) $\quad \alpha_1 = \dfrac{1}{1 + \dfrac{[\mathrm{H^+}]}{K_1}}$

For ammonium, $K_1 = 10^{-9.26}$. In this problem C_T represents the total equilibrium nitrogen species ($\mathrm{NH_4^+}$—$\mathrm{NH_3}$—N) concentration.

3. Develop the mass balance equation which describes the total soluble copper concentration:

$$[\mathrm{Cu}]_T = [\mathrm{Cu^{2+}}] + [\mathrm{Cu_2(OH)_2^{2+}}] + [\mathrm{Cu(OH)_3^-}] + [\mathrm{Cu(OH)_4^{2-}}]$$
$$+ [\mathrm{CuNH_3^{2+}}] + [\mathrm{Cu(NH_3)_2^{2+}}]$$

Substituting from the appropriate equilibrium constant expressions, it is possible to write

$$[\mathrm{Cu}]_T = [\mathrm{Cu^{2+}}] + \dfrac{10^{-20.6}}{[\mathrm{OH^-}]^2} + (10^{-3.6})[\mathrm{OH^-}] + (10^{-2.7})[\mathrm{OH^-}]^2$$
$$+ (10^{4.31})[\mathrm{Cu^{2+}}][\mathrm{NH_3}] + (10^{7.98})[\mathrm{Cu^{2+}}][\mathrm{NH_3}]^2$$

or

$$[\mathrm{Cu}]_T = \dfrac{10^{-18.8}}{[\mathrm{OH^-}]^2}\Big[1 + (10^{4.31})[\mathrm{NH_3}] + (10^{7.98})[\mathrm{NH_3}]^2\Big] + \dfrac{10^{-20.6}}{[\mathrm{OH^-}]^2}$$
$$+ (10^{-3.6})[\mathrm{OH^-}] + (10^{-2.7})[\mathrm{OH^-}]^2$$

Substituting $(10^{-14}/[\mathrm{H^+}])$ for $[\mathrm{OH^-}]$ and $\alpha_1 C_T$ for $[\mathrm{NH_3}]$ gives

$$[\mathrm{Cu}]_T = (10^{9.2})[\mathrm{H^+}]^2\,[1 + (10^{4.31})(\alpha_1 C_T) + (10^{7.98})(\alpha_1 C_T)^2]$$
$$+ (10^{7.4})[\mathrm{H^+}]^2 + \dfrac{10^{-17.6}}{[\mathrm{H^+}]} + \dfrac{10^{-30.7}}{[\mathrm{H^+}]^2}$$

4. Convert the equilibrium total nitrogen species concentration to moles/ℓ.

$$C_T = \dfrac{(100)(10)^{-3}}{14} = 7.1 \times 10^{-3}\ \mathrm{mole}/\ell$$

5. Develop a log $[\mathrm{Cu}]_T$ vs. pH diagram for a C_T value of 7.1×10^{-3}. Such a plot will indicate the pH value at which $\mathrm{Cu(OH)_2}$ has its minimum solubility (see Figure 3-7).

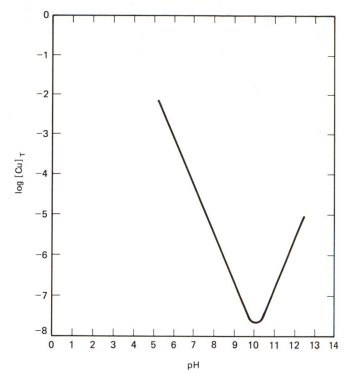

FIGURE 3-7 *Solubility diagram for the $Cu(OH)_2$-NH_3-H_2O system at equilibrium C_T value of 7.1×10^{-3} M.*

6. Figure 3-7 shows that at a pH of 10.2 the minimum soluble copper concentration is $10^{-7.6}$ mole/ℓ, or 0.0016 mg/ℓ.

It should be understood that the calculations presented in example problems 3-4, 3-5, and 3-6 have certain limitations. Probably the most important limitation is that the calculations are based on the assumption that equilibrium is attained within the retention time provided by the reaction units, whereas in reality the rate of formation of chemical precipitates can be so slow that the theoretical equilibrium concentrations are not reached within the relatively short reaction time provided in most treatment units.

Other limitations which may result in a difference between the calculated values and field results are failure to include in the equilibrium model all the soluble species which affect the solubility of the precipitate and the failure to include ionic strength and temperature corrections.

Competing Solid Phase Equilibria

In the previous section solubility diagrams were constructed for systems in which only one solid phase was possible. However, in process chemistry it is not uncommon to encounter a situation in which two or more precipitation reactions are competing. In such cases the ion-ratio method may be used to predict the order of precipitation of solids (Freiser and Fernando, 1963).

Thermodynamically, the solid phase that is most stable at a given pH is the one which gives the smallest soluble metal ion concentration. To illustrate the ion-ratio method for determining the controlling solid phase, the zinc-carbonate-water system is considered. For this system the solubility equilibrium reactions are

(a) $\qquad ZnCO_{3(s)} \;\rightleftharpoons\; Zn^{2+}_{(aq)} + CO^{2-}_{3(aq)}$ (3-89)

where

$$K_s = 10^{-10.8} = [Zn^{2+}][CO_3^{2-}] \qquad (3\text{-}90)$$

(b) $\qquad Zn(OH)_{2(s)} \;\rightleftharpoons\; Zn^{2+}_{(aq)} + 2OH^-_{(aq)}$ (3-91)

where

$$K_s = 10^{-16} = [Zn^{2+}][OH^-]^2 \qquad (3\text{-}92)$$

(c) $\qquad 10^{-11.9} = [Zn(OH)^+][OH^-]$ (3-93)

(d) $\qquad 10^{-5.9} = [Zn(OH)_2^0]$ (3-94)

(e) $\qquad 10^{-1.8} = \dfrac{[Zn(OH)_3^-]}{[OH^-]}$ (3-95)

(f) $\qquad 10^{-0.5} = \dfrac{[Zn(OH)_4^{2-}]}{[OH^-]^2}$ (3-96)

The total soluble zinc concentration is given by the following mass balance expression:

$$[Zn]_T = [Zn^{2+}] + [Zn(OH)^+] + [Zn(OH)_2^0] + [Zn(OH)_3^-] + [Zn(OH)_4^{2-}] \qquad (3\text{-}97)$$

Substituting from the appropriate equilibrium constant expression, it is possible to write

$$[Zn]_T = [Zn^{2+}] + \frac{10^{-11.9}}{[OH^-]} + 10^{-5.9} + (10^{-1.8})[OH^-] + (10^{-0.5})[OH^-]^2 \qquad (3\text{-}98)$$

However, at this point the problem arises as to whether to substitute $(10^{-10.8}/[CO_3^{2-}])$ or $(10^{-16}/[OH^-]^2)$ for $[Zn^{2+}]$. To overcome this problem, consider that the ratio of the $Zn(OH)_{2(s)}$ and $ZnCO_{3(s)}$ solubility products and the ratio of the concentrations of the individual ions will determine which solid phase controls the soluble Zn^{2+} concentration under a given set of conditions. Thus,

$$\frac{[Zn^{2+}][OH^-]^2}{[Zn^{2+}][CO_3^{2-}]} = \frac{(K_s)_{Zn(OH)_2}}{(K_s)_{ZnCO_3}} \qquad (3\text{-}99)$$

or

$$\frac{[OH^-]^2}{[CO_3^{2-}]} = \frac{(K_s)_{Zn(OH)_2}}{(K_s)_{ZnCO_3}} \qquad (3\text{-}100)$$

Since the ratio $(K_s)_{Zn(OH)_2}/(K_s)_{ZnCO_3}$ is a constant, it can be represented by the constant R. Hence,

$$\frac{(K_s)_{Zn(OH)_2}}{(K_s)_{ZnCO_3}} = R \qquad (3\text{-}101)$$

or for the Zn—CO$_3$—H$_2$O system,

$$\frac{10^{-16}}{10^{-10.8}} = 10^{-5.2} \qquad \qquad \textbf{(3-102)}$$

If the quotient of the actual equilibrium concentrations of the hydroxyl and carbonate ions, $[OH^-]^2/[CO_3^{2-}]$, are computed to be greater than $10^{-5.2}$, the $Zn(OH)_2$ solid phase controls the soluble Zn^{2+} concentration and $(10^{-16}/[OH^-]^2)$ would be substituted for the $[Zn^{2+}]$ term in equation 3-98. However, if $[OH^-]^2/[CO_3^{2-}]$ is less than $10^{-5.2}$, the $ZnCO_3$ solid phase will control the soluble Zn^{2+} concentration and $(10^{-10.8}/[CO_3^{2-}])$ would be substituted for the $[Zn^{2+}]$ term in equation 3-98. If $[OH^-]^2/[CO_3^{2-}]$ equals $10^{-5.2}$, either $(10^{-16}/[OH^-]^2)$ or $(10^{-10.8}/[CO_3^{2-}])$ can be substituted for $[Zn^{2+}]$ in equation 3-98. Thus, it can be seen that the boundary of the solid phase for the Zn-CO$_3$-H$_2$O system is actually described by two different equations. The first equation describes the boundary within the pH range where the carbonate solid phase controls the soluble Zn^{2+} concentration, and the second equation describes the solid phase boundary within the pH range where the hydroxide solid phase controls the soluble Zn^{2+} concentration.

Figure 3-8 shows only the log $[Zn^{2+}]$ vs. pH plot for the zinc hydroxide and a 10^{-3} M zinc carbonate system. Note that $ZnCO_3$ shows the lowest solubility up to a pH value of approximately 9.4 and, as a result, it will control the soluble Zn^{2+} concentration in the pH range of 0 to 9.4. Above a pH value of 9.4, $Zn(OH)_2$ shows the lowest solubility and will control the soluble Zn^{2+} concentration when the equilibrium pH is greater than 9.4.

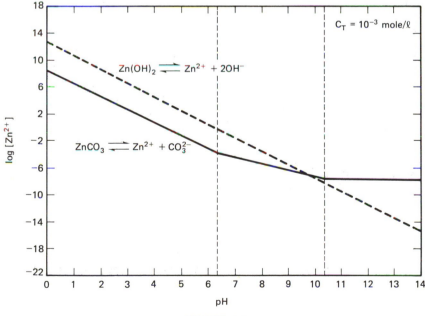

FIGURE 3-8

The pH value at which control of the soluble Zn^{2+} concentration passes from the carbonate system to the hydroxide system can be computed mathematically. The relationship required to compute the changeover pH can be developed by noting that the concentration of the carbonate species, $[CO_3^{2-}]$, is given by

$$[CO_3^{2-}] = \alpha_2 C_T \qquad \text{(2-99)}$$

where α_2 is the third ionization fraction for the carbonic acid system, and C_T represents the total equilibrium carbonic specie concentration. Equation 3-100 can therefore be expressed as

$$[OH^-] = [R\alpha_2 C_T]^{1/2} \qquad \text{(3-103)}$$

or

$$[H^+] = \frac{K_w}{[R\alpha_2 C_T]^{1/2}} \qquad \text{(3-104)}$$

Equation 2-106 shows that for the carbonic acid system (assuming an ideal solution) α_2 is equal to

$$\alpha_2 = \frac{1}{1 + \dfrac{[H^+]}{K_2} + \dfrac{[H^+]^2}{K_1 K_2}} \qquad \text{(2-106)}$$

Therefore, equation 3-104 can be written as

$$[H^+]\left[\frac{1}{1 + \dfrac{[H^+]}{K_2} + \dfrac{[H^+]^2}{K_1 K_2}}\right]^{1/2} = \frac{K_w}{(RC_T)^{1/2}} \qquad \text{(3-105)}$$

or

$$\left[\frac{[H^+]^2}{1 + \dfrac{[H^+]}{K_2} + \dfrac{[H^+]^2}{K_1 K_2}}\right]^{1/2} = \frac{K_w}{(RC_T)^{1/2}} \qquad \text{(3-106)}$$

or

$$\frac{[H^+]^2}{1 + \dfrac{[H^+]}{K_2} + \dfrac{[H^+]^2}{K_1 K_2}} = \left[\frac{K_w}{(RC_T)^{1/2}}\right]^2 \qquad \text{(3-107)}$$

or

$$\frac{[H^+]^2}{1 + \dfrac{[H^+]}{K_2} + \dfrac{[H^+]^2}{K_1 K_2}} = \frac{(K_w)^2}{RC_T} \qquad \text{(3-108)}$$

If

$$B = \frac{(K_w)^2}{RC_T} \qquad \text{(3-109)}$$

it is possible to write

$$[H^+]^2\left[\frac{B}{K_1 K_2} - 1\right] + \frac{B}{K_2}[H^+] + B = 0 \tag{3-110}$$

Therefore,

$$[H^+] = \frac{-\left(\frac{B}{K_2}\right) \pm \left[\left(\frac{B}{K_2}\right)^2 - 4\left(\frac{B}{K_1 K_2} - 1\right)(B)\right]^{1/2}}{2\left(\frac{B}{K_1 K_2} - 1\right)} \tag{3-111}$$

3-3 REMOVAL OF HEAVY METALS FROM COMPLEX WATER AND WASTEWATER SYSTEMS

Heavy metals such as zinc, copper, nickel, and cadmium are present in many industrial wastewaters and must be removed prior to discharge. Various methods are available for removal, including neutralization/precipitation, sorption, and ion exchange. If the metals are present in wastewaters that contain hexavalent chromium, this chromium must be reduced prior to metal removal. Likewise, metals are normally removed from cyanide-bearing wastes only after cyanide destruction.

Removal by Neutralization/Precipitation

The concentration of a metal in solution can be lowered by precipitation to a point dictated by the solubility of the various species of that metal. Most metals are relatively insoluble as hydroxides, carbonates, and sulfides and are usually precipitated in one of these forms. Metals can be precipitated as hydroxides simply by raising the pH of the wastewater to the range of pH 8 to 11. Precipitation as a metal carbonate can be achieved by adding dolomite, and as a metal sulfide by adding inorganic sulfide or hydrogen sulfide gas (Rouse, 1976). The precipitate that yields the desired residual metal concentration at the lowest cost should be selected as the target for treatment. Theoretical solubility calculations and plots such as Figure 3-8 can be used to help determine the insoluble species that will control the solubility of a given metal.

In general, hydroxides usually prove to be the controlling species for precipitating metal from industrial waste. Many metal hydroxides are amphoteric in nature and exhibit an optimum pH for removal by precipitation. This optimum pH value and the associated minimum solubility can be determined from solubility calculations previously discussed in this chapter. Unfortunately, actual solubilities and optimum pH values for metals in industrial wastewaters are seldom the same as indicated by theoretical calculations. In most cases, actual solubilities will be greater than theoretical solubilities due to incomplete reactions, poor separation of colloidal precipitates and the formation of soluble metal complexes not considered in the equilibrium model. Solubilities lower than the theoretical values may occasionally occur as a result of coprecipitation.

The difficulties encountered in predicting residual metal concentrations for industrial waste are evidenced in the studies of Yost and Scarfi (1979a). These researchers compared metal hydroxide precipitation from freshly prepared zinc cyanide plating solutions with precipitation from solutions that had been used for plating (working solutions). Residual zinc solubility in the freshly prepared solutions was found to be independent of initial zinc concentration and averaged 0.078 mg/ℓ in the pH range of 8.5 to 9.5. Solubility in the working zinc cyanide solutions was much higher than in the fresh solutions and increased with initial zinc concentration. When the working zinc cyanide solution was mixed with a working copper cyanide solution prior to treatment, residual solubilities increased dramatically and the optimum pH for precipitation varied with the initial zinc concentration. Furthermore, the allowable zinc concentration specified in current pretreatment standards was not met for solutions having initial zinc concentrations exceeding 200 mg/ℓ. This increase in solubility was attributed to the formation of mixed polynuclear complexes in which both zinc and copper appeared to be part of the same metal-ligand complex molecule (Yost and Scarfi, 1979a).

When zinc cyanide plating solutions were mixed with nickel plating solutions or solution of reduced hexavalent chromium, an improvement in zinc removal was noted (Yost and Scarfi, 1979a). This observation suggests the possibility of coprecipitation of zinc by trivalent chromium and divalent and/or tetravalent nickel.

In a related study Yost and Scarfi (1979b) investigated the solubility of copper in in copper cyanide plating solutions following treatment by cyanide destruction and pH adjustment. The study also included copper cyanide solutions mixed with either zinc cyanide, cadmium cyanide, chromate, or nickel plating solutions.

Residual copper solubilities in treated copper cyanide plating solutions were higher than for comparable copper sulfate solutions. This increased solubility was thought to be due to formation of a complex ion, $Cu(CO_3)_3^{4-}$, formed between copper and the carbonate produced during cyanide destruction. Likewise, when copper cyanide plating solutions were mixed with zinc or cadmium plating solutions, a large increase in soluble copper was noted in the treated solutions. Mixing copper and nickel plating solutions prior to treatment resulted in substantially lower copper concentrations for pH values below 10.

Chromium appeared to have a variable effect on copper solubility. Mixtures of dilute (50 mg/ℓ of each metal) copper and chromium plating solutions showed higher concentrations of residual copper than did individually treated copper cyanide solutions. Conversely, residual copper concentrations were reduced when copper and chromium plating solutions containing 200 to 500 mg/ℓ of each metal were mixed prior to treatment.

The preceding discussion points out the need for laboratory studies when selecting a treatment scheme for plating wastes. Waste streams should be studied individually and in various combinations in order to determine feasible treatment alternatives. Some general guidelines can be observed so as to eliminate certain options and minimize the testing process. For example, cyanide and chromium plating solutions should be segregated until cyanide has been destroyed and hexavalent chromium has been reduced. Furthermore, untreated cyanide waste must always be separated from acid waste to avoid formation of toxic HCN gas.

Waste streams should not be combined prior to treatment if they contain components that are known to form complex ions which might interfere with treatment or, being highly soluble, might increase residual metal concentrations. For example, cyanide binds very tightly to iron, and cyanide wastes should be segregated from iron-bearing wastes prior to the cyanide destruction step. A number of complex ions associated with metals commonly found in plating wastes are listed in Appendix V. This list may serve as a guide in determining wastes that should not be mixed because of potential complex formation.

PROBLEMS

3-1. Plot a log $[Zn^{2+}]$ vs. pH diagram for the zinc-sulfide-carbonate-water system, using total sulfide and carbonic species concentrations of 10^{-3} mole/ℓ. From this diagram estimate the pH range over which each solid (ZnS, $ZnCO_3$, $Zn(OH)_2$) controls the soluble Zn^{2+} concentration. The applicable reactions are

(a) $ZnCO_{3(s)} \rightleftharpoons Zn^{2+}_{(aq)} + CO^{2-}_{3(aq)}$
$10^{-10.8} = [Zn^{2+}][CO_3^{2-}]$

(b) $Zn(OH)_{2(s)} \rightleftharpoons Zn^{2+}_{(aq)} + 2OH^-_{(aq)}$
$10^{-16.3} = [Zn^{2+}][OH^-]^2$

(c) $ZnS_{(s)} \rightleftharpoons Zn^{2+}_{(aq)} + S^{2-}_{(aq)}$
$10^{-22.8} = [Zn^{2+}][S^{2-}]$

3-2. Construct the log [species] vs. pH diagram for the $Fe(OH)_2$-H_2O system. From this diagram determine the minimum solubility of $Fe(OH)_{2(s)}$ and the pH where the minimum solubility occurs.

2-3. Evaluate the effect of the equilibrium carbonate concentration on the total soluble calcium concentration. This is accomplished by constructing a log $[Ca]_T$ vs. pH diagram for equilibrium total carbonic species concentrations (C_T values) of 10^{-3} and 10^{-5} mole/ℓ.

2-4. Determine whether $Ni(OH)_2$ or $NiCO_3$ will control the soluble Ni^{2+} concentration if the equilibrium pH is 8.0 and the equilibrium total alkalinity is 400 mg/ℓ expressed as $CaCO_3$. Assume that the applicable equilibrium constant expressions are

(a) $NiCO_{3(s)} \rightleftharpoons Ni^{2+}_{(aq)} + CO^{2-}_{3(aq)}$
$10^{-6.9} = [Ni^{2+}][CO_3^{2-}]$

(b) $Ni(OH)_{2(s)} \rightleftharpoons Ni^{2+}_{(aq)} + 2OH^-_{(aq)}$
$10^{-15.8} = [Ni^{2+}][OH^-]^2$

(c) $10^{-11.3} = [Ni(OH)^+][OH^-]$

(d) $10^{-4.5} = [Ni(OH)_2^0]$

(e) $10^{-1.7} = \dfrac{[Ni(OH)_3^-]}{[OH^-]}$

3-5. Evaluate the effect of the equilibrium carbonate concentration on the total soluble cadmium concentration for the cadmium-carbonate-water system. This is accomplished by constructing a log $[Cd]_T$ vs. pH diagram for equilibrium total carbonic species concentrations (C_T values) of 10^{-2}, 10^{-3}, and 10^{-4} mole/ℓ. Assume that the applicable equilibrium constant expressions are

(a) $CdCO_{3(s)} \rightleftharpoons Cd^{2+}_{(aq)} + CO^{2-}_{3(aq)}$
$$10^{-11.3} = [Cd^{2+}][CO_3^{2-}]$$

(b) $Cd(OH)_{2(s)} \rightleftharpoons Cd^{2+}_{(aq)} + 2OH^-_{(aq)}$
$$10^{-13.7} = [Cd^{2+}][OH^-]^2$$

(c) $10^{-9.5} = [Cd(OH)^+][OH^-]$

(d) $10^{-5.3} = [Cd(OH)_2^0]$

(e) $10^{-4.6} = \dfrac{[Cd(OH)_3^-]}{[OH^-]}$

(f) $10^{-4.9} = \dfrac{[Cd(OH)_4^{2-}]}{[OH^-]^2}$

REFERENCES

BARD, A.J., *Chemical Equilibrium*, Harper and Row, Publishers, Inc. New York (1966).

BLACK, A.P., and CHEN, C., "Electrophoretic Studies of Coagulation and Flocculation of River Sediment Suspensions with Aluminum Sulfate," *J. Am. Water Works Assoc.*, **57,** 354 (1965).

BOLT, G.H., and BRUGGENWERT, M.G.M., *Soil Chemistry: Basic Elements*, Elsevier Scientific Publishing Company, Amsterdam (1976).

FREISER, H., and FERNANDO, Q., *Ionic Equilibria in Analytical Chemistry*, John Wiley & Sons, Inc., New York (1963).

RINGBOM, A., *Complexation in Analytical Chemistry*, Wiley-Interscience, New York, (1963).

ROUSE, J.V., "Removal of Heavy Metals from Industrial Effluents," *Journal Environmental Engineering Division*, ASCE, **102,** 929 (1976).

RUBIN, A.J., and KOVAC, T.W., "Effects of Aluminum(III) Hydrolysis on Alum Coagulation," in *Chemistry of Water Supply, Treatment, and Distribution*, [Ed.] A.J. RUBIN, Ann Arbor Science Publishers, Ann Arbor, MI (1974).

SILLEN, L.G., and MARTELL, A.C., *Stability Constants of Metal-ion Complexes*, 2nd ed. Special Publication No. 17, The Chemical Society, London (1964).

STUMM, W., and O'MELIA, C.R., "Stoichiometry of Coagulation," *J. Am. Water Works Assoc.*, **60,** 514 (1968).

YOST, K.J., and MASARIK, D.R., "A Study of Chemical-Destruct Waste Treatment Systems in the Electroplating Industry," *Plating and Surface Finishing*, **35,** (January, 1977).

YOST, K.J., and SCARFI, A., "Factors Affecting Zinc Solubility in Electroplating Waste," *Journal Water Pollution Control Federation*, **51,** 1878 (1979a).

YOST, K.J., and SCARFI, A., "Factors Affecting Copper Solubility in Electroplating Waste," *Journal Water Pollution Control Federation*, **51,** 1887 (1979b).

4

OXIDATION-REDUCTION EQUILIBRIA

4-1 OXIDATION-REDUCTION PROCESSES

A solute that can be dissolved in water to become a solution which conducts electricity is called an *electrolyte*. The electrical conducting property of an electrolyte solution can be understood by noting that such a solution contains both positive and negative ions. When an electrical field is applied to such a solution, the positive ions will migrate toward the electrode with the negative charge, and the negative ions will migrate toward the electrode with the positive charge. This movement of charged particles constitutes an electrical current. (Metal materials contain charge carriers of only one sign, whereas solutions contain charge carriers of both signs.) In the simplest case, a positive ion accepts one or more electrons from the electrode to which it moves, and a negative ion releases one or more electrons to the electrode to which it moves. Such a transfer of electrons to or from a chemical species changes the oxidation state (or oxidation number) of that species, thus constituting an oxidation-reduction chemical reaction.

The electrode where the positive ion accepts electrons (is reduced) is called the *cathode*, whereas the electrode at which the negative ion releases electrons (is oxidized) is called the *anode*. A number of different types of reactions are possible at each electrode. Some of the more common ones are (Brescia et al., 1970):

1. *Anode Reactions:*
   (a) The electrode material may be oxidized to positive ions:

$$Zn_{(s)} \longrightarrow Zn^{2+}_{(aq)} + 2\,e^- \qquad\qquad \textbf{(4-1)}$$

   (b) Solutes may be oxidized and remain in solution:

$$Fe^{2+}_{(aq)} \longrightarrow Fe^{3+}_{(aq)} + e^- \qquad\qquad \textbf{(4-2)}$$

(c)   Negatively charged ions may be oxidized and form neutral molecules which may escape from solution:

$$2Cl^-_{(aq)} \longrightarrow Cl_{2(g)} + 2\,e^- \qquad\qquad \textbf{(4-3)}$$

(d)   A gas in contact with the electrode may be oxidized to a positive ion:

$$H_{2(g)} \longrightarrow 2H^+_{(aq)} + 2\,e^- \qquad\qquad \textbf{(4-4)}$$

(e)   In an aqueous solution, water may be oxidized to molecular oxygen:

$$2H_2O_{(aq)} \longrightarrow O_{2(g)} + 4H^+_{(aq)} + 4\,e^- \qquad\qquad \textbf{(4-5)}$$

2. *Cathode Reactions:*
   (a)   Positive ions in solution may be reduced and deposited on the electrode:

$$Cu^{2+}_{(aq)} + 2\,e^- \longrightarrow Cu_{(s)} \qquad\qquad \textbf{(4-6)}$$

(b)   Solutes may be reduced and remain in solution:

$$Fe^{3+}_{(aq)} + e^- \longrightarrow Fe^{2+}_{(aq)} \qquad\qquad \textbf{(4-7)}$$

(c)   A gas may be reduced to a negative ion:

$$Cl_{2(g)} + 2\,e^- \longrightarrow 2Cl^-_{(aq)} \qquad\qquad \textbf{(4-8)}$$

(d)   In an aqueous solution, water may be reduced to molecular hydrogen:

$$2H_2O + 2\,e^- \longrightarrow H_{2(g)} + 2OH^-_{(aq)} \qquad\qquad \textbf{(4-9)}$$

It has been observed that an electric current can cause certain chemical reactions to occur and, conversely, that electricity can be obtained from certain chemical reactions. Based on these observations, two kinds of electrochemical cells have been defined: (a) the *galvanic cell*, i.e., an electrochemical cell where the spontaneous occurrence of electrode reactions produces electrical energy which can be converted to useful work; (b) the *electrolytic cell*, i.e., an electrochemical cell where nonspontaneous electrode reactions are forced to proceed when an external voltage is impressed or connected across the two electrodes, which means that electrical energy is consumed rather than produced during the electrode reactions. In this chapter the discussion is limited to galvanic cells and oxidation-reduction reactions occurring in aqueous solutions.

4-2 GALVANIC CELLS AND CHEMICAL THERMODYNAMICS

If elemental zinc and copper(II) sulfate are mixed together in an aqueous solution the following reaction will occur:

$$Zn_{(s)} + Cu^{2+}:SO^{2-}_{4(aq)} \longrightarrow Cu_{(s)} + Zn^{2+}:SO^{2-}_{4(aq)} \qquad\qquad \textbf{(4-10)}$$

In this reaction, elemental zinc loses two electrons and is oxidized to Zn^{2+}, whereas

Cu^{2+} gains two electrons and is reduced to elemental copper. Because the chemical species exist together in the same solution, there will be a direct transfer of electrons between zinc and copper and, as a result, it is not possible to produce electrical work from the reaction. However, it is possible to produce electrical work from the reaction if the chemical species are separated so that during the reaction the electrons are transferred from zinc to copper through an external connecting wire. To understand how this can be accomplished, recall that an oxidation-reduction reaction can be considered as two half-reactions:

Oxidation:

$$Zn_{(s)} \longrightarrow Zn^{2+}_{(aq)} + 2\,e^- \tag{4-11}$$

Reduction:

$$Cu^{2+}_{(aq)} + 2\,e^- \longrightarrow Cu_{(s)} \tag{4-12}$$

Figure 4-1 shows a solution of $ZnSO_4$ containing a metallic zinc electrode connected by an external wire to a solution of $CuSO_4$ containing a metallic copper electrode. In this system zinc will go into solution as Zn^{2+}, and the electrons left behind will flow through the wire to the copper electrode, where Cu^{2+} ions from solution will accept the electrons and form elemental copper. Because electroneutrality must be maintained in the solution, negatively charged ions migrate from the copper cell through the salt bridge toward the zinc cell, whereas positively charged ions migrate from the zinc cell toward the copper cell.

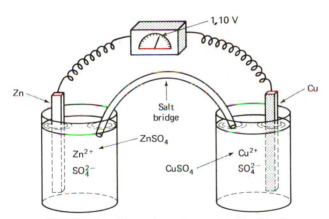

FIGURE 4-1 *Measuring cell potential (after Moeller and O'Connor, 1972).*

At the instant that the external wire is first connected between the zinc electrode and copper electrode, an initial voltage (potential difference) can be measured. However, the release of Zn^{2+} ions in the left cell and the removal of Cu^{2+} ions in the right cell will change the original concentrations of these ions. This suggests that oxidation-reduction reactions establish an equilibrium state where there is no net electron transfer. For the zinc-copper galvanic cell the equilibrium reaction can be written as

$$Zn_{(s)} + Cu^{2+}_{(aq)} \rightleftharpoons Zn^{2+}_{(aq)} + Cu_{(s)} \tag{4-13}$$

for which

$$(K_a)_{eq.} = \frac{(Zn^{2+})}{(Cu^{2+})} \qquad \textbf{(4-14)}$$

Thus, the thermodynamic treatment of chemical equilibrium presented in Chapter 1 may be applied to oxidation-reduction equilibria.

The sign (positive or negative) of the free energy change ΔG for any chemical reaction indicates the spontaneity of the reaction, and the value of ΔG represents the maximum amount of useful work that can be obtained from the process. However, voltage is also a measure of the maximum work which can be obtained from a chemical reaction. Hence, the following equation has been developed to describe the relationship between ΔG and voltage, E:

$$\Delta G_{reaction} = -nFE_{cell} \qquad \textbf{(4-15)}$$

where $\Delta G_{reaction}$ = Gibbs free energy change (calories/mole)

n = number of moles of electrons transferred, as indicated in the balanced chemical equation

F = Faraday constant, which has a value of 23,061 cal/mole-volt

E_{cell} = potential of a cell (volts).

If the cell reaction occurs under standard state conditions, equation 4-15 would be expressed as

$$\Delta G^{\circ}_{reaction} = -nFE^{\circ}_{cell} \qquad \textbf{(4-16)}$$

Substituting for $\Delta G^{\circ}_{reaction}$ from equation 1-54 gives

$$-nFE^{\circ}_{cell} = -RT \ln (K_a)_{eq.}$$

or

$$E^{\circ}_{cell} = \frac{RT \ln (K_a)_{eq.}}{nF} \qquad \textbf{(4-17)}$$

In Chapter 1 it is shown that the free energy change for a chemical reaction under nonstandard state conditions is given by the relationship

$$\Delta G_{reaction} = \Delta G^{\circ}_{reaction} + RT \ln Q \qquad \textbf{(1-57)}$$

Substituting for $\Delta G_{reaction}$ from equation 4-15 and for $\Delta G^{\circ}_{reaction}$ from equation 4-16, equation 1-57 becomes

$$-nFE_{cell} = -nFE^{\circ}_{cell} + RT \ln Q$$

or

$$E_{cell} = E^{\circ}_{cell} - \frac{RT}{nF} \ln Q \qquad \textbf{(4-18)}$$

Equation 4-18 is commonly referred to as the *Nernst equation* and is very useful in relating half-reaction potentials and cell potentials to concentrations of reacting substances. When the equation is applied to half-reactions, it is standard practice to represent E_{cell} by E_h to denote that the half-reaction potential is based on the hydrogen electrode.

Standard Electrode Potentials

The electrodes of a galvanic cell must have different potentials if a voltage measurement is to be made, which suggests that it is not possible to compute the potential for a single half-cell. What is done in electrochemistry is to compute the cell potential of many different half-cells against an arbitrarily selected reference half-cell. The reference standard is the hydrogen electrode which is assigned a potential of 0 volts under standard conditions:

$$2H^+_{(aq)} + 2\,e^- \longrightarrow H_{2(g)} \qquad : E^\circ = 0.00 \text{ V} \quad \textbf{(4-19)}$$

Thus, it is possible to measure cell voltages for many different half-cells against the hydrogen electrode and, by comparison, to assign a standard reduction potential to the half-cell reactions. A list of reduction potentials developed in this manner is presented in Appendix VI. The following observations can be made regarding Appendix VI:

1. The standard hydrogen half-cell is assigned a reduction potential of 0.0 V.

2. All species in standard half-cells are at standard state. For example,
   gases: $P = 1$ atm
   liquid: $a = 1$
   solid: $a = 1$
   ions: $a = 1$

3. The most negative reduction potential has the least tendency to exist in the reduced state. For example,

$$Li^+ + e^- \longrightarrow Li \qquad : E^\circ = -3.03 \text{ V}$$

   This reaction has little tendency to go to the right. In other words, Li prefers the ionic form.

4. The reduction potential of a metal ion, nonmetal, or compound is a measure of the tendency of the substance to accept an electron from an electron donor. The higher the potential (as expressed in volts), the greater the affinity of the acceptor for the electron.

Standard reduction potentials may be used to compute a cell potential, E°_{cell}, for any electrochemical cell. The E°_{cell} value can then be used to determine the spontaneity of the overall cell reaction under standard conditions. To illustrate how this may be done, consider the following half-cell reactions:

$$Pb^{2+} + 2\,e^- \longrightarrow Pb \qquad : E^\circ = -0.126 \text{ V} \quad \textbf{(4-20)}$$
$$Sn^{4+} + 2\,e^- \longrightarrow Sn^{2+} \qquad : E^\circ = +0.150 \text{ V} \quad \textbf{(4-21)}$$

These two half-cell reactions can produce two possible situations: (a) Pb can be oxidized and Sn^{4+} reduced or (b) Pb^{2+} can be reduced and Sn^{2+} oxidized. These two situations are represented by the following overall cell reactions:

(a) $\quad Pb + Sn^{4+} \longrightarrow Pb^{2+} + Sn^{2+}$ (4-22)

$$E^{\circ}_{cell} = E^{\circ}_{ox} + E^{\circ}_{red} = +0.126 + 0.150 = +0.276 \text{ V} \qquad \text{(4-23)}$$

Note: Since lead is being oxidized in the cell reaction and the half-cell potential is given as a reduction potential, the sign of the potential is reversed.

(b) $\quad Sn^{2+} + Pb^{2+} \longrightarrow Sn^{4+} + Pb$ (4-24)

$$E^{\circ}_{cell} = -0.150 + (-0.126) = -0.276 \text{ V}$$

The relationship between standard state cell potential and free energy change is given by equation 4-16 as

$$\Delta G^{\circ}_{reaction} = -nFE^{\circ}_{cell} \qquad \text{(4-16)}$$

Hence, if E°_{cell} is negative, that means $\Delta G^{\circ}_{reaction}$ will be positive and the reaction will not be spontaneous. However, if E°_{cell} is positive, that means $\Delta G^{\circ}_{reaction}$ will be negative and the reaction will be spontaneous. Therefore, the cell reaction will occur in the manner represented by equation 4-22.

Applications of the Nernst Equation

The Nernst equation can be used to calculate cell potentials for conditions other than standard state. Hence, spontaneity predictions may be applied to reactions under conditions other than those of standard state by using equation 4-15.

When equilibrium is attained for any cell reaction, the cell potential becomes zero and the value of Q from the Nernst equation will then be that of the equilibrium constant for the cell reaction. Thus, at equilibrium

$$0 = E^{\circ}_{cell} - \frac{RT}{nF} \ln (K_a)_{eq.}$$

or

$$\ln (K_a)_{eq.} = \frac{nFE^{\circ}_{cell}}{RT} \qquad \text{(4-25)}$$

Therefore, it is possible to compute equilibrium constants for oxidation-reduction reactions from standard half-cell reaction potentials such as those presented in Appendix VI.

Examples of calculations which involve standard potential data and the Nernst equation are presented in example problems 4-1, 4-2, and 4-3. To simplify the calculations in these problems, analytical concentrations are used instead of activities. However, it should be kept in mind that the activities of ions in dilute solutions are normally somewhat less than analytical concentration values because of interionic forces.

EXAMPLE PROBLEM 4-1: Assume the following reaction at standard state

$$5H_2S_{(g)} + 6H^+_{(aq)} + 2MnO^-_{4(aq)} \rightleftharpoons 2Mn^{2+}_{(aq)} + 5S_{(s)} + 8H_2O_{(1)}$$

Calculate the cell potential, the free energy change and the equilibrium constant for the reaction.

Solution:

1. Obtain the half-cell potentials from Appendix VI.

$$MnO^-_4 + 8H^+ + 5e^- \longrightarrow Mn^{2+} + 4H_2O \qquad : E^\circ = +1.51$$
$$S + 2H^+ + 2e^- \longrightarrow H_2S \qquad : E^\circ = +0.14$$

2. Write the cell reaction from the half-cell reactions. Insure that the same number of electrons appear in both half-cell reactions.

Oxidation:
$$5H_2S \longrightarrow 5S + 10H^+ + 10e^- \qquad : E^\circ = -0.14$$

Reduction:
$$\underline{2MnO^-_4 + 16H^+ + 10e^- \longrightarrow 2Mn^{2+} + 8H_2O \qquad : E^\circ = +1.51}$$
$$5H_2S + 6H^+ + 2MnO^-_4 \longrightarrow 2Mn^{2+} + 5S + 8H_2O$$

3. Compute the standard cell potential from equation 4-23:

$$E^\circ_{cell} = E^\circ_{ox} + E^\circ_{red} = (-0.14) + (+1.51) = 1.37$$

Note: Do not multiply the half-cell potentials by the multiplication factors required for electron balance because all E° values are normalized to the transfer of one equivalent of electrons.

4. Calculate the free energy change of the reaction from equation 4-16.

$$\Delta G^\circ_{reaction} = -nFE^\circ_{cell}$$
$$= -(10)(23,061)(1.37)$$
$$= -315,936 \, cal/mole$$

5. Calculate the equilibrium constant for the reaction from equation 4-25:

$$\ln (K_a)_{eq.} = \frac{nFE^\circ_{cell}}{RT} = \frac{(10)(23,061)(1.37)}{(1.98)(298)} = 535.4$$

EXAMPLE PROBLEM 4-2: Compute the cell potential, the free energy change, and the equilibrium constant for the reaction given in example problem 4-1 if the chemical species are present in the following concentrations:

$$H_2S = 0.2 \, atm$$
$$H^+ = 10^{-7} \, M$$
$$MnO^-_4 = 10^{-4} \, M$$
$$Mn^{2+} = 10^{-5} \, M$$
$$S = 1.0 \, M$$
$$H_2O = 1.0 \, M$$

The reaction temperature is 25°C.

Solution:

1. Compute the cell potential from equation 4-18, using the E°_{cell} value from step 3 of example problem 4-1:

$$E_{cell} = E^\circ_{cell} - \frac{RT}{nF} \ln Q$$

$$= 1.37 - \frac{(1.98)(298)(2.3)}{(10)(23,061)} \log \frac{[10^{-5}]^2}{[0.2]^5[10^{-7}]^6[10^{-4}]^2}$$

$$= 1.28 \text{ V}$$

2. Calculate the free energy change for the reaction from equation 4-15:

$$\Delta G_{reaction} = -nFE_{cell}$$

$$= -(10)(23,061)(1.28)$$

$$= -295,181 \text{ cal/mole}$$

3. Compute the equilibrium constant from equation 1-59:

$$\Delta G_{reaction} = RT \ln \left[\frac{Q}{(K_a)_{eq.}} \right]$$

or

$$\log (K_a)_{eq.} = 203$$

It should be noted that the same information about a chemical reaction can be obtained with either standard free energy of formation data or standard half-cell potentials. The only difference is that the use of half-cell potentials is limited to redox reactions.

EXAMPLE PROBLEM 4-3: To assess the effect of precipitation of a chemical species involved in an oxidation-reduction reaction on the cell potential, determine the cell potential at a pH of 7 for the oxidation of iron(II) with oxygen, and compare it to the cell potential when iron(III) is in equilibrium with iron(III) hydroxide. The overall cell reaction has the form

$$4Fe^{2+}_{(aq)} + O_{2(g)} + 4H^+_{(aq)} \rightleftharpoons 4Fe^{3+}_{(aq)} + 2H_2O_{(l)}$$

Assume standard state conditions for all chemical species whose concentrations are not specified.

Solution:

1. Obtain the half-cell potentials from Appendix VI:

$$Fe^{3+} + e^- \longrightarrow Fe^{2+} \qquad : E^\circ = +0.771$$
$$O_2 + 4H^+ + 4e^- \longrightarrow 2H_2O \qquad : E^\circ = +1.229$$

2. Write the cell reaction from the half-cell reactions. Ensure that the same number of electrons appear in both half-cell reactions.

Oxidation: $4Fe^{2+} \longrightarrow 4Fe^{3+} + 4e^-$ $\qquad : E^\circ = -0.771$
Reduction: $O_2 + 4H^+ + 4e^- \longrightarrow 2H_2O$ $\qquad : E^\circ = +1.229$

$4Fe^{2+} + O_2 + 4H^+ \rightleftharpoons 4Fe^{3+} + 2H_2O$

3. Compute the standard cell potential from equation 4-23:

$$E_{\text{cell}}^{\circ} = E_{\text{ox}}^{\circ} + E_{\text{red}}^{\circ} = (-0.77) + (+1.23) = +0.46 \text{ V}$$

4. Calculate the cell potential from equation 4-18:

$$E_{\text{cell}} = E_{\text{cell}}^{\circ} - \frac{RT}{nF} 2.3 \log Q$$

$$= (+0.46) - \frac{(1.98)(298)(2.3)}{(4)(23,061)} \log \frac{1}{[10^{-7}]^4}$$

$$= +0.05 \text{ V}$$

5. Develop the solubility product constant expression for iron(III) hydroxide, using data from Appendix IV.

$$4 \times 10^{-38} = [Fe^{3+}][OH^-]^3$$

Substituting for $[OH^-]$ from equation 2-3 gives

$$4 \times 10^{-38} = [Fe^{3+}]\left[\frac{10^{-14}}{10^{-7}}\right]^3$$

Solving for $[Fe^{3+}]$ gives

$$[Fe^{3+}] = 4 \times 10^{-17}$$

6. Compute the cell potential from equation 4-18, using the Fe^{3+} ion concentration given in step 5:

$$E_{\text{cell}} = (+0.46) - \frac{(1.98)(298)(2.3)}{(4)(23,061)} \log \frac{[4 \times 10^{-17}]^4}{[10^{-7}]^4}$$

$$= +1.07 \text{ V}$$

Thus, the presence of iron(III) hydroxide increases the cell potential, which tends to shift the equilibrium position toward the right. Complexation of Fe^{3+} would have the same effect.

4-3 STABILITY DIAGRAMS

An important aspect of redox reactions is that many of them include H^+ or OH^- ions, and in all such cases the half-cell reactions and cell potentials will vary with pH. A number of investigators have presented stability field diagrams (sometimes called *Pourbaix diagrams*) to describe the relationship between E_h (cell potential with respect to the standard hydrogen electrode), pH, and stable chemical species for a particular chemical system. These diagrams have little use in process chemistry, but are discussed here because they are helpful when studying metallic corrosion or redox equilibria in natural water systems.

A stability field diagram is constructed by plotting E_h on the ordinate and pH on the abscissa. Each diagram is constructed for a particular chemical system at a specified temperature and pressure. When standard conditions are specified, the pressure is 1 atm, the temperature is 25°C, and species are at unit activity. Before the construction of a stability field diagram is begun, it is necessary to establish the stability of water within limits of pH and E_h possible in aqueous systems. In this regard, molecular oxygen is generally the strongest oxidizing agent found in natural water systems. Oxidizing agents stronger than molecular oxygen cannot exist for prolonged periods because they will react with water to liberate O_2. Thus, the oxidation limit for the stability field diagram is controlled by this reaction:

$$2H_2O_{(1)} \rightleftharpoons O_{2(g)} + 4H^+_{(aq)} + 4\,e^- \qquad \text{(4-26)}$$

or, using the standard reduction convention form,

$$O_{2(g)} + 4H^+_{(aq)} + 4\,e^- \rightleftharpoons 2H_2O_{(1)} \qquad : E° = +1.23 \quad \text{(4-27)}$$

Reducing agents stronger than molecular hydrogen are generally not found in natural water systems because stronger agents cause the breakdown of water by reduction to form hydroxyl ions and hydrogen gas:

$$2H_2O_{(1)} + 2\,e^- \rightleftharpoons H_{2(g)} + 2OH^-_{(aq)} \qquad : E° = -0.828 \quad \text{(4-28)}$$

The plotting equations used to establish the oxidation and reduction boundary of the stability field diagram can be developed from the Nernst equation (4-18) and the half-reactions represented by equations 4-27 and 4-28.

1. Oxidizing Boundary:
Applying the Nernst equation to equation 4-27 gives

$$E_h = 1.23 - \left[\frac{(1.98)(2.3)(T)}{(4)(23,061)}\right] \log\left[\frac{(H_2O)^2}{(O_2)(H^+)^4}\right]$$

Assuming a 25°C temperature, an activity of 1 for H_2O, and a pressure of 0.21 atm for O_2 (air is 21 % oxygen) and solving for E_h in terms of pH gives

$$E_h = 1.23 - (0.0147)[\log(1) - 4\log(H^+) - \log(0.21)]$$
$$= 1.23 - [0.059\,\text{pH} + 0.01]$$
$$= 1.22 - 0.059\,\text{pH} \qquad \text{(4-29)}$$

2. Reducing Boundary:
Applying the Nernst equation to equation 4-28 gives

$$E_h = -0.828 - \left[\frac{(1.98)(23)(298)}{(2)(23,061)}\right] \log\left[\frac{(H_2)(OH^-)^2}{(H_2O)^2}\right]$$

Assuming an activity of 1 for H_2O, and a pressure of 1 atm for H_2 and solving for E_h in terms of pH gives

$$E_h = -0.828 - (0.029)[\log(1) + 2\log(OH^-)]$$
$$= -0.828 + 0.059\,\text{pOH} \qquad \text{(4-30)}$$

Substituting pK_w-pH for pOH and 14 for pK_w, equation 4-30 takes the form

$$E_h = -0.828 + (0.059)(14\ \text{pH}) = -0.059\ \text{pH} \qquad \textbf{(4-31)}$$

At equilibrium, most aqueous solutions will have E_h and pH values within the boundaries shown in this figure.

Three types of equilibria may be encountered when constructing a stability field diagram. These are

1. Equilibrium state independent of pH. This situation is described by the general equation

$$\text{a OX} + \text{b e}^- \rightleftharpoons \text{c RED}$$

2. Equilibrium state independent of electrical potential. This situation is described by the general equation

$$\text{d A} + \text{f H}^+ \rightleftharpoons \text{g B} + \text{h H}_2\text{O}$$

3. Equilibrium state dependent on pH and electrical potential. This situation is described by the general equation

$$\text{a OX} + \text{f H}^+ + \text{b e}^- \rightleftharpoons \text{c RED} + \text{h H}_2\text{O}$$

Each of the three types of equilibria are illustrated graphically in Figure 4-2.

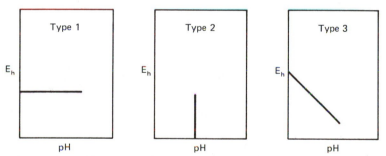

FIGURE 4-2 *Relationship between* E_h *and pH for various types of equilibrium.*

The first step in constructing a stability field diagram is to identify the pertinent reactions which affect the equilibrium state of the system of interest. Keep in mind that these reactions are not limited to redox reactions but should include all reactions which affect the equilibrium state. Once the pertinent reactions have been identified, the necessary plotting equations must be developed. This method will be illustrated by constructing a stability field diagram for iron. For convenience, unit activity will be assumed for all soluble iron species, and the total carbonic species will be taken as $10^{-5}\ M$. It should be understood that if the actual soluble iron concentration were used, a more accurate diagram could be constructed. However, the use of actual concentrations will not alter the shape of the E_h vs. pH diagram to any great extent.

The reactions which will be used in the construction of the stability field diagram for iron are as follows:

1) $Fe^{2+}_{(aq)} + 2\,e^- \rightleftharpoons Fe_{(s)}$ $: E^\circ = -0.44\ V$ **(4-32)**

2) $Fe^{3+}_{(aq)} + e^- \rightleftharpoons Fe^{2+}_{(aq)}$ $: E^\circ = 0.77\ V$ **(4-33)**

3) $Fe(OH)_{3(s)} \rightleftharpoons FeOH^{2+}_{(aq)} + 2OH^-$ $: K = 10^{-24.2}$ **(4-34)**

4) $Fe(OH)_{2(s)} \rightleftharpoons Fe^{2+}_{(aq)} + 2OH^-_{(aq)}$ $: K = 10^{-14.6}$ **(4-35)**

5) $FeCO_{3(s)} \rightleftharpoons Fe^{2+}_{(aq)} + CO^{2-}_{3(aq)}$ $: K = 10^{-10.5}$ **(4-36)**

6) $Fe(OH)_{3(s)} \rightleftharpoons Fe^{3+}_{(aq)} + 3OH^-$ $: K = 10^{-38.6}$ **(4-37)**

7) $Fe(OH)_{3(s)} + H^+_{(aq)} + e^- \rightleftharpoons Fe(OH)_{2(s)} + H_2O_{(l)}$

$: E^\circ = 0.26\ V$ **(4-38)**

8) $Fe(OH)_{3(s)} + 3H^+_{(aq)} + e^- \rightleftharpoons Fe^{2+}_{(aq)} + 3H_2O_{(l)}$

$: E^\circ = 1.04\ V$ **(4-39)**

9) $Fe(OH)_{2(s)} + 2H^+_{(aq)} + 2\,e^- \rightleftharpoons Fe_{(s)} + 2H_2O_{(l)}$

$: E^\circ = -0.049\ V$ **(4-40)**

The plotting equation for each of these equations is developed as follows:

1. *Plotting equation 1:*

   (a) Redox equilibrium reaction:

   $$Fe^{2+}_{(aq)} + 2\,e^- \rightleftharpoons Fe_{(s)} \qquad : E^\circ = -0.44\ V$$

   (b) E_h calculation:

   $$E_h = -0.44 - \frac{(1.98)(298)(2.3)}{(2)(23,061)} \log \frac{[Fe]}{[Fe^{2+}]}$$

   It is assumed that the activity of all soluble iron species is unity, and since the activity of any species in the solid phase is assumed to be unity, the above equation reduces to the form

   $$E_h = -0.44$$

   (c) Plotting equation:

   $$E_h = -0.44 \qquad\qquad \textbf{(4-41)}$$

   Thus, the equilibrium state for equation 4-32 is independent of pH.

2. *Plotting equation 2:*

   (a) Redox equilibrium reaction:

   $$Fe^{3+}_{(aq)} + e^- \rightleftharpoons Fe^{2+}_{(aq)} \qquad : E^\circ = 0.77\ V$$

(b) E_h calculation:

$$E_h = 0.77 - \frac{(1.98)(298)(2.3)}{(1)(23,061)} \log \frac{[Fe^{2+}]}{[Fe^{3+}]} = 0.77$$

(c) Plotting equation:

$$E_h = 0.77 \qquad\qquad \textbf{(4-42)}$$

Thus, the equilibrium state for equation 4-33 is independent of pH.

3. *Plotting equation 3:*

(a) Solubility equilibrium reaction:

$$Fe(OH)_{3\,(s)} \;\rightleftharpoons\; FeOH^{2+}_{(aq)} + 2OH^-_{(aq)} \qquad : K = 10^{-24.2}$$

(b) Equilibrium constant expression:

$$10^{-24.2} = [FeOH^{2+}][OH^-]^2$$

Take the logarithm of both sides of this expression and solve for log $[FeOH^{2+}]$.

$$\log [FeOH^{2+}] = -24.2 + 2\,pOH$$

or

$$\log [FeOH^{2+}] = -24.2 + 2(14\,pH)$$

Since the soluble iron species are assumed to be at unit activity, this equation reduces to

$$0 = -24.2 + 28 - 2\,pH$$

or

$$pH = 1.9$$

(c) Plotting equation:

$$pH = 1.9 \qquad\qquad \textbf{(4-43)}$$

Thus, the equilibrium state for equation 4-34 is independent of E_h.

4. *Plotting equation 4:*

(a) Solubility equilibrium reaction:

$$Fe(OH)_{2\,(s)} \;\rightleftharpoons\; Fe^{2+}_{(aq)} + 2OH^-_{(aq)} \qquad : K = 10^{-14.6}$$

(b) Equilibrium constant expression:

$$10^{-14.6} = [Fe^{2+}][OH^-]^2$$

Take the logarithm of both sides of this expression and solve for log Fe^{2+}.

$$\log [Fe^{2+}] = -14.6 - 2 \log [OH^-]$$

Since Fe^{2+} is assumed to be at unit activity, this equation reduces to

$$0 = -14.6 + 2\,pOH$$

Substitute the term (14-pH) for pOH and solve the resulting equation for pH.

$$pH = 6.7$$

(c)   Plotting equation:

$$pH = 6.7 \qquad\qquad\qquad \textbf{(4-44)}$$

Thus, the equilibrium state for equation 4-35 is independent of E_h.

5.   *Plotting equation 5:*

(a)   Solubility equilibrium reaction:

$$FeCO_{3\,(s)} \; \rightleftharpoons \; Fe^{2+}_{(aq)} + CO^{2-}_{3\,(aq)} \qquad : K = 10^{-10.5}$$

(b)   Equilibrium constant expression:

$$10^{-10.5} = [Fe^{2+}][CO_3^{2-}]$$

Take the logarithm of both sides of this expression and solve for log Fe^{2+}.

$$\log [Fe^{2+}] = -10.5 - \log [CO_3^{2-}]$$

or

$$0 = -10.5 - \log [CO_3^{2-}]$$

Substitute $\alpha_2 C_T$ for $[CO_3^{2-}]$ and 10^{-5} for C_T and solve for α_2.

$$0 = -5.5 - \log \alpha_2$$

or

$$\alpha_2 = 0.0000032$$

when $\alpha_2 = 0.0000032$ the pH $\simeq 5.6$. Hence,

$$pH = 5.6$$

(c)   Plotting equation:

$$pH = 5.6 \qquad\qquad\qquad \textbf{(4-45)}$$

Thus, the equilibrium state for equation 4-36 is independent of E_h.

6.   *Plotting equation 6:*

(a)   Solubility equilibrium reaction:

$$Fe(OH)_{3\,(s)} \; \rightleftharpoons \; Fe^{3+}_{(aq)} + 3OH^{-}_{(aq)} \qquad : K = 10^{-38.6}$$

(b) Equilibrium constant expression:

$$10^{-38.6} = [Fe^{3+}][OH^-]^3$$

Take the logarithm of both sides of this expression and solve for $\log [Fe^{3+}]$.

$$\log [Fe^{3+}] = -38.6 - 3 \log [OH^-]$$

or

$$0 = -38.6 + 3 \text{ pOH}$$

Substitute the term (14-pH) for pOH and solve the resulting equation for pH.

$$pH = 1.1$$

(c) Plotting equation:

$$pH = 1.1 \qquad\qquad \textbf{(4-46)}$$

Thus, the equilibrium state of equation 4-37 is independent of E_h.

7. *Plotting equation 7:*

(a) Redox equilibrium reaction:

$$Fe(OH)_{3\,(s)} + H^+_{(aq)} + e^- \;\rightleftharpoons\; Fe(OH)_{2\,(s)} + H_2O_{(l)} \quad : E^\circ = 0.262 \text{ V}$$

(b) E_h calculation:

$$E_h = 0.262 - \frac{(1.98)(298)(2.3)}{(1)(23,061)} \log \frac{[Fe(OH)_2][H_2O]}{[Fe(OH)_3][H^+]}$$

or

$$E_h = 0.262 - 0.059 \text{ pH}$$

(c) Plotting equation:

$$E_h = 0.262 - 0.059 \text{ pH} \qquad\qquad \textbf{(4-47)}$$

Thus, the equilibrium state of equation 4-38 is dependent on both pH and electrical potential.

8. *Plotting equation 8:*

(a) Redox equilibrium reaction:

$$Fe(OH)_{3\,(s)} + 3H^+_{(aq)} + e^- \;\rightleftharpoons\; Fe^{2+}_{(aq)} + 3H_2O_{(l)} \quad : E^\circ = 1.04 \text{ V}$$

(b) E_h calculation:

$$E_h = 1.04 - \frac{(1.98)(298)(2.3)}{(1)(23,061)} \log \frac{[Fe^{2+}][H_2O]^3}{[Fe(OH)_3][H^+]^3}$$

or

$$E_h = 1.04 - 0.18 \text{ pH}$$

(c) Plotting equation:

$$E_h = 1.04 - 0.18 \text{ pH} \qquad \text{(4-48)}$$

Thus, the equilibrium state of equation 4-39 is dependent on both pH and electrical potential.

9. *Plotting equation 9:*

(a) Redox equilibrium reaction:

$$Fe(OH)_{2(s)} + 2H^+_{(aq)} + 2\,e^- \; \rightleftharpoons \; Fe_{(s)} + 2H_2O_{(l)} \quad : E_0 = -0.049 \text{ V}$$

(b) E_h calculation:

$$E_h = -0.049 - \frac{(1.98)(298)(2.3)}{(2)(23{,}061)} \log \frac{[Fe][H_2O]^2}{[Fe(OH)_2][H^+]^2}$$

or

$$E_h = -0.049 - 0.059 \text{ pH}$$

(c) Plotting equation:

$$E_h = -0.049 - 0.059 \text{ pH} \qquad \text{(4-49)}$$

Thus, the equilibrium state of equation 4-40 is dependent on both pH and electrical potential.

The plotting equations for species separation are summarized as follows:

Species Separated	*Plotting equation*
$Fe^{2+}_{(aq)} : Fe_{(s)}$	$E_h = -0.44$
$Fe^{3+}_{(aq)} : Fe^{2+}_{(aq)}$	$E_h = 0.77$
$FeOH^{2+}_{(aq)} : Fe(OH)_{3(s)}$	$\text{pH} = 1.9$
$Fe(OH)_{2(s)} : Fe^{2+}_{(aq)}$	$\text{pH} = 6.7$
$FeCO_{3(s)} : Fe^{2+}_{(aq)}$	$\text{pH} = 5.6$
$Fe(OH)_{3(s)} : Fe^{3+}_{(aq)}$	$\text{pH} = 1.1$
$Fe(OH)_{3(s)} : Fe(OH)_{2(s)}$	$E_h = 0.262 - 0.059 \text{ pH}$
$Fe(OH)_{3(s)} : Fe^{2+}_{(aq)}$	$E_h = 1.04 - 0.18 \text{ pH}$
$Fe(OH)_{2(s)} : Fe_{(s)}$	$E_h = -0.049 - 0.059 \text{ pH}$

A first-trial construction of the stability field diagram is shown in Figure 4-3. The

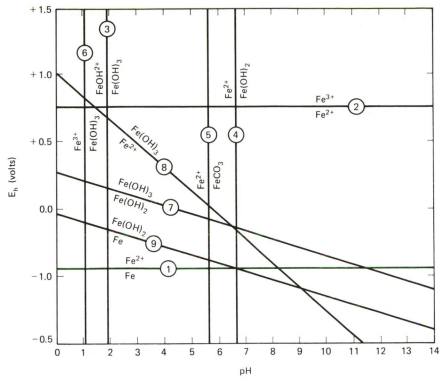

FIGURE 4-3 *First-trial* E_h *vs pH diagram.*

following rules are helpful when deciding on which side of a specific boundary to place a particular chemical species:

1. Iron species of higher oxidation numbers are above iron species with lower oxidation numbers.
2. Aqueous phase iron species lie to the left of solid phase iron species.
3. Solid carbonate species lie to the left of solid hydroxide species.

To reduce Figure 4-3 to a finished stability field diagram, consider that

1. The $FeOH^{2+}$ species increases the solubility of the $Fe(OH)_3$ solid phase.
2. The presence of carbonate causes $FeCO_3$ to precipitate at a lower pH than $Fe(OH)_2$.

The construction lines (lines 1–9) are then partially erased to give the proper domain for each iron species. The completed diagram is shown in Figure 4-4, where the dashed lines represent the stability limits for water. This figure shows that elemental iron is thermodynamically unstable in contact with water; i.e., Fe causes the evolution of H_2 from water, which is the reason why iron will rust in the presence of water. Figure 4-4 also indicates that in waters where the pH is greater than 2, the predominant iron

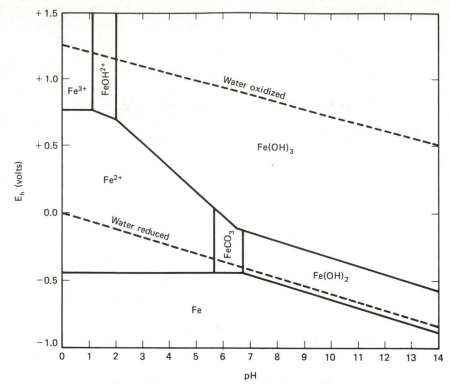

FIGURE 4-4 E_h *vs* pH *diagram for the iron system.*

species will be Fe^{2+}, $FeCO_3$, $Fe(OH)_2$, and $Fe(OH)_3$. Furthermore, above pH 5 in waters containing an appreciable amount of dissolved oxygen (having a relatively large positive E_h value), $Fe(OH)_3$ will be the predominant iron species. If soluble iron is present in such water, it will be in the form of a complex.

4-4 MEASURING REDOX POTENTIALS

According to the Nernst equation, the cell potential or redox potential of a galvanic cell depends on the activities (concentrations in infinitely dilute solutions) of the reduced chemical species and oxidized chemical species in the solution. The Nernst equation also makes it possible in certain situations to relate the cell potential of a galvanic cell to the activity of a single chemical species. This type of use is known as *direct potentiometry.*

Ion-Selective Electrodes

The galvanic cell used for making potentiometric measurements is comprised of two electrodes. One of the electrodes is called a *reference electrode* and the other an *indicator electrode*. The half-cell potential of the reference electrode remains insensitive to changes in solution composition during the measurement, whereas the response of the indicator electrode is dependent on changes in the solution environment.

Probably the most commonly used reference electrode is the *saturated calomel electrode*, which is actually an electrode and a salt bridge encased in a single unit. The inner tube of the unit contains mercury metal and calomel (Hg_2Cl_2) paste which are in contact with a saturated potassium chloride solution. Contact between the test solution and the salt bridge (the saturated KCl solution) is made through an asbestos fiber. The external circuit connection is made with an amalgamated platinum wire from the inner tube. The calomel electrode reaction is given by the equation

$$Hg_2Cl_{2(s)} + 2\,e^- \;\rightleftharpoons\; 2Hg_{(s)} + 2Cl^-_{(aq)} \qquad : E° = 0.242\ V \quad \textbf{(4-50)}$$

The silver-silver chloride electrode in $1\,M$ or $4\,M$ KCl solution are also popular reference electrodes. The half-cell potentials of these reference electrodes are indicated in Table 4-1.

TABLE 4-1 *Potential of the silver—silver chloride reference electrodes relative to the standard hydrogen electrode at 25°C.*

Half-Cell Reaction	Standard Potential
$AgCl + e^- \longrightarrow Ag + Cl^-$ (1 M KCl)	$E° = +0.235\ V$
$AgCl + e^- \longrightarrow Ag + Cl^-$ (4 M KCl)	$E° = +0.199\ V$

pH determination is the most common potentiometric measurement, and normally a *glass electrode* is used as the indicator electrode for this measurement; This electrode is composed of a silver-silver chloride reference electrode inside a tube which is filled with a solution of fixed pH (usually $0.1\,M$) hydrochloric acid. The bottom of the tube is a thin-walled bulb of special glass which is sensitive to hydrogen ion activity. The potential of the silver-silver chloride reference electrode remains constant, but a potential difference develops across the glass membrane. The magnitude of the potential difference depends on the difference between the hydrogen ion activity of the test solution and the HCl solution inside the electrode.

It has been found experimentally that the electrical potential of the glass electrode varies linearly with changes in the pH of the test solution. This means that a plot of E_{cell} vs. pH will produce a straight line of slope $\Delta E_{cell}/\Delta pH$. The slope $\Delta E_{cell}/\Delta pH$ will have a value of 0.059 V per pH unit at 25°C if the glass electrode and the potentiometer circuit are operating properly.

The valuable property of the glass electrode is its highly selective response to hydrogen ions. In recent years extensive research has resulted in a better understanding of the selectivity of glass membranes as well as the development of membranes which are selective of ions other than hydrogen. As a result, membrane electrodes are now available for ions such as NO_3^-, Na^+, F^-, and many others. The interested reader is referred to Skoog and West (1971) and Durst (1969) for detailed discussions of membrane electrodes.

Redox Potentials in Biological Systems

Most wastewaters are extremely complex solutions. Normally, during biological treatment a variety of chemical and biological reactions are occurring simultaneously, which means that there are several redox pairs present in the medium. According to Kjaergaard (1977), the redox reactions occurring in such a system can be formulated as

$$\sum_{i=1}^{M_1} OX_{1i} + b_1 e^- \rightleftharpoons \sum_{i=1}^{M_1} Red_{1i} \qquad (4\text{-}51)$$

$$\sum_{j=1}^{M_2} Red_{2j} \rightleftharpoons \sum_{j=1}^{M_2} OX_{2j} + b_2 e^- \qquad (4\text{-}52)$$

where b_1 and b_2 are the numbers of electrons needed for the reduction of

$$\sum_{i=1}^{M_1} OX_{1i} \quad \text{and} \quad \sum_{j=1}^{M_2} OX_{2j}$$

respectively. This suggests that the redox potential measured for such a system is related to all the pairs of reducible and oxidizable compounds found in the medium.

The redox potential of a biological system can be measured by an electrode because the potential of an immersed electrode is dependent on the activity of the electrons in the solution (the higher the electron activity, the lower the redox potential). The saturated calomel electrode is commonly used as the reference electrode for redox potential measurements in biological systems. To complete the circuit of the galvanic cell, an indicator electrode is required. Inert platinum electrodes are generally used as indicator electrodes. The inert electrode does not take part in the redox reactions which occur during treatment but acts only as a neutral acceptor or donor of electrons to the ions in solution. According to the Nernst equation, the redox potential which is measured is proportional to the activity ratios of the oxidized and reduced chemical forms existing at equilibrium:

$$E_{\text{cell}} \propto \ln \left[\frac{\text{oxidized material}}{\text{reduced material}} \right] \qquad (4\text{-}53)$$

The redox potentials measured in anaerobic environments normally have negative values, whereas the redox potentials measured in aerobic environments normally have positive values. For example, Molof (1960) observed that under normal digestion conditions the redox potential measured in anaerobic digesters containing municipal wastewater sludges remained fairly constant within an operating range of -460 to -530 mV. It was also observed that a movement toward the positive millivolt range indicated inhibited or decreasing levels of digester performance.

Redox Potentials in Chemical Systems

In many chemical systems a nonspecific redox measurement is made by using the same kind of galvanic cell that is employed in measuring the redox potential of biological systems, e.g., an inert platinum or gold indicator electrode and a saturated calomel or silver-silver chloride reference electrode. This type of redox potential measurement is nonspecific in that it does not indicate the presence or absence of a particular ion but, rather, indicates the activity ratio of total oxidizing species present to that of total reducing species present.

For any nonspecific galvanic cell redox potential measurement at 25°C the Nernst equation for any redox reaction may be written as (assuming ideal conditions)

$$E_{meter} = E^{\circ}_{adjusted} - \frac{0.0591}{n} \log Q \qquad (4\text{-}54)$$

where $E_{meter} = $ potential read on a millivolt meter

$E^{\circ}_{adjusted} = $ the standard cell potential adjusted for the reference electrode used in the galvanic cell.

The value of $E^{\circ}_{adjusted}$ to be used in equation 4-54 is indicated by the following relationship:

$$E^{\circ}_{adjusted} = (E^{\circ}_{ox} + E^{\circ}_{red}) - E^{\circ}_{reference} \qquad (4\text{-}55)$$

or

$$E^{\circ}_{adjusted} = E^{\circ}_{cell} - E^{\circ}_{reference} \qquad (4\text{-}56)$$

EXAMPLE PROBLEM 4-4: What meter reading (in millivolts) will indicate that oxygenation of iron(II) is 95% complete if a galvanic cell composed of a platinum indicator electrode and a 1 M KCl silver-silver chloride reference electrode is used to monitor the reaction. Assume that the reaction pH is held constant at 7, the oxygen partial pressure is 0.2 atm, and the initial iron(II) is 10^{-4} mole/ℓ. The overall cell reaction is

$$4Fe^{2+}_{(aq)} + O_{2(g)} + 4H^{+}_{(aq)} \rightleftharpoons 4Fe^{3+}_{(aq)} + 2H_2O_{(l)}$$

Also assume that iron(III) is in equilibrium with iron(III) hydroxide.

Solution:

1. Obtain the half-cell potentials from Appendix VI. See step 1, example problem 4-3.

2. Write the cell reaction from the half-cell reactions. See step 2, example problem 4-3.

3. Compute the standard cell potential from equation 4-23. See step 3, example problem 4-3.

4. Determine $E^{\circ}_{adjusted}$ from equation 4-56:
$$E^{\circ}_{adjusted} = +0.46 - 0.235 = +0.225 \text{ V}$$

5. Compute the equilibrium iron(III) concentration based on the saturated equilibrium reaction of iron(III) hydroxide. See Step 5, example problem 4-3.

6. Calculate the concentration of iron(II) when the oxygenation reaction is 95% complete:
$$[Fe^{2+}] = (1 \times 10^{-4}) - (0.95)1 \times 10^{-4} = 5 \times 10^{-6} \text{ mole/}\ell$$

7. Compute the expected meter reading from equation 4-54:

$$E_{meter} = +0.225 - \frac{0.0591}{4} \log \left[\frac{[4 \times 10^{-17}]^4}{[10^{-7}]^4 [5 \times 10^{-6}]^4 [0.21]} \right]$$

$$= +0.457 \text{ V}$$

or

$$E_{meter} = +457 \text{ mV}$$

It is important to remember that the computed E_{meter} value is based on ideal conditions and a solution temperature of 25°C. The actual value recorded in complex waste treatment solutions will differ from this value.

PROBLEMS

4-1. Compute the equilibrium constant for the ionization of water at 25°C.

$$2H_2O + 2e^- \rightleftharpoons H_2 + 2OH^- \qquad : E° = -0.828$$

$$2H^+ + 2e^- \rightleftharpoons H_2 \qquad : E° = 0.000$$

4-2. Is the following reaction feasible under standard conditions?

$$Fe_{(s)} + 2Fe^{3+}_{(aq)} \rightleftharpoons 3Fe^{2+}_{(aq)}$$

4-3. Determine the feasibility of the following reaction under standard conditions:

$$2Fe^{2+}_{(aq)} + Cl_{2(g)} \rightleftharpoons 2Fe^{3+}_{(aq)} + 2Cl^-_{(aq)}$$

4-4. Construct the E_h vs. pH diagram for the Pb-H$_2$O system under standard conditions.

$$PbO_{(s)} + 2H^+_{(aq)} \rightleftharpoons Pb^{2+}_{(aq)} + H_2O_{(l)} \qquad : (K_a)_{eq.} = 10^{12.7}$$

$$Pb^{2+}_{(aq)} + 2e^- \rightleftharpoons Pb_{(s)} \qquad : E° = -0.126$$

$$PbO_{2(s)} + 4H^+_{(aq)} + 2e^- \rightleftharpoons 2H_2O_{(l)} + Pb^{2+}_{(aq)} \qquad : E° = +1.455$$

$$PbO_{(s)} + 2H_2O_{(l)} \rightleftharpoons Pb(OH)^-_{3(aq)} + H^+_{(aq)} \qquad : (K_a)_{eq.} = 10^{-15.4}$$

4-5. Calculate the equilibrium constant for the following reaction at 25°C:

$$5Fe^{2+} + MnO_4^- + 8H^+ \rightleftharpoons 5Fe^{3+} + Mn^{2+} + 4H_2O$$

Assume unit activity for all species except H$^+$, for which a value of 10^{-7} M is assumed.

4-6. Construct an E_{meter} vs. percent reaction completion diagram for the reaction outlined in example problem 4-5. To construct this diagram, calculate E_{meter} values at 0, 20, 30, 50, 70, 90, 95, 98 and 99%.

4-7. Repeat Problem 4-6, using a reaction pH of 5.

4-8. A solution contains 10^{-3} mole of iron(II) ions. Iron is to be converted from iron(II) to iron(III) by the addition of a solution containing cerium(IV) ions. The appropriate half-reactions are

$$Fe^{3+} + e^- \longrightarrow Fe^{2+} \qquad : E° = +0.771 \text{ V}$$

$$Ce^{4+} + e^- \longrightarrow Ce^{3+} \qquad : E° = +1.61 \text{ V}$$

The overall reaction has the form:

$$Fe^{2+} + Ce^{4+} \rightleftharpoons Fe^{3+} + Ce^{3+}$$

A platinum indicator electrode and a calomel reference electrode are used for the galvanic cell. Construct an E_{meter} vs. moles of Ce^{4+} ions added, using Ce^{4+} ion concentrations of 0.1×10^{-3}, 0.3×10^{-3}, 0.5×10^{-3}, 0.7×10^{-3}, and 0.9×10^{-3} moles.

REFERENCES

BRESCIA, F., ARENTS, J., MEISLICH, H., and TURK, A., *Fundamentals of Chemistry: A Modern Introduction*, Academic Press, Inc., New York (1970).

DURST, R., *Ion-Selective Electrodes*, National Bureau of Standards Special Publication No. 314, U.S. Government Printing Office, Washington, DC (1969).

FISCHER, R.B., and PETERS, D.G., *Chemical Equilibrium*, W. B. Saunders Company, Philadelphia, (1970).

JURINAK, J.J., "Chemistry of Aquatic Systems," Lecture Notes, Utah State University (1976).

KJAERGAARD, L., "The Redox Potential: Its Use and Control in Biotechnology," in *Advances in Biochemical Engineering*, 7, [Ed.] T. K. GHOSE, A. FIECHTER, and N. BLAKEBROUGH, Springer-Verlag New York, Inc., New York (1977).

MOELLER, T., and O'CONNOR, R., *Ions in Aqueous Systems*, McGraw-Hill Book Company, New York (1972).

MOLOF, A.H., "A Study of Oxidation-Reduction Potentials Applied to Sewage Sludge Digestion," Ph.D. Dissertation, University of Michigan, Ann Arbor, MI (1960).

ROCHOW, E.G., FLECK, G., and BLACKBURN, T.R., *Chemistry: Molecules That Matter*, Holt, Rinehart and Winston, New York (1974).

SKOOG, D.A., and WEST, D.M., *Principles of Instrumental Analysis*, Holt, Rinehart and Winston, New York (1971).

WHITE, G.C., *Disinfection of Wastewater and Water for Reuse*, Van Nostrand Reinhold Company, New York (1978).

WILLIAMS, V.R., and WILLIAMS, H.B., *Basic Physical Chemistry for the Life Sciences*, W.H. Freeman & Company Publishers San Francisco, (1967).

5

FUNDAMENTALS OF PROCESS KINETICS

5-1 INTRODUCTION

Two basic questions require answers when one is concerned with the reactions which occur in a particular chemical process. These questions are

1. Where is the relative equilibrium position of the reaction?
2. At what rate does the reaction approach equilibrium?

Chemical thermodynamics answers the first question, whereas chemical kinetics answers the second.

5-2 REACTION RATES

The rate of a chemical change is usually described in terms of the decrease in the concentration of reactant, $-dC_r/dt$, or the increase in the concentration of product, $+dC_p/dt$ (in process chemistry the former term is normally used). The rate of chemical change defined in this manner does not apply to the entire chemical reaction but only to one particular chemical species involved in the reaction. For example, consider the following reaction:

$$NH_4^+ + 1.5O_2 \longrightarrow NO_3^- + 4H^+ \qquad \textbf{(5-1)}$$

The relative rate of change for each species is defined by the molar coefficient in the balanced chemical equation (e.g., mole for mole, the hydrogen ion is being formed four times faster than the ammonium ion is disappearing). Thus, the specific numerical value for the rate of the chemical change depends on whether it is expressed in terms of ammonia, oxygen, nitrate, or hydrogen. Because of this, there are researchers in the field who suggest that the overall rate of chemical change be defined as the rate of change in concentration of a particular chemical species divided by its molar coefficient in the balanced chemical equation. Following this suggestion, the rate of change for each species shown in equation 5-1 would be as follows:

$$\begin{bmatrix} \text{rate of change in} \\ \text{ammonium ion concentration} \end{bmatrix} = \frac{-dC_{NH_4^+}/dt}{1}$$

$$\begin{bmatrix} \text{rate of change in} \\ \text{oxygen concentration} \end{bmatrix} = \frac{-dC_{O_2}/dt}{1.5}$$

$$\begin{bmatrix} \text{rate of change in} \\ \text{nitrate ion concentration} \end{bmatrix} = \frac{+dC_{NO_3^-}/dt}{1}$$

$$\begin{bmatrix} \text{rate of change in} \\ \text{hydrogen ion concentration} \end{bmatrix} = \frac{+dC_{H^+}/dt}{4}$$

Note that most kinetic data are analyzed on the basis of the rate of change for a particular chemical species and not in terms of the rate for the overall chemical change.

Rate and Order

Chemical reactions may be classified in one of the following ways:

1. On the basis of the number of molecules that must react to form the reaction product.

2. On a kinetic basis by reaction order.

It is the latter classification which is useful in describing the kinetics of many chemical treatment processes.

The concept of classifying reactions on the basis of order is generally applicable to the following types of reactions:

1. Essentially irreversible reactions, where a stoichiometric combination of reactant chemical species is totally or almost totally converted to product (see Figure 5-1).

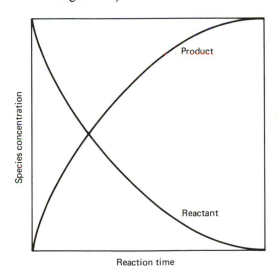

FIGURE 5-1 *Material balance for reaction going to completion.*

2. The initial stages of most reversible reactions, where the reactant concentration-to-product concentration ratio is large.

3. Reversible reactions, where the position of equilibrium lies far to the right, i.e., where the equilibrium product concentration-to-equilibrium reactant concentration ratio is large.

A large number of reactions encountered in process chemistry will be one of these three types.

For a particular chemical reaction, the order, n, with respect to a specific reactant species, A, may be defined by equation 5-2 in terms of the concentration of A, C_A, and its rate of change with respect to time, $-dC_A/dt$:

$$-\frac{dC_A}{dt} = K(C_A)^n \tag{5-2}$$

where K is a proportionality constant and is referred to as the *rate constant*. Taking the logarithm of both sides of equation 5-2 one obtains

$$\log\left[\frac{-dC_A}{dt}\right] = \log K + n \log C_A \tag{5-3}$$

By applying equation 5-3, experimental results may be interpreted to establish a reaction order and rate. For any constant order reaction, if the log of the instantaneous rate of change of reactant concentration at any time is plotted as a function of the log of the reactant concentration at that instant, a straight line will result and the slope of the line will be the order of the reaction (see Figure 5-2). The zero-order reaction results in a horizontal line, and the rate of reaction is concentration independent or the same at any reactant concentration. For the first-order reaction, the rate of reaction is directly proportional to the reactant concentration, and with second-order equations the rate is proportional to the concentration squared. Fractional reaction orders are

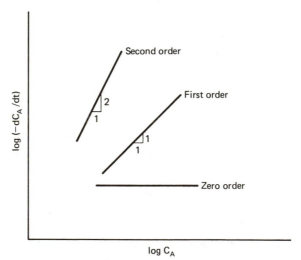

FIGURE 5-2 *Determining reaction order by log plotting.*

possible, but for the solution of many rate problems an integer value for the reaction order is determined or assumed. With this condition, a more detailed evaluation of integer order rate equations can be made as a function of reaction elapsed time.

In most situations it is difficult to determine numerical values for the instantaneous rate of change of reactant concentration at specific times. To circumvent this problem, Swinbourne (1971) noted that at the midpoint of the time interval, Δt, $dC/dt \simeq \Delta C/\Delta t$. If $\bar{C}$ represents the midpoint concentration for the concentration interval, ΔC, equation 5-3 may be represented in the following form (Swinbourne, 1971):

$$\log\left(\frac{-\Delta C_A}{\Delta t}\right) \simeq \log K + n \log \bar{C}_A \tag{5-4}$$

If the reactant concentrations are measured at equal time intervals in a batch reactor, t is a constant, and equation 5-4 reduces to

$$\log(-C_A) \simeq \text{constant} + n \log \bar{C}_A \tag{5-5}$$

A reaction order determined by using either equations 5-4 or 5-5 is termed *the order with respect to time*.

EXAMPLE PROBLEM 5-1: Using the experimental measurements obtained from a batch reactor and given below, determine the reaction order.

Time (min)	Reactant Concentration, C_A (mg/ℓ)
10	132
22	68
31	42
40	26

Solution:

1. Arrange the experimental data in tabular form suitable for plotting.

t	C_A	ΔC_A	Δt	$-(\Delta C_A/\Delta t)$	$\bar{C}_A$	$\log(-\Delta C_A/\Delta t)$	$\log(\bar{C}_A)$
10	132						
		−64	12	5.4	100	0.73	2.00
22	68						
		−26	9	2.9	55.0	0.46	1.74
31	42						
		−16	9	1.8	34.0	0.25	1.53
40	26						

2. Plot data according to equation 5-4:
   The slope of the linear trace has a value of 1, indicating a first-order reaction.

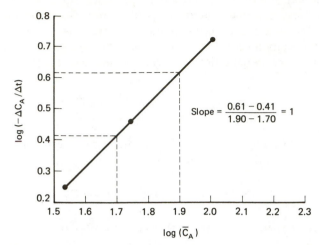

The order of a reaction may also be determined by plotting the logarithm of the initial reaction rate vs. the logarithm of the initial reactant concentration for a number of different initial concentrations of reactant. The order of the reaction is given by the slope of the linear trace. A reaction order determined by this method is termed *the order with respect to concentration.*

EXAMPLE PROBLEM 5-2: Using the experimental measurements obtained from a batch reactor and tabulated as follows, determine the reaction order.

Time (min)	C_A: Run 1 (mg/ℓ)	C_A: Run 2 (mg/ℓ)	C_A: Run 3 (mg/ℓ)	C_A: Run 4 (mg/ℓ)
0	139.0	233.6	330.8	486.4
1	133.6	224.8	318.2	468.6

Solution:

1. Arrange the experimental data in tabular form suitable for plotting.

Run	$-(\Delta C_A/\Delta t)_0$	$\log(-\Delta C_A/\Delta t)_0$	$\log(C_A)_0$
1	5.4	0.73	2.14
2	8.8	0.94	2.37
3	12.6	1.10	2.52
4	17.8	1.25	2.68

2. Plot log $(-\Delta C_A/\Delta t)_0$ vs. log $(C_A)_0$, and determine the slope of the linear trace:

    The slope of the linear trace has a value of 0.95 or approximately 1, indicating a first-order reaction.

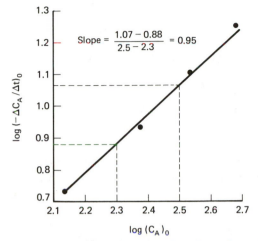

Swinbourne (1971) noted that for a given reaction the order with respect to time is not always the same as the order with respect to concentration. For example, when there is autocatalysis the order with respect to concentration will be greater than the order with respect to time (the concept of autocatalysis is discussed in a later section).

Zero-Order Reactions

Zero-order reactions are those reactions which proceed at a rate independent of the concentration of any reactant. As an example, consider the conversion of a single reactant to a single product.

$$A \longrightarrow P$$
$$\text{reactant} \qquad \text{product}$$

If such a conversion follows zero-order kinetics, the rate of disappearance of A is described by the rate equation

$$-\frac{d[A]}{dt} = K[A]^0 = K$$

where $\qquad -\dfrac{d[A]}{dt} =$ rate of disappearance of A

$\qquad\qquad K =$ reaction rate constant.

If C represents the concentration of A at any time, t, then the rate equation can be expressed as

$$-\frac{dC}{dt} = K \qquad\qquad\qquad \textbf{(5-6)}$$

where $-\dfrac{dC}{dt}$ = rate of change in concentration of A with time (mass volume^{-1} time^{-1})

K = reaction rate constant, (mass volume^{-1} time^{-1}).

Integrating equation 5-6 gives the formulation

$$C = -Kt + \text{constant of integration} \tag{5-7}$$

The constant of integration is evaluated by letting $C = C_0$ at $t = 0$, which implies that

$$C_0 = \text{constant of integration}$$

and shows that the integrated rate law has the form

$$C - C_0 = -Kt \tag{5-8}$$

A plot of concentration (obtained from a batch reactor) vs. time for a zero-order reaction is illustrated in Figure 5-3. Note that the response is linear when the plot is made on arithmetic paper.

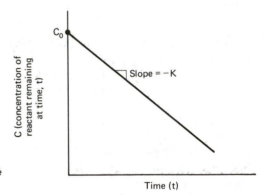

FIGURE 5-3 A plot of the course of a zero-order reaction.

First-Order Reactions

First-order reactions are those reactions which proceed at a rate directly proportional to the concentration of one reactant. Since the rate of the reaction depends on the concentration of the reactant and since the concentration of the reactant changes with time, an arithmetic plot of the variation in the concentration of the reactant with time will not give a linear response as it did for a zero-order reaction. Such a graph is presented in Figure 5-4.

Again, considering the conversion of a single reactant to a single product,

$$A \longrightarrow P$$
$$\text{reactant} \quad \text{product}$$

If first-order kinetics are followed, the rate of disappearance of A is described by the rate equation

$$-\dfrac{dC}{dt} = K(C)^1 = KC \tag{5-9}$$

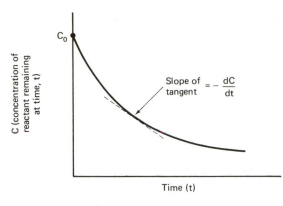

FIGURE 5-4 *An arithmetic plot of the course of a first-order reaction.*

where $-dC/dt$ = rate of change in the concentration of A with time (mass volume^{-1} time^{-1})

C = concentration of A at any time, t (mass volume^{-1})

K = reaction rate constant (time^{-1}).

Integrating equation 5-9 and letting $C = C_0$ at $t = 0$ gives an integrated rate law of the form

$$\ln\left[\frac{C_0}{C}\right] = Kt \qquad \textbf{(5-10)}$$

or, in the more familiar form,

$$\log\left[\frac{C_0}{C}\right] = \frac{Kt}{2.3} \qquad \textbf{(5-11)}$$

Equation 5-11 suggests that a plot of log C (obtained from a batch reactor) vs. time for a first-order reaction will give a linear trace as shown in Figure 5-5.

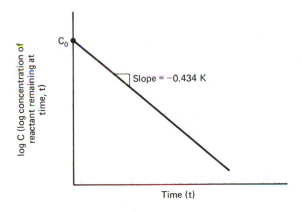

FIGURE 5-5 *A semi-log plot of the course of a first-order reaction.*

Second-Order Reactions

Reaction order is defined as the sum of the exponents of the concentrations of reactants in the differential rate law. The numerical value of a particular exponent indicates the order of the reaction with respect to the associated chemical species.

For example, consider the following rate law:

$$-\frac{dC_A}{dt} = KC_A(C_B)^2 \tag{5-12}$$

The reaction is third-order, since the sum of the exponents is three. However, on the basis of individual components, the reaction is first-order with respect to reactant A and second-order with respect to reactant B, since the exponent of the B concentration is 2. Thus, with regard to a reaction involving a single reactant, a second-order reaction will proceed at a rate proportional to the second power of the reactant concentration. This suggests that, for the second-order reaction describing the conversion of a single reactant to a single product, two molecules of reactant A will react to yield product P.

$$2A \quad \longrightarrow \quad P \tag{5-13}$$
$$\text{reactant} \qquad \text{product}$$

Hence, the rate of disappearance of A for the reaction described by equation 5-13 is defined by the differential rate law

$$-\frac{dC_A}{dt} = K(C_A)^2 \tag{5-14}$$

where K = reaction rate constant, mass^{-1} volume time^{-1}.

The integrated rate law for a second-order reaction has the form

$$\frac{1}{C} - \frac{1}{C_0} = Kt \tag{5-15}$$

Figure 5-6 indicates that an arithmetic plot of $1/C$ vs. time will give a linear trace, the slope of which yields the value of K.

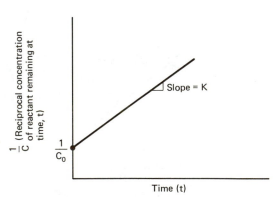

FIGURE 5-6 *A plot of the course of a second-order reaction.*

For a given set of experimental values of C obtained from a batch reactor system and t, equations 5-8, 5-11, and 5-15 can be used to test for a particular reaction order. This is accomplished by making the appropriate concentration vs. time plot and noting any deviation from linearity.

EXAMPLE PROBLEM 5-3: Glucose was added to a batch culture of micro-organisms and removal was measured over time. The following data were obtained:

Glucose Concentration Measured as COD (mg/ℓ)	Time (min)
180	0
155	5
95	12
68	22
42	31
26	40

Determine the reaction order of the removal process by curve fitting.

Solution:

1. Make the appropriate concentration vs. time plots and note any deviation from linearity:

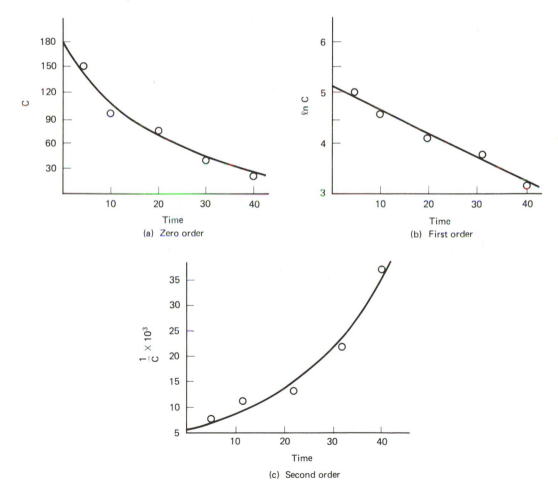

(a) Zero order

(b) First order

(c) Second order

Because curve (b) gives the best fit to the experimental data, glucose removal is assumed to follow first-order kinetics.

Although any differential rate law may be integrated to give a graphable time-dependent function of concentration which may be tested for fit against the data from a single progress curve, care must be exercised in the use of integrated rate law plots to determine reaction orders. There will always be some experimental error in the data, and small but important deviations from linearity can be obscured by such random error of the measurements. This is especially true if the data are taken only over a small part of the extent of the reaction (e.g., the first 10%) where the rate may be very nearly constant and the data can be deceptively well-fitted by any of the simple integrated equations.

It is possible to have a reaction in which two different compounds A and B react mole for mole to form product P.

$$\underset{\text{reactant}}{A} + \underset{\text{reactant}}{B} \longrightarrow \underset{\text{product}}{P} \tag{5-16}$$

The rate of disappearance of either A or B is described by the differential rate law

$$-\frac{dC_A}{dt} = -\frac{dC_B}{dt} = KC_A C_B \tag{5-17}$$

For the boundary conditions, $C_A = C_{A_0}$ and $C_B = C_{B_0}$ at $t = 0$, integration of equation 5-17 yields

$$\frac{1}{C_{A_0} - C_{B_0}} \ln\left[\frac{C_{B_0} C_A}{C_{A_0} C_B}\right] = Kt \tag{5-18}$$

or

$$\log\left(\frac{C_A}{C_B}\right) = \left[K\left(\frac{C_{A_0} - C_{B_0}}{2.303}\right)\right] t - \log\left(\frac{C_{B_0}}{C_{A_0}}\right) \tag{5-19}$$

A plot of $\log(C_A/C_B)$ vs. t will give a linear trace with a slope numerically equal to $K(C_{A_0} - C_{B_0})/2.303$. Equation 5-19 is valid only when the initial concentrations of reactants A and B are not equal and the data are obtained from a batch reactor system. If $C_{A_0} = C_{B_0}$, the differential rate law described by equation 5-17 is mathematically equivalent to the differential rate law described by equation 5-14. For this situation the integrated rate law has the same form as equation 5-15.

The stoichiometry of certain second-order reactions is such that reactants A and B do not react mole for mole. For example, consider the case where one mole of A reacts with b moles of B.

$$\underset{\text{reactant}}{A} + \underset{\text{reactant}}{bB} \longrightarrow \underset{\text{product}}{P} \tag{5-20}$$

The rate of disappearance of either A or B is described by the differential rate law

$$-\frac{dC_A}{dt} = -\frac{dC_B}{bdt} = KC_A C_B \tag{5-21}$$

In this case, the integrated rate law has the following form:

$$\frac{1}{bC_{A_0} - C_{B_0}} \ln\left(\frac{C_{B_0}C_A}{C_{A_0}C_B}\right) = Kt \tag{5-22}$$

Pseudo First-Order Reactions

In some situations where the reactants A and B react to form a single product, one of the reactants is present in great excess over the other. The concentration of the reactant present in great excess will change very little during the course of the reaction and, as a result, can be considered constant. This means that the rate of the reaction is proportional only to the concentration of the reactant which is not in excess. As an example, assume that B is the reactant present in great excess. The differential rate law for this situation has the form

$$-\frac{dC_A}{dt} = KC_AC_B \tag{5-23}$$

However, since C_B is constant, a second constant term can be defined as shown in equation 5-24.

$$K_{obs} = KC_B \tag{5-24}$$

Substituting from equation 5-24 for KC_B in equation 5-23 gives

$$-\frac{dC_A}{dt} = K_{obs}C_A \tag{5-25}$$

The differential rate law described by this equation is mathematically equivalent to the differential rate law described by equation 5-9. Thus, the reaction will appear to follow first-order kinetics but is in reality a second-order reaction. Such a reaction is referred to as a *pseudo first-order reaction*, and the rate constant, K_{obs}, determined for this reaction is a pseudo first-order rate constant. The second-order rate constant may be evaluated by plotting K_{obs} vs. C_B for several different concentrations of B obtained from a batch reactor system. The slope of the linear trace obtained from such a plot will have a value numerically equal to K.

Rate, Order and, Stoichiometry

For the general chemical reaction

$$aA + bB \rightleftharpoons cD + dD$$

in which a, b, c, and d represent the stoichiometric number of particles of A, B, C and D, respectively, involved in the balanced reaction, the equilibrium constant may be expressed by the relationship

$$(K_c)_{eq} = \frac{[C]^c[D]^d}{[A]^a[B]^b}$$

regardless of the actual reaction mechanism. Even though the actual reaction may be quite different from the stoichiometry of the overall process, the equilibrium properties of a chemical system do not depend on the pathway by which equilibrium is attained.

It is often mistakenly assumed that the order of a chemical reaction can be predicted directly from its stoichiometry. However, it must be understood that the exponents of concentration in a differential rate law expression are not necessarily equal to the corresponding molar coefficients in the balanced chemical equation. The order of a reaction can only be determined by experimentation; there is no precise way of predicting it directly from the balanced chemical equation. This can be illustrated by considering the following chemical equation:

$$A + B \longrightarrow P$$

If the differential rate law is based solely on stoichiometry, it would be written as

$$-\frac{dC_A}{dt} = KC_A C_B$$

indicating a second-order reaction with respect to reactant concentration. Yet, laboratory studies might show the rate to be independent of the concentration of reactant, B, over a wide range of concentrations. In this case, the correct differential rate law would be

$$-\frac{dC_A}{dt} = KC_A(C_B)^0$$

or

$$-\frac{dC_A}{dt} = KC_A$$

indicating a pseudo first-order reaction with respect to reactant concentration.

In the majority of chemical treatment processes the differential rate laws describing zero-order, first-order, pseudo first-order, and second-order reactions have the widest applications. This includes many complex reactions which involve a number of different steps. In such cases, one step is normally slower than all the rest. This step controls the overall rate of the reaction (the rate-limiting step), and it can generally be described by one of the rate laws previously mentioned.

5-3 CATALYSIS

In Chapter 1 it is noted that when the reactant material contains large amounts of energy relative to the products, the chemical reaction will occur spontaneously. The driving force for the reaction is a combination of enthalpy change and entropy change and is called the *Gibbs free energy change* of the reaction. Experimental observations have shown that a decrease in free energy (a negative free energy change) is associated with reactions which occur spontaneously. Yet, even though a negative free energy change indicates that a reaction has a tendency to occur, the rate of certain spontaneous reactions is so slow that it may take many years to detect any change in the reactant concentration. This type of behavior may be explained on the basis of the theory of *absolute reaction rates* developed by Eyring and his collaborators. This theory proposes that before molecules react they must pass through a configuration known as the *transition state or activated complex*, which has an energy content greater than either the reactant or product. Energy may have to be added to the system before the reaction can proceed. The added energy is required to elevate the reactant to the transition state, and this energy is called the *free energy of activation*. An inverse relationship exists between the free energy of activation and the rate of the reaction; i.e., the faster the rate of reaction, the lower the free energy of activation is, and vice versa. Another important point to note here is that, according to the absolute reaction rate theory, the reaction rate is a function of the concentration of the activated complex, which means that kinetics gives the stoichiometry of the transition state and not the stoichiometry of the reactants. However, in most reactions the concentration of the activated complex may be expressed as a function of the concentrations of the reactants.

A catalyst is generally defined as a substance that increases the rate of a chemical reaction but does not appear in the balanced chemical equation. From a thermodynamic point of view a catalyst speeds up a reaction by lowering the activation energy. It does so by increasing the number of molecules that are activated and therefore reactive. This concept is illustrated in Figure 5-7. This figure also indicates that, regardless of the path the reaction takes, both the uncatalyzed and catalyzed reaction have the same ΔG of reaction. Thus, it can be seen that a catalyst does not change the final state of a reaction but reduces the energy of activation and increases the rate at which

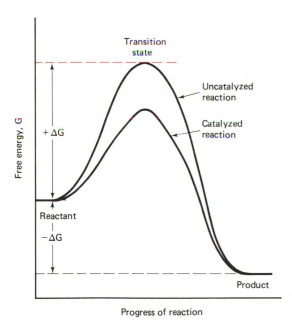

FIGURE 5-7 Energy diagram illustrating transition state for a catalyzed and an uncatalyzed reaction.

the final state is attained. At the end of the reaction the catalyst itself remains unchanged.

Catalysis by acids and bases is commonly encountered in process chemistry. The appropriate rate laws (as developed by Piszkiewicz, 1977) governing hydrogen ion, hydroxyl ion, and weak acid catalysis are presented in the following three sections.

Catalysis by Hydrogen Ions

Many chemical reactions which occur in aqueous solutions are observed to be catalyzed by the hydrogen ion. A general reaction describing such a situation may be written as

$$A + n\,H^+ \longrightarrow P + n\,H^+ \qquad (5\text{-}26)$$

The differential rate law for the reaction described by equation 5-26 is given by

$$-\frac{dC_A}{dt} = K_H[H^+]^n C_A \qquad (5\text{-}27)$$

Equation 5-27 shows that the H^+ ion is not consumed in the reaction. Thus, the hydrogen ion concentration may be treated as a constant. Since $[H^+]$ is constant, a second constant term can be defined as

$$K_{obs} = K_H[H^+]^n \qquad (5\text{-}28)$$

Substituting from equation 5-28 for $K[H^+]$ in equation 5-27 gives

$$-\frac{dC_A}{dt} = K_{obs} C_A \qquad (5\text{-}29)$$

which is the same as the differential rate law defined by equation 5-25. Thus, the H^+ ion-catalyzed reaction is a pseudo first-order reaction.

Taking the logarithm of both sides of equation 5-28 gives

$$\log K_{obs} = \log K_H + n \log [H^+] \qquad (5\text{-}30)$$

If ideal conditions are assumed,

$$pH = - \log [H^+]$$

Thus, equation 5-30 may be expressed as

$$\log K_{obs} = \log K_H - n\,pH \qquad (5\text{-}31)$$

Equation 5-31 indicates that by using data obtained from a batch reactor system a plot of $\log K_{obs}$ vs. pH will give a straight line with a slope of $-n$.

Catalysis by Hydroxyl Ions

A general reaction describing hydroxyl ion catalysis is given by equation 5-32.

$$A + n\,OH^- \longrightarrow P + n\,OH^- \tag{5-32}$$

A development similar to that presented for hydrogen ion catalysis will yield the following relationships:

$$-\frac{dC_A}{dt} = K_{OH}[OH^-]^n C_A \tag{5-33}$$

$$K_{obs} = K_{OH}[OH^-]^n \tag{5-34}$$

$$-\frac{dC_A}{dt} = K_{obs}C_A \tag{5-35}$$

and

$$\log K_{obs} = \log K_{OH} + n\,(pH - 14) \tag{5-36}$$

Equation 5-36 suggests that by using data obtained from a batch reactor system a plot of $\log K_{obs}$ vs. pH will give a straight line with a slope of n.

Catalysis by a Weak Monoprotic Acid

A reaction catalyzed by a weak monoprotic acid may be written as follows:

$$\underset{\text{monoprotic acid}}{HA} + \underset{\text{reactant}}{B} \longrightarrow \underset{\text{product}}{P} \tag{5-37}$$

The differential rate law for this reaction has the form

$$-\frac{dC_B}{dt} = K_{HA}[HA]C_B \tag{5-38}$$

Although a hydrogen ion term is not shown in the balanced chemical equation, the rate of the reaction will vary as a function of pH. This happens because the rate is dependent on the concentration of the unionized acid, and this quantity does vary with pH.

$$HA \rightleftharpoons A^- + H^+$$

The equilibrium constant expression for this reaction (assuming ideal conditions) is

$$K_a = \frac{[A^-][H^+]}{[HA]} \tag{5-39}$$

A mass balance expression for the A specie may be written as

$$C_T = [HA] + [A^-] \tag{5-40}$$

Solving equation 5-39 for [A⁻] and substituting for [A⁻] in equation 5-40 gives

$$C_T = [\text{HA}] + \frac{[\text{HA}]K_a}{[\text{H}^+]} \tag{5-41}$$

or

$$C_T = [\text{HA}]\left(\frac{[\text{H}^+] + K_a}{[\text{H}^+]}\right) \tag{5-42}$$

Equation 5-42 may be rearranged into the form

$$[\text{HA}] = C_T\left(\frac{[\text{H}^+]}{K_a + [\text{H}^+]}\right) \tag{5-43}$$

Substituting for [HA] in equation 5-38 from equation 5-43 gives

$$-\frac{dC_B}{dt} = K_{\text{HA}}C_B C_T\left(\frac{[\text{H}^+]}{K_a + [\text{H}^+]}\right) \tag{5-44}$$

If, during the course of the reaction the pH remains constant, an observed rate constant can be defined as

$$K_{\text{obs}} = K_{\text{HA}}C_T\left(\frac{[\text{H}^+]}{K_a + [\text{H}^+]}\right) \tag{5-45}$$

For this situation the differential rate law expression reduces to

$$-\frac{dC_B}{dt} = K_{\text{obs}}C_B \tag{5-46}$$

Equation 5-45 shows that the pseudo first-order rate constant will vary with both pH and total acid concentration; i.e., at a constant pH several values of K_{obs} may be determined by varying C_T. Using such data, K_{HA} may be evaluated by plotting K_{obs} vs. C_T. Such a plot will give a linear trace with a slope numerically equal to $K_{\text{HA}}([\text{H}^+]/K_a + [\text{H}]^+)$.

Equation 5-45 may be rearranged to give

$$\frac{K_{\text{obs}}}{C_T} = K_{\text{HA}}\left(\frac{[\text{H}^+]}{K_a + [\text{H}^+]}\right) \tag{5-47}$$

Equation 5-48 is obtained by taking the logarithm of both sides of equation 5-47.

$$\log\left(\frac{K_{\text{obs}}}{C_T}\right) = \log K_{\text{HA}} + \log[\text{H}^+] - \log(K_a + [\text{H}^+]) \tag{5-48}$$

or

$$\log\left(\frac{K_{\text{obs}}}{C_T}\right) = \log K_{\text{HA}} - \text{pH} - \log(K_a + [\text{H}^+]) \tag{5-49}$$

There are two limiting cases to consider when evaluating equation 5-49 on the basis of a log (K_{obs}/C_T) vs. pH plot.

Case 1:

$$[H^+] \gg K_a \quad \text{or} \quad pK_a > pH$$

For this case, K_a can be neglected so that equation 5-49 reduces to

$$\log\left(\frac{K_{obs}}{C_T}\right) = \log K_{HA} - pH + pH \tag{5-50}$$

or

$$\log\left(\frac{K_{obs}}{C_T}\right) = \log K_{HA} \tag{5-51}$$

Equation 5-51 will be applicable when $[H^+]$ is at least 10 times greater than K_a, i.e., when the pH is at least one unit less than the pK_a. The slope of the line described by equation 5-51 is

$$\frac{d \log d(K_{obs}/C_T)}{d\,pH} = 0 \tag{5-52}$$

Thus, a plot of log (K_{obs}/C_T) vs. pH in the region where the pH value is at least one unit less than pK_a will give a horizontal trace which will intersect the ordinate at the value given by log K_{HA}.

Case 2:

$$[H^+] \ll K_a \quad \text{or} \quad pH > pK_a$$

For this case, $[H^+]$ can be neglected in the $\{K_a + [H^+]\}$ sum term in equation 5-49 so that the equation reduces to

$$\log\left(\frac{K_{obs}}{C_T}\right) = \log K_{HA} + pK_a - pH \tag{5-53}$$

Equation 5-53 will be applicable when $[H^+]$ is at least 10 times smaller than K_a, i.e., when the pH is at least one unit larger than the pK_a. The slope of the line described by equation 5-53 is

$$\frac{d \log (K_{obs}/C_T)}{d\,pH} = -1 \tag{5-54}$$

A plot of log (K_{obs}/C_T) vs. pH in the region where the pH is at least one unit larger than the pK_a will give a linear trace of slope -1.

Figure 5-8 shows a plot of log (K_{obs}/C_T) vs. pH by using data obtained from a batch reactor system. A vertical line constructed through the point of intersection of the straight line defined by equations 5-51 and 5-53 will intersect the abscissa at a pH value numerically equal to pK_a.

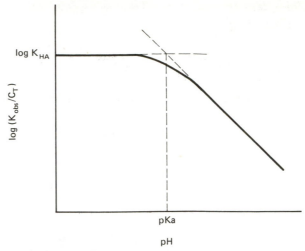

FIGURE 5-8 *Relationship between* K_{obs}/C_T *and* pH *for a weak monoprotic acid catalyzed reaction.*

Autocatalysis

Autocatalysis occurs when the reaction is catalyzed by a product of the reaction. For this situation the reaction rate will increase with time. The simplest type of autocatalytic reaction may be represented by the following chemical equation:

$$A + P \longrightarrow P + P \tag{5-55}$$

The differential rate law for this reaction has the form

$$-\frac{dC_A}{dt} = KC_A C_P \tag{5-56}$$

During the course of the reaction the algebraic sum of the moles/ℓ of product remains constant; i.e.,

$$C_T = C_A + C_P = C_{A_0} + C_{P_0} = \text{constant} \tag{5-57}$$

where C_{A_0} = initial reactant concentration (moles/ℓ)

         C_{P_0} = initial product concentration (moles/ℓ)

         C_T = moles/ℓ of reactant plus moles/ℓ of product at any time during the course of the reaction.

Equation 5-57 can be solved for C_P to give

$$C_P = C_T - C_A \tag{5-58}$$

Substituting for C_P in equation 5-56 from equation 5-58 results in equation 5-59:

$$-\frac{dC_A}{dt} = KC_A(C_T - C_A) \tag{5-59}$$

Equation 5-59 may be arranged into the form

$$-\frac{dC_A}{C_A(C_T - C_A)} = K\,dt \tag{5-60}$$

For the boundary conditions, $C_A = C_{A_0}$ at $t = 0$, integration of equation 5-60 yields

$$\ln\left[\frac{C_{A_0}(C_T - C_A)}{C_A(C_T - C_{A_0})}\right] = C_T K t \tag{5-61}$$

or

$$\log\left[\frac{C_{A_0}(C_T - C_A)}{C_A(C_T - C_{A_0})}\right] = \frac{C_T K}{2.303}t \tag{5-62}$$

Since $C_T = C_{A_0} + C_{P_0}$, equation 5-61 may also be expressed as

$$\ln\left[\frac{C_{A_0}(C_T - C_A)}{C_A(C_T - C_{A_0})}\right] = (C_{A_0} - C_{P_0})Kt \tag{5-63}$$

Equation 5-62 suggests that a plot of $\log[C_{A_0}(C_T - C_A)/C_A(C_T - C_{A_0})]$ (using data obtained from a batch reactor system) vs t, will give a straight line with a slope numerically equal to $C_T K/2.3$.

5-4 CRYSTAL GROWTH KINETICS

Probably the most common method employed to remove soluble metal ions from water and wastewater is precipitating them in the form of sparingly soluble salts. Precipitation can be thought of as a two-step process: nucleation and crystal growth. Nucleation is the generation (birth) of crystals in the solution (the formation of very small nuclei which generally contain between 10–100 atoms or molecules per nucleus), whereas growth is the process where atoms or molecules are transported to the individual crystal surface and are then oriented into the crystal lattice.

The driving force for crystallization can be expressed mathematically as (Nancollas and Reddy, 1976):

$$\mathrm{DF} = \frac{C - C_S}{C_S} \tag{5-64}$$

where DF = driving force or degree of oversaturation

C = concentration of the supersaturated solution

C_S = saturation concentration of the solution.

The effect of the degree of oversaturation on the precipitation process in a batch reactor system is illustrated in Figure 5-9. For high degrees of oversaturation (the region of Figure 5-9 labeled as "labile"), the solution will experience rapid precipitation through spontaneous nucleation and crystal growth. For solutions with low degrees of oversaturation (the region of Figure 5-9 labeled as "metastable"), no noticeable precipitation will occur for long periods of time. According to Nancollas and Reddy (1976), the upper boundary of the metastable region will vary from salt to salt and has

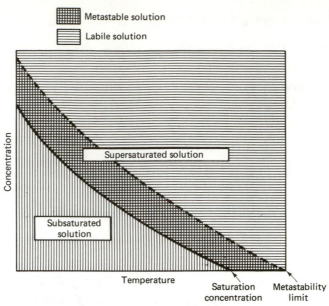

Metastable solution

Labile solution

Supersaturated solution

Subsaturated solution

Temperature

Saturation concentration

Metastability limit

FIGURE 5-9 *Solubility diagram showing subsaturated, labile and metastable supersaturated solutions (after Nancollas and Reddy, 1976).*

been observed to be as high as 10 times the solubility value. In the metastable region rapid precipitation can be initiated by the addition of seed crystals.

According to Nancollas and Reddy (1976), the differential rate law describing the *growth* of seed crystals from the metastable region of many supersaturated salt solutions has the general form

$$\text{rate of crystal growth} = -\frac{dC_A}{dt} = KS(C_A - C_{A_e})^n \qquad \textbf{(5-65)}$$

where

$C_A =$ concentration of species A in solution at time, t

$K =$ crystal growth rate constant

$S =$ crystal surface area available for growth. This term is generally taken as the weight concentration of added seed crystal.

$C_{A_e} =$ equilibrium concentration of A based on the solubility of the precipitate

$(C_A - C_{A_e}) =$ amount of precipitate remaining to be precipitated from the solution before equilibrium is reached

$n =$ constant.

The rate of crystal growth during precipitation is controlled by one of two different mechanisms: (1) the diffusion of ions from the liquid phase to the crystal surface or (2) the surface reaction. According to the step dislocation theory, the surface reaction involves the following steps: (a) adsorption of the depositing ion to the crystal surface,

(b) diffusion of the adsorbed ion along the crystal surface to a step edge, and (c) diffusion of the adsorbed ion along the step edge to a kink site (its final position in the crystal lattice). These steps are illustrated in Figure 5-10.

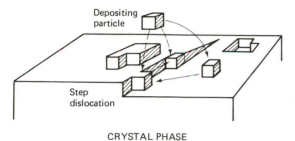

FIGURE 5-10 Steps involved in crystal growth (after Nancollas and Reddy, 1976).

Equation 5-65 is applicable to those situations where the surface reaction rather than bulk diffusion controls the rate of crystal growth. In this regard, when the growth rate is observed to be independent of the reactor mixing intensity, a surface reaction controlled precipitation process is indicated. In other words, a surface controlled reaction is independent of mixing speed.

5-5 EFFECT OF TEMPERATURE ON REACTION RATE

The rate of most simple chemical reactions is increased when the temperature is elevated, provided the higher temperature does not produce alterations in the reactants or catalyst.

To describe the variation in reaction rate with temperature, a frequently quoted approximation known as the *van't Hoff rule* states that the reaction rate doubles for a 10° temperature rise.

Arrhenius proposed that the effect of temperature on the reaction rate constant in a chemical reaction may be described by equation 5-66:

$$\frac{d(\ln K)}{dT} = \frac{E_a}{R} \cdot \frac{1}{T^2} \qquad (5\text{-}66)$$

where K = reaction rate constant

E_a = activation energy (calories mole^{-1})

R = ideal gas constant (1.98 calories mole^{-1} degree^{-1})

T = reaction temperature (°K)

Equation 5-66 can be integrated to give the expression

$$\ln K = -\frac{E_a}{R} \cdot \frac{1}{T} + \ln B \qquad (5\text{-}67)$$

where B represents a constant. Plotting, according to equation 5-67, experimental data obtained from a batch reactor system, is useful when one desires to determine the activation energy associated with a particular reaction. The plot necessary for such

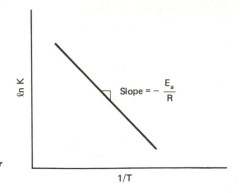

FIGURE 5-11 *Arrhenius plot for determining activation energy.*

a determination is shown in Figure 5-11. This figure shows that the slope of the line given when ln K is plotted vs. $1/T$ is $-E_a/R$. The straight line response will result as long as E_a and B remain constant.

When equation 5-66 is integrated between the limits of T_1 and T_2, equation 5-68 is obtained.

$$\ln\left[\frac{K_2}{K_1}\right] = \frac{E_a}{R}\left[\frac{T_2 - T_1}{T_2 T_1}\right] \tag{5-68}$$

If K_1 is known for T_1 then, using equation 5-68, it is possible to compute K_2 for the temperature change from T_1 to T_2 if the activation energy for the reaction is known. The equation has been used to estimate the effect of temperature over a limited range for many reactions.

For a number of situations encountered in process chemistry, the quantity E_a/RT_1T_2 located on the right side of equation 5-68 may be considered constant. Thus, equation 5-68 may be approximated by the expression

$$\ln\left[\frac{K_2}{K_1}\right] = \text{constant}\,(T_2 - T_1) \tag{5-69}$$

Equation 5-69 may then be written in the form

$$\frac{K_2}{K_1} = e^{\text{constant}\,(T_2 - T_1)} \tag{5-70}$$

It is common practice to introduce a temperature characteristic term, θ which has a value equal to that given by e^{constant}. For this situation equation 5-70 becomes

$$\frac{K_2}{K_1} = \theta^{(T_2 - T_1)} \tag{5-71}$$

When equation 5-71 is applied, T_1 and T_2 may be expressed as degrees Celsius rather than degrees Kelvin, since the difference will be the same for both temperature scales.

Note that the rate-temperature response of many chemical reactions does not follow the Arrhenius law stated by equation 5-66. In such cases, the reaction mechanism is generally very complex or else the mechanism changes over the temperature range studied.

EXAMPLE PROBLEM 5-4: Given, the following experimental data obtained from a batch reactor system:

Temperature (°C)	Reaction Rate Constant, K (days⁻¹)
15.0	0.53
20.5	0.99
25.5	1.37
31.5	2.80
39.5	5.40

Determine the value of the temperature characteristic, θ.

Solution:

1. Plot data according to equation 5-67 and determine E_a/R. To simplify the graphical construction, all K values will be multiplied by 10. Such a manipulation will not change the slope of the line.

$K \times 10$	$ln\,(K \times 10)$	$1/T$
5.3	1.67	0.00347
9.9	2.29	0.00340
13.7	2.62	0.00335
28.0	3.33	0.00328
54.0	3.99	0.00320

$$\text{slope} = -\frac{E_a}{R} = \frac{3.99 - 1.67}{0.00320 - 0.00347} = -8593$$

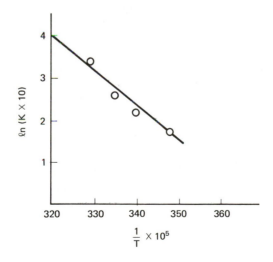

2. Estimate θ, using the value of 8593 for E_a/R. Assume that θ is valid over the temperature range of 15 to 40°C. Then for any temperature variation within this range θ is indicated by

$$\theta = e^{(8593/T_1 T_2)}$$

where T_1 and T_2 are the respective temperatures in degrees Kelvin.

However, one value of θ is generally used for a range of temperatures. In this case, if $T_1 = 15 + 273 = 288°K$ and $T_2 = 40 + 273 = 313°K$, then

$$\theta = e^{[8593/(288)(313)]} = 1.100$$

At this point it should be understood that θ is not really a constant, but that its value will actually vary depending on the two temperatures over which it is computed.

PROBLEMS

5-1. The rate of kill of *E. coli* with chlorine was investigated and the surviving organisms were determined by a plate count. The following data were obtained:

Contact Time (min)	Number of Organisms Remaining
0	40,800
2	36,000
4	23,200
6	16,000
8	12,800
10	10,400
15	5,600
20	3,000
25	1,500

Determine the reaction order of the disinfection process and the value of the associated reaction rate constant, K.

5-2. A certain reaction was found to have an activation energy of 20,000 calories per mole and a specific reaction rate constant of 2.0 min⁻¹ at 20°C. Calculate the specific reaction rate constant at 0°C.

5-3. Determine K_{OH}, using the kinetic data shown in Figure P5-3.

5-4. The oxygenation of Mn(II) in aqueous solution is an autocatalytic reaction. If C_{A_0} and C_T are taken as 10^{-4} mole/ℓ and 10^{-3} mole/ℓ, respectively, determine the reaction rate constant, K. Use the experimental data shown in Figure P5-4.

5-5. The effect of substrate concentration on the initial rate of the following reaction,

$$A + H_2O \longrightarrow P$$

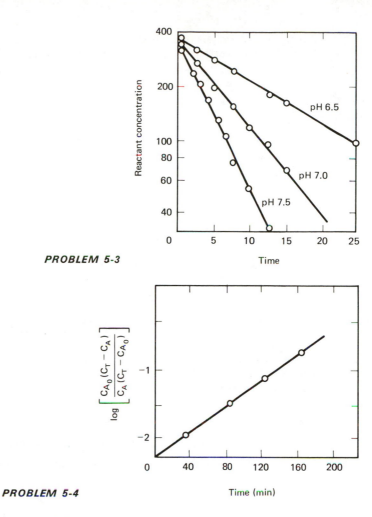

PROBLEM 5-3

Reactant concentration

Time

$$\log \left[\frac{C_{A_0}(C_T - C_A)}{C_A(C_T - C_{A_0})} \right]$$

PROBLEM 5-4

Time (min)

has been investigated in both directions at a pH of 7.0. The experimental data obtained in this investigation are as follows:

Concentration (moles/ℓ)	Initial Rate $\times 10^5$, moles/sec	
	A	P
0.40	5.0	2.8
0.20	6.0	3.2
0.10	5.9	3.2
0.05	5.6	4.0
0.025	5.4	2.5
0.0125	6.0	3.0

Determine the order of the reaction in the initial stages.

5-6. The hydrolysis of a particular chemical compound is catalyzed by phosphate buffers. Experimental data obtained for this reaction are as follows:

pH	K_{obs}/C_T (moles^{-1} min^{-1})
5.4	0.5
5.9	1.0
6.2	1.4
6.7	2.4
7.0	3.3
7.6	4.8
8.0	5.2
8.5	5.4

What is the pK_a of the phosphate catalyst?

REFERENCES

NANCOLLAS, G.H., and REDDY, M.M., "Crystal Growth Kinetics of Minerals Encountered in Water Treatment Processes," in *Aqueous-Environmental Chemistry of Metals*, [Ed.] A.J. RUBIN, Ann Arbor Science Publishers, Ann Arbor, MI (1976).

PISZKIEWICZ, D., *Kinetics of Chemical and Enzyme-Catalyzed Reactions*, Oxford University Press, New York (1977).

SEGEL, I.H., *Biochemical Calculations*, John Wiley & Sons, Inc., New York (1968).

SWINBOURNE, E.S., *Analysis of Kinetic Data*, Appleton-Century-Crofts, New York (1971).

6

FUNDAMENTALS OF SURFACE
AND COLLOIDAL CHEMISTRY

Surface chemistry is the study of the interfaces between two bulk phases in contact. Considering the three states of matter, the following types of interfaces can exist:

Liquid-Liquid
Liquid-Gas
Solid-Solid
Solid-Gas
Solid-Liquid

An interface cannot exist between two gases, since mixtures of gases always form true solutions.

Surface chemistry can be used to explain or predict the behavior of many well-known interfacial systems. The spreading of a drop of oil on water and the evaporation of vapor from the surface of a liquid are examples of liquid-liquid and liquid-gas interfaces. The adhesion of two solid substances creates a solid-solid interface and the adsorption of a gas onto a solid gives rise to a solid-gas interface.

The type of interface of primary interest to the environmental engineer is the interface between a solid and a liquid. Solid-liquid interfaces play a vital role in stabilizing colloidal impurities found in water and wastewater, and a knowledge of surface chemistry is essential to an understanding of the mechanisms by which such systems may be destroyed. Surface chemistry is also of interest when considering the removal of dissolved impurities by adsorption onto a solid adsorption medium such as activated carbon.

The purpose of this chapter is to present some fundamentals of surface chemistry related to colloidal stability and adsorption. Applications of these fundamentals to water and wastewater treatment are presented in Chapters 7 and 11 respectively.

6-1 COLLOIDAL SYSTEMS

A colloidal system is defined as a system in which particles, in a finely divided state, are dispersed in a continuous medium. The particles are called the *dispersed phase*, and the medium in which they exist is called the *dispersing phase*.

The particles that form a colloidal dispersion are sufficiently large for a definite surface of separation, or interface, to exist between them and the medium in which they are contained. At this interface there is no sharp transition between one bulk phase and the other, but a transition region exists which shows properties differing from either of the two bulk phases. The transition region is very thin but plays an important role in determining the behavior of such systems.

Colloidal particles are not limited to any particular group of substances but are defined by size. The colloidal size range is generally regarded to extend from 1 nanometer (nm) to 1 micrometer (μm)* although some authors consider the colloidal range to extend up to a size of 10 μm. It is frequently quite difficult to distinguish between colloids and solutions at the lower end of the scale and colloids and suspensions at the upper end. The size relationships are shown in Figure 6-1.

Colloidal systems may exist in which either the dispersed phase or the dispersing phase is a solid, liquid, or gas. Since gases always form true solutions, eight types of colloidal systems are possible as shown in Table 6-1.

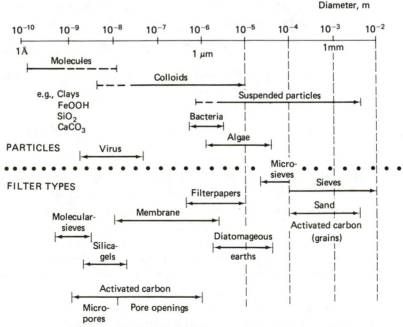

FIGURE 6-1 *Size spectrum of waterborne particles and filter pores (after Stumm, 1978).*

* 1 nm = 10^{-9} m; 1 μm = 10^{-6} m.

TABLE 6-1 Types of colloidal dispersions (after Shaw, 1970).

Dispersed Phase	Dispersion Medium	Name	Examples
Liquid	Gas	Liquid aerosol	Fog, liquid sprays
Solid	Gas	Solid aerosol	Smoke, dust
Gas	Liquid	Foam	Foam on soap solutions, fire-extinguisher foam
Liquid	Liquid	Emulsion	Milk, mayonnaise
Solid	Liquid	Sol. colloidal suspension: Paste (high solid concentration)	Au sol, AgI sol: toothpaste
Gas	Solid	Solid foam	Expanded polystyrene
Liquid	Solid	Solid emulsion	Opal, pearl
Solid	Solid	Solid suspension	Pigmented plastics

Colloidal dispersions commonly encountered in environmental engineering include emulsified oils (liquid-in-liquid dispersion), which are found in many industrial wastes, and foams (gas-in-liquid dispersion), which frequently develop during wastewater aeration. The colloidal system that is perhaps most often encountered in environmental engineering is the solid-in-liquid dispersion (sol) formed by clay particles present in surface waters. These clay particles impart turbidity to the water, and their removal is required if the water is to be used for human consumption.

Colloidal systems may be classified by the affinity of the dispersed phase for the dispersing medium. Systems are referred to as *hyophobic* (liquid hating) when a weak affinity exists and as *hyophilic* (liquid loving) when there is a strong affinity. The terms *hydrophobic* and *hydrophilic* are used when water is the dispersing medium. Hydrophobic colloids include such things as clay, gold, and other metals, whereas hydrophilic colloids are characterized by proteins, soaps, and synthetic detergents.

6-2 SURFACE CHARGE ON COLLOIDAL PARTICLES

If a colloidal sol is placed in an electrical field, the particles will move toward one of the electrodes. This phenomenon, called *electrophoresis*, indicates that the colloidal particles carry an electric charge. The sign of this charge (positive or negative) and its magnitude will depend on the nature of the colloidal material. Since like charges repel, similarly charged colloids are held apart from each other by their electric charges and thus are prevented from aggregating into larger particles. This electric charge, which is the primary factor responsible for colloidal stability, can be acquired by the particle in a number of ways.

(a) *Imperfections in the Crystal Structure:* In certain cases, a colloidal particle may acquire a charge as a result of isomorphic replacements within the crystal lattice, as represented schematically in Figure 6-2(a) for a SiO_2 tetrahedron in which an Si atom is replaced by an Al atom, resulting in a net negative charge on the particle (Fair,

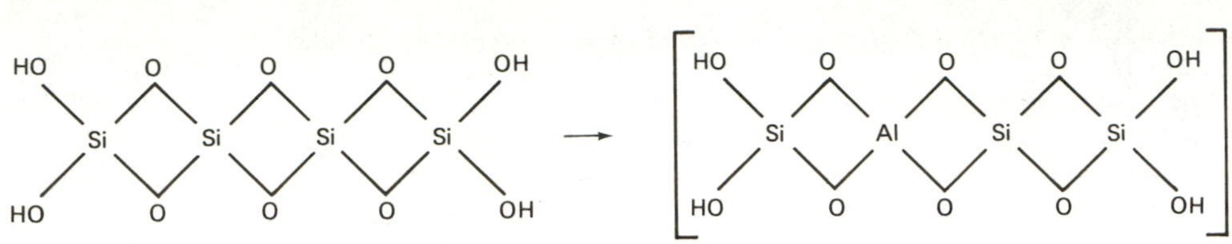

(a) Charge acquisition through isomorphic replacement of Al for Si. (after Fair, Geyer, and Okun, 1968.)

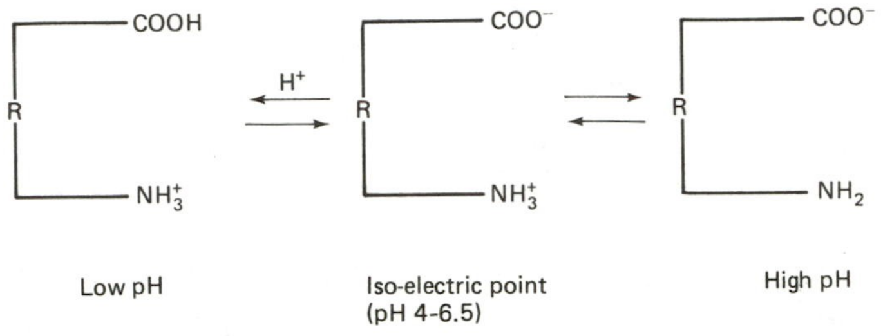

(b) Effect of pH on the ionization of a protein particle.

FIGURE 6-2 *Charge acquisition by clay and protein particles.*

Geyer, and Okun, 1968). Although this situation is comparatively rare, the clay particles responsible for turbidity in surface waters acquire their negative charge in this manner (van Olphen, 1977).

(*b*) *Adsorption of Ions Onto the Particle Surface:* Many colloidal particles acquire a charge as a result of the preferential adsorption of either positive or negative ions on their surface. The adsorbed ions are called *peptizing ions.* Colloidal particles in aqueous media usually adsorb anions and acquire a negative charge. This is because cations are generally more hydrated than anions and are separated from the colloids by the shield of hydrated water. Gas bubbles and oil droplets frequently acquire their charge by preferential adsorption.

(*c*) *Ion Dissolution:* Certain colloidal substances acquire an electric charge if the oppositely charged ions of which they are composed do not dissolve equally.

(*d*) *Ionization of Surface Sites:* Many colloidal particles acquire an electrical charge through the ionization of surface functional groups. For example, proteins acquire their charge as a result of the ionization of carboxyl or amino groups. Ionization of these groups is pH dependent, and particles may exhibit a net positive charge at low pH, a net negative charge at high pH, and a net charge of zero at some intermediate pH, *known as the iso-electric point.* The ionization of a protein particle is represented schematically in Figure 6-2(b).

An important difference between hydrophobic and hydrophilic colloids is that the former are stabilized almost exclusively by electrical repulsion forces, whereas the latter may be partly stabilized by particle solvation. This difference is so significant that it is convenient to divide colloidal dispersions into the two types and to discuss their behavior separately. Since clay particles that cause turbidity in water supplies are of prime importance, and since they are hydrophobic particles, this discussion will emphasize the behavior of hydrophobic systems.

6-3 THE ELECTRICAL DOUBLE LAYER

Although individual hydrophobic colloids have an electrical charge, a colloidal dispersion, like an ionic solution, does not have a net electrical charge. For electroneutrality to exist, the charge on the colloidal particle must be counterbalanced by ions of opposite charge (counter-ions) contained in the dispersing phase. The ions involved in this electroneutrality are arranged in such a way as to constitute what is called the *electrical double layer*. The concept of an electrical double layer was first proposed by Helmholtz and later refined by Gouy (1910), Chapman (1913), and Stern (1924).

To discuss the electrical double layer, it is convenient to start by considering the charge on the colloidal particle. As previously mentioned, some hydrophobic colloids acquire a charge by the preferential adsorption of certain ions on their surface. Once adsorbed, these ions, called *peptizing ions*, constitute the *inner coating* of the electrical double layer (van Olphen, 1977). For other colloids such as clay particles, the electrical double layer originates from charges caused by imperfections in the crystal lattice. In either case the charge on the particle will be offset by ions of opposite charge which accumulate around the particle to form the *outer coating* of the double layer. The high electrical potential on the particle surface will decrease to zero at some distance from the particle as a result of the accumulation of counter-ions.

A model proposed by Stern (1924) can be used to describe the distribution of the electrical potential in the vicinity of a colloidal particle. Consider a colloidal clay particle with a net negative surface charge as shown in Figure 6-3. The electrical potential created by this charge at the particle surface will attract counter-ions toward the particle. According to Stern, the distance of closest approach of these counter-ions to the particle will be limited by the size of the ions. He proposed that the center of the closest counter-ions are separated from the surface charge by a layer of thickness Ω (a distance approximately equal to the hydrated radius of the ion) in which there is no charge. This layer is referred to as the *Stern layer*. The electrical potential drops linearly across the Stern layer from a value of ψ_0 (Nernst potential) at the particle surface to a value ψ_Ω, which is called the *Stern potential*. Beyond this point, in what is called the *diffuse (Gouy) layer*, the electrical potential decreases exponentially with increasing distance from the particle.

Since like charges repel, similarly charged colloids are held apart as a result of their electrical charge. The magnitude of this charge on a colloidal particle cannot be measured directly; however, a value of the potential at some distance from the particle can be calculated from a measurement of the electrophoretic mobility of the particle. This potential, called the *zeta potential*, Z_p, is defined as

$$Z_p = \frac{4\pi q \delta}{D} \tag{6-1}$$

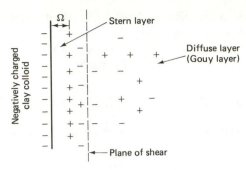

(a) Distribution of charges in the vicinity of a colloidal particle

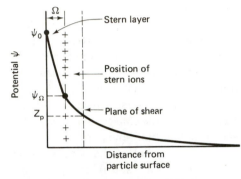

FIGURE 6-3 *Stern's model for the electrical double layer (after Van Olphen, 1977).*

(b) Distribution of potential in the electrical double-layer

where Z_p = zeta potential

q = charge on the particle

δ = thickness of the zone of influence of the charge on the particle

D = dielectric constant of the liquid.

The zeta potential is the potential that exists at the *plane of shear* between the bulk liquid and an envelope of water that moves with the particle. The exact location of this shearing plane is not known, but it is generally thought to lie in the diffuse layer, some distance beyond the Stern layer as shown in Figure 6-3. The magnitude of the zeta potential is a rough measure of the stability of a colloidal particle. High zeta potentials suggest strong forces of separation and stable colloidal systems, whereas lower zeta potentials are associated with less stable systems. The use of the zeta potential as a measure of particle stability is discussed in more detail in section 7-1.

6-4 STABILITY OF HYDROPHOBIC COLLOIDS

The individual particles in a hydrophobic sol are acted upon by both repulsive and attractive forces. The repulsive forces result from the electrical double layer, and the principle attractive forces result from van der Waals' forces of intermolecular attraction. Interactions between these forces contribute to the overall stability of a colloidal dispersion, according to theories developed by Derjaguin and Landau (1941) and Verwey and Overbeek (1948).

Colloidal particles in suspension are constantly moving as a result of Brownian motion. As two similarly charged particles approach each other, their diffuse counterion atmospheres begin to interfere and cause the particles to be repulsed. The amount of work required to overcome this repulsion and bring the particles from infinite separation to a given distance apart is called the *repulsive energy* or the *repulsive potential*, V_R, at that distance. The electrical potential for an individual particle decreases with distance from the particle as shown in Figure 6-3. Consequently, the repulsive energy between two particles decreases roughly exponentially with increasing particle separation as shown in Figure 6-4.

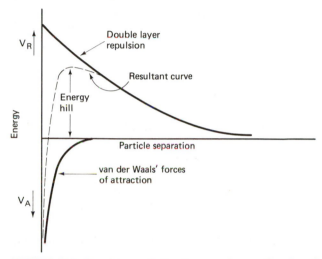

FIGURE 6-4 *Repulsive and attractive energies as a function of particle separation (after Van Olphen, 1977).*

The van der Waals' forces of attraction, which oppose the repulsive forces, are largely due to universal attractive forces (called *dispersion forces*), first explained by London. The London-van der Waals' attractive energy of interaction, V_A, between two particles is inversely proportional to the second power of the distance separating the particles and thus decreases very rapidly with increasing intermolecular distance (O'Melia, 1969). The variation in particle attractive energy with particle separation distance is shown in Figure 6-4.

The repulsion and attraction curves can be combined to form a curve representing the resultant energy of interaction. This curve indicates that repulsion forces predominate at certain distances of separation, but that if the particles can be brought close enough together, the van der Waals' attractive forces will predominate and the particles will coalesce. To come together, the particles must possess enough kinetic energy to overcome the so-called energy hill on the total energy curve.

The forces stabilizing colloidal particles must be overcome and the individual particles must aggregate if they are to be separated from suspension. Aggregation can be brought about by the addition of selected chemicals that act in various ways to overcome the stabilizing forces. This process, referred to as *coagulation*, is discussed in Chapter 7.

6-5 ADSORPTION: SOLID-LIQUID INTERFACE

Adsorption is defined as the taking up of molecules by the external or internal surface of solids or by the surface of liquids. Adsorption occurs on these surfaces because of attractive forces of the atoms and molecules that make up the surfaces. When impurities are adsorbed from a liquid onto a surface, the adsorption process occurs at the solid-liquid interface, and reactions occurring at the interface determine the rate and extent of adsorption.

There are many applications of adsorption in both industrial and pollution control processes and many solid adsorbents, including activated alumina, silica gel, and activated carbon, are used. This discussion is limited to the use of activated carbon to adsorb materials from liquid, since this is the most common situation encountered in water and wastewater treatment.

6-6 PRODUCTION OF ACTIVATED CARBON

Activated carbon has a great capacity for the adsorption of organic molecules. It is produced by exposing selected carbonaceous materials to a series of treatment processes referred to as *dehydration*, *carbonization*, and *activation*. These processes burn away impurities in the raw material and leave a highly porous residue that has an extremely large surface area per unit volume. Since adsorption is a surface phenomenon, the large surface/volume ratio of activated carbon is its primary attribute as an adsorbent.

Activated carbon has been produced from many different carbonaceous materials, including coal, wood, coconut shells, and peat. The first step in production, dehydration, is accomplished by heating the material at temperatures up to 170°C to remove excess water. In some cases, zinc chloride or phosphoric acid may be used in addition to heat as a dehydrating agent. Following dehydration, heating is continued, usually in the absence of air, at temperatures up to 400–600°C. This treatment, known as *carbonization*, causes decomposition of the material and drives off impurities such as tars and methanol. The escape of these volatile substances causes pores to form within the material and leaves a product usually referred to as a *char*. Although surface area is opened up during carbonization, chars possess relatively little adsorptive power due to the presence of amorphous residues (tars), which block their pores. These residues are removed and the pores are cleaned and enlarged by the process of activation, which is the final step in activated carbon production. Activation is normally achieved by treating the carbon product with mixtures of CO_2, air, and steam at temperatures of 750–950°C so as to burn off the amorphous residues.

Pore spaces within the carbon are well developed following activation and can be classified according to size as follows: macropores, transitional pores, and micropores. According to Dubinin et al. (1964), macropores have an effective radius of 5000 to 20,000 Å and open directly to the outer surface of the carbon particle. Transitional pores with radii of 40–200 Å develop off the macropores. Micropores develop off the transitional pores and have an effective radii of 18–20 Å or less.

6-7 FACTORS INFLUENCING ADSORPTION

Molecules of solute are removed from solution and taken up by the adsorbent during the process of adsorption. The majority of molecules are adsorbed onto the large surface area within the pores of a carbon particle and relatively few are adsorbed on the outside surface of the particle. The transfer of solute from solution to adsorbent continues until the concentration of solute remaining in solution is in equilibrium with the concentration of solute adsorbed by the adsorbent. When equilibrium is reached, the transfer of solute stops and the distribution of solute between the liquid and solid phases is measurable and well-defined.

The equilibrium distribution of solute between the liquid and solid phases is an important property of adsorption systems and helps define the capacity of a particular system. Of equal importance to the engineer are the kinetics of the system which describe the rate at which this equilibrium is reached. The rate of adsorption determines the detention time required for treatment and thus the size of carbon contacting systems.

Process kinetics describe the rate at which molecules are transferred from solution to the pores of the carbon particle. Three distinct steps must take place for adsorption to occur:

1. The adsorbed molecule must be transferred from the bulk phase of the solution to the surface of the adsorbent particle. In so doing, it must pass through a film of solvent that surrounds the adsorbent particle. This process is referred to as *film diffusion.*

2. The adsorbate molecule must be transferred to an adsorption site on the inside of the pore. This process is referred to as *pore diffusion.*

3. The absorbate must become attached to the surface of the absorbent, i.e., be adsorbed.

Many factors influence the rate at which adsorption reactions occur and the extent to which a particular material can be adsorbed. Several of the more important factors are now discussed.

Agitation

The rate of adsorption is controlled by either film diffusion or pore diffusion, depending on the amount of agitation in the system. If relatively little agitation occurs between the carbon particle and the fluid, the surface film of liquid around the particle will be thick and film diffusion will likely be the rate-limiting step. If adequate mixing is provided, the rate of film diffusion will increase to the point that pore diffusion becomes the rate-limiting step. According to Weber (1972), pore diffusion is generally rate-limiting for batch-type contacting systems which provide a high degree of agitation. Film diffusion will most likely be the rate-limiting step for continuous-flow systems at flow rates of 10 gal/min per square foot or less.

Characteristics of the Adsorbent (Activated Carbon)

Particle size and surface area are important properties of an activated carbon with respect to its use as an adsorbent. The size of a carbon particle influences the rate at which adsorption occurs: adsorption rates increase as particle sizes decrease. Thus, adsorption rates are faster for powdered carbon than for granular carbon.

The total adsorptive capacity of a carbon depends on its total surface area. The size of a carbon particle does not have a great effect on total surface area, since most of the surface area lies within the pores of a carbon particle. Thus, equal weights of the same carbon would have essentially the same capacity in the granular form as in the powdered form.

Solubility of the Adsorbate

For adsorption to occur, a molecule must be separated from the solvent and become attached to the carbon surface. Soluble compounds have a strong affinity for their solvent and thus are more difficult to adsorb than insoluble compounds. However, there are exceptions, since many compounds that are slightly soluble are difficult to adsorb, whereas some very soluble compounds may be adsorbed readily (Hassler, 1974). Efforts to find a quantitative relationship between adsorbability and solubility have met with only limited success.

Size of Adsorbate Molecules

Molecular size would logically be important in adsorption, since the molecules must enter the micropores of a carbon particle so as to be adsorbed. Research studies have shown that within a homologous series of aliphatic acids, aldehydes, or alcohols, adsorption usually increases as the size of the molecule becomes greater (Hassler, 1974). This can partly be explained by the fact that the forces of attraction between a carbon and a molecule are greater the closer the size of the molecule is to the size of the pores in the carbon (Culp and Culp, 1971). Adsorption is strongest when the pores are just large enough to permit the molecules to enter.

Most wastewaters contain a mixture of compounds representing many different sizes of molecules. In this situation there would appear to be a danger of molecular screening, i.e., large molecules blocking the pores to prevent the entrance of small molecules. However, the irregular shape of both the molecules and the pores, as well as the constant motion of the molecules, prevents such blockage from occurring (Culp and Culp, 1971). Furthermore, the greater mobility of the small molecules allows them to diffuse faster and to enter the pores ahead of the large molecules. The relative position of various sizes of molecules during adsorption is illustrated in Figure 6-5.

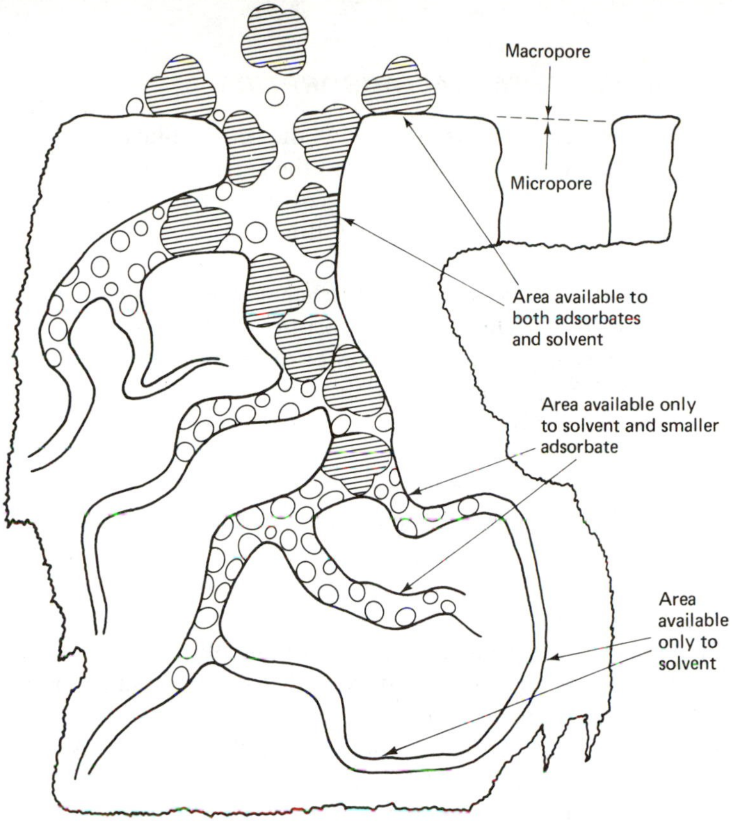
Macropore
Micropore
Area available to both adsorbates and solvent
Area available only to solvent and smaller adsorbate
Area available only to solvent

FIGURE 6-5 Concept of molecular screening in micropores (Courtesy, *Calgon Corporation).*

pH

The pH at which adsorption is carried out has been shown to have a strong influence on the extent of adsorption. This is partly due to the fact that hydrogen ions themselves are strongly adsorbed and partly that pH influences the ionization, and thus the adsorption, of many compounds. Organic acids are more adsorbable at low pH, whereas the adsorption of organic bases is favored by high pH. The optimum pH for any adsorption process must be determined by laboratory testing.

Temperature

The temperature at which an adsorption process is conducted will affect both the rate of adsorption and the extent to which adsorption occurs. Adsorption rates increase with increased temperature and decrease with decreased temperature. However, since adsorption is an exothermic process, the degree of adsorption will increase at lower temperature and decrease at higher temperatures.

Many other factors in addition to those mentioned above could influence adsorption reactions. Thus, adsorption is an extremely complex process and one that must be carefully analyzed in order to ensure optimum process design.

6-8 PHYSICAL VS. CHEMICAL ADSORPTION

After an adsorbent molecule has passed through the surface film and entered the pores of a carbon particle, it must become attached to the particle if it is to be removed from solution. This process is referred to as *physical adsorption* when attachment is by physical forces and *chemical adsorption* (chemisorption) when chemical forces cause attachment.

Chemical adsorption occurs as a result of a chemical reaction between the adsorbate molecule and the adsorbent. This type of adsorption is thought to occur on the sides and corners of the microcrystallites that comprise the activated carbon, since these sites are characterized by the presence of various types of functional groups which participate in adsorption through electron-sharing reactions (Snoeyink and Weber, 1967).

Particles removed by physical adsorption are held to the adsorbent by relatively weak van der Waals' forces of attraction or perhaps by π-bonding under certain conditions. Physically adsorbed particles are assumed to be free to move on the surface of the adsorbent, and adsorption is assumed to be multilayered with each new layer of molecules forming on top of previously adsorbed layers.

Physical adsorption is thought to take place on the planar surfaces of carbon particles. These surfaces are uniform in nature and do not contain functional groups because the electrons of the carbon atoms are involved in covalent bonding. The large amount of surface area found in the micropores of carbon particles is probably of the planar surface type (Snoeyink and Weber, 1967).

Most adsorption processes in wastewater treatment are neither purely physical or chemical processes but are a combination of the two. It is frequently difficult to distinguish between these two processes and, fortunately, such distinction is not necessary in order to be able to analyze and design adsorption processes.

6-9 ADSORPTION ISOTHERMS

Data collected during an adsorption test will describe the performance of the carbon and will yield valuable information if properly interpreted. Several mathematical relationships have been developed to describe the equilibrium distribution of solute between the solid and liquid phases and thus aid in the interpretation of adsorption data. These relationships apply when the adsorption tests are conducted at constant temperature and are referred to as *adsorption isotherms*. Three of the most common are the Langmuir isotherm, the Fruendlich isotherm, and the Brunaur-Emmett-Teller (BET) isotherm.

Langmuir's isotherm is based on the assumption that points of valency exist on the surface of the adsorbent and that each of these sites is capable of adsorbing one molecule; thus, the adsorbed layer will be one molecule thick. Furthermore, it is assumed that all the adsorption sites have equal affinities for molecules of the adsorbate and that the presence of adsorbed molecules at one site will not affect the adsorption of molecules at an adjacent site. The Langmuir equation is commonly written as follows (Langmuir, 1918):

$$\frac{x}{m} = \frac{abc}{1 + ac} \tag{6-2}$$

where x = amount of material adsorbed (mg or g)

m = weight of adsorbent (mg or g)

c = concentration of material remaining in solution after adsorption is complete (mg/ℓ)

a and b = constants.

Taking the reciprocal of both sides of equation 6-2 yields

$$\frac{1}{(x/m)} = \frac{1 + ac}{abc} \tag{6-3}$$

or

$$\frac{1}{(x/m)} = \frac{1}{b} + \frac{1}{abc} \tag{6-4}$$

If adsorption follows the Langmuir isotherm, a linear trace should result when the quantity $1/(x/m)$ is plotted against $1/c$ as shown in Figure 6-6. Values of the constants a and b can be determined from the slope and intercept of the plot.

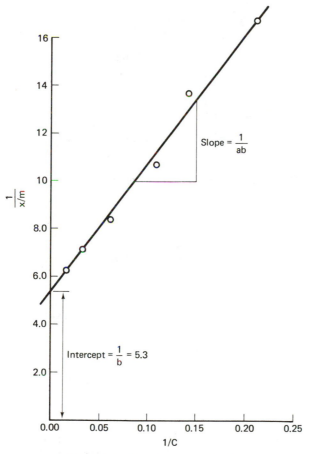

FIGURE 6-6 Plot of Langmuir equation.

EXAMPLE PROBLEM 6-1: An adsorption study is set up in the laboratory by adding a known amount of activated carbon to six flasks which contain 200 mℓ of an industrial waste. An additional flask containing 200 mℓ of waste but no carbon is run as a blank. Plot the data according to the Langmuir equation, and determine the values of the constants a and b.

Flask No.	Wt. of Carbon (mg) (m)	Volume in Flask (mℓ)	Final COD (mg/ℓ) (C)	Wt. of Adsorbate Adsorbed (mg)	$\frac{x}{m}$ (mg/mg)
1	804	200	4.70	49.06	0.061
2	668	200	7.0	48.6	0.073
3	512	200	9.31	48.1	0.094
4	393	200	16.6	46.7	0.118
5	313	200	32.5	43.5	0.139
6	238	200	62.8	37.4	0.157
7	0	200	250	0	0

Solution:

1. Calculate the values of x and x/m from the data:
   For flask #1:

   $$x = (250 \text{ mg/}\ell - 4.70 \text{ mg/}\ell) \frac{200 \text{ m}\ell \text{ in flask}}{1000 \text{ m}\ell/\text{liter}}$$

   $$= 49.06 \text{ mg}$$

   $$\frac{x}{m} = \frac{49.06 \text{ mg}}{804 \text{ mg}} = 0.061 \text{ mg/mg}$$

2. Plot values of $(1/x/m)$ vs. $1/c$ as shown in Figure 6-6.

3. Determine the values of the constants a and b.

   (a) Read value of intercept $= 1/b = 5.3$:

   $$b = \frac{1}{5.3} = 0.189$$

   (b) Calculate slope:

   $$\frac{13.42 - 10.0}{0.15 - 0.085} = \frac{3.42}{0.065} = 52.615$$

   (c) Since slope $= 1/ab$ and $b = 0.189$,

   $$a = \frac{1}{52.615(0.189)} = 0.101$$

4. The Langmuir equation is

   $$\frac{1}{x/m} = \frac{1}{0.189} - \frac{1}{(0.101)(0.189)c}$$

   $$= 5.3 - \frac{52.615}{c}$$

Freundlich (1926) developed an empirical equation to describe the adsorption process. His development was based on the assumption that the adsorbent had a heterogeneous surface composed of different classes of adsorption sites, with adsorption on each class of site following the Langmuir isotherm. Freundlich offered the following equation:

$$\frac{x}{m} = KC^{1/n} \tag{6-5}$$

where
x = amount of solute adsorbed (mg, g)

m = weight of adsorbent (mg, g)

C = concentration of solute remaining in solution after adsorption is complete (mg/ℓ)

K and n = constants that must be evaluated for each solute and temperature.

The Freundlich equation can be put in a useful form by taking the log of both sides.

$$\log\left(\frac{x}{m}\right) = \log K + \frac{1}{n}\log C \tag{6-6}$$

Thus, a plot of log x/m vs. log C should yield a straight line for adsorption data which follow the Freundlich theory. Values of the constants n and K can be determined from the plot as shown in Figure 6-7.

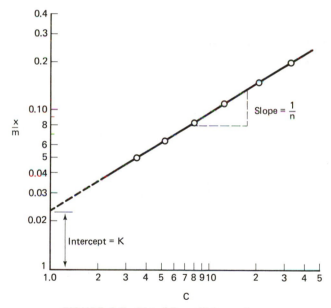

FIGURE 6-7 *Plot of Freundlich equation.*

EXAMPLE PROBLEM 6-2: The following laboratory data were collected in a batch adsorption study. Plot the data according to the Freundlich isotherm and determine the values for the constants n and k. A volume of 500 mℓ is placed in each flask, and the waste has an initial COD of 100 mg/ℓ.

Flask No.	Wt. of Carbon (mg) (m)	Volume in Flask (ml)	Final COD mg/l (C)	Wt. of Adsorbate Adsorbed (mg) (x)	$\frac{x}{m}$ (mg/mg)
1	965	500	3.5	48.25	0.05
2	740	500	5.2	47.40	0.064
3	548	500	8.0	46.0	0.084
4	398	500	12.5	43.75	0.11
5	265	500	20.5	39.75	0.15
6	168	500	33	33.5	0.20
7	0	500	100	0	0

Solution:

1. Calculate the values of x and x/m from the data. Refer to example problem 6-1.

2. Plot values of x/m vs. C on log-log paper as shown in Figure 6-7.

3. Determine the values of the constants n and K.

   (a) To determine the intercept of a line on a log-log plot, the value of the intercept must be read at the point where the value of the abscissa is equal to 1.0.

   To determine K from Figure 6-7, locate a value of $C = 1.0$ and read

   $$K = 0.023$$

   (b) The slope of the line will yield a value of $1/n$. The slope of a log-log plot can be determined by scaling or by the following calculations.

   $$\log \frac{x}{m} = \log K + \frac{1}{n} \log C$$

   $$\frac{1}{n} = \frac{\log \frac{x}{m} - \log K}{\log C}$$

   For a point on the line at $x/m = 0.07$, $C = 6$,

   $$\frac{1}{n} = \frac{\log 0.07 - \log 0.023}{\log 6} = \frac{-1.154 - (-1.638)}{0.778}$$

   $$\frac{1}{n} = 0.622$$

   $$n = 1.607$$

   (c) The Freundlich equation then becomes

   $$\log \frac{x}{m} = \log 0.023 + \frac{1}{1.607} \log C$$

   $$\log \frac{x}{m} = -1.638 + 0.622 \log C$$

Brunauer, Emmett, and Teller (1938) derived an adsorption isotherm based on the assumption that molecules could be adsorbed more than one layer thick on the surface of the adsorbent. Their equation, like the Langmuir equation, assumes that the adsorbent surface is composed of uniform, localized sites and that adsorption at one site does not affect adsorption at neighboring sites. Moreover, it was assumed that the energy of adsorption holds the first monolayer but that the condensation energy of the adsorbate is responsible for adsorption of successive layers. The equation, known as the *BET equation*, is commonly written as follows:

$$\frac{x}{m} = \frac{ACx_m}{(C_s - C)\left[1 + (A - 1)\dfrac{C}{C_s}\right]} \tag{6-7}$$

where

$x =$ amount of solute adsorbed (mg, moles)

$m =$ weight of adsorbent (mg, g)

$x_m =$ amount of solute adsorbed in forming a complete monolayer (mg/g, moles/g)

$C_s =$ saturation concentration of solute (mg/ℓ, moles/ℓ)

$C =$ concentration of solute in solution at equilibrium (mg/ℓ, moles/ℓ)

$A =$ a constant to describe the energy of interaction between the solute and the adsorbent surface.

Rearranging the BET equation yields

$$\frac{C}{(C_s - C)\dfrac{x}{m}} = \frac{1}{A(x_m)} + \frac{A - 1}{Ax_m}\left(\frac{C}{C_s}\right) \tag{6-8}$$

Data from adsorption processes that conform to the BET equation will yield a straight line when the left-hand side of equation 6-8 is plotted against C/C_s as shown in Figure 6-8. The resulting line has a slope of $(A - 1)/Ax_m$ and an intercept of $1/Ax_m$.

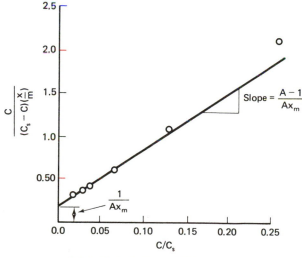

FIGURE 6-8 *Plot of BET equation.*

EXAMPLE PROBLEM 6-3: Plot the data from example problem 6-1, using the BET equation, and determine the values of the constants A and x_m.

Flask No.	Wt. of Carbon (mg) (m)	Final COD (mg/ℓ) (C)	$\dfrac{x}{m}$ (mg/mg)	$(C_s - C)$	$\dfrac{C}{(C_s - C)\dfrac{x}{m}}$	$\dfrac{C}{C_s}$
1	804	4.70	0.061	245.3	0.314	0.0188
2	668	7.0	0.073	243	0.395	0.028
3	512	9.31	0.094	240.7	0.412	0.037
4	393	16.6	0.118	233.4	0.603	0.066
5	313	32.5	0.139	217.5	1.075	0.13
6	238	62.8	0.157	187.2	2.136	0.251
7	0	250	0	250	—	—

Solution:

1. Calculate the values of $\dfrac{C}{(C_s - C)x/m}$ and C/C_s and plot as shown in Figure 6-8.

2. Read the value of the intercept:

$$\frac{1}{Ax_m} = 0.18$$

3. Calculate the value of the slope of the line:

$$\text{slope} = \frac{1.58 - 1.30}{0.21 - 0.165} = \frac{0.28}{0.045} = 6.22$$

4. Determine the value of A and x_m.

   (a) The value of A can be calculated:

$$\frac{A - 1}{Ax_m} = \text{slope} = 6.22$$

   Since $Ax_m = 0.18$,

$$(A - 1) = 6.22(5.55)$$
$$A = 6.22(5.55) + 1 = 35.55$$

   (b) The value of x_m can be calculated:

$$\frac{1}{Ax_m} = 0.18$$

$$x_m = \frac{1}{A(0.18)} = \frac{1}{(35.55)(0.18)} = 0.16$$

All three of the above equations have been successfully used to analyze adsorption data from wastewater studies. In general, the Langmuir and BET equations do not apply as well as the Freundlich equation to mixed solutes or to dilute solutions. Thus, the Freundlich equation finds wide application in environmental engineering.

Usually, the isotherm equation that yields a linear plot of the data is selected for use. In some cases, none of the equations will yield linear plots. The situation with nonlinear isotherms is discussed in section 11-3.

PROBLEMS

6-1. An adsorption study was conducted by adding varying amounts of activated carbon to a series of six flasks containing 500 mℓ of an industrial waste having an initial TOC of 150 mg/ℓ. The flasks were agitated for 4 hr, and the residual, steady-state, TOC concentrations were determined. Plot the Langmuir and Freundlich isotherms for the data presented below and determine the values of the appropriate constants.

Flask No.	Carbon Dosage (mg)	Final TOC (mg/ℓ)
1	0	150
2	75	105
3	175	70
4	250	54.5
5	500	28.3
6	1000	12.5

6-2. The following data were obtained in a batch adsorption study of an industrial waste with an initial COD of 150 mg/ℓ. Six flasks were set up to contain 250 mℓ of the sample and a measured amount of activated carbon. A seventh flask contained no carbon and served as a blank. Determine the values of the constants in the BET equation for these data.

Flask No.	Wt. of Carbon (mg) (m)	Final COD (mg/ℓ) (C)
1	2837	3.6
2	1285	6.0
3	465	10.5
4	287	13.5
5	103	24.0
6	52	33
7	0	150

REFERENCES

BRUNAUR, J., EMMETT, P.H., and TELLER, E., "Adsorption of Gases in Multimolecular Layers", *J. Am. Chem. Soc.*, **60**, 309 (1938).

CHAPMAN, D.L., "A Contribution to the Theory of Electrocapillarity," *Phil. Mag.*, **25**, 6, 475–481.

CULP, R.L., and CULP, G.L., *Advanced Wastewater Treatment*, Van Nostrand Reinhold Company, New York, (1971).

DERJAGUIN, B., and LANDAU, L.D., *Acta. Physicochem.* (U.S.S.R.), **14**, 635 (1941), *V. Exp. Theor. Phys.* (U.S.S.R.) **11**, 802 (reprinted **15**, 662, 1945).

DUBININ, M.M., PLAVNIK, G.M., and ZAVERINA, E.F., "Integrated Study of the Porous Structure of Activated Carbon from Carbonized Sucrose," *Carbon*, **2**, 261 (1964).

FAIR, G.M., GEYER, J.C., and OKUN, D.A., *Water and Wastewater Engineering*, Vol. 2, *Water Purification and Wastewater Treatment and Disposal*, John Wiley & Sons, Inc., New York (1968).

FREUNDLICH, H., *Colloid and Capillary Chemistry*, Metheun, London (1926).

GOUY, G., "Sur la Constitution de la Charge E'lectrique a'la Surface D'um E'lectrolyte", *Ann. Phys.* (Paris) Serie 4, 9, 457–468.

HASSLER, J.W., *Purification with Activated Cardon*, Chemical Publishing Company, New York, (1974).

LANGMUIR, I.J., "The Adsorption of Gases on Plane Surfaces of Glass, Mica and Platinum," *J. Am. Chem. Soc.*, **40**, 1361 (1918).

O'MELIA, C.R., "A Review of the Coagulation Process," *Public Works* (May, 1969).

SHAW, D.J., *Introduction to Colloid and Surface Chemistry*, 2nd ed., Butterworths, London (1970).

SNOEYINK, V.L., and WEBER, W.J., Jr., "The Surface Chemistry of Active Carbon," *Environmental Science and Technology*, **1**, 3, 228 (March, 1967).

STERN, O., "Zur Theorie der, Elektrolytischem Doppelschicht," *A. Electrochem.*, **30**, 508 (1924).

STUMM, W., "Chemical Interactions in Particle Separation," in *Chemistry of Wastewater Technology*, [Ed.] A.J. RUBIN, Ann Arbor Science Publishers, Ann Arbor, MI (1978).

VAN OLPHEN, H., *An Introduction to Clay Colloid Chemistry* 2nd ed., Wiley-Interscience, New York (1977).

VERWEY, E.J.W., and OVERBECK, J.T.G., *Theory of the Stability of Lyophobic Colloids*, Elsevier Scientific Publishing Company, Amsterdam (1948).

WEBER, W.J., Jr., "*Physicochemical Processes for Water Quality Control*," John Wiley & Sons, Inc., New York, (1972).

7

COAGULATION IN WATER TREATMENT

Surface waters generally contain a wide variety of colloidal impurities that may cause the water to appear turbid or may impart color. Turbidity is most often caused by colloidal clay particles produced by soil erosion. Color may result from colloidal forms of iron and manganese or, more commonly, from organic compounds contributed by decaying vegetation. It is of course possible for a water to be high in color and low in turbidity or low in color and high in turbidity.

Colloidal particles that cause color and turbidity are difficult to separate from water because the particles will not settle by gravity and are so small that they pass through the pores of most common filtration media. To be removed, the individual colloids must aggregate and grow in size. Aggregation is complicated not only by the small size of the particles but more importantly by the fact that physical and electrical forces keep the particles separated from each other and prevent the collisions that would be necessary for aggregation to occur.

Chemical agents can be used to promote colloid aggregation by destroying the forces that stabilize colloidal particles. Mechanisms responsible for destabilization of inorganic clay colloids have been identified through extensive research studies and are well understood. The process of destroying the stabilizing forces and causing aggregation of clay colloids is referred to as *chemical coagulation*. The mechanisms responsible for destabilization of organic color colloids differ from those for clay and may be more properly regarded as chemical precipitation rather than chemical coagulation (AWWA, 1971).

The origins and distribution of forces contributing to the stability of colloidal clay particles are discussed in Chapter 6. The use of chemical coagulation for destruction of these stabilizing forces is discussed in the following pages. The reader should be familiar with the contents of Chapter 6 before proceeding further.

211

7-1 DESTABILIZATION OF COLLOIDAL DISPERSIONS

Colloidal systems are discussed in Chapter 6, where the overall stability of a colloidal particle is shown to be controlled by double-layer repulsion forces and van der Waals' forces of attraction. A resultant curve, representing the combined repulsion and attractive forces for a typical colloidal particle, is shown in Figure 6-4.

To induce colloidal particles to aggregate, two distinct steps must occur: (1) the repulsion forces must be reduced (i.e., the particle must be destabilized), and (2) particle transport must be achieved to provide contacts between the destabilized particles (Stumm and O'Melia, 1968). Particle destabilization can be achieved by four mechanisms: (1) double-layer compression, (2) adsorption and charge neutralization, (3) enmeshment in a precipitate, and (4) adsorption and interparticle bridging. It is absolutely essential that each of these mechanisms be understood in order to interpret coagulation data and to select the proper type and dosage of coagulant for a particular application.

At this point, it is necessary to define certain terms used in connection with colloid destabilization. Many authors use the term *coagulation* to describe processes by which the charge on particles is destroyed and *flocculation* to describe the aggregation of particles into larger units (LaMer and Healy, 1963). In this sense double-layer compression and charge neutralization would be classified as coagulation, while enmeshment and bridging would be considered to be flocculation. Chemicals used to destabilize colloids would be referred to as *coagulants* or *flocculants*, depending on their mode of operation.

According to common engineering usage, the term *flocculation* describes the particle transport step, while the term *coagulation* is used to describe the overall process of aggregation, including both destabilization and transport (Weber, 1972). This terminology is used in this text.

Reduction of Surface Potential by Double-Layer Compression

It has long been known that colloidal systems could be destabilized by the addition of ions having a charge opposite to that of the colloid. Schulze, in 1882, noted that the coagulating power of cations increased in the ratio of $1:10:1000$ as the valence of the ions increased from 1 to 2 to 3. A similar observation for anions was noted by Hardy (1900), who formulated the Schulze-Hardy rule:

> The coagulating power of a salt is determined by the valency of one of its ions. The prepotent ion is either the negative or positive ion, according to whether the colloidal particles move down or up the potential gradient. The coagulating ion is always of the opposite electrical sign to the particle.

The Schulze-Hardy rule is valid for indifferent electrolytes, those that do not engage in any sort of reaction with the sol.

The phenomenon observed by Schulze and Hardy is attributable to compression of the electrical double layer surrounding the colloidal particle and can be explained by the theory developed by Verwey and Overbeek (1948). The diffuse layer contains a quantity of counter ions sufficient to balance the electrical charge on the particle. The charge distribution in the diffuse layer of a negatively charged colloid can be represented by the curve $ABCD$ in Figure 7-1. The line BD represents the concentration of both cations and anions at a large distance from the surface. The average local concentration of ions of opposite sign is given by curve DA, and the concentration of ions with the same sign as the surface charge is indicated by curve DC. The surface area CAD represents the total, net, diffuse layer charge, which is equivalent to the surface charge (van Olphen, 1977).

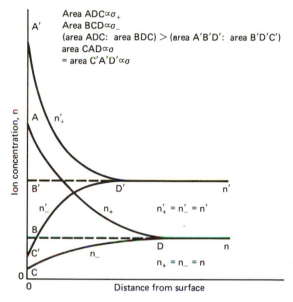

FIGURE 7-1 *Charge distribution in the diffuse double layer of a negative particle surface at two electrolyte concentrations for a constant surface charge (after Van Olphen, 1977).*

If an electrolyte is added to a colloidal dispersion, the surface charge on the particles will remain unchanged if that charge originates from crystal imperfections (i.e., clay particles). However, the added electrolyte will increase the charge density in the diffuse layer and result in less volume of the diffuse layer being required to neutralize the surface charge. Thus, the diffuse layer is *compressed* toward the particle surface as shown by the curve $A'B'C'D'$ in Figure 7-1. The total net charge in the diffuse layer has not changed (area CAD = area $C'A'D'$), but the thickness of the layer has been reduced. The effect of this compression is to change the distribution of double-layer repulsion forces in the vicinity of the colloid and cause a reduction in surface potential with increasing electrolyte concentration, which allows the van der Waals' attractive forces to be more dominant, thus enhancing particle aggregation.

The effect on particle stability of increasing the electrolyte concentration is shown in Figure 7-2. At low electrolyte concentrations the resultant of the repulsion and attraction forces shows a high energy hill that must be overcome if agglomeration is to occur. At a high electrolyte concentration the double layer has been compressed, and the resultant curve shows no repulsion at any distance. Particle agglomeration occurs very rapidly.

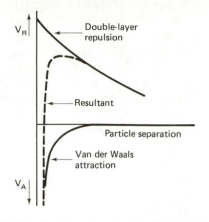

(a) Low electrolyte concentration with normal double-layer thickness. System is stable and agglomeration is imperceptable.

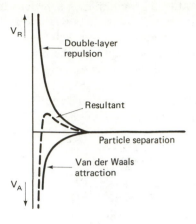

(b) Intermediate electrolyte concentration causes some double-layer compression. Slow agglomeration can occur.

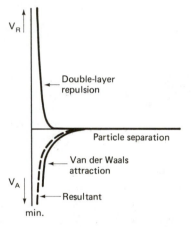

(c) High electrolyte concentration causes severe double-layer compression. Rapid agglomeration can occur.

FIGURE 7-2 *Effect of electrolyte concentration on double layer compression (after Van Olphen, 1977).*

The decrease in surface potential resulting from double-layer compression will be reflected by an associated decrease in zeta potential. It has been suggested that coagulation by double-layer compression is usually optimum when the value of the zeta potential is near zero. However, because of uncertainties involved in calculating the zeta potential and because the location of the plane of shear is not known exactly, the zeta potential is not a useful quantitative criterion of colloidal stability (van Olphen, 1977).

A mathematical model developed independently by Derjaguin and Landau and Verwey and Overbeek (1948) makes it possible to calculate the concentration of an indifferent electrolyte required to cause coagulation. This model predicts that the coagulation concentration for counter-ions with charge numbers 1, 2, and 3 should be in the ratio of (Shaw, 1970):

$$\frac{1}{1^6} : \frac{1}{2^6} : \frac{1}{3^6} \quad \text{or equivalent ratios of } 1000 : 16 : 1.3$$

Thus, the theoretical model is in good agreement with the empirical Schulze-Hardy rule.

The relative coagulating power of aluminum, calcium, and sodium ions is shown in Figure 7-3(a). It is important to understand that all metal cations are hydrated in water and that such simple ions as Al^{3+}, Ca^{2+}, and Na^+ do not actually exist (Weber, 1972). Instead, these ions do exist as aquocomplexes such as $Al(H_2O)_6^{3+}$. Well-hydrated ions usually are not adsorbed to the colloidal surface but remain in the diffuse part of the double layer. Two interesting aspects of double-layer compression are (1) the amount of electrolyte required to achieve coagulation by double-layer compression is practically independent of the concentration of colloids in the dispersion, and (2) it is not possible to cause a *charge reversal* on a colloid by double-layer compression, regardless of how much electrolyte is added.

Reduction of Surface Potential by Adsorption and Charge Neutralization

Some chemical species are capable of being adsorbed at the surface of colloidal particles. If the adsorbed species carry a charge opposite to that of the colloids, such adsorption causes a reduction of surface potential and a resulting destabilization of the colloidal particle. Reduction of surface charge by adsorption is a much different mechanism than reduction by double-layer compression, and the behavior of systems destabilized in this manner cannot be explained on the basis of the Verwey-Overbeek model.

Destabilization by adsorption differs from destabilization by double-layer compression in three very important ways. First, sorbable species are capable of destabilizing colloids at much lower dosages than nonsorbable, "double-layer compressing" ions. Tamamushi and Tamaki (1959) reported that a dosage as low as 6×10^{-5} mole/ℓ of the sorbable dodecylammonium ion ($C_{12}H_{25}NH_3^+$) would produce destabilization of a negatively charged silver iodide sol. In contrast, the nonsorbable Na^+ with a similar charge is only effective as a coagulant in concentrations above 10^{-1} moles/ℓ [Figure 7-3(a)].

Secondly, destabilization by adsorption is stoichiometric. Thus, the required dosage of coagulant increases as the concentration of colloids (more specifically, the total surface area of the colloids) increases. As previously mentioned, the amount of electrolyte required to achieve coagulation by double-layer compression is not stoichiometric and is practically independent of colloid concentration.

Thirdly, it is possible to overdose a system with an adsorbable species and cause restabilization as a result of a reversal of charge on the colloidal particle. The effect of overdosing with dodecylammonium ion is shown in Figure 7-3(b). The fact that ions can be adsorbed beyond neutralization to the point of charge reversal suggests that

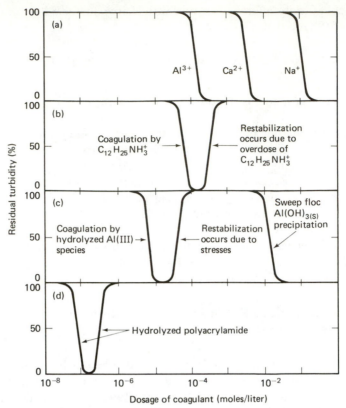

FIGURE 7-3 *Schematic coagulant curves for various modes of destabilization (after Weber, 1972).*

specific chemical interactions can outweigh electrostatic repulsion effects in some cases (Stumm and O'Melia, 1968).

Hydrolyzed species of Al(III) and Fe(III) can cause coagulation by adsorption. Hydrolyzed aluminum(III) species are shown in Figure 7-3(c) to be effective at a somewhat lower dosage than dodecylammonium ions; again, restabilization due to charge reversal occurs if the system is overdosed.

Destabilization of Colloids by Enmeshment in a Precipitate

If certain metal salts are added to water or wastewater in sufficient amounts, rapid formation of precipitates will occur. Colloids may serve as condensation nuclei for these precipitates or may become enmeshed as the precipitates settle. Coagulants such as $Al_2(SO_4)_3$, $FeCl_3$, $MgCO_3$, and $Ca(OH)_2$ can induce coagulation through the formation of insoluble $Al(OH)_{3(s)}$, $Fe(OH)_{3(s)}$, $Mg(OH)_{2(s)}$, and $CaCO_{3(s)}$. Removal of colloids in this manner is frequently referred to as *sweep-floc* coagulation.

Several characteristics that distinguish sweep-floc coagulation from double-layer compression and adsorption have been reported by Packham (1965). For example, an inverse relationship exists between the optimum coagulant dosage and the concentration of colloids to be removed. This can be explained as follows: at low colloid concentrations a large excess of coagulant is required to produce a large amount of precipitate that will enmesh the relatively few colloidal particles as it settles. At high colloid concentrations, coagulation will occur at a lower chemical dosage because the colloids serve as nuclei to enhance precipitate formation.

Since sweep-floc coagulation does not depend on neutralization of surface charge, the conditions for optimum coagulation do not correspond to a minimum zeta potential. However, an optimum pH does exist for each coagulant, depending on its solubility-pH relationship.

Destabilization by Adsorption and Interparticle Bridging

Many different natural compounds such as starch, cellulose, polysaccharide gums, and proteineous materials, as well as a wide variety of synthetic polymeric compounds are known to be effective coagulating agents. These materials are characterized by a large molecular size and most have multiple electrical charges along a molecular chain of carbon atoms.

Research has revealed that both positive (cationic) and negative (anionic) polymers are capable of destabilizing negatively charged colloidal particles. Neither the double-layer compression model nor the charge neutralization model can be used to explain these results. Ruehrwein and Ward (1952) and LaMer and Healy (1963) have developed a chemical bridging theory that is consistent in explaining the observed behavior of these polymeric compounds.

The chemical bridging theory proposes that a polymer molecule will become attached to a colloidal particle at one or more sites as shown by reaction 1 in Figure 7-4. Attachment may result from coulombic attraction if the polymer and particle are of opposite charge, or from ion exchange, hydrogen bonding, or van der Waals' forces if they are of similar charge (O'Melia, 1969). The "tail" of the adsorbed polymer will extend out into the bulk of the solution and can become attached to vacant sites on the surface of another particle to form a chemical bridge as shown by reaction 2. This bridging action results in the formation of a floc particle having favorable settling characteristics. If the extended segment fails to contact another particle, it may fold back and attach to other sites on the original surface, thus restabilizing the particle as shown by reaction 3.

Inefficient coagulation may result from an overdose of polymer to the system or from intense or prolonged agitation. If excessive polymer is added, the segments may saturate the surfaces of colloidal particles so that no sites are available for the formation of polymer bridges (see reaction 4, Figure 7-4). This can restabilize the particles and may or may not be accompanied by charge reversal. Intense or prolonged mixing may destroy previously formed bridges and lead to restabilization as shown in reactions 5 and 6.

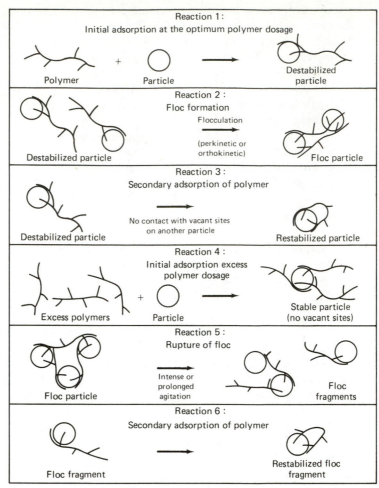

FIGURE 7-4 *Schematic of reactions between colloidal particles and polyelectrolytes (after O'Melia, 1969).*

7-2 COAGULATION IN WATER TREATMENT

As mentioned previously, coagulation is a two-step process involving particle destabilization followed by particle transport to promote collisions between the destabilized particles. Destabilization is induced by the addition of a suitable coagulant, and particle contact is ensured through appropriate mixing devices. In a typical water treatment plant coagulation occurs in the rapid-mixing and flocculation units.

The characteristics and capabilities of the most popular coagulants are discussed in the following pages. Factors of interest in the design of rapid mixing and flocculation units are presented in section 7-3.

Coagulation with Al(III) and Fe(III)

The salts of Al(III) and Fe(III) listed in Table 7-1 are commonly used as coagulants in water and wastewater treatment. Both Al(III) and Fe(III) are hydrolyzing metal ions, and a knowledge of the chemistry of these ions is essential to an understanding of their role in coagulation.

TABLE 7-1 *Common water treatment coagulants (after Parsons, 1965).*

Chemical Name, Formula, and Trade Name	Molecular Weight	Common Forms and Commercial Strength		Bulk Density (lbs/cu ft)	Solubility (g/100 g Water, °C)
Aluminum sulfate $Al_2(SO_4)_3 \cdot 14H_2O$ Alum	594.4	Tan to gray-green: powder granules } 17% Al_2O_3 lump } minimum liquid 8.3% Al_2O_3		38–45 60–63 62–67	65.3 at 10°C 71 at 20°C 78.8 at 30°C
Ferric Chloride $FeCl_3$	162.2	Anhydrous:	green-black powder 96–97% $FeCl_3$	65–70	74.4 at 0°C
		Heptahydrate:	yellow-brown lump 60% $FeCl_3$	60–64	536 at 100°C
		Liquid:	dark brown solution 37–47% $FeCl_3$		
Ferric Sulfate $Fe_2(SO_4)_3$	399.9	Dihydrate:	red-brown granules 20.5% Fe	70–72	300 at 20°C
		Trihydrate:	red-gray granules 18.5% Fe	70–72	

* Reagent grade alum is $Al_2(SO_4)_3 \cdot 18H_2O$. Commercial alum or filter alum differs from reagent grade and has no exact composition. Filter alum is usually written as $Al_2(SO_4)_3 \cdot 14H_2O$.

When a salt of Al(III) or Fe(III) is added to water, it will dissociate to yield trivalent Al^{3+} or Fe^{3+} ions, which hydrate to form the aquometal complexes $Al(H_2O)_6^{3+}$ and $Fe(H_2O)_6^{3+}$. These complexes then pass through a series of hydrolytic reactions in which H_2O molecules in the hydration shell are replaced by OH^- ions. This gives rise to the formation of a variety of soluble species, including mononuclear species (one aluminum ion) such as $Al(OH)^{2+}$ and $Al(OH)_2^+$, and polynuclear species (several aluminum ions) such as $Al_8(OH)_{20}^{4+}$. Despite the fact that some of these products have only one or two positive charges, they are quite effective as coagulants because they adsorb very strongly onto the surface of most negative colloids. Hydrolysis schemes for aluminum(III) and iron-(III) are shown in Figure 7-5.

The equilibria for the various species shown in Figure 7-5 are a function of pH and can be represented by a log [species] vs. pH solubility diagram. Diagrams for aluminum(III) and iron(III) have been previously presented as Figures 3-1 and 3-2.

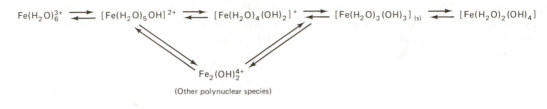

(a) Hydrolysis scheme for iron (III)

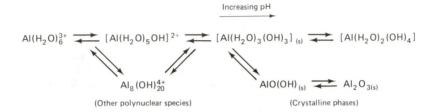

(b) Hydrolysis scheme for aluminum (III)

FIGURE 7-5 *Hydrolysis schemes for iron (III) and aluminum (III) (after Rubin and Kovac, 1974).*

According to Stumm and O'Melia (1968), aluminum(III) and iron(III) accomplish destabilization by two mechanisms: (1) adsorption and charge neutralization and (2) enmeshment in a sweep floc. If an aluminum(III) or iron(III) salt is added to water in concentrations less than the solubility limit of the metal hydroxide, the hydrolysis products will form and adsorb onto the particles, causing destabilization by charge neutralization. When the amount of aluminum(III) or iron(III) added to water is sufficient to exceed the solubility of the metal hydroxide, the hydrolysis products will form as kinetic intermediates in the formation of the metal hydroxide precipitate. In this situation charge neutralization and enmeshment in the precipitate both contribute to coagulation. Interrelations between pH, coagulant dosage, and colloid concentration determine the mechanism responsible for coagulation.

The charge on hydrolysis products and the precipitation of metal hydroxides are both controlled by pH. Referring to Figures 3-1 and 3-2, the hydrolysis products possess a positive charge at pH values below the iso-electric point of the metal hydroxide (O'Melia, 1969). These positively charged species can cause destabilization of negatively charged colloids by adsorption and charge neutralization. Negatively charged species ($Al(OH)_4^-$ and $Fe(OH)_4^-$), which predominate above the iso-electric point, are ineffective for the destabilization of negatively charged colloids.

Precipitation of amorphous metal hydroxide is necessary for sweep-floc coagulation. The solubility of $Al(OH)_{3(s)}$ and $Fe(OH)_{3(s)}$ is minimal at a particular pH and increases as the pH increases or decreases from that value.

Obviously, pH must be controlled to establish optimum conditions for coagulation.

Control is complicated by the fact that the aquometal ions of aluminum(III) and iron(III) are acidic in nature, as shown in the following equations (Weber, 1972):

$$Al(H_2O)_6^{3+} + H_2O \rightleftharpoons Al(H_2O)_5(OH)^{2+} + H_3O^+ \qquad \textbf{(7-1)}$$

$$Fe(H_2O)_6^{3+} + H_2O \rightleftharpoons Fe(H_2O)_5(OH)^{2+} + H_3O^+ \qquad \textbf{(7-2)}$$

Hydrogen ions liberated by the addition of alum will react with natural alkalinity in the water as follows:

$$Al_2(SO_4)_3 \cdot 14H_2O + 3Ca(HCO_3)_2 \longrightarrow$$
$$2Al(OH)_3 + 3CaSO_4 + 14H_2O + 6CO_2 \qquad \textbf{(7-3)}$$

Theoretically, each mg/ℓ of alum will consume approximately 0.50 mg/ℓ (as $CaCO_3$) of alkalinity and produce 0.44 mg/ℓ of carbon dioxide. If the natural alkalinity is not sufficient to react with the alum and buffer the pH, it may be necessary to add alkalinity to the water in the form of lime or soda ash.

$$Al_2(SO_4)_3 \cdot 14H_2O + 3Ca(OH)_2 \longrightarrow 2Al(OH)_3 + 3CaSO_4 + 14H_2O \qquad \textbf{(7-4)}$$

$$Al_2(SO_4)_3 \cdot 14H_2O + 3Na_2CO_3 + 3H_2O \longrightarrow$$
$$2Al(OH)_3 + 3Na_2SO_4 + 3CO_2 + 14H_2O \qquad \textbf{(7-5)}$$

Example problem 7-1 shows how the above equations can be used to estimate the volume of sludge produced during coagulation. The reader is cautioned that the quantities determined by such calculations are only approximate because equation 7-3 does not accurately describe what actually occurs during the coagulation process.

EXAMPLE PROBLEM 7-1: A raw water supply is treated with an alum dosage of 25 mg/ℓ. Calculate the following:

1. The amount of alum required to treat a flow of 1.0 MGD.

2. The amount of natural alkalinity required to react with the alum added.

3. The volume of $Al(OH)_3$ sludge produced per MGD if it is collected at 2.0% solids. Assume that the dry solids have a specific gravity of 2.2.

Solution:

1. The dosage of alum for a 1.0 MGD flow is

$$\frac{lb}{day} = 8.34 \frac{lb/day}{(mg/\ell)(MGD)} \times 25\ mg/\ell \times 1.0\ MGD = 208.5$$

2. The reaction between alum and natural alkalinity is

$$\underset{594.4}{Al_2(SO_4)_3 \cdot 14H_2O} + \underset{3(162)}{3Ca(HCO_3)_2} \longrightarrow$$
$$\underset{155.96}{2Al(OH)_3} + 3CaSO_4 + 14H_2O + 6CO_2 \qquad \textbf{(7-3)}$$

One mole of alum reacts with 3 moles of $Ca(HCO_3)_2$ as shown in equation 7-3. Expressing the reaction on a weight basis,

$$1 \text{ mg/}\ell \text{ of alum will react with } \frac{3(162)}{594.4} = 0.818 \text{ mg/}\ell \text{ } Ca(HCO_3)_2$$

Expressing $Ca(HCO_3)_2$ in mg/ℓ as $CaCO_3$,

$$\text{mg/}\ell \text{ as } CaCO_3 = 0.818 \times \frac{\text{eq. wt. } CaCO_3}{\text{eq. wt. } Ca(HCO_3)_2} = 0.818\left(\frac{50}{81}\right) = 0.505 \text{ mg/}\ell$$

$$\begin{array}{l} \text{Amount of natural} \\ \text{alkalinity reacting} \\ \text{with alum} \end{array} = 25 \text{ mg/}\ell \text{ alum} \times 0.5 \frac{\text{mg/}\ell \text{ alkalinity}}{\text{mg/}\ell \text{ alum}} = 12.5 \text{mg/}\ell$$

3. The volume of $Al(OH)_3$ sludge produced per day:

(a) Referring to equation 7-3, 1 mole of alum produces 2 moles of $Al(OH)_3$. Expressing the reaction on a weight basis,

$$1 \text{ lb of alum will produce } \frac{155.96}{594.4} = 0.262 \text{ lb } Al(OH)_3$$

Thus, 208.5 lb alum will produce

$$0.262(208.5) = 54.627 \text{ lb/day } Al(OH)_3$$

(b) Calculate the specific gravity of sludge at 2% solids:

$$\frac{1}{S_s} = \sum_{i=1}^{n} \left(\frac{W_i}{S_i}\right)$$

where S_s = specific gravity of sludge

W_i = weight fraction of ith component of sludge

S_i = specific gravity of ith component of sludge.

$$\frac{1}{S_s} = \left(\frac{0.98}{1.0} + \frac{0.02}{2.2}\right) = 0.98 + 0.009 = 0.989$$

$$S_s = 1.011$$

(c) Calculate the volume of sludge:

$$\text{volume} = 54.627 \frac{\text{lb}}{\text{day}} \times \frac{100}{2} \times \frac{1}{8.34 \text{ lb/gal}} \times \frac{1}{1.011} = 323.94 \text{ gal}$$

Alternatively,

$$\text{water in sludge} = 54.627 \frac{\text{lb}}{\text{day}} \times \frac{98}{2} = 2676.72 \text{ lb/day}$$

$$\text{volume of wet sludge} = \frac{54.627 \text{ lb/day}}{8.34 \text{ lb/gal} \times 2.2} + \frac{2676.72 \text{ lb/day}}{8.34 \text{ lb/gal}}$$

$$= 2.97 \text{ gal} + 320.965$$

$$= 323.94 \text{ gal}$$

In practice, most water treatment plants utilizing alum operate between a pH of 6.0 and 7.5 and with alum dosages of 5 to 50 mg/ℓ. This combination of pH and alum dosage is in the range of insolubility of $Al(OH)_3$ as shown by the shaded area in Figure 3-1. It is likely that under these conditions destabilization is brought about by a combination of adsorption and enmeshment (Stumm and O'Melia, 1968).

Actually, the mechanism of destabilization depends on the concentration of colloids (more specifically, the surface area of the colloids) and the amount of alum added. Figure 7-6(a) represents the results of typical coagulation studies with various concentrations of colloids at a constant pH. In zone 1 insufficient coagulant has been applied, and destabilization does not occur. Increasing the coagulant dosage results in destabilization in zone 2 and can lead to restabilization of the dispersion (zone 3). If a sufficient amount of coagulant is added, sweep-floc coagulation will occur in zone 4.

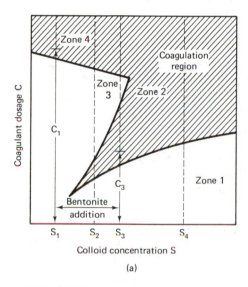

(a)

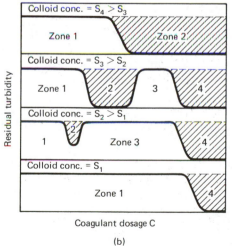

Coagulant dosage C

(b)

FIGURE 7-6 Schematic coagulation curves at constant pH. Shaded areas represent regions in which coagulation occurs (after Stumm and O'Melia, 1968).

Figure 7-6(b) is a plot of sections through the curve of Figure 7-6(a) at the points indicated by the vertical lines. The relationship between colloid concentration, coagulant dosage, and destabilization mode is evident from this figure.

(a) *Colloid Concentration S_1:* At this low colloid concentration limited opportunity exists for contact between hydrolysis products and the surface of the colloids and between destabilized colloids. Coagulation does not occur until the coagulant concentration is high enough to cause sweep-floc coagulation.

(b) *Colloid Concentration $S_2 > S_1$:* At a higher colloid concentration, increased contact opportunity leads to destabilization by adsorption and charge neutralization at a relatively low coagulant dosage. Charge reversal and destabilization occur as the coagulant dosage is increased. Finally, removal by sweep-floc coagulation occurs when the solubility of the metal hydroxide is exceeded.

(c) *Colloid Concentration $S_3 > S_2$:* The same relationships are evident here as in the previous case. Note the stoichiometric relationship associated with adsorption and charge neutralization and the inverse relationship associated with sweep-floc coagulation. The destabilization zone (zone 2) widens, indicating that a larger dosage of coagulant is required to cause a charge reversal on the increased surface area.

Coagulation with Polymers

Synthetic organic polymers can be effective as coagulants or coagulant aids. These polymers are long-chain molecules comprised of many subunits called *monomers*. A polymer that contains only one type of monomer is referred to as a *homopolymer*; those containing different monomers are called *copolymers*. The number and type of subunits can be varied to yield a wide range of polymers with different characteristics and molecular weights. A polymer is referred to as a *polyelectrolyte* if its monomers contain ionizable groups. Cationic polyelectrolytes acquire a positive charge upon ionization, whereas anionic polyelectrolytes acquire a negative charge. Polymers that contain no ionizable groups are termed *nonionic*.

Ruehrwein and Ward (1952) studied the mechanism by which clay particles were aggregated by polyelectrolytes and proposed the concept of chemical bridging to explain the behavior of polymer-clay systems. Their results showed cationic polymers to be effective coagulants for negatively charged clay particles. It was suggested that bridges were formed when the polymers became attached to the surfaces of clay particles by the process of electrostatic attraction or ion exchange. Anionic polymers were found to be ineffective coagulants for negatively charged clay particles when used alone, but their effectiveness was increased when used in conjunction with an extraneous electrolyte such as NaCl or $CaCl_2$ or a coagulant such as alum (Ruehrwein and Ward, 1952; Black, Birkner, and Morgan, 1965).

The size and shape of a polymer in solution is important in determining its effectiveness and is influenced by ionic strength, ionic valences, and pH. Polymers typically have a helical molecular structure comprised of carbon chains with ionizing groups attached. When the groups are ionized in solution, an electrical repulsion is created which causes the polymer to assume the shape of an extended rod (Shea, 1972). As the ionized groups become attached to colloidal particles the charges are neutralized and the polymer starts to coil and form a dense floc with favorable settling properties. An overdose of polymer may result in an excess of unneutralized sites and cause the stable, straight-chain structure to persist. A low ionic strength or a pH that favors a high degree of ionization may also cause the polymer to remain extended and thus adversely affect coagulation.

Dosages of only 0.5 to 1.5 mg/ℓ of cationic polymer are often effective for coagulation, whereas as much as 10 to 20 times that amount of alum would be required to achieve the same results. Another important difference between the use of polymers and alum is in dosage control. Referring to Figure 7-3(d), it is obvious that a very narrow optimum exists for the polymer and that overdosing or underdosing will result in restabilization of the colloids. Figure 7-3(c) indicates that a system may also be stabilized by overdosing with hydrolyzing Al(III) ions but that removal by sweep-floc precipitation will eventually occur at extremely high dosages. Thus, dosage control is more difficult to achieve when using polymers because of their narrow range of effectiveness. In the absence of precise dosage control, polymers will not yield satisfactory performance.

Polymers are not acidic and do not lower the pH of water as alum does; thus, their use offers a decided advantage for treating low alkalinity waters similar to those found in the southeastern United States, particularly if the waters are high in turbidity. The large alum dosages required to treat such waters would liberate a significant amount of hydrogen ions and necessitate the addition of lime or soda ash to maintain the proper pH level. Other advantages of polymers over alum include reducing the volume of sludge produced, providing a sludge that is easier to dewater, preventing the carryover of soluble aluminum into the distribution system, and preventing the carryover of light floc (Beardsley, 1973).

Coagulation with Magnesium

Thompson, Singley, and Black (1972) have shown that magnesium precipitated as $Mg(OH)_2$ can be an effective coagulant for removal of color and turbidity from natural waters. Coagulation occurs as a result of colloidal particles becoming enmeshed in the gelatinous hydroxide precipitate.

For waters high in naturally occurring magnesium, coagulation can be achieved by simply adding a sufficient amount of lime to precipitate $Mg(OH)_2$. For waters low in natural magnesium it is necessary to provide magnesium through the addition of a suitable salt. One advantage of the process is that both magnesium and lime can be recovered from the sludge and recycled. This simplifies sludge disposal problems and improves the overall economics of treatment.

The chemistry of magnesium coagulation is a combination of water softening and coagulation chemistry. Sufficient lime must be added to satisfy the CO_2 demand, to convert all bicarbonate alkalinity to carbonate alkalinity, and to precipitate the $Mg(OH)_2$.

Reactions are as follows:

$$CO_2 + Ca(OH)_2 \longrightarrow CaCO_3 + H_2O \qquad \text{(7-6)}$$

$$Ca(HCO_3)_2 + Ca(OH)_2 \longrightarrow 2CaCO_3 + 2H_2O \qquad \text{(7-7)}$$

$$Mg(HCO_3)_2 + Ca(OH)_2 \longrightarrow MgCO_3 + CaCO_3 + 2H_2O \qquad \text{(7-8)}$$

$$MgCO_3 + Ca(OH)_2 \longrightarrow Mg(OH)_2 + CaCO_3 \qquad \text{(7-9)}$$

Figure 9-8 is a solubility diagram for magnesium based on equation 9-31. Obviously, precipitation is enhanced at high pH values. In practice, sufficient lime is added to the water to raise the pH to 10.7 and to sustain it at that level during treatment.

Recovery of magnesium is achieved by recarbonating the coagulant sludge to selectively dissolve $Mg(OH)_2$.

$$Mg(OH)_2 + CO_2 \longrightarrow Mg(HCO_3)_2 \qquad \text{(7-10)}$$

The solubilized magnesium can be separated from the remaining sludge by vacuum filtration and returned to the coagulation process. The remaining sludge can be treated by flotation to separate turbidity and then centrifuged to dewater $CaCO_3$, which can be recalcined for lime recovery.

$$CaCO_3 + heat \longrightarrow CaO + CO_2 \qquad \text{(7-12)}$$

A flow diagram for such a process is shown in Figure 7-7.

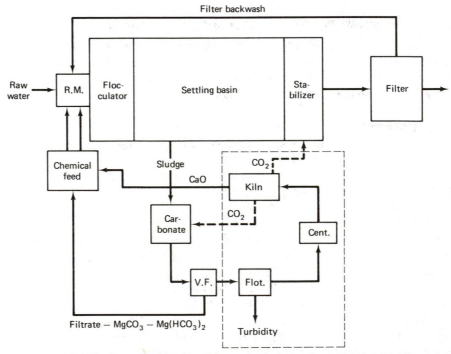

FIGURE 7-7 *Flow diagram of turbidity removal plant using magnesium and lime recovery (after Thompson, Singley, and Black, 1972).*

Although the process is one of recovery and reuse, some source of makeup chemical is required for waters lacking in natural magnesium. Economics usually dictate the choice of chemicals for this purpose. The so-called basic carbonate, $4MgCO_3 \cdot Mg(OH)_2 \cdot 5H_2O$, has been used in some cases but is somewhat expensive and relatively insoluble. Dolomitic lime is a promising source of magnesium and is currently being studied on a pilot basis (Thompson, 1979). Magnesium carbonate trihydrate could be used as a coagulant but is not commercially available. If necessary, this product could be recovered from sludges produced at water treatment plants softening high magnesium waters (Thompson, Singley, and Black, 1972a).

The use of magnesium coagulation reportedly offers the following advantages (Thompson, Singley, and Black, 1972a, b):

1. In soft surface waters the process results in increased alkalinity in the finished water, which makes the water easier to stabilize with respect to corrosion control. When alum is used as a coagulant for soft surface waters, alkalinity is consumed and the finished water is frequently difficult to stabilize.

2. Waters high in carbonate hardness are softened by the addition of lime required for magnesium coagulation. Hardness of soft waters is not reduced because of the solubility limitation of calcium and magnesium.

3. The high pH required for precipitation of $Mg(OH)_2$ will provide for improved disinfection and may eliminate the need for pre-chlorination in many plants.

4. The high pH provides for essentially complete removal of iron and manganese when these are present in the raw water.

Coagulant Aids

Flocs produced during coagulation should settle rapidly and be resistant to destruction by shearing forces. Unfortunately, in many instances the flocs produced do not possess these characteristics, particularly in the case of waters low in turbidity or low mineralized waters that are high in color.

The situation for low turbidity can be explained by referring to Figure 7-6(a) and considering a water with a colloid concentration (S_1). Because of the low concentration of colloids and the low collision potential, destabilization is achieved only at high coagulant concentrations through the sweep-floc mechanism. This results in a light-weight, fragile, slow-settling floc consisting primarily of $Al(OH)_3$. In such cases, certain materials referred to as *coagulant aids* can be added to the water to improve floc properties and enhance coagulation. Among the coagulant aids are clays, activated silica, and polymers.

Bentonitic clays are frequently used as coagulant aids for waters low in turbidity. Their use can reduce the amount of coagulant required to treat a water and can improve the nature of the flocs produced.

The method by which bentonite reduces coagulant demand can be explained with the aid of Figure 7-6(a). Consider a water with a colloid concentration S_1, to which bentonite is added to raise the colloid concentration to S_3. The shaded portion of

Figure 7-6(a) indicates that the effective coagulant dosage has been reduced from C_1 to C_3. This change is brought about because the increased concentration of colloids provides greater contact opportunity and leads to destabilization by charge neutralization rather than by sweep-floc coagulation.

The second advantage of using bentonite, improving floc properties, results from the fact that it weights the flocs and causes them to settle more rapidly than flocs containing mostly $Al(OH)_3$. Bentonite dosages in the range of 10 to 50 mg/ℓ are generally sufficient, but the exact dosage for a given water must be determined by laboratory testing.

Activated silica has long been used in water treatment as a primary coagulant in water softening operations or as a coagulant aid. When used as a coagulant aid in combination with alum it is capable of increasing the rate of flocculation, improving floc toughness, and increasing settleability.

Activated silica consists of negatively charged particles of hydrous silicon dioxide which are formed by acid neutralization of a dilute solution of sodium silicate. The mechanisms by which activated silica is formed have been described by Stumm, Hüper, and Champlin (1967) and can be explained with the aid of Figure 7-8. The

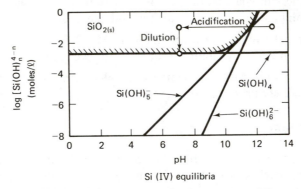

FIGURE 7-8 *Solubility equilibria for amorphous SiO$_{2\,(s)}$ (after Weber, 1972).*

solubility of silica is independent of pH below pH 8.0 but increases dramatically with increasing pH above that point. When a saturated sodium silicate solution at pH 12 is neutralized it will become oversaturated and polymeric silicates will begin to form as kinetic intermediates to the precipitation of amorphous silica (Weber, 1972). These polymeric species represent the activated silica particles, and if the solution is diluted to relieve supersaturation these species will be preserved and precipitation of amorphous silica will not occur. If the partially neutralized solution is aged for more than 2 hr prior to dilution, it is possible that the entire mixture may solidify by gelation (AWWA, 1971). Once diluted, the polysilicate may retain coagulative properties for up to several weeks. Because of the relatively short lifetime of activated silica it is best to produce the material at the point of use. Production must be carefully controlled to yield an "active" solution that will not gel.

When used in conjunction with alum, the optimum dose of activated silica usually falls between 7 and 11 % of the alum dose. Jar tests are required to identify the optimum combination of chemicals to use.

Anionic and nonionic polymers, while ineffective as primary coagulants, have been proven effective as coagulant aids. Although incapable of providing charge neutralization for negatively charged colloids, these polymers help to promote large floc particles by a bridging mechanism once the colloids have been destabilized by coagulants such as alum. The effectiveness of anionic polymers has been reported to improve dramatically with increases in the concentration of divalent metal ions (Ca^{2+}, Mg^{2+}, etc.) in the water. It has been suggested that this improvement can be attributed to the following factors (Black et al. 1965).

1. The divalent metal ions compress the thickness of the double layer of negative colloids, thereby reducing interparticle repulsive forces.

2. They reduce the repulsive forces between the anionic polymer and the colloids.

3. They reduce interactions between polymer molecules adsorbed on the surface of colloidal particles.

4. The addition of divalent metal ions reduces the size of the polymer coil, thus allowing more polymer to adsorb to the particle surface.

The reaction of divalent metal ions to form chemical complexes with certain ionogenic groups on the polymer and on the particle surface has also been suggested as a possible mode of action (Sommerauer et al., 1968).

Operating experience indicates that the use of only 0.1 to 0.5 mg/ℓ of anionic polymer in conjunction with alum can significantly improve floc settleability and toughness compared to the use of alum alone. Overdosing the system with anionic polymer can seriously inhibit coagulation and must be avoided (AWWA, 1971).

Determination of Coagulant Dosage

Coagulant dosages cannot be calculated but must be determined experimentally, usually by simple jar test procedures. Unfortunately, the optimum dose determined by the jar test is frequently not the same as that for the actual plant. One reason for this discrepancy is that the jar test is a batch test, while rapid mixing units in actual water treatment plants are continuous flow processes. A batch system is inherently more efficient than a completely mixed, continuous-flow system for any reaction which has a reaction order greater than zero. Consequently, a jar test will usually indicate an optimum efficiency at a lower dose than that needed for actual plant operation. Agreement between jar tests and plant performance will be better for plants which have efficient rapid-mixing units.

Four classifications of water, based on turbidity and alkalinity concentrations, have been characterized with respect to ease of coagulation in Table 7-2. The classifications will serve as a guide for selection of coagulants to be studied by jar test procedures.

TABLE 7-2 *Comparison of various coagulants (after Weber, 1972).*

Type of Water	Alum	Ferric Salts	Polymer	Magnesium
High turbidity. High alkalinity. (easiest to coagulate). (Type 1)	Effective for pH 5–7. Addition of alkalinity and coagulant aid not required.	Effective for pH 6–7. Addition of alkalinity and coagulant aid usually not required.	Cationic polymers very effective. Anionic and non-ionic may also be effective. High molecular weight materials are best.	Effective due to precipitation of $Mg(OH)_2$.
High turbidity. Low alkalinity. (Type 2)	Effective for pH 5–7. May need to add alkalinity if pH drops during treatment.	Effective for pH May need to add alkalinity if pH drops during treatment.	Same as above.	Effective and results in increased alkalinity, which makes water easier to stabilize.
Low turbidity. High alkalinity (Type 3)	Effective in relatively large dosages, which promote precipitation of $Al(OH)_{3(s)}$. Coagulant aid may be needed to weight floc and improve settling.	Effective in relatively large dosages, which promote precipitation of $Fe(OH)_{3(s)}$. Coagulant aid should be added to weight floc and improve settling.	Cannot work alone due to low turbidity. Coagulant aids such as clay should be added ahead of polymer.	Effective due to precipitation of $Mg(OH)_2$.
Low turbidity. Low alkalinity (most difficult to coagulate). (Type 4)	Effective only by sweep-floc formation, but resulting dosage will destroy alkalinity. Must add alkalinity to produce type 3 or clay to produce type 2 water.	Effective only by sweep-floc formation, but resulting dosage will destroy alkalinity. Must add alkalinity to produce type 3 or clay to produce type 2 water.	Will not work alone due to low turbidity. Coagulant aids such as clay should be added ahead of polymer.	Results in increased alkalinity, which makes water easier to stabilize.

Low turbidity < 10 JTU, low alkalinity < 50 mg/ℓ (as $CaCO_3$).
High turbidity > 100 JTU, high alkalinity > 250 mg/ℓ (as $CaCO_3$).

7-3 RAPID MIXING AND FLOCCULATION

Coagulation has previously been described as a two-stage process involving particle destabilization and particle transport. Destabilization results from the addition of chemical coagulants; particle transport is accomplished by physical mixing. Actually, both of these stages are interdependent. Coagulants cannot cause destabilization unless they are properly distributed and contacted with the colloids. Likewise, destabilized colloids will not aggregate unless sufficient contact opportunities are provided.

Rapid-mixing and flocculation units are installed in water treatment plants for the purposes of chemical mixing and particle transport. These units are vitally important and must be properly designed and operated if the coagulation process is to be successful.

Rapid Mixing

The function of a rapid-mixing unit is to provide complete mixing of the coagulant and raw water. Recent advances in coagulation chemistry have resulted in an increased awareness of the importance of rapid mixing in the overall coagulation process. For example, it is now known that destabilization and the early stages of floc formation occur during rapid mixing. Proper design of rapid-mixing units can optimize these processes, resulting in reduced coagulant demands and improved aggregation in subsequent flocculation units (Letterman et al., 1973).

Many devices have been used for rapid mixing, including baffle chambers, hydraulic jumps, and mechanically mixed tanks. Mechanically mixed tanks, commonly called *completely mixed* or *back-mixed units*, have been the most popular devices, and have typically been designed for detention times of 10 to 30 sec and velocity gradients of about 300 (sec)$^{-1}$ (AWWA, 1969).

Although completely mixed units are widely used, they do not provide uniform mixing because of short circuiting of flow and mass rotation of water (Kawamura, 1976). As a result, some portions of the water may have a high concentration of chemicals, while other portions may have a very low concentration.

This nonuniform mixing is acceptable when destabilization is the result of double-layer compression. During double-layer compression there is no chemical interaction between ions and colloids as the ions accumulate around the particles. If unequal concentrations of chemicals initially occur due to back-mixing, continued mixing may redistribute the chemicals until a uniform chemical concentration occurs and particle destabilization results. Thus, for double-layer compression only minimal or even "slow" rapid mixing is all that is required to achieve good results (Vrale and Jorden, 1971).

The existence of nonuniform mixing is of more concern when hydrolyzing metal ions are used as coagulants. The metal ion concentration will be uniform throughout the water if the coagulant is uniformly distributed; however, local concentrations of metal ions will vary if mixing is nonuniform. Furthermore, the pH may be lowered in regions of high coagulant concentration because of the acidic nature of the metal ions (equations 7-1 and 7-2). Variations in pH and metal ion concentration may cause a great variation in the number and type of hydrolysis products formed [see Figure 7-6(a)] and will influence the rate of hydrolysis, adsorption, and precipitation.

Nonuniform mixing of metal ion coagulants will be acceptable if sweep-floc precipitation is desired. The regions of high chemical concentration produced by back-mixing apparently stimulate formation of the hydroxide flocs necessary for this mode of destabilization (Vrale and Jorden, 1971).

However, nonuniform mixing will be quite undesirable if destabilization by adsorption is desired. Destabilization will be poor in areas of low chemical concentration because of insufficient adsorption and low chemical potential. Destabilization may also be poor in areas of high chemical concentration because of over-adsorption and charge reversal. Once charge reversal occurs, continued mixing will not improve the situation because the adsorption process is essentially irreversable. For this reason, completely mixed units are not efficient if destabilization by adsorption of hydrolyzed metal ions is desired (Vrale and Jorden, 1971). The same holds true for the use of polyelectrolytes, which also function by adsorption.

Coagulation is thought to result from a combination of adsorption and enmeshment when hydrolyzing metal ions are used in water treatment plants. Because completely mixed units have been shown to be detrimental to the adsorption mechanism, they are no longer recommended by many engineers. These units are being replaced by in-line blenders, which have been found to be very efficient rapid-mixing units. In addition to providing good mixing, in-line blenders are inexpensive to construct and operate. Design procedures have been presented by Kawamura (1976) and Chao and Stone (1979).

Flocculation

Contacts between destabilized particles are essential for agglomeration to occur. These contacts can be achieved by three separate mechanisms: (1) thermal motion (Brownian motion), (2) bulk liquid motion (stirring), and (3) differential settling (Weber, 1972). When contacts are produced by Brownian motion, the process is termed *perikinetic* flocculation and, when produced by stirring or settling, as *orthokinetic* flocculation.

The rate of change of particle concentration by perikinetic flocculation can be expressed as follows (Swift and Friedlander, 1964):

$$J_{pk} = \frac{dN^\circ}{dt} = \frac{-4\eta\bar{k}T}{3\mu}(N^\circ)^2 \tag{7-12}$$

where N° = total concentration of particles in suspension at time t

η = a collision efficiency factor, fraction of collisions that produce aggregates

$\bar{k}$ = Boltzmann's constant (1.38×10^{-16} erg/degree)

T = absolute temperature (°K)

μ = fluid viscosity $\left(\frac{gram}{cm\text{-}sec}\right)$.

The rate of change of particle concentration as a result of orthokinetic flocculation is indicated by Weber (1972):

$$J_{ok} = \frac{dN^\circ}{dt}\frac{-2\eta(\bar{G})d^3(N^\circ)^2}{3} \tag{7-13}$$

where $\bar{G}$ = velocity gradient

          d = particle diameter (cm)

and the other terms are as previously defined.

The ratio of the rate of orthokinetic flocculation to perikinetic flocculation can be obtained by dividing equation 7-13 by 7-12:

$$\frac{J_{ok}}{J_{pk}} = \frac{-2\eta(\bar{G})d^3(N^{\circ})^2/3}{-4\eta\bar{k}T(N^{\circ})^2/3\mu}$$

Thus,

$$\frac{J_{ok}}{J_{pk}} = \frac{\mu(\bar{G})d^3}{2\bar{k}T} \tag{7-14}$$

Referring to Equation 7-14, it is obvious that perikinetic flocculation dominates (i.e., $J_{ok}/J_{pk} < 1.0$) for the aggregation of small particles. Even for particles as large as 0.10 micrometer (μm) in diameter, a velocity gradient of 10,000/sec would be necessary for the rate of orthokinetic flocculation to equal the rate of perikinetic flocculation. In practice such high velocity gradients cannot feasibly be achieved. However, if particles are 1.0 μm in diameter, a G value of only 10/sec is necessary for the two rates to be equal. It can be concluded that small particles must aggregate to a size of approximately 1.0 μm by Brownian motion, but once they reach that size, stirring must be provided to promote further aggregation. Particles that are 1 μm in diameter do not settle very rapidly; thus, orthokinetic flocculation must be utilized in water treatment plants to produce settleable-size particles. Flocculation basins in which velocity gradients are produced by baffles, compressed air, or mechanical mixing are commonly employed.

Velocity gradients vary considerably throughout a flocculation basin, and the value at any given point, $\bar{G}$, is difficult to determine. Camp and Stein (1943) defined a mean velocity gradient, G, that can be used to describe the average conditions within the basin. The mean velocity gradient, which serves as an estimate of $\bar{G}$, is a function of the energy dissipation per unit volume of tank and can be expressed as follows:

$$G = \sqrt{\frac{P}{V\mu}} \tag{7-15}$$

where G = mean velocity gradient $\left(\dfrac{1}{sec}\right)$

          P = power input to the basin $\left(\dfrac{ft\text{-}lb}{sec}\right)$

          V = volume of basin (ft^3)

          μ = absolute viscosity $\left(\dfrac{lb\text{-}sec}{ft^2}\right)$.

To calculate G values, it is necessary to evaluate the power input to the flocculation basin. For mixing chambers equipped with over-and-under or around-the-end baffles, the useful power input is as follows (Fair, Geyer, and Okun, 1968):

$$P = Q\gamma h \qquad \text{(7-16)}$$

where $Q =$ flow rate of water (ft³/sec)

$\gamma =$ weight of water (lb/ft³)

$h =$ head loss through unit (ft).

The head loss should include the head loss due to friction and that caused by the presence of the baffles. The head loss due to $(n - 1)$ equally spaced baffles can be calculated from

$$h = \frac{nV_1^2}{2g} + (n - 1)\frac{V_2^2}{2g} \qquad \text{(7-17)}$$

where $V_1 =$ velocity in the channels between baffles (ft/sec)

$V_2 =$ velocity around the end of the baffles (ft/sec).

Head loss due to friction must be added to this value to get the total head loss through the unit.

When compressed air flocculators are used the mixing action is provided by the rising air bubbles. The useful power can be calculated from

$$P = 81.5(Q_a) \log \left(\frac{h + 34}{34}\right) \qquad \text{(7-18)}$$

where $Q_a =$ ft³/min of free air injected into water

$h =$ depth of air release below the surface (ft).

The useful power input to a mechanically mixed flocculation basin can be calculated by the following expression (Fair, Geyer, and Okun, 1968):

$$P = (1.44 \times 10^{-4})C_D(\rho)[(1 - k)n]^3 b \sum (r^4 - r_0^4) \qquad \text{(7-19)}$$

where $C_D =$ drag coefficient (see Table 7-3)

$\rho =$ mass density of the fluid $\left(\frac{\text{lbs-sec}}{\text{ft}^4}\right)$

$k =$ ratio of fluid velocity to impeller velocity usually taken to be 0.25

$n =$ paddle speed (rpm)

$r =$ radius to outside edge of paddle (ft)

$r_0 =$ radius to inside edge of paddle (ft)

$b =$ length of paddle (ft).

Mechanically mixed tanks are the most popular type of flocculation device. Paddles consist of boards attached to horizontal shafts that rotate at 2 to 15 rpm. Stator blades should be used to prevent rotation of the water and improve the overall mixing process.

Velocity gradients in flocculation basins must be high enough to achieve the particle contacts necessary to promote aggregation; however, they must not be so high that they shear the flocs apart. Flocs produced by organic polymers may be stronger than those produced from iron(III) and aluminum(III) salts and can withstand higher shearing stresses (O'Melia, 1969). However, if flocs are ruptured, reaggregation is more likely to occur for iron(III) and aluminum(III) flocs than for polymer flocs because the polymer segments, once detached by mixing, may fold back and restabilize the particle as shown in Figure 7-4.

Design values for the velocity gradient depend on the coagulant used. Gradients between 25/sec and 100/sec have been found to be acceptable for iron(III) and aluminum(III) flocs, but lower values, 15/sec to 20/sec, are better when polymers are used.

In the past, most flocculation basins consisted of single, uniformly mixed tanks. However, both the degree of flocculation and the subsequent clarification step have been found to improve if flocculation basins are compartmentalized. Current design favors the use of basins with three or more compartments, with the G value decreasing from compartment to compartment (e.g., 90/sec, 70/sec, 40/sec). This plan provides maximum mixing to enhance aggregation at the influent end but promotes larger flocs by reducing mixing and shear at the effluent end.

The detention time in a flocculation basin is also an important parameter because it determines the amount of time that particles are exposed to the velocity gradient and thus is a measure of contact opportunity in the basin. A detention time of 30 min is commonly used for flocculator design; however, many engineers consider the dimensionless product of G in $(\text{sec})^{-1}$ and t in sec to be more important than the detention time. $G(t)$ values in the range of 10^4 to 10^5 are frequently recommended for design. Typical values for design of mechanical flocculation basins are summarized in Table 7-3.

TABLE 7-3 *Typical range of flocculator basin design criteria for water treatment, mechanical mixing (after Humenick, 1977).*

Detention time t	30 to 60 min
Mixing Intensity G	10 to 75 fps/ft (25 to 65 is most common.)
Gt	20,000 to 200,000
Paddle tip speed	Less than 2 ft/sec, weak floc; less than 4 ft/sec, strong floc
Paddle area	Less than 15 to 20% of the area in the plane of paddle rotation
Relative velocity between water and paddle	70 to 80 percent of paddle speed without stators (75% is most common); with stators, 100% of paddle speed is approached.
Drag coefficient for flat paddles, C_d	For Reynolds numbers greater than 1,000 and flat plates, $C_d = 1.16, 1.20, 1.50,$ and 1.90 for length-to-width ratios of 1, 5, 20, and ∞ respectively.

PROBLEMS

7-1. Jar test results indicate that an alum $(Al_2(SO_4)_3 \cdot 14H_2O)$ dosage of 20 mg/ℓ will be required to treat a raw water which has an alkalinity of 50 mg/ℓ as $CaCO_3$. Calculate the pounds per day of alum that must be fed to treat 5 MGD of water. Will the natural alkalinity of the water be sufficient to satisfy the reaction of the alum?

7-2. What concentration of alkalinity is theoretically required to react with an alum feed of 200 lb/MG?

7-3. Jar test results reveal that 8 mℓ of a 10 mg/ℓ alum solution produced good coagulation in a 1500 mℓ sample of water. What amount of chemical, in pounds per million gallons, would be required at a water treatment plant to yield this same dosage?

7-4. Calculate the specific gravity of a sludge produced from coagulation if it is collected at 2.1% solids and the dry solids have a specific gravity of 2.4.

7-5. Calculate the daily volume of sludge produced in Problem 7-1 if the water contains 12 mg/ℓ of suspended solids and it is assumed that all of the alum and suspended solids are removed during settling. Assume that the sludge settles to 1.9% solids and that the dry solids have a specific gravity of 2.1.

7-6. Why is an understanding of coagulation-flocculation helpful in water treatment plant design and operation?

7-7. The curve shown in Figure P7-7 was determined from jar tests. To reduce the overall cost of chemicals and sludge disposal it is desired to reduce the optimal alum dosage. Describe what laboratory tests you would conduct, and what you would look for in these tests, to determine if a lower alum dosage could be used.

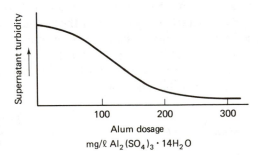

PROBLEM 7-7

REFERENCES

AWWA, *Water Quality and Treatment*, 3rd ed., McGraw-Hill Book Company, New York, (1971).

AWWA, ASCE, and Conference of State Sanitary Engineers, *Water Treatment Plant Design*, McGraw-Hill Book Company, New York (1969).

BEARDSLEY, J.A., "Use of Polymers in Municipal Water Treatment," *J. Am.* Water Works Assoc., **65**, 1, 85 (1973).

BLACK, A.P., BIRKNER, F.B., and MORGAN J.J., "Destabilization of Dilute Clay Suspensions with Labeled Polymers," *J. Am. Water Works Assoc.*, **57**, 12, 1547 (1965).

CAMP, T.R., and STEIN, P.C., "Velocity Gradients and Internal Work in Fluid Motion," *J. Boston Soc. Civil Engineers*, **30**, 219–237 (October, 1943).

COHEN, J.M., ROURKE, G.A., and WOODWARD, R.L., "Natural and Synthetic Polyelectrolytes As Coagulant Aids," *J. Am. Water Works Assoc.*, **50**, 4, 463 (April, 1958).

CHOA, J.L., and STONE, B.G., "Initial Mixing by Jet Injection Blending," *J. Am. Water Works Assoc.*, **71**, 10, 570 (1979).

FAIR, G.M., GEYER, J.C., and OKUN, D.A., *Water and Wastewater Engineering*, Vol. 2, *Water Purification and Wastewater Treatment and-Disposal*, John Wiley & Sons, Inc., New York (1968).

HARDY, W.B., "A Preliminary Investigation of the Conditions Which Determine the Stability of Irreversible Hydrosols," *Proc. Roy. Soc.* (London) **66**, 110–125 (1900).

HUMENICK, M.J., Jr., *Water and Wastewater Treatment: Calculations for Chemical and Physical Processes*, Marcel Dekker, Inc., New York (1977).

KAWAMURA, S., "Considerations on Improving Flocculation", *J. Am. Water Works Assoc.*, **68**, 6, 328 (1976).

LAMER, V.K., and HEALY, T.W., "Adsorption-Flocculation Reactions of Macromolecules at the Solid-Liquid Interface", *Rev. Pure App. Chem.*, **13**, 112–132 (1963).

LETTERMAN, R.D., QUON, J.E., and GEMMELL, R.S., "Influence of Rapid-Mix Parameters on Flocculation," *J. Am. Water Works Assoc.*, **65**, 11, 716 (1973).

O'MELIA, C.R., "A Review of the Coagulation Process," *Public Works* (May, 1969).

PACKHAM, R.F., "Some Studies of the Coagulation of Dispersed Clays with Hydrolyzing Salts," *J. Coll. Sci.*, **20**, 81–92 (1965).

PARSONS, W.A., "Chemical Treatment of Sewage and Industrial Waste," *Bulletin No. 215* National Lime Association, Washington, DC, (1965).

RUBIN, A.J., and KOVAC, T.W., "Effect of Aluminum III Hydrolysis on Alum Coagulation," in *Chemistry of Water Supply, Treatment and Distribution*, [Ed.] A.J. RUBIN, Ann Arbor Science Publishers, Ann Arbor, MI (1974).

RUEHRWEIN, R.A., and WARD, D.W., "Mechanisms of Clay Aggregation by Polyelectrolytes," *Soil Science*, **73**, 485–492 (1952).

SCHULZE, H.J., *J. Prakt. Chem.*, (2) **25**, 431 (1882).

SHAW, D.J., *Introduction to Colloid and Surface Chemistry*, 2nd ed., Butterworths London (1970).

SHEA, T.G., "Use of Polymers As a Primary Coagulant," *Proc. Am. Water Works Seminar* (June 4, 1972).

SOMMERAUER, A., SUSSMAN, D.L., and STUMM, W., "The Role of Complex Formation in the Flocculation of Negatively Charged Sols with Anionic Polyelectrolytes," *Koll, Z. Z. Polymere*, **225**, 147 (1968).

STUMM, W., HÜPER, H., and CHAMPLIN, R.L., "Formation of Polysilicates As Determined by Coagulation Effects," *Environmental Science and Technology*, **1**, 3, 221 (1967).

STUMM, W., and O'MELIA, C.R., "Stoichiometry of Coagulation," *J. Am. Water Works Assoc.*, **60**, 5, 514 (1968).

SWIFT, D.L., and FRIEDLANDER, S.K., "The Coagulation of Hydrosols by Brownian Motion and Laminar Shear Flow," *J. Coll, Sci.*, **19,** 621 (1964).

TAMAMUSHI, B., and TAMAKI, K., "The Action of Long-Chain Cations on Negative Silver Iodide Sol," *Koll-Z*, **163,** 122–126 (1959).

THOMPSON, C.G., personal communication (1979).

THOMPSON, C.G., SINGLEY, J.E., and BLACK, A.P., "Magnesium Carbonate—A Recycled Coagulant," *J. Am. Water Works Assoc.*, **64,** 1, 11 (1972a).

THOMPSON, C.G., SINGLEY, J.E., and BLACK, A.P., "Magnesium Carbonate—A Recycled Coagulant," Part II, *J. Am. Water Works Assoc.*, **64,** 2, 93 (1972b).

VAN OLPHEN, H., *An Introduction to Clay Colloid Chemistry*, 2nd ed., Wiley-Interscience, New York (1977).

VERWEY, E.J.W., and OVERBEEK, J.Th.G., "Theory of the Stability of Lyophobic Colloids," Elsevier Scientific Publishing Company, Amsterdam, (1948).

VRALE. H., and JORDEN, R.M., "Rapid Mixing in Water Treatment," *J. Am. Water Works Assoc.*, **63,** 1, 52 (1971).

WEBER, W.J., Jr., *Physicochemical Processes for Water Quality Control*, John Wiley & Sons, Inc., New York (1972).

8

WATER STABILIZATION

For the purpose of this discussion corrosion is defined as the destruction of metals by interaction with the environment. However, it should be understood that nonmetal material such as concrete can also corrode. In this regard, a more general definition of corrosion would be the destruction or deterioration of a material because of reaction with its environment.

Corrosion is part of a cycle of events in which metals tend to return to their natural state. Most metals occur in nature in the form of a hydroxide or oxide, and the metal is produced through reduction of the ore. If a metal is placed in the proper environment, it will corrode and return to its more stable oxidized state. Because a change in oxidation state is involved, corrosion is electrochemical in nature ; i.e., the destruction of the metal is the result of an oxidation-reduction chemical reaction.

In the past the concept of local anode and cathode areas (local-element hypothesis) has been the basis of corrosion theory. However, in recent years this has been replaced by the mixed-potential theory, which provides a more general approach to the study of corrosion. These two theories are not conflicting but simply represent two different approaches to treating corrosion. Since the local-element approach is the one most used in environmental engineering literature, it will be used in this discussion.

8-1 ELECTROCHEMICAL ASPECTS OF CORROSION

For corrosion of a metal to occur, an electrochemical cell (generally a galvanic cell) must be established. For this to occur, a potential difference must exist between one part of the corroding structure and another. According to Singley (1978), such a situation may develop for several reasons:

1. from the connection of dissimilar metals,
2. from concentration variations in either the metal structure or the solution adjacent to the metal structure, or
3. from surface heterogeneities in the metal structure.

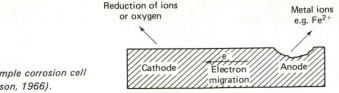

FIGURE 8-1 Simple corrosion cell
(after Butler and Ison, 1966).

For any of these situations, Figure 8-1 shows the electrochemical cell which is established.

The anode reaction is an oxidation reaction where electrons are released by the metal; e.g.

$$Fe \longrightarrow Fe^{2+} + 2\,e^- \tag{8-1}$$

The electrons released by the metal at the anode travel through the metal structure to the cathode, where they participate in a surface reaction involving the reduction of either oxygen or aqueous phase ions. The two most common cathode reactions have been presented by Butler and Ison (1966): these are:

1. *Reduction of hydrogen:*

$$2H^+ + 2\,e^- \longrightarrow 2H \longrightarrow H_2 \tag{8-2}$$

2. *Reduction of oxygen:*

$$O_2 + 2H_2O + 4\,e^- \longrightarrow 4OH^- \tag{8-3}$$

These reactions are illustrated in Figure 8-2. When a metal corrodes in an acidic solution or under anoxic conditions, hydrogen reduction with the evolution of hydrogen gas is the primary cathode reaction. This reaction can be considered as a two-step process. In the first reaction hydrogen ions are reduced to atomic hydrogen, which tends to plate out on the metal surface in the area of the cathode.

$$2H^+ + 2\,e^- \longrightarrow 2H \tag{8-4}$$

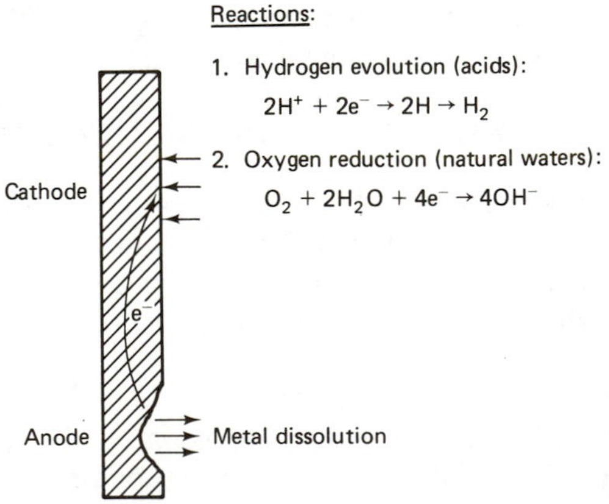

Reactions:

1. Hydrogen evolution (acids):
   $$2H^+ + 2e^- \rightarrow 2H \rightarrow H_2$$

2. Oxygen reduction (natural waters):
   $$O_2 + 2H_2O + 4e^- \rightarrow 4OH^-$$

Cathode

Anode — Metal dissolution

FIGURE 8-2 Cathode reactions on the metal
surface in a simple corrosion cell (after Butler
and Ison, 1966).

If the atomic hydrogen remained on the metal surface, the transfer of electrons to the solution would be inhibited and the metal would ultimately cease to corrode. However, atomic hydrogen generally forms molecular hydrogen, which is evolved as a gas, thereby removing the atomic hydrogen layer.

$$2H \longrightarrow H_2 \tag{8-5}$$

The formation of the atomic hydrogen layer is called *polarization*; the removal of the layer is referred to as *depolarization*.

In most natural waters or waters with a near neutral pH which contain an appreciable amount of oxygen, the primary cathode reaction is oxygen reduction.

$$O_2 + 2H_2O + 4\,e^- \longrightarrow 4OH^- \tag{8-6}$$

This reaction results in the production of alkaline conditions in the vicinity of the cathode. The metal ions going into solution at the anode react with the hydroxyl ions to form metal hydroxides. For example, iron(II) reacts with hydroxyl ions to form iron-(II) hydroxide.

$$Fe^{2+} + 2OH^- \rightleftharpoons Fe(OH)_2 \tag{8-7}$$

When the solubility of $Fe(OH)_2$ is exceeded, solid $Fe(OH)_2$ will precipitate. In the presence of oxygen, $Fe(OH)_2$ is rapidly oxidized to $Fe(OH)_3$.

$$4Fe(OH)_2 + O_2 + 2H_2O \rightleftharpoons 4Fe(OH)_3 \tag{8-8}$$

These reactions will result in the deposition of solid metal hydroxides and the growth of a tubercule on the metal structure.

In general, corrosion can be viewed as the process of loss of metal ions from one point on a metal surface and a simultaneous reaction of an equivalent quantity of electrons at some other point on the metal surface. To retard or halt the corrosion process, it is necessary to prevent the reactions which occur at either the anode or cathode. One way to accomplish this is by covering the metal surface with a material which will prevent passage of electrons through the coating. This is one of the objectives in treating water for stability control.

8-2 WATER STABILIZATION

In the general sense, a stable water is defined as a water which exhibits neither scale-forming nor corrosion properties. However, in the water utility industry a stable water is normally considered to be one which will neither dissolve nor deposit calcium carbonate. The intent of water stabilization, as practiced by the water utility industry, is to provide a thin protective coating of calcium carbonate on the interior of the piping in a water distribution system. To be effective, film deposition must be controlled or else excessive buildup of scale deposits in the system will drastically reduce the carrying capacity of the water pipes.

Langelier Saturation Index

If a water is oversaturated with respect to $CaCO_3$, solid $CaCO_3$ will precipitate. On the other hand, if the water is undersaturated with respect to $CaCO_3$, it will dissolve solid $CaCO_3$. The solubility equilibrium constant expression for calcium carbonate has the form

$$K'_s = [Ca^{2+}][CO_3^{2-}] \tag{8-9}$$

According to equation 2-99, $[CO_3^{2-}]$ can be expressed as $C_T\alpha_2$, where C_T represents the total carbonic species concentration, and α_2 represents the fraction of the total carbonic species concentration which is in the CO_3^{2-} form. Hence, equation 8-9 can be written as

$$K'_s = [Ca^{2+}]\alpha_2 C_T \tag{8-10}$$

$$\alpha_2 = \cfrac{1}{\dfrac{[H^+]^2}{K'_1 K'_2} + \dfrac{[H^+]}{K'_2} + 1} \tag{2-106}$$

Substituting for α_2 in equation 8-10 from equation 2-106 gives

$$K'_s = \cfrac{[Ca^{2+}]C_T}{\dfrac{[H^+]^2}{K'_1 K'_2} + \dfrac{[H^+]}{K'_2} + 1} \tag{8-11}$$

This equation illustrates the relationship between calcium carbonate solubility and pH and indicates that the solubility will increase with decreasing pH.

If the alkalinity and pH of a water are measured, the total carbonic species concentration can be computed from equation 2-113. The equilibrium concentration of calcium required for saturation can then be calculated from equation 8-11. If the actual calcium concentration (measured by one of the available analytical techniques) is less than the calculated value, the water will be undersaturated with respect to $CaCO_3$; i.e., it will dissolve calcium carbonate. On the other hand, if the calculated value is less than the measured calcium value, the water will be oversaturated with respect to $CaCO_3$; i.e., $CaCO_3$ will precipitate. Applying this principle, Langelier (1936) developed a method for predicting the saturation pH of a water with a specified alkalinity and calcium concentration. A control parameter called the *Langelier index* is defined on the basis of the saturation pH.

$$\text{Langelier index (LI)} = pH - pH_s \tag{8-12}$$

where   pH = measured pH of the water

               pH_s = calculated saturation pH of the water.

A negative LI indicates a water which is undersaturated with respect to $CaCO_3$; a positive LI indicates a water that is oversaturated with respect to $CaCO_3$. An LI of zero indicates a water which is just saturated with respect to $CaCO_3$.

Loewenthal and Marais (1976) have presented an excellent development of the equation required for the pH_s caculation, and it is this development which is summarized in this section.

On the assumption that alkalinity is due to the presence of a strong base in a carbonic acid solution, alkalinity may be defined mathematically as

$$[Alk]_e = [CO_3^{2-}]_e + [HCO_3^-]_e + [OH^-]_e - [H^+]_e \qquad (2\text{-}74)$$

where $[\ \ ]_e$ represents concentration in terms of equivalents per liter. To reformulate this equation in terms of moles per liter, it is noted that univalent species are numerically equal in terms of moles or equivalent, whereas both alkalinity (thought of in terms of equivalents of $CaCO_3$) and carbonate show a 2:1 ratio of equivalents to moles (i.e., 1 mole of CO_3^{2-} represents two equivalents). Therefore, equation 2-74 can be modified as follows:

$$2[Alk] = 2[CO_3^{2-}] + [HCO_3^-] + [OH^-] - [H^+] \qquad (2\text{-}107)$$

where $[H^+]$ is the molar hydrogen ion concentration. This value is computed from the measured pH value by employing equation 2-93.

$$[H^+] = \frac{10^{-pH}}{\gamma_m} \qquad (2\text{-}93)$$

Solving equation 2-107 for $[CO_3^{2-}]$ gives

$$[CO_3^{2-}] = [Alk] - \tfrac{1}{2}[HCO_3^-] - \tfrac{1}{2}[OH^-] + \tfrac{1}{2}[H^+] \qquad (8\text{-}13)$$

Substituting for $[HCO_3^-]$ in equation 8-13 from equation 2-90 gives

$$[CO_3^{2-}] = [Alk] - \frac{[H^+][CO_3^{2-}]}{2K_2'} - \tfrac{1}{2}[OH^-] + \tfrac{1}{2}[H^+]$$

or

$$[CO_3^{2-}] = \frac{[Alk] - \tfrac{1}{2}[OH^-] + \tfrac{1}{2}[H^+]}{\left[1 + \dfrac{[H^+]}{2K_2'}\right]} \qquad (8\text{-}14)$$

When the denominator term in equation 8-14 is multiplied by $2K_2'[H^+]/2K_2'[H^+]$, equation 8-15 is obtained.

$$[CO_3^{2-}] = \frac{K_2'\{2[Alk] - [OH^-] + [H^+]\}}{[H^+]\left[\dfrac{2K_2'}{[H^+]} + 1\right]} \qquad (8\text{-}15)$$

For any equilibrium calcium concentration the carbonate concentration required for saturation equilibrium may be computed by rearranging the solubility equilibrium constant expression into the form

$$[CO_3^{2-}]_s = \frac{K_s'}{[Ca^{2+}]} \tag{8-16}$$

where $[CO_3^{2-}]_s$ = carbonate concentration required for saturation at a specified calcium concentration (moles/ℓ).

For a specific total carbonic species concentration, the carbonate concentration is dependent only on pH. This means that equation 8-15 can be written as

$$[CO_3^{2-}]_s = \frac{K_2'\{2\,[\text{Alk}] - [\text{OH}^-]_s + [\text{H}^+]_s\}}{[\text{H}^+]_s\left[\dfrac{2K_2'}{[\text{H}^+]_s} + 1\right]} \tag{8-17}$$

where $[\text{H}^+]_s$ and $[\text{OH}^-]_s$ represent the saturation hydrogen ion and hydroxyl ion concentrations. Substituting for $[CO_3^{2-}]_s$ in equation 8-17 from equation 8-16 gives

$$\frac{K_s'}{[Ca^{2+}]} = \frac{K_2'\{2\,[\text{Alk}] - [\text{OH}^-]_s + [\text{H}^+]_s\}}{[\text{H}^+]_s\left[\dfrac{2K_2'}{[\text{H}^+]_s} + 1\right]} \tag{8-18}$$

or

$$K_s' = \frac{[Ca^{2+}]K_2'\{2\,[\text{Alk}] - [\text{OH}^-]_s + [\text{H}^+]_s\}}{[\text{H}^+]_s\left[\dfrac{2K_2'}{[\text{H}^+]_s} + 1\right]} \tag{8-19}$$

Rearranging and taking the logarithm of both sides of equation 8-19 produces equation 8-20:

$$-\log[\text{H}^+]_s = \log K_s' - \log K_2' - \log[Ca^{2+}]$$
$$- \log\{2\,[\text{Alk}] - [\text{OH}^-]_s + [\text{H}^+]_s\} + \log\left[\frac{2K_2'}{[\text{H}^+]_s} + 1\right] \tag{8-20}$$

Since

$$pH_s = -\log[\gamma_m[\text{H}^+]_s]$$
$$pK_s' = -\log K_s'$$
$$pK_2' = -\log K_2'$$
$$pCa^{2+} = -\log[Ca^{2+}]$$

equation 8-20 reduces to

$$pH_s = pK_2' + pCa^{2+} - pK_s' - \log\{2\,[\text{Alk}] - [\text{OH}^-]_s + [\text{H}^+]_s\}$$
$$+ \log\left[\frac{2K_2'}{[\text{H}^+]_s} + 1\right] - \log\gamma_m \tag{8-21}$$

Using equation 8-21, the saturation pH can be computed for a water at any measured alkalinity and calcium concentrations. However, because $[H^+]_s$ appears on the right-hand side of this equation, a direct solution for pH_s cannot be made, and the equation will have to be solved by a trial-and-error method. To simplify the solution, Loewenthal and Marais (1976) noted that in the pH range of 6.5 to 9.5, the $[H^+]_s$ and $[OH^-]_s$ terms in the sum $\{2\,[Alk] - [OH^-]_s + [H^+]_s\}$ and the $2K_2'/[H^+]_s$ term in the sum $[2K_2'/[H^+]_s + 1]$ are negligible. Thus, within this pH range equation 8-21 reduces to

$$pH_s = pK_2' + pCa^{2+} - pK_s' - \log\,(2\,[Alk]) - \log \gamma_m \qquad (8\text{-}22)$$

Since the pH of most natural waters falls within the pH range of 6.5 to 9.5, equation 8-22 is commonly used to compute pH_s.

EXAMPLE PROBLEM 8-1: Compute the Langelier index for a water with the following characteristics:

$$\text{calcium} = 240 \text{ mg}/\ell \text{ as } CaCO_3$$
$$\text{alkalinity} = 190 \text{ mg}/\ell \text{ as } CaCO_3$$
$$pH = 6.83$$
$$\text{temperature} = 21°C$$
$$TDS = 500 \text{ mg}/\ell$$

Will this water have a tendency to dissolve or deposit $CaCO_3$?

Solution:

1. Compute the ionic strength of the water from equation 1-41:

$$I = (2.5 \times 10^{-5})(500) = 0.0125 \text{ } M$$

2. Determine γ_m from the Davies relationship (equation 1-47):

$$\log \gamma_m = -(1.82 \times 10^6)[(78.3)(294)]^{-3/2}(1)^2 \left[\frac{[0.0125]^{1/2}}{1 + [0.0125]^{1/2}} - 0.3(0.0125)\right]$$

$$= -0.0515$$

$$\gamma_m = 0.89$$

3. Convert calcium concentration to moles/ℓ.

(a) Apply equation 1-14 and convert to concentration as Ca^{2+}:

$$Ca^{2+}(\text{mg}/\ell) = 240\,[\tfrac{20}{50}] = 96 \text{ mg}/\ell$$

(b) Apply equation 1-11 and convert mg/ℓ to mole/ℓ:

$$Ca^{2+}(\text{moles}/\ell) = \frac{96 \times 10^{-3}}{40} = 2.4 \times 10^{-3} \text{ mole}/\ell$$

4. Convert alkalinity concentration to moles/ℓ as $CaCO_3$ by applying equation 1-11:

$$Alk = \frac{190 \times 10^{-3}}{100} = 1.9 \times 10^{-3} \text{ mole}/\ell$$

5. Compute pK'_2.

   (a) Calculate pK_2 from equation 2-95:

   $$pK_2 = \frac{2902.39}{(294)} + 0.02379(294) - 6.498 = 10.36$$

   (b) Convert the pK_2 value to the K_2 value:

   $$pK_2 = \log \frac{1}{K_2}$$

   therefore,

   $$K_2 = 10^{-10.36}$$

   (c) Calculate γ_D from the Davies relationship (equation 1-47):

   $$\log \gamma_D = -(1.82 \times 10^6)[(78.3)(294)]^{-3/2}(2)^2\left[\frac{[0.0125]^{1/2}}{1 + [0.0125]^{1/2}}\right.$$
   $$\left. - 0.3(0.0125)\right]$$

   $$= -0.2$$
   $$\gamma_D = 0.63$$

   (d) Compute K'_2 from equation 2-90:

   $$K'_2 = \frac{10^{-10.36}}{0.63} = 10^{-10.16}$$

   (e) Calculate pK'_2:

   $$pK'_2 = \log \frac{1}{10^{-10.16}} = 10.16$$

6. Compute pK'_s:

   (a) Loewenthal and Marais (1976) have presented the following equation for computing pK_s for $CaCO_3$:

   $$pK_s = 0.01183t + 8.03$$

   where t is the water temperature in °C. Thus, for $t = 21$°C,

   $$pK_s = 0.01183(21) + 8.03 = 8.28$$

   (b) Convert the pK_s value to the K_s value:

   $$pK_s = \log \frac{1}{K_s}$$

   therefore,

   $$K_s = 10^{-8.28}$$

(c) Using the value of γ_D obtained in step 5(c), calculate K_s':

$$K_s = [\gamma_D Ca^{2+}][\gamma_D CO_3^{2-}]$$

or

$$K_s' = \frac{K_s}{(\gamma_D)^2} = [Ca^{2+}][CO_3^{2-}]$$

therefore,

$$K_s' = \frac{10^{-8.28}}{(0.63)^2} = 10^{-7.88}$$

(d) Compute pK_s':

$$pK_s' = \log \frac{1}{10^{-7.88}} = 7.88$$

7. Calculate pCa^{2+}:

$$pCa^{2+} = \log \frac{1}{2.4 \times 10^{-3}} = 2.62$$

8. Compute pH_s from equation 8-22:

$$pH_s = 10.16 + 2.62 - 7.88 - \log[2(1.9 \times 10^{-3})] + 0.05 = 7.37$$

9. Determine the LI, using equation 8-12:

$$LI = 6.83 - 7.37 = -0.54$$

Since the Langelier Index is negative, the water is undersaturated and will have a tendency to dissolve $CaCO_3$.

When the Langelier index is used to indicate water stability, the optimum pH (pH_s) of the treated water is computed by using equation 8-22. The plant operator then adjusts the chemical feed rate to give the desired pH.

According to Merrill and Sanks (1977), the oversaturated water (the water described by point 2 on Figure 8-3) should have the following characteristics to ensure uniform coverage of the calcium carbonate film:

1. The precipitation potential of the oversaturated water should be in the range of 4–10 mg/ℓ of $CaCO_3$; i.e., 4–10 mg/ℓ of $CaCO_3$ should have precipitated when the water is at saturation equilibrium.

2. Both the alkalinity and calcium concentration should be greater than 40 mg/ℓ as $CaCO_3$.

3. When all concentrations are expressed in terms of mg/ℓ as $CaCO_3$, the following condition should be satisfied:

$$\frac{Alk}{Cl^- + SO_4^{2-}} \geq 5$$

4. The pH should be within the range of 6.8–7.3.

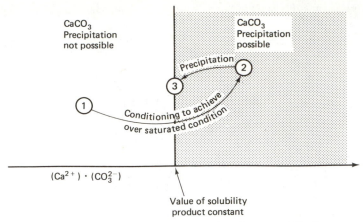

FIGURE 8-3 *Schematic representation of water stabilization (after Merrill and Sanks, 1977).*

In most cases, all four of these conditions cannot economically be achieved. Merrill and Sanks (1977) recommended that when it is not possible to achieve all four conditions during the treatment process, conditions 1, 2, and 3 should be preserved. However, note that when the Langelier index is used to indicate water stability, it is not possible to evaluate any of these conditions directly. To circumvent this problem, both Loewenthal and Marais (1976) and Merrill and Sanks (1977) recommended the use of a graphic method employing Caldwell-Lawrence diagrams to solve water stability problems.

Caldwell-Lawrence Diagrams

A Caldwell-Lawrence (C-L) diagram for a temperature of 15°C and an ionic strength of 0.03 M is shown in Figure 8-4. Theoretically, such diagrams may be constructed for any combination of temperature and ionic strength; however, in many cases this is not practiced. Under these conditions a reasonable answer can usually be obtained by using a diagram which was constructed for a water with characteristics fairly similar to the actual water to be treated.

A C-L diagram is a graphical representation of saturation equilibrium for $CaCO_3$. Any point on the diagram indicates the pH, soluble calcium concentration, and alkalinity required for $CaCO_3$ saturation. The coordinate system for the diagram is defined as follows:

$$C_1 = - \text{ acidity} \tag{8-23}$$

$$C_2 = (\text{Alk} - \text{Ca}) \tag{8-24}$$

where   acidity = acidity concentration expressed as mg/ℓ $CaCO_3$

        Alk = alkalinity concentration as mg/ℓ $CaCO_3$

        Ca = calcium concentration expressed as mg/ℓ $CaCO_3$.

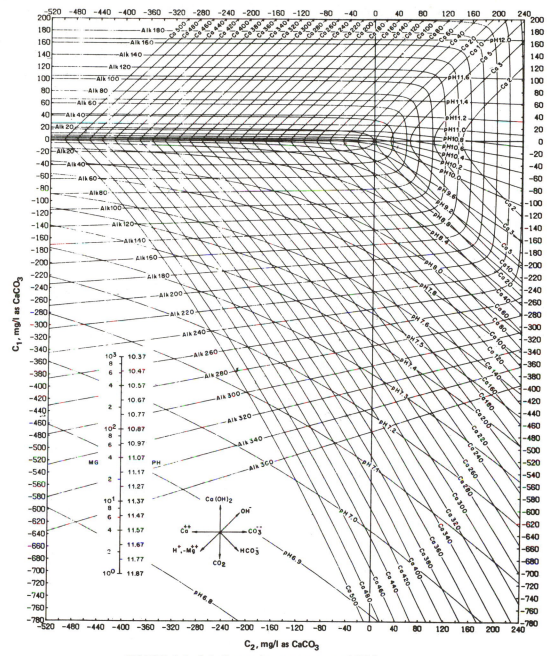

FIGURE 8-4 C-L *diagram for a temperature of 15°C and an Ionic strength of 0.03.*

Note that, by definition, a positive movement along the ordinate is given by a negative acidity value and vice versa.

Before the use of the C-L diagram is discussed, the reader should become familiar with how the addition of various chemicals to a water will affect the alkalinity, acidity, and total carbonic species (C_T) concentrations. Such changes are presented in Table 8-1 for a number of different chemicals. The table is based on the change induced on a

TABLE 8-1 *Changes in alkalinity, acidity, and total carbonic species concentrations associated with the addition of specific chemicals (after Lowenthal and Marais, 1976).*

X mg/ℓ of Chemical Added (expressed as CaCO₃)	Alkalinity Change (mg/ℓ as CaCO₃)	Acidity Change (mg/ℓ as CaCO₃)	C_T Change (mg/ℓ as CaCO₃)
$Ca(OH)_2$	$+X$	$-X$	0
$NaOH$	$+X$	$-X$	0
HCl	$-X$	$+X$	0
$NaHCO_3$	$+X$	$+X$	$+X$
Na_2CO_3	$+X$	0	$+X/2*$
CO_2	0	$+X$	$+X/2*$
$CaCO_3$ (added)	$+X$	0	$+X/2*$
$CaCO_3$ (precipitated)	$-X$	0	$-X/2*$
$CaCl_2$	0	0	0

* The 1/2 values are obtained because these chemical species show a 2:1 ratio of equivalents to moles, and the total carbonic species concentration is based on a mass balance given in terms of moles/ℓ.

specified parameter by the addition of X amount expressed in terms of mg/ℓ as $CaCO_3$ of a specific chemical. For example, the addition of 10 mg/ℓ of $Ca(OH)_2$ as $CaCO_3$ will increase alkalinity by 10 mg/ℓ, will decrease acidity by 10 mg/ℓ, but will have no change on the total carbonic species concentration. These numbers are obtained from the basic mathematical expressions for alkalinity, acidity, and total carbonic species concentrations.

$$[\text{Alk}]_e = [CO_3^{2-}]_e + [HCO_3^-]_e + [OH^-]_e - [H^+]_e \qquad \textbf{(2-74)}$$

$$[\text{Acd}]_e = [H_2CO_3^*]_e + [HCO_3^-]_e + [H^+]_e - [OH^-]_e \qquad \textbf{(2-78)}$$

$$C_T = [H_2CO_3^*] + [HCO_3^-] + [CO_3^{2-}] \qquad \textbf{(2-88)}$$

Equation 2-74 shows that the addition of one equivalent of OH^- ions will increase the alkalinity by one equivalent, whereas equation 2-78 indicates that one equivalent of OH^- ions will decrease acidity by one equivalent. Since OH^- ions do not appear in the mass balance expression for C_T, this parameter will remain unchanged upon OH^- addition to a water.

There are six general steps involved in solving water stabilization problems with C-L diagrams. These are as follows:

1. Measure the pH, alkalinity, and soluble calcium concentration of the water to be treated.

2. Evaluate the equilibrium state of the untreated water with respect to $CaCO_3$ precipitation. This is done by locating the point of intersection of the measured pH and alkalinity lines. Determine the value of the calcium line which passes through this point. Compare this value to the measured calcium value. If the measured value is greater, the water is oversaturated with respect to $CaCO_3$. On the other hand, if the measured value is less than the value obtained from the C-L diagram, the water is undersaturated with respect to $CaCO_3$. The latter is generally the case.

3. Compute the initial acidity of the untreated water. To do so, simply draw a horizontal line through the intersection of the initial pH and alkalinity lines. Determine the C_1 value at the point where this line intersects the ordinate. Calculate acidity from equation 8-23.

$$C_1 = - \text{(acidity)}$$

4. Determine the state of the system with respect to conditions 1, 2, and 3 desired for the saturated water if some given amount of chemical is added and $CaCO_3$ is assumed to be infinitely soluble. To locate this point on the C-L diagram, compute the change in calcium, alkalinity, and acidity which will occur as a result of the chemical addition.

$$[Alk]_2 = [Alk]_1 + [Alk]_{added} \qquad \textbf{(8-25)}$$

$$[Ca]_2 = [Ca]_1 + [Ca]_{added} \qquad \textbf{(8-26)}$$

$$[Acd]_2 = [Acd]_1 + [Acd]_{added} \qquad \textbf{(8-27)}$$

Next, determine the C_1 coordinate of the system after chemical addition.

$$C_1 = - [Acd]_2$$

The intersection of the $[Alk]_2$ line and a horizontal line through C_1 defines the system after chemical addition. If conditions 1, 2, and 3 are satisfied at this point, remove the condition of infinite $CaCO_3$ solubility and allow $CaCO_3$ to precipitate. When evaluating condition 1, note that the difference between $[Ca]_2$ and the Ca value given on the C-L diagram is the maximum amount available for $CaCO_3$ precipitation. Cadena, et al., (1974) indicate that the $CaCO_3^\circ$ species accounts for 13.5 mg/ℓ as $CaCO_3$. Their work is based in part on the relationship given by Martynova (1971) for the variation in the dissociation constant for $CaCO_3^\circ$ with temperature:

$$\log K = \frac{2280}{T} - 12.10$$

where T represents the temperature in degrees Kelvin. The concentration of $CaCO_3^\circ$ may be estimated by dividing the solubility equilibrium expression for calcium carbonate by the equilibrium constant expression for $CaCO_3^\circ$. This gives

$$[CaCO_3^\circ] = \frac{K_{CaCO_3}}{K_{CaCO_3^\circ}}$$

A graphical representation of the variation in the $CaCO_3^\circ$ concentration with temperature is presented in Figure 8-5. Merrill (1980) and Trussell, et al., (1977)

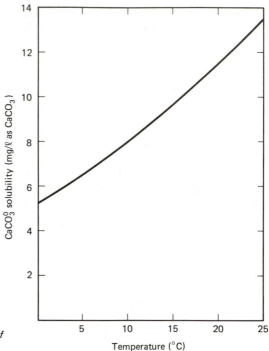

FIGURE 8-5 *Variation in the solubility of the CaCO$_3$ complex ion with temperature.*

do not consider the $CaCO_3^\circ$ species to be important. Trussell, et al., (1977) indicate that the concentration of $CaCO_3^\circ$ in a saturated solution of calcium carbonate will be about $0.17 \, mg/\ell$ as $CaCO_3$ rather than $13.5 \, mg/\ell$. This implies that the value of the dissociation constant for $CaCO_3^\circ$ at $25°C$ is about 6.02×10^{-4} rather than 3.58×10^{-5}.

If conditions 1, 2 and 3 are not satisfied at the new system point, add more chemical.

5. Determine the final equilibrium state of the system after the condition of infinite $CaCO_3$ solubility is removed. This can be accomplished by considering the following two conditions:

(a) Acidity remains constant during $CaCO_3$ precipitation.

(b) During $CaCO_3$ precipitation equivalent amounts of calcium and alkalinity are removed. Therefore, the quantity Alk-Ca will remain constant during precipitation.

On the basis of (a) and (b) above, the coordinates of the equilibrium state after $CaCO_3$ precipitation may be computed from these expressions:

$$C_1 = - [Acd]_2$$
$$C_2 = ([Alk]_2 - [Ca]_2)$$

The intersection of a vertical line through C_2 and a horizontal line through C_1 will define the equilibrium state after $CaCO_3$ precipitation.

6. Check the accuracy of the computations by comparing the change in alkalinity and calcium during $CaCO_3$ precipitation. These values should be the same if the computations are correct.

$$\Delta Alk = [Alk]_2 - [Alk]_{final}$$
$$\Delta Ca = [Ca]_2 - [Ca]_{final}$$
$$\Delta Alk = \Delta Ca$$

The use of a C-L diagram for computing the required chemical dose for water stabilization is illustrated in example problem 8-2. At this point, the reader's attention is directed to Figure 8-4, where it can be seen that the figure contains both a directional format diagram and a pH-magnesium nomograph. These are applicable only to water softening and neutralization problems which are presented in Chapter 9; therefore, their use will be discussed at that time.

EXAMPLE PROBLEM 8-2: Analysis of a water gives an alkalinity of 200 mg/ℓ, calcium of 100 mg/ℓ (both expressed as $CaCO_3$), and a pH of 7.3. Determine if a lime dose of 31 mg/ℓ as $CaCO_3$ will provide a good finished water. Assume a water temperature of 5°C and an ionic strength of 0.01 M, and also assume that chlorides and sulfates are present in negligible concentrations.

Solution:

1. Evaluate the equilibrium state of the untreated water.

(a) Locate the intersection of initial pH and initial alkalinity lines (shown as point 1 in Figure 8-6).

(b) Since point 1 is beyond the maximum calcium concentration line of 500, it is safe to assume that the water, which contains 100 mg/ℓ Ca, is undersaturated.

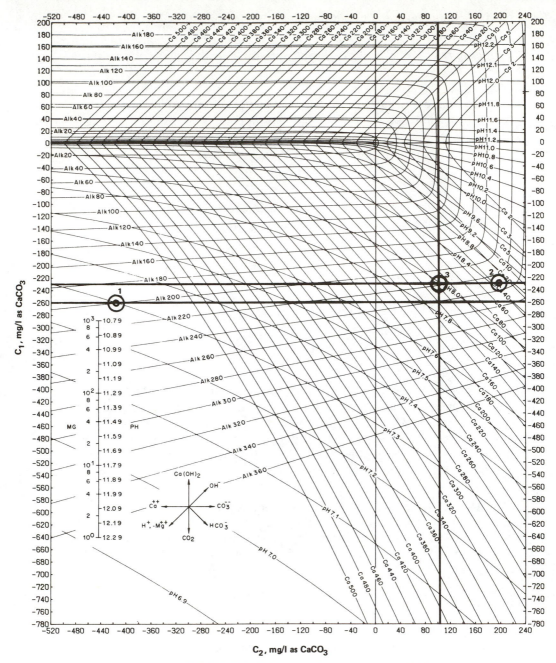

FIGURE 8-6 C-L diagram for a temperature of 5°C and ionic strength of 0.01.

2. Compute the initial acidity of the untreated water.

   (a) Construct a horizontal line through point 1 and read the C_1 value at the point where that line intersects the ordinate:

   $$C_1 = -260$$

   therefore,

   $$-260 = -(Acd)$$

   or

   $$Acd = 260 \text{ mg}/\ell \text{ as } CaCO_3$$

3. Determine the state of the system with respect to conditions 1, 2, and 3 after 31 mg/ℓ of lime as $CaCO_3$ have been added.

   (a) Compute the alkalinity, acidity, and calcium content at state 2:

   $$[Alk]_2 = 200 + 31 = 231 \text{ mg}/\ell \text{ as } CaCO_3$$
   $$[Acd]_2 = 260 - 31 = 229 \text{ mg}/\ell \text{ as } CaCO_3$$
   $$[Ca]_2 = 100 + 31 = 131 \text{ mg}/\ell \text{ as } CaCO_3$$

   (b) Determine the C_1 coordinate of the system after chemical addition:

   $$C_1 = -[Acd]_2 = -229$$

   (c) Locate the intersection of the $[Alk]_2$ line and a horizontal line through $C_1 = -229$. (shown as point 2 in Figure 8-6).

   (d) *Condition 1:* precipitation potential between 4–10 mg/ℓ of $CaCO_3$.

   The soluble calcium concentration at point 2 is 35 mg/ℓ. Since this is less than $[Ca]_2$ concentration, the water is oversaturated and precipitation will occur. The precipitation potential is the difference between $[Ca]_2$ and $[Ca]_{final}$ and will be evaluated in step 4.

   (e) *Condition 2:* Both $[Alk]_2$ and $[Ca]_2$ greater than 40 mg/ℓ as $CaCO_3$.

   This condition is satisfied, since $[Alk]_2 = 231$ and $[Ca]_2 = 131$.

   (f) *Condition 3:*

   $$\frac{Alk}{Cl^- + SO_4^{2-}} > 5$$

   Since Cl^- and SO_4^{2-} concentrations are very small, this condition is satisfied.

4. Determine the final equilibrium state of the system after the condition of infinite $CaCO_3$ solubility is removed.

   (a) Determine the C_2 coordinate of the system after precipitation:

   $$C_2 = 231 - 131 = 100$$

   (b) Locate the intersection of a vertical line through C_2 and the horizontal line through C_1 (shown as point 3 in Figure 8-6).

(c)   Estimate the calcium and alkalinity at point 3:

$$[Ca]_{final} = 118 \text{ mg}/\ell \text{ as } CaCO_3$$
$$[Alk]_{final} = 218 \text{ mg}/\ell \text{ as } CaCO_3$$

(d)   Compute the precipitation potential:

$$\Delta\ CaCO_3 = ([Ca]_2 - [Ca]_{final})$$
$$= 131 - 118$$
$$= 13 \text{ mg}/\ell \text{ as } CaCO_3$$

Note: The final equilibrium pH is established at point 3 and is slightly less than 8. Although this pH is outside the suggested range of 6.8 to 7.3, it may be acceptable since it is often uneconomical to achieve equilibrium pH values this low. Two criteria should be followed when evaluating the acceptability of the final pH: these are (1) avoid excessively high pH values, which may cause scaling of hot-water heaters because of $Mg(OH)_2$ precipitation. Figure 8-7 is presented as a guide to acceptable pH values to avoid this

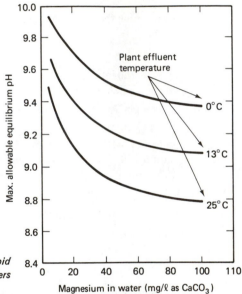

FIGURE 8-7 Approximate pH *limits to avoid* $Mg(OH)_2$ *scaling in domestic water heaters (after Merrill and Sanks, 1977).*

problem. (2) Avoid final pH values within the 8.0 to 8.5 range. This is a range of minimum buffer intensity for the carbonic acid system (see Figure 2-24), and as a result the water will have little ability to resist pH changes induced by products released from the corrosion process.

5.   Check the accuracy of the computations:

$$\Delta\ Alk = 231 - 218 = 13$$
$$\Delta\ Ca = 131 - 118 = 13$$

Therefore,

$$\Delta\ Alk = \Delta\ Ca$$

Some waters, such as those from snow-melt or glaciers, are low in total dissolved solids (TDS). These are referred to as *carbonic species deficient waters*. Such waters generally have a fairly low pH and very little buffer capacity. Stabilization of carbonic species deficient waters is accomplished with two chemicals. Lime is usually added to increase the calcium and alkalinity concentrations to some desired level. Because of the initial low calcium and alkalinity values, a substantial lime dose is required. However, because of the low buffer capacity, lime addition alone produces a pH which is inordinately high. As a result, CO_2 is added to reduce the pH to some desired precipitation value (this value is usually arrived at by field experimentation).

There are eight general steps involved in solving water stabilization problems for carbonic-species deficient waters. These are as follows:

1. Measure the pH, alkalinity, and soluble calcium concentration of the water to be treated.

2. Determine the initial acidity of the untreated water. Since the intersection of the initial alkalinity line and the initial pH line for carbonic-species deficient waters occurs at some point to the left of the C-L diagram, these diagrams cannot be used to determine the initial acidity. Instead, initial acidity is calculated from equation 8-28:

$$\text{initial acidity} \atop (\text{mg}/\ell \text{ as } CaCO_3) = \frac{\left([\text{Alk}] - \frac{K_w}{[H^+]} + [H^+]\right)\left(1 + \frac{[H^+]}{K_1}\right)}{1 + \frac{K_2}{[H^+]}} + [H^+] - \frac{K_w}{[H^+]} \qquad (8\text{-}28)$$

where K_w, K_1, and K_2 = equilibrium constants for the ionization of H_2O, $H_2CO_3^*$, and HCO_3^-, respectively. These values are given in Table 8-2 for different TDS concentrations and temperatures.

[Alk] = initial alkalinity of untreated water (mg/ℓ as $CaCO_3$)

[H^+] = initial hydrogen ion concentration of untreated water (mg/ℓ as $CaCO_3$). Values for H^+ may be obtained from Figure 8-8 for different TDS concentrations and initial pH values. Temperature corrections are compensated for in the pH measurement.

3. Establish the approximate pH and alkalinity values which are desired for the oversaturation state. Typical pH values lie within the 9.0 to 10.5 range; typical alkalinity values are between 30 and 100 mg/ℓ as $CaCO_3$.

4. Compute the lime dose required to adjust the alkalinity to the desired value. Since each mg/ℓ of $Ca(OH)_2$ (as $CaCO_3$) added to the water will increase both

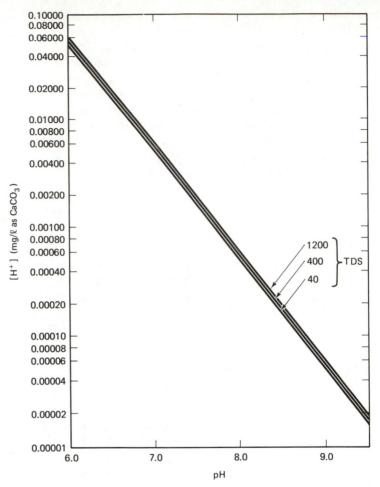

FIGURE 8-8 *Hydrogen ion concentration in mg/l as* $CaCO_3$ *vs pH at different* TDS *values (after Merrill and Sanks, 1977).*

alkalinity and calcium by 1 mg/ℓ (as $CaCO_3$) and decrease acidity by 1 mg/ℓ (as $CaCO_3$), the following relationships are valid:

$$\frac{\text{lime dose}}{(\text{mg}/\ell \text{ as } CaCO_3)} = [\text{Alk}]_{\text{desired}} - [\text{Alk}]_{\text{initial}} \qquad (8\text{-}29)$$

$$[\text{Alk}]_2 = [\text{Alk}]_{\text{desired}} = \text{lime dose} + [\text{Alk}]_{\text{initial}} \qquad (8\text{-}30)$$

$$[\text{Ca}]_2 = \text{lime dose} + [\text{Ca}]_{\text{initial}} \qquad (8\text{-}31)$$

$$[\text{Acd}]_2 = [\text{Acd}]_{\text{initial}} - \text{lime dose} \qquad (8\text{-}32)$$

In this step $CaCO_3$ is assumed to be infinitely soluble.

TABLE 8-2 Values for K_w, K_1 and K_2 at various temperatures and TDS concentrations which are to be used with equation 8-23 (after Merrill and Sanks, 1977).

	Temperature (°C)			
TDS	2°	5°	15°	25°
		K_w		
40	0.00000373	0.00000496	0.0000120	0.0000268
400	0.00000386	0.00000567	0.0000138	0.0000307
1200	0.00000482	0.00000640	0.0000155	0.0000346
		K_1		
40	0.00739	0.00774	0.00996	0.01150
400	0.00845	0.00917	0.01140	0.01320
1200	0.00954	0.01030	0.01290	0.01490
		K_2		
40	0.00000290	0.00000320	0.00000426	0.00000537
400	0.00000380	0.00000419	0.00000558	0.00000703
1200	0.00000484	0.00000534	0.00000711	0.00000896

5. Determine the pH of the water after lime addition. To do this, determine the C_1 coordinate of the system after lime addition:

$$C_1 = - [Acd]_2$$

The intersection of the $[Alk]_2$ line and a horizontal line through C_1 will define the system after lime addition; this point will be referred to as *point 2*. The pH line through point 2 will give the pH of the system after lime addition; this will be referred to as $[pH]_2$.

6. Compute the CO_2 dose required to adjust the pH from $[pH]_2$ to the desired oversaturation pH established in step 3. This can be accomplished by considering the following two conditions:

(a) When CO_2 is added to a water, acidity increases by an amount equal to the quantity of CO_2 added.

(b) As long as $CaCO_3$ does not precipitate, the addition of CO_2 does not affect the alkalinity or calcium content of a water.

   On the basis of (a) and (b) above, the point on the C-L diagram attained after CO_2 addition is located at the intersection of the [pH] desired line and the

[Alk]$_2$ line; this point will be referred to as *point 3*. The required CO_2 dose is given by equation 8-33:

$$\frac{CO_2 \text{ dose}}{(mg/\ell \text{ as } CaCO_3)} = [Acd]_3 - [Acd]_2 \qquad (8\text{-}33)$$

The [Acd]$_3$ value is obtained by constructing a horizontal line through point 3. Determine the C_1 value at the point where this line intersects the ordinate. Calculate acidity from the relationship:

$$C_1 = - \text{ (acidity)}$$

7. Determine the final equilibrium state of the system after the condition of infinite $CaCO_3$ solubility is removed.

   (a) Evaluate the saturation state of the system. If enough lime has been added, the system will be oversaturated. This is determined by comparing the [Ca]$_2$ value to the value of the calcium line through point 3. If [Ca]$_2$ is greater than [Ca]$_3$, the water is oversaturated.

   (b) Determine the C_2 coordinate of the system after precipitation:

   $$C_2 = [Alk]_2 - [Ca]_2$$

   (c) Locate the intersection of a vertical line through C_2 and the horizontal line through point 3; this point will be referred to as *point 4*.

   (d) Estimate the values of the calcium and alkalinity lines passing through point 4.

   (e) Compute the precipitation potential:

   $$\Delta CaCO_3 = ([Ca]_2 - [Ca]_4)$$

8. Check the accuracy of the computations by ensuring that Δ Alk and Δ Ca are equal

$$\Delta \text{ Alk} = [Alk]_2 - [Alk]_4$$
$$\Delta \text{ Ca} = [Ca]_2 - [Ca]_4$$

The chemical dose calculations for the stabilization of a carbonic-species deficient water are illustrated in example problem 8-3.

EXAMPLE PROBLEM 8-3: The state of a water entering a treatment plant was found to contain alkalinity of 2 mg/ℓ, calcium of 2 mg/ℓ (both as $CaCO_3$), and pH of 6.0. Determine the lime and carbon dioxide dose required to produce the oversaturation state of pH of 9.1 and alkalinity of 40 mg/ℓ. Does the precipitation potential of the treated water fall within the 4–10 mg/ℓ range? Assume a water temperature of 25°C and an ionic strength of 0.001 M (40 mg/ℓ TDS).

Solution:

1. Calculate the initial acidity of the untreated water, using equation 8-28:

$$[Acd]_{initial} = \frac{\left(2.0 - \dfrac{0.0000268}{0.052} + 0.052\right)\left(1 + \dfrac{0.052}{0.01150}\right)}{1 + \dfrac{0.0000053}{0.052}} + 0.052 - \frac{0.0000268}{0.052}$$

$$= 11.0 \text{ mg}/\ell \text{ as } CaCO_3$$

2. Compute the lime dose required to adjust the alkalinity to 40 mg/ℓ. Also, compute the $[Ca]_2$ and $[Acd]_2$ values.

   lime dose = $40 - 2 = 38$ mg/ℓ as $CaCO_3$

   $[Alk]_2 = 40$ mg/ℓ as $CaCO_3$

   $[Ca]_2 = 38 + 2 = 40$ mg/ℓ as $CaCO_3$

   $[Acd]_2 = 7.6 - 38 = -27$ mg/ℓ as $CaCO_3$

3. Locate point 2 on the C-L diagram shown in Figure 8-9. To do this, first compute the C_1 coordinate value:

$$C_1 = -[-27] = 27$$

   Construct a horizontal line through C_1 and locate the point where this line intersects the $[Alk]_2$ line. This intersection is shown as point 2 on Figure 8-9. The pH line through point 2 has an approximate value of 10.8.

4. Locate point 3 on the C-L diagram shown in Figure 8-9. This point is located at the intersection of the $[pH]_{desired}$ line and the $[Alk]_2$ line.

5. Compute the CO_2 dose required to adjust the pH from 10.8 to 9.1. To make this computation, first determine the acidity at point 3. This is accomplished by constructing a horizontal line through point 3 and determining the C_1 value at the point where this line intersects the ordinate.

$$C_1 = -34$$

   Therefore,

$$-34 = -(acidity)$$

   or

$$acidity = 34 \text{ mg}/\ell \text{ as } CaCO_3$$

   Thus,

$$CO_2 \text{ dose} = 34 - (-27) = 61 \text{ mg}/\ell \text{ as } CaCO_3$$

6. Determine the final equilibrium state of the system after the condition of infinite $CaCO_3$ solubility is removed.

   (a) Compute the C_2 coordinate of the system after precipitation:

$$C_2 = 40 - 40 = 0 \text{ mg}/\ell$$

   (b) Locate point 4 at the intersection of the vertical line constructed through C_2 and the horizontal line constructed through point 3.

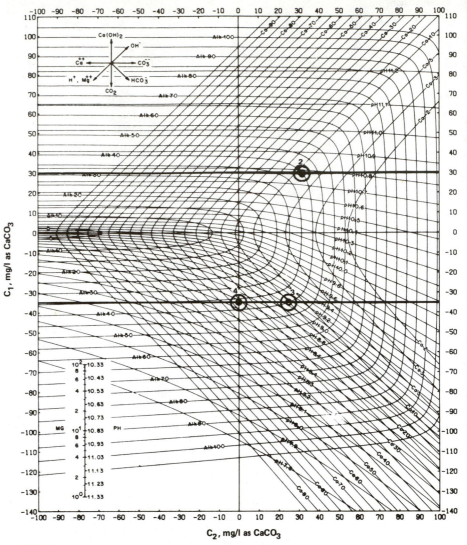

FIGURE 8-9 C-L *diagram for a temperature of 25°C and an ionic strength of 0.001.*

(c) Estimate the value of the calcium and alkalinity lines through point 4:

$$[Alk]_4 = 36$$
$$[Ca]_4 = 36$$

(d) Compute the precipitation potential:

$$\Delta CaCO_3 = (40 - 36) = 4 \text{ mg}/\ell$$

8. Check the accuracy of the computations by ensuring the Δ Alk and Δ Ca are equal:

$$\Delta \text{Alk} = 40 - 36 = 4$$

$$\Delta \text{Ca} = 40 - 36 = 4$$

It is important that a water utility produce a stable or well-conditioned water to reduce or prevent corrosion of the pipes in the distribution system, since there are a number of undesirable effects associated with this problem: these include (a) contamination of the finished water by corrosion products, (b) reduction in carrying capacity of the pipes, and (c) pipe failure. Lane (1978) reported that when corrosive water was allowed to stand overnight in household plumbing, copper concentrations as high as 2 mg/ℓ were measured in copper tubing, and zinc concentrations up to 5.5 mg/ℓ and iron concentrations as great as 5.4 mg/ℓ were measured in galvanized piping. Primary regulations, which are enforceable, described in the Safe Drinking Water Act (PL 93–523), establish the maximum levels for specific contaminants that are allowed in drinking water. These levels are listed in Table 8-3 for inorganic chemicals other than fluoride. Proposed secondary regulations are presented in Table 8-4. Note that the

TABLE 8-3 *Maximum contaminant levels for inorganic chemicals other than fluoride.*

Inorganic Contaminant	Level (mg/ℓ)
Arsenic	0.05
Barium	1.
Cadmium	0.010
Chromium	0.05
Lead	0.05
Mercury	0.002
Nitrate (as N)	10.
Selenium	0.01
Silver	0.05

TABLE 8-4 *Proposed secondary regulations from PL 93-523.*

Contaminant	Level
Chloride	250 mg/ℓ
Color	15 color Units
Copper	1 mg/ℓ
Corrosivity	Noncorrosive
Foaming agents	0.5 mg/ℓ
Hydrogen sulfide	0.05 mg/ℓ
Iron	0.3 mg/ℓ
Manganese	0.05 mg/ℓ
Odor	3 threshold odor number
pH	6.5–8.5
Sulfate	250 mg/ℓ
Total dissolved solids	500 mg/ℓ
Zinc	5 mg/ℓ

question of corrosion is addressed in these regulations. At this time, secondary regulations are not enforceable; however, they may be in the future.

Corrosion of water distribution piping will result in a decrease in the carrying capacity of the system. The change in hydraulic capacity of a pipe can be evaluated in terms of a change in the value of the Hazen-Williams coefficient. The numerical value of this coefficient is dependent on the roughness of the pipe surface. Most new pipes have C values between 130 and 140. However, as the interior pipe surface corrodes it becomes rough, and the C value may decrease to as low as 60. The effect of decreasing C on the carrying capacity of a pipe can be computed from the Hazen-Williams formula for circular conduits:

$$Q = 405(C)(d)^{2.63}(S)^{0.54} \tag{8-34}$$

where $Q = $ pipe flow (gal/day)

$d = $ pipe diameter (ft)

$S = $ slope of hydraulic gradient (ft per ft of pipe length)

$C = $ Hazen-Williams coefficient.

Annual economic losses resulting from corrosion of water distribution systems have been estimated by Hudson and Gilcreas (1976) to be near 375 million dollars. They suggest that such losses could be avoided if the water utility industry as a whole practiced water stability control. The annual lime cost for such conditioning was estimated to be 27 million dollars.

It is important to remember that, although the deposition of a calcium carbonate coating will generally protect against corrosion, the continuous deposition of such a coating will form a heavy scale which will restrict flow in the piping system. Therefore, $CaCO_3$ precipitation should be discontinued after the protective coating has been established. At that point the chemical dose should be adjusted to produce a stable water, i.e., a water that will neither precipitate nor dissolve a $CaCO_3$ film. Since the actual chemical dose depends on the chemical characteristics of the water to be treated, the type of pipe in the distribution system, and the flow regimes in the pipelines, the water utility must continually evaluate the effectiveness of any water conditioning program. This is normally done by the coupon test, which measures the effects of water on small sections of metal pipe placed in the distribution system. Corrosivity is assessed on the basis of weight loss by the coupon insert, whereas scale buildup is indicated by weight gain. Adjustments in chemical dose can be made if continuous evaluation of pipe inserts indicate that it is necessary. Keep in mind that conditioning chemicals should be added after the disinfection step to avoid pH changes resulting from chlorine addition.

Methods other than calcium carbonate deposition are also available for corrosion control. Sodium silicate dosages between 4 and 8 mg/ℓ as SiO_2 have been used to inhibit corrosion in building pipe systems. Lane (1978) reported that when applied to waters with hardness less than 100 mg/ℓ and a near neutral pH, a few mg/ℓ of zinc polyphosphate or zinc orthophosphate reduces corrosion in steel, galvanized steel, and copper piping.

PROBLEMS

8-1. Calculate the Langelier index for a raw water with the following characteristics:

$$pH = 7.4$$
$$Alk = 160 \text{ mg}/\ell \text{ as } CaCO_3$$
$$Ca = 300 \text{ mg}/\ell \text{ as } CaCO_3$$
$$TDS = 400 \text{ mg}/\ell$$
$$temperature = 5°C$$

Is this water oversaturated or undersaturated with respect to $CaCO_3$ precipitation?

8-2. The chemical characteristics of a snow-melt water are as follows:

$$temperature = 25°C$$
$$TDS = 40 \text{ mg}/\ell$$
$$alkalinity = 8 \text{ mg}/\ell \text{ as } CaCO_3$$
$$calcium = 5 \text{ mg}/\ell \text{ as } CaCO_3$$
$$pH = 6.5$$
$$Cl^- + SO_4^{2-} = negligible$$

Does this water require conditioning for corrosion control? If so, determine the amount of $Ca(OH)_2$ and CO_2 that should be used for conditioning.

8-3. Analysis of a water gives an alkalinity of 200 mg/ℓ, calcium of 100 mg/ℓ (both expressed as $CaCO_3$), and a pH of 7.3. Determine if a NaOH dose of 33 mg/ℓ as $CaCO_3$ will provide a good finished water. Assume a water temperature of 5°C and an ionic strength of 0.01 M, and also that chlorides and sulfates are present in negligible concentrations.

8-4. A raw water has the following characteristics:

$$temperature = 15°C$$
$$alkalinity = 220 \text{ mg}/\ell \text{ as } CaCO_3$$
$$calcium = 300 \text{ mg}/\ell \text{ as } CaCO_3$$
$$TDS = 400 \text{ mg}/\ell$$

What is the final equilibrium pH after 20 mg/ℓ of $Ca(OH)_2$ (expressed as $CaCO_3$) have been added to the water? Assume that the water is initially saturated with respect to calcium carbonate.

8-5. Calculate the Langelier index of the water described in Problem 8-1 for a pH of 7.6 and a temperature of 15°C.

8-6. Rework Problem 8-2, where it is required to determine the amount of H_2SO_4 and powdered $CaCO_3$ that should be used for conditioning. In other words sulfuric acid and powdered $CaCO_3$ rather than lime and carbon dioxide are to be used as the conditioning chemicals.

8-7. The finished water from a water treatment plant has the following characteristics:

$$pH = 7.6$$
$$temperature = 15°C$$
$$Ca^{2+} = 65 \text{ mg}/\ell \text{ as } Ca^{2+}$$
$$Mg^{2+} = 17 \text{ mg}/\ell \text{ as } Mg^{2+}$$
$$Na^{+} = 20 \text{ mg}/\ell \text{ as } Na^{+}$$
$$K^{+} = 10 \text{ mg}/\ell \text{ as } K^{+}$$
$$HCO_3^{-} = 125 \text{ mg}/\ell \text{ as } HCO_3^{-}$$
$$SO_4^{2-} = 80 \text{ mg}/\ell \text{ as } SO_4^{2-}$$
$$Cl^{-} = 20 \text{ mg}/\ell \text{ as } Cl^{-}$$

Calculate the Langelier index of this water.

REFERENCES

BUTLER, G., and ISON, H.C.K., *Corrosion and Its Prevention in Waters*, Van Nostrand Reinhold Company, New York (1966).

CADENA, F., MIDKIFF, W.S., and O'CONNOR, G.A., "The Calcium Carbonate Ion-Pair as a Limit to Hardness Removal," *J. Am. Water Works Assoc.*, **66**, 524 (1974).

CULVER, R.H., "Soft Water: A Distribution Problem," *Water and Sewage Works*, **52** (February, 1975).

HUDSON, H.E., and GILCREAS, F.W., "Health and Economic Aspects of Water Hardness and Corrosiveness," *J. Am. Water Works Assoc.*, **68**, 20 (1976).

LANE, R.W., "Stabilization Needed to Control Pipe Corrosion and Scale," *Water and Sewage Works*, **60** (July, 1978).

LANGELIER, W.F., "The Analytical Control of Anti-Corrosion Water Treatment," *J. Am. Water Works Assoc.*, **28**, 1500 (1936).

LOEWENTHAL, R.E., and MARAIS, G.V.R., *Carbonate Chemistry of Aquatic Systems: Theory and Application*. Ann Arbor Science Publishers, Ann Arbor, MI (1976).

MARTYNOVA, O.J., et al., "The Determination of the Dissociation Constants of $CaOH^{+}$, $CaHCO_3^{+}$ and $CaCO_3^{o}$ Ion-Pairs in the 22-98°C Temperature Range." Report of the Academy of Science, USSR (1971).

MERRILL, D.T., "Chemical Conditioning for Water Softening and Corrosion Control," paper presented at the Fifth Environmental Engineers Conference, Big Sky of Montana (June, 1976).

MERRILL, D.T., and SANKS, R.L., "Corrosion Control by Deposition of $CaCO_3$ Films: Part 1, A Practical Approach for Plant Operators," *J. Am. Water Works Assoc.*, **69**, 592 (1977).

MERRILL, D.T., Personal Communication (1980).

SINGLEY, J.E., "Principles of Corrosion," *Proc, AWWA Seminar on Controlling Corrosion Within Water Systems*, Atlantic City, NJ (1978).

TRUSSELL, R.R., RUSSELL, L.L., and THOMAS, J.F., "The Langelier Index," in *Water Quality in the Distribution System*, Fifth Annual AWWA Water Quality Technology Conference, Kansas City, MO (1977).

9

WATER SOFTENING AND NEUTRALIZATION

A water may be referred to as *hard* or *soft*, depending on the concentration of polyvalent metallic cations contained in the water. Two major disadvantages of using hard water are the increased amount of soap required to produce a lather when bathing and the formation of scale on heat exchange surfaces such as cooking utensils, hot-water pipes, hot-water heaters, and boilers. Although public tolerance of hardness varies, a level ranging from 60 to 120 mg/ℓ expressed as $CaCO_3$ is generally acceptable. The classification of water supplies according to their degree of hardness is summarized in Tables 9-1 and 9-2, and a map showing the general character of the water in various states in Figure 9-1.

Although hardness is caused by the presence of any polyvalent metallic cation, the most prevalent of these species are the divalent cations calcium and magnesium. As a result, hardness is usually defined as the total concentration of calcium and magnesium ions and is usually expressed as mg/ℓ of $CaCO_3$. It is emphasized that the salts formed from soap and the monovalent ions in water are soluble over the normal concentration range of these ions in water and hence do not contribute to hardness (e.g., the sodium ion is not considered in hardness calculations).

TABLE 9-1 United States classification of hard waters (*after Kunin, 1972*).

Hardness Range (mg/ℓ CaCO$_3$)	Hardness Description	Percent of Population of 100 Large Cities
0–60	Soft	35.0
61–120	Moderately hard	25.0
121–180	Hard	26.6
> 180	Very hard	13.4

TABLE 9-2 *International classification of hard waters (after Kunin, 1972).*

Hardness Range (mg/ℓ CaCO₃)	Hardness Description	Percent of Population				
		France	England	Netherlands	Sweden	Switzerland
0–50	Soft	7.9	20.5	13.1	37.9	3
50–100	Moderately soft	6.6	14.1	34.0	33.8	2
100–150	Slightly hard	6.4	11.9	35.0	4.7	
150–200	Moderately hard	7.2	8.5	16.7	5.1	49
200–300	Hard	55.6	39.7	1.1	8.0	41
> 300	Very hard	16.3	5.3	0.1	10.5	5

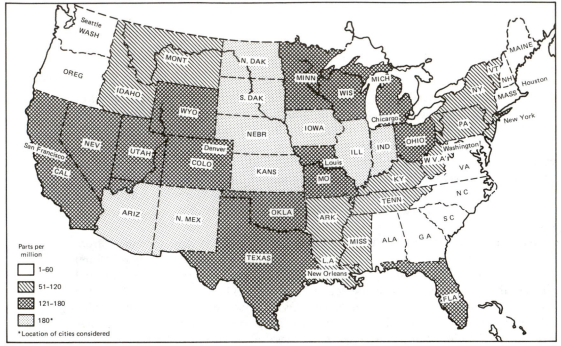

WEIGHTED AVERAGE HARDNESS, BY STATES, OF WATER FURNISHED IN 1932 BY PUBLIC SUPPLY SYSTEMS IN OVER 600 CITIES IN THE UNITED STATES.

FIGURE 9-1 *Hardness characteristics of various state water supplies (from U.S. Geological Survey Paper 658).*

Hardness can be classified in a number of different ways:

1. Total hardness: the sum of the calcium and magnesium ion concentrations expressed as mg/ℓ of CaCO₃.

2. Carbonate hardness: hardness equal to or less than the total alkalinity.

3. Noncarbonate hardness: hardness exceeding carbonate hardness, i.e., the difference between total hardness and carbonate hardness.

4. Calcium hardness: the portion of the total hardness due to the calcium ion.

5. Magnesium hardness: the portion of the total hardness due to the magnesium ion.

In this discussion it is assumed that alkalinity is due to the carbonic acid system. However, remember that hydroxide, silicate, phosphate, or borate ions may also contribute to alkalinity.

Two commonly used processes for reducing hardness in water are precipitation and ion exchange. Both of these methods are discussed in subsequent sections of this chapter.

9-1 CHEMICAL PRECIPITATION

The precipitation of calcium and magnesium ions is usually affected by the addition of lime, $Ca(OH)_2$, and soda ash, Na_2CO_3, to the water. This results in the formation of the sparingly soluble compounds of calcium carbonate, $CaCO_3$, and magnesium hydroxide, $Mg(OH)_2$, at elevated pH levels. These precipitates are separated by gravity settling, leaving a clarified, softened water with an elevated pH. The clarified effluent is then generally neutralized by adding carbon dioxide, CO_2. In some instances sodium hydroxide, NaOH, is used in the softening process, and a strong acid such as sulfuric acid, H_2SO_4, is used for neutralization.

The effect of pH on the species distribution for the carbonic acid system is shown in Figure 9-2. Since the pH of most natural waters is in the neutral range, the alkalinity

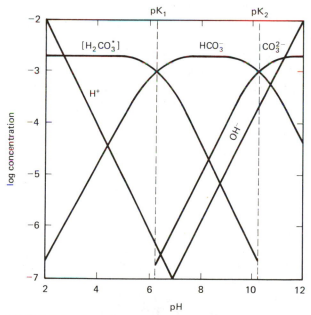

FIGURE 9-2 *Effect of* pH *on the species composition of a carbonic acid solution with a total alkalinity equal to 200 mg/l as* $CaCO_3$ *(after Singley, 1972).*

(assuming that alkalinity is due mainly to the carbonic acid system) is in the form of bicarbonate alkalinity.

The solubility equilibrium for $CaCO_3$ is described by equation 9-1:

$$CaCO_{3(s)} \rightleftharpoons Ca^{2+}_{(aq)} + CO^{2-}_{3(aq)} \tag{9-1}$$

The addition of $Ca(OH)_2$ to a water increases the hydroxyl ion concentration and elevates the pH which, according to Figure 9-2, shifts the equilibrium of the carbonic acid system in favor of the carbonate ion, CO^{2-}_3 (i.e., it changes the alkalinity from the bicarbonate form to the carbonate). This increases the concentration of the CO^{2-}_3 ion and, according to Le Châtelier's principle, shifts the equilibrium described by equation 9-1 to the left. Such a response results in the precipitation of $CaCO_3$ and a corresponding decrease in the soluble calcium concentration.

The solubility equilibrium for $Mg(OH)_2$ is described by equation 9-2:

$$Mg(OH)_2 \rightleftharpoons Mg^{2+}_{(aq)} + 2OH^-_{(aq)} \tag{9-2}$$

According to Le Châtelier's principle, the addition of hydroxyl ions shifts the equilibrium described by equation 9-2 to the left, resulting in the precipitation of $Mg(OH)_2$ and a corresponding decrease in the soluble magnesium concentration.

The amount of calcium which can be removed through lime addition is limited by the number of carbonate ions that can be formed by pH adjustment, which is, in turn, limited by the amount of alkalinity initially present in the water. What this means in terms of calcium removal is that only calcium carbonate hardness can be reduced by adding lime. If calcium removal beyond this limit is required (i.e., if the removal of calcium noncarbonate hardness is required), it is necessary to provide a source of carbonate ions other than natural alkalinity, which is normally accomplished by the addition of soda ash, Na_2CO_3. Magnesium can theoretically be reduced to the solubility limits of $Mg(OH)_2$ by using only lime. Hence, magnesium hardness and calcium carbonate hardness can be removed with lime, whereas soda ash is required to remove calcium noncarbonate hardness.

The reactions that occur when a water containing both carbonate and noncarbonate hardness is softened by the addition of lime and soda ash are summarized as follows (Faust and McWhorter, 1976):

$$1) \quad H_2CO_3 + Ca(OH)_2 \rightleftharpoons CaCO_{3(s)} + 2H_2O \tag{9-3}$$

Equation 9-3 represents the reaction of lime with carbonic acid. Although this reaction does not produce any softening, it must be considered because it creates a lime demand. The reaction results from the fact that bases react with acids. Most natural waters contain free CO_2, which reacts with water to form carbonic acid. Because of the equilibrium position of this reaction, the amount of H_2CO_3 present in the system is generally assumed to be equal to the free CO_2 concentration.

$$2) \quad Ca^{2+} + 2HCO^-_3 + Ca(OH)_2 \rightleftharpoons 2CaCO_{3(s)} + 2H_2O \tag{9-4}$$

WATER SOFTENING AND NEUTRALIZATION   CHAPTER 9

Equation 9-4 represents the removal of calcium carbonate hardness, i.e., the calcium which can be removed by converting bicarbonate ions to carbonate ions and precipitating calcium as calcium carbonate. This reaction shows that one equivalent of lime will remove one equivalent of carbonate hardness.

$$3) \quad Ca^{2+} + SO_4^{2-} + Na_2CO_3 \rightleftharpoons CaCO_{3(s)} + Na_2SO_4 \qquad (9\text{-}5)$$

Equation 9-5 represents the removal of calcium noncarbonate hardness, i.e., the calcium which must be removed by adding carbonate ions from an external source. In this case, calcium ions are associated with anions other than bicarbonate (e.g. SO_4^{2-}, Cl^-). This reaction shows that one equivalent of soda ash will remove one equivalent of noncarbonate hardness.

$$4) \quad Mg^{2+} + 2HCO_3^- + Ca(OH)_2 \rightleftharpoons CaCO_{3(s)} + MgCO_3 + 2H_2O \quad (9\text{-}6)$$

$$MgCO_3 + Ca(OH)_2 \rightleftharpoons CaCO_{3(s)} + Mg(OH)_{2(s)} \qquad (9\text{-}7)$$

$$overall: Mg^{2+} + 2HCO_3^- + 2Ca(OH)_2 \rightleftharpoons 2CaCO_{3(s)} + Mg(OH)_{2(s)}$$
$$+ 2H_2O \qquad (9\text{-}8)$$

Equation 9-8 represents the removal of magnesium carbonate hardness. Because this reaction is actually the sum of two reactions (reactions 9-6 and 9-7), it indicates that two equivalents of lime are required to remove one equivalent of magnesium carbonate hardness. In effect, what equation 9-6 represents is a situation where any bicarbonate alkalinity in excess of the amount associated with calcium carbonate hardness has to be neutralized before the hydroxyl ion concentration can be increased significantly for magnesium removal. Therefore, equation 9-6 could be represented as follows:

$$2HCO_3^- + Ca(OH)_2 \rightleftharpoons CaCO_{3(s)} + 2H_2O + CO_3^{2-} \qquad (9\text{-}9)$$

$$5) \quad Mg^{2+} + SO_4^{2-} + Ca(OH)_2 + Na_2CO_3 \rightleftharpoons CaCO_{3(s)} + Mg(OH)_{2(s)}$$
$$+ Na_2SO_4 \qquad (9\text{-}10)$$

Equation 9-10 represents the removal of magnesium noncarbonate hardness. This reaction shows that one equivalent of lime and one equivalent of soda ash is required to remove one equivalent of magnesium noncarbonate hardness. Lime is required to supply the hydroxyl ions necessary for $Mg(OH)_2$ precipitation, whereas soda ash is required to remove the calcium derived from lime addition. Unless an equivalent amount of soda ash is added no softening is achieved by this reaction because, for each equivalent of magnesium noncarbonate hardness removed, an equivalent amount of calcium noncarbonate hardness is added.

In the reactions involving lime addition, lime was represented as $Ca(OH)_2$ (calcium hydroxide, generally referred to as *slaked lime* or *hydrated lime*). Lime is usually purchased in the CaO (calcium oxide, generally referred to as *unslaked lime* or *quick lime*) form. However, when CaO is added to water it forms $Ca(OH)_2$, and since this is the lime form present at the time of softening, it is felt that the use of calcium hydroxide in the softening reactions is preferable to the use of calcium oxide.

Three different methods are presented which may be employed in the calculation of the chemical dosages required for softening. These methods are stoichiometric calculations, Caldwell-Lawrence diagrams, and equilibrium calculations.

Stoichiometry

The assumptions made when using this method are that the softening reactions represented by equations 9-3, 9-4, 9-5, 9-8, and 9-10 go to completion and that the reactants combine to form products in fixed, constant proportions. Based on these assumptions and the appropriate softening reactions, the lime and soda ash dosages required to soften a water may be computed from the concentrations of free carbon dioxide, total alkalinity, calcium, and magnesium in the raw water. A procedure for making these calculations is illustrated in example problem 9-1.

EXAMPLE PROBLEM 9-1: The following raw water analysis has been provided for Jacksonville, Florida:

Parameter	Concentration (mg/ℓ)
TDS	373
Calcium (Ca)	70
Magnesium (Mg)	22
Sodium (Na)	14
Potassium (K)	2
Bicarbonate (HCO)₃	187
Sulfate (SO₄)	118
Chloride (Cl)	17
pH	7.5

Determine the lime and soda ash dose required to soften this water.

Solution:

1. Compute the carbonic acid concentration of the water.

    (a) Convert the bicarbonate concentration into moles/ℓ:

$$[HCO_3^-] = \frac{\text{concentration (mg/}\ell\text{)}}{1000} \times \frac{1 \text{ mole}}{61 \text{ grams}}$$

$$= \frac{187}{(1000)(61)} = 3.06 \times 10^{-3} \text{ mole/}\ell$$

    (b) Calculate the total carbonic species concentration from a rearrangement of equation 2-98:

$$C_T = \frac{[HCO_3^-]}{\alpha_1}$$

Calculate the value of α_1 at pH 7.5 and then compute C_T:

$$C_T = \frac{3.06 \times 10^{-3}}{0.93} = 3.29 \times 10^{-3} \text{ mole/}\ell$$

(c) Compute the carbonic acid concentration from a rearrangement of equation 2-88:

$$[H_2CO_3^*] = C_T - [HCO_3^-] - [CO_3^{2-}]$$

At a pH of 7.5, the carbonate term can be neglected. Hence,

$$[H_2CO_3^*] = C_T - [HCO_3^-]$$
$$= 3.29 \times 10^{-3} - 3.06 \times 10^{-3}$$
$$= 0.23 \times 10^{-3} \text{ mole}/\ell$$

(d) Convert the carbonic acid concentration to mg/ℓ:

$$H_2CO_3^* \text{ (mg}/\ell) = 0.23 \times 10^{-3} \times 1000 \times 62 = 14.3 \text{ mg}/\ell$$

2. Construct an equivalence table for the chemical species of interest.

Species	Mole. wt.	No. eq./mole	Concentration		
			(mg/ℓ)	(moles/ℓ)	(eq./ℓ)
Ca^{2+}	40	2	70	1.75×10^{-3}	3.50×10^{-3}
Mg^{2+}	24.3	2	22	0.91×10^{-3}	1.82×10^{-3}
Na^+	23	1	14	0.61×10^{-3}	0.61×10^{-3}
K^+	29.1	1	2	0.05×10^{-3}	0.05×10^{-3}
HCO_3^-	61	1	187	3.06×10^{-3}	3.06×10^{-3}
SO_4^{2-}	96.1	2	118	1.23×10^{-3}	2.46×10^{-3}
Cl^-	35.5	1	17	0.48×10^{-3}	0.48×10^{-3}
$H_2CO_3^*$	62	2	14.3	0.23×10^{-3}	0.46×10^{-3}

3. Construct an equivalence diagram for the cationic and anionic chemical species. When constructing this diagram, the calcium equivalents should be placed first on the cationic scale, and immediately following calcium the magnesium equivalents should be shown. The bicarbonate equivalents should be placed first on the anionic scale, and immediately following bicarbonate should be the divalent species, which would, in turn, be followed by other monovalent species.

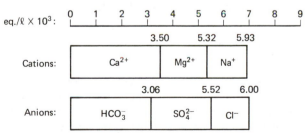

4. From the equivalence diagram determine the types of hardness present in the water.

Types of Hardness	Concentration (eq./ℓ)
Calcium carbonate	3.06×10^{-3}
Magnesium carbonate	0
Calcium noncarbonate	0.44×10^{-3}
Magnesium noncarbonate	1.82×10^{-3}

5. Based on the chemical requirement indicated by the softening reactions, compute the lime and soda ash dose.

Constituent to be Removed	Pertinent Chemical Reaction	Amount to be Removed (eq./ℓ)	$Ca(OH)_2$ Required (eq./ℓ)	Na_2CO_3 Required (eq./ℓ)
Calcium carbonate hardness	9–4	3.06×10^{-3}	3.06×10^{-3}	0
Magnesium carbonate hardness	9–8	0	$2 \times 0 = 0$	0
Calcium noncarbonate hardness	9–5	0.44×10^{-3}	0	0.44×10^{-3}
Magnesium noncarbonate hardness	9–10	1.82×10^{-3}	1.82×10^{-3}	1.82×10^{-3}
Carbonic acid	9–3	0.46×10^{-3}	0.46×10^{-3}	0
Total chemical requirement			5.34×10^{-3}	2.26×10^{-3}

6. Convert the lime and soda ash requirements into lb/MG of water treated.

   (a) Convert the dose requirements from eq/ℓ to mg/ℓ.

$$Ca(OH)_2 = \frac{eq.}{\ell} \times \frac{moles}{eq} \times \frac{g}{mole} \times \frac{mg}{g}$$
$$(mg/\ell)$$

$$= (5.34 \times 10^{-3})(0.5)(74)(1000)$$

$$= 198 \text{ mg}/\ell$$

$$Na_2CO_3 = (2.26 \times 10^{-3})(0.5)(106)(1000) = 120 \text{ mg}/\ell$$
$$(mg/\ell)$$

   (b) Convert the dose requirements to lb/MG.

$$Ca(OH)_2 = 198 \times 8.34 = 1651 \text{ lb/MG}$$
$$(lb/MG)$$

$$Na_2CO_3 = 120 \times 8.34 = 1001 \text{ lb/MG}$$
$$(lb/MG)$$

To accelerate the rate of the softening reactions, a 5 to 10% excess of the stoichiometric lime and soda ash is often added. Furthermore, the chemical dose calculations have been based on pure chemical compounds. To obtain the actual chemical dose, the calculated requirement should be divided by the fractional purity of the chemical used. For example, if the lime used was 70% pure, the actual chemical dose would be 1651/0.7, or 2359 lb/MG.

The residual soluble calcium and magnesium concentrations are controlled by the calcium carbonate and magnesium hydroxide solid phases, respectively. Under conditions normally encountered in water softening processes, the theoretical residual hardness is generally about 25 mg/ℓ as $CaCO_3$. However, hardness levels less than 50 mg/ℓ as $CaCO_3$ are seldom achieved during plant operation.

Sodium hydroxide (caustic soda) can be used to remove both carbonate and noncarbonate hardness and may be used in place of lime and as a substitute for part or all of the soda ash requirement. The reactions which occur when a water containing both carbonate and noncarbonate hardness is softened by the addition of NaOH are summarized as follows:

1) $$H_2CO_3 + 2NaOH \rightleftharpoons Na_2CO_3 + 2H_2O \qquad \text{(9-11)}$$

Equation 9-11 represents the reaction of caustic soda with carbonic acid. Although this reaction does not produce any softening, it must be considered because it creates a chemical demand. This reaction shows that 1 mole of H_2CO_3 will react with 2 moles of NaOH to form 1 mole of Na_2CO_3 or one equivalent of H_2CO_3 will react with one equivalent of NaOH to form one equivalent of Na_2CO_3.

2) $$Ca^{2+} + 2HCO_3^- + 2NaOH \rightleftharpoons CaCO_{3(s)} + Na_2CO_3 + 2H_2O \qquad \text{(9-12)}$$

Equation 9-12 represents the removal of calcium carbonate hardness. This reaction shows that one equivalent of NaOH will react with one equivalent of calcium carbonate hardness to form one equivalent of calcium carbonate and one equivalent of soda ash.

3) $$Mg^{2+} + 2HCO_3^- + 2NaOH \rightleftharpoons Na_2CO_3 + MgCO_3 + 2H_2O \qquad \text{(9-13)}$$
$$MgCO_3 + 2NaOH \rightleftharpoons Mg(OH)_2 + Na_2CO_3 \qquad \text{(9-14)}$$

overall: $$Mg^{2+} + 2HCO_3^- + 4NaOH \rightleftharpoons Mg(OH)_2 + 2Na_2CO_3$$
$$+ 2H_2O \qquad \text{(9-15)}$$

Equation 9-16 represents the removal of magnesium carbonate hardness. This reaction shows that two equivalents of NaOH will react with one equivalent of magnesium carbonate hardness to form one equivalent of magnesium hydroxide and two equivalents of soda ash.

4) $$Mg^{2+} + SO_4^{2-} + 2NaOH \rightleftharpoons Mg(OH)_2 + Na_2SO_4 \qquad \text{(9-16)}$$

Equation 9-16 represents the removal of magnesium noncarbonate hardness. This reaction shows that one equivalent of NaOH will react with one equivalent of magnesium noncarbonate hardness to form one equivalent of magnesium hydroxide.

The soda ash formed in reactions 9-11, 9-12, and 9-15 will react with calcium noncarbonate hardness according to equation 9-5. If the amount of soda ash formed by caustic soda addition is insufficient to remove the calcium noncarbonate hardness, then soda ash addition will be required. Example problem 9-2 illustrates the procedure for computing the chemical dose requirements for caustic soda softening. The same raw water characteristics as in example problem 9-1 are used.

EXAMPLE PROBLEM 9-2:   The following raw water analysis has been provided for Jacksonville, Florida:

Parameter	Concentration (mg/ℓ)
TDS	373
Calcium (Ca)	70
Magnesium (Mg)	22
Sodium (Na)	14
Potassium (K)	2
Bicarbonate (HCO_3)	187
Sulfate (SO_4)	118
Chloride (Cl)	17
pH	7.5

Determine the caustic soda and soda ash dose required to soften this water.

Solution:

1. See step 1, example problem 9-1.
2. See step 2, example problem 9-1.
3. See step 3, example problem 9-1.
4. See step 4, example problem 9-1.
5. Based on the chemical requirements indicated by the softening reactions 9-11, 9-12, 9-15, 9-16, and 9-5, compute the caustic soda and soda ash dose.

Constituent to be Removed	Pertinent Chemical Reaction	Amount to be Removed $(eq.\ell)$	NaOH Required $(eq./\ell)$	Na_2CO_3 Required $(eq./\ell)$	Na_2CO_3 Produced $(eq./\ell)$
CCH	9–12	3.06×10^{-3}	3.06×10^{-3}	0	3.06×10^{-3}
MCH	9–15	0	$2 \times 0 = 0$	0	$2 \times 0 = 0$
CNH	9–5	0.44×10^{-3}	0	0.44×10^{-3}	0
MNH	9–16	1.82×10^{-3}	1.82×10^{-3}	0	0
CA	9–11	0.46×10^{-3}	0.46×10^{-3}	0	0.46×10^{-3}

Total caustic soda requirement	5.34×10^{-3}

Total soda ash requirement	0.44×10^{-3}

Soda ash produced	3.52×10^{-3}

Net soda ash requirement	$0.44 \times 10^{-3} - 3.52 \times 10^{-3} = -3.08 \times 10^{-3}$

A negative value for the net soda ash requirement indicates that no soda ash addition will be required.

6. Convert the caustic soda requirement into lb/MG of water treated.

(a) Convert the dose requirements from eq./ℓ to mg/ℓ:

$$NaOH = (5.34 \times 10^{-3})(40)(1000) = 214 \text{ mg}/\ell$$
(mg/ℓ)

(b) Convert the dose requirement to lb/MG:

$$NaOH = 214 \times 8.34 = 1785 \text{ lb/MG}$$
(lb/MG)

At this point, the theoretical dose should be corrected for chemical purity to obtain the actual dose required.

Merrill (1976) listed the following advantages for using caustic soda in the softening process: (1) in many cases only one chemical feed system is required, (2) it is easier to handle and feed than lime, (3) it does not deteriorate in storage, and (4) less $CaCO_3$ sludge is produced. Considering these advantages, the choice of chemicals to use in the softening process should be based on an economic comparison which considers the delivered costs of the three chemicals, cost of equipment for chemical feed, and the cost of sludge handling and disposal.

EXAMPLE PROBLEM 9-3: Compare the amount of $CaCO_3$ sludge formed in the lime-soda ash softening process, presented in example problem 9-1, to the amount of $CaCO_3$ sludge formed in the caustic soda softening process, presented in example problem 9-2.

Solution:

1. Construct a table to show the amount of $CaCO_3$ sludge produced by the softening reactions.

Constituent Removed	Lime-Soda Ash Softening		Caustic Soda Softening	
	Pertinent Reaction	*$CaCO_3$ Formed (eq./ℓ)*	*Pertinent Reaction*	*$CaCO_3$ Formed (eq./ℓ)*
CA	9–3	0.46×10^{-3}		
CCH	9–4	$2(3.06 \times 10^{-3})$	9–12	3.06×10^{-3}
CNH	9–5	0.44×10^{-3}	9–5	0.44×10^{-3}
MCH	9–8	0		
MNH	9–10	1.82×10^{-3}		
Total Sludge		8.84×10^{-3}		3.5×10^{-3}

2. Convert the calcium carbonate sludge production into lb/MG of water treated.

(a) Convert the sludge production from eq./ℓ to mg/ℓ:

Lime-soda ash:

$$CaCO_3 = (8.84 \times 10^{-3})(0.5)(100(1000) = 442 \text{ mg}/\ell$$
(mg/ℓ)

Caustic soda:

$$CaCO_3 = (3.5 \times 10^{-3})(0.5)(100)(1000) = 175 \text{ mg}/\ell$$
(mg/ℓ)

(b) Convert the sludge production to lb/MG:

Lime-soda ash:

$$CaCO_3 = 492 \times 8.34 = 3686 \text{ lb/MG}$$
(lb/MG)

Caustic soda:

$$CaCO_3 = 175 \times 8.34 = 1460 \text{ lb/MG}$$
(lb/MG)

Caldwell-Lawrence Diagrams

Caldwell-Lawrence diagrams are based on equilibrium principles and can be used effectively in solving water softening problems. The diagrams provide a means to determine rapidly the chemical requirement with a relatively high degree of accuracy. The major disadvantage of using such diagrams is that unless the user fully understands their development, little understanding of the mechanisms which control the water softening process is gained.

A general explanation relating to the use of C-L diagrams appears in Chapter 8. The reader is advised to review that section thoroughly before continuing with this section.

When C-L diagrams are employed to estimate chemical dosages for water softening, it is necessary to use both the direction format diagram and the Mg-pH nomograph which are located on each diagram. The general steps involved in solving water softening problems with C-L diagrams are as follows:

1. Measure the pH, alkalinity, soluble calcium concentration, and soluble magnesium concentration of the water to be treated.

2. Evaluate the equilibrium state with respect to $CaCO_3$ precipitation of the untreated water. This is done by locating the point of intersection of the measured pH and alkalinity lines. Determine the value of the calcium line which passes through that point. Compare that value to the measured calcium value. If the measured value is greater, the water is oversaturated with respect to $CaCO_3$. If the measured value is less than the value obtained from the C-L diagram, the water is undersaturated with respect to $CaCO_3$.

WATER SOFTENING AND NEUTRALIZATION CHAPTER 9

3. To use the direction format diagram, the water must be saturated with $CaCO_3$. The procedure for establishing this point for waters which are not saturated is as follows:

   (a) *Raw water oversaturated:* Locate the point of $CaCO_3$ saturation by allowing $CaCO_3$ to precipitate until equilibrium is established. This point is located at the point of intersection of a horizontal line through the ordinate value $C_1 = - [\text{acidity}]_{\text{initial}}$ and a vertical line through the abscissa value $C_2 = ([\text{Alk}]_{\text{initial}} - [\text{Ca}]_{\text{initial}})$.

   (b) *Raw water undersaturated:* Locate the point of $CaCO_3$ saturation by allowing recycled $CaCO_3$ particles to dissolve until equibrium is established. This point is located by the same procedure followed in step 3(a).

4. Establish the pH required to produce the desired residual soluble magnesium concentration. This is accomplished by simply noting the pH associated with the desired concentration on the Mg-pH nomograph.

5. On a C-L diagram, $Mg(OH)_2$ precipitation produces the same response as the addition of a strong acid. This response is indicated on the direction format diagram as downward and to the left at 45°. When using a C-L diagram for softening calculations, the effect of $Mg(OH)_2$ precipitation should be accounted for before the chemical dose is computed. The starting point for the chemical dose calculation is located as follows:

   (a) Compute the change in the magnesium concentration as a result of $Mg(OH)_2$ precipitation:

   $$\Delta \, Mg = [Mg]_{\text{initial}} - [Mg]_{\text{desired}} \qquad (9\text{-}17)$$

   (b) From the initial saturation point construct a vector downward and to the left at 45°. The magnitude of this vector should be such that the horizontal and vertical projections have a magnitude equal to $\Delta \, Mg$.
   The starting point for the chemical dose calculation is located at the head of the vector.

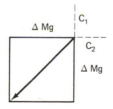

6. From the point located in step 5(b), construct a vector whose head intersects the pH line established in step 4. The direction of this vector will depend on the chemical selected for the softening process. If lime is selected, the direction format diagram shows that the vector will move vertically. On the other hand, if sodium hydroxide is added the diagram shows that the vector will move

upward and to the right at 45°. The required $Ca(OH)_2$ dose is given by the magnitude of the projection of the lime addition vector on the C_1 axis, and the required NaOH dose is given by either the magnitude of the horizontal projection on the C_2 axis or by the magnitude of the vertical projection on the C_1 axis of the sodium hydroxide vector.

7. Evaluate the calcium line which passes through the point located at the intersection of the pH line established in step 4 and the chemical addition vector established in step 6. If this line indicates that the soluble calcium concentration is too high, soda ash addition is necessary. The required soda dose is determined as follows:

    (a) Locate the calcium line representing the desired residual calcium concentration.

    (b) Construct a vector from the point established in step 6 to intersect the desired calcium line. The direction format diagram indicates that the direction of this vector is horizontal and to the right.

    (c) The required soda ash dose is given by the magnitude of the projection of the soda ash vector on the C_2 axis.

    The saturation state defined by the intersection of the desired calcium line and the soda ash vector describes the characteristics of the softened water.

The use of a C-L diagram for computing the required chemical dose for water softening is illustrated in example problem 9-4.

EXAMPLE PROBLEM 9-4: A raw water has the following characteristics:

$$pH = 7.5$$
$$Ca^{2+} = 380 \text{ mg}/\ell \text{ as } CaCO_3$$
$$Mg^{2+} = 80 \text{ mg}/\ell \text{ as } CaCO_3$$
$$Alk = 100 \text{ mg}/\ell \text{ as } CaCO_3$$
$$temperature = 15°C$$
$$I = 0.01 \text{ } M$$

Determine the chemical dose requirement for both lime-soda softening and caustic soda softening if the finished water is to contain 40 mg/ℓ calcium and 10 mg/ℓ magnesium (both as $CaCO_3$).

Solution:

A. *Lime-Soda Ash Softening*

1. Evaluate the equilibrium state of the untreated water.

    (a) Locate the intersection of the initial pH and initial alkalinity lines (shown as point 1 in Figure 9-3).

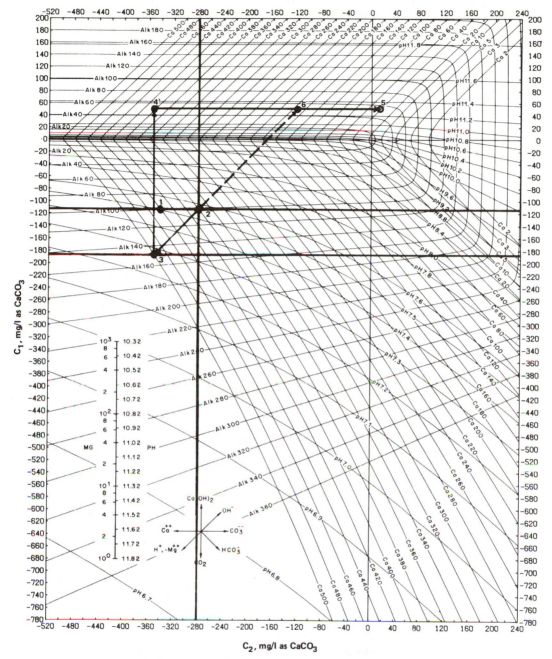

FIGURE 9-3 C-L *diagram for a temperature of 15°C and an ionic strength of 0.01.*

(b) The calcium line which passes through point 1 is 440. Since this represents a concentration greater than 380, the water is undersaturated with respect to $CaCO_3$.

2. Compute the initial acidity of the untreated water.

   (a) Construct a horizontal line through point 1 and read the C_1 value at the point where that line intersects the ordinate:

   $$C_1 = -117$$

   Therefore,

   $$-117 = - \text{(acidity)}$$

   or

   $$\text{acidity} = 117 \text{ mg/}\ell$$

3. Compute the C_2 value for the untreated water:

   $$C_2 = ([\text{Alk}] - [\text{Ca}]) = (100 - 380) = -280$$

4. Locate the system equilibrium point at the intersection of a horizontal line through $C_1 = -117$ and a vertical line through $C_2 = -280$ (shown as point 2 in Figure 9-3). The saturation condition is achieved by contacting the raw water with recycled $CaCO_3$ sludge.

5. Establish the pH required to produce the desired residual soluble magnesium concentration. This is obtained from the Mg-pH nomograph, which shows that a pH of 11.32 is required to reduce the soluble magnesium concentration to 10 mg/ℓ as $CaCO_3$.

6. Compute the change in the magnesium concentration as a result of $Mg(OH)_2$ precipitation:

   $$\Delta \text{Mg} = [\text{Mg}]_{\text{initial}} - [\text{Mg}]_{\text{desired}}$$
   $$= 80 - 10$$
   $$= 70 \text{ mg/}\ell \text{ as } CaCO_3$$

7. Construct a downward vector from point 2, which will account for the effects of $Mg(OH)_2$ precipitation.

   (a) Draw a horizontal line through the C_1 value of $-(117 + 70) = -187$.

   (b) Beginning at point 2, construct a vector downward and to the left at 45° until it intersects the horizontal line through $C_1 = -187$ (shown as point 3 in Figure 9-3).

8. Construct a vertical vector beginning at point 3 to intersect the pH = 11.32 line (shown as point 4 in Figure 9-3). The lime dose is equal to the magnitude of the projection of this vector onto the C_1 axis.

   $$\text{lime dose} = 187 + 50 = 237 \text{ mg/}\ell \text{ as } CaCO_3$$
   $$(\text{mg/}\ell \text{ as } CaCO_3)$$

9. Construct a horizontal vector beginning at point 4 to intersect the $Ca = 40$ line (shown as point 5 in Figure 9-3). The soda ash dose is equal to the magnitude of the projection of this vector onto the C_2 axis.

$$\text{soda ash dose} = 350 + 18 = 368 \text{ mg/}\ell \text{ as } CaCO_3$$
$$(\text{mg/}\ell \text{ as } CaCO_3)$$

B. Caustic Soda Softening

1. Same as step A-1.
2. Same as step A-2.
3. Same as step A-3.
4. Same as step A-4.
5. Same as step A-5.
6. Same as step A-6.
7. Same as step A-7.
8. Beginning at point 3, Figure 9-3, construct a vector upward and to the right at 45° until it intersects the pH = 11.32 line (shown as point 6 on Figure 9-3, and the vector is shown as a dashed line). The NaOH dose is equal to the magnitude of the vertical projection of this vector onto the C_1 axis.

$$\text{caustic soda dose} = 187 + 50 = 237 \text{ mg/}\ell \text{ as } CaCO_3$$
$$(\text{mg/}\ell \text{ as } CaCO_3)$$

9. Construct a horizontal vector beginning at point 6 to intersect the $Ca = 40$ line (shown as point 5 in Figure 9-3). The soda ash dose is equal to the magnitude of the projection of this vector onto the C_2 axis.

$$\text{soda ash dose} = 120 + 18 = 138 \text{ mg/}\ell \text{ as } CaCO_3$$
$$(\text{mg/}\ell \text{ as } CaCO_3)$$

Neutralization

The product water from a chemical precipitation softening process has a very high pH, and because equilibrium is generally not attained during the reaction time provided in the softening units, the water is usually supersaturated with calcium carbonate. This condition causes afterprecipitation on the filter sand and in the distribution system. To prevent afterprecipitation from occurring, the pH of the water is reduced by the addition of a strong acid (usually H_2SO_4) or by the addition of a weak acid (usually CO_2 and referred to as *recarbonation*) to the water. Keep in mind that the prevention of afterprecipitation is not the sole objective of neutralization. It is also important that the finished water not be corrosive. Hence, there is a limit to the amount of acid which should be added to the water.

There are basically two types of neutralization processes. One type is referred to as *two-stage recarbonation*. In this process enough CO_2 is added to produce the minimum calcium carbonate solubility. The water is then passed to a clarifier where $CaCO_3$ is precipitated, and as a result additional hardness is removed. After clarification more

CO_2 is added to the water to adjust the pH to an acceptable level (generally to a value between 8 and 9). The flow schematic for a typical two-stage recarbonation process is presented in Figure 9-4.

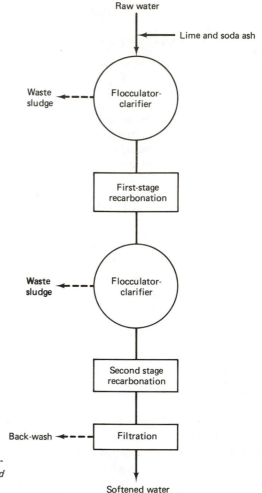

FIGURE 9-4 *Typical flow schematic for two-stage recarbonation (after Clark, Viessman, and Hammer, 1977).*

The second type of neutralization process is referred to as *one-stage neutralization* when a strong acid is used and as one-stage recarbonation when CO_2 is used. In one-stage treatment enough strong acid or CO_2 is added to the water to directly reduce the pH of the softened water to the desired value. The flow schematic for a typical one-stage recarbonation process is shown in Figure 9-5.

Because most softening plants employ CO_2 for neutralization an example problem will be worked to illustrate the procedure for computing the required carbon dioxide dose for both two-stage and one-stage carbonation.

WATER SOFTENING AND NEUTRALIZATION CHAPTER 9

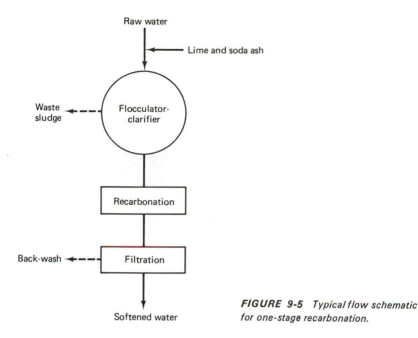

Raw water

Lime and soda ash

Waste sludge ← Flocculator-clarifier

Recarbonation

Back-wash ← Filtration

Softened water

FIGURE 9-5 *Typical flow schematic for one-stage recarbonation.*

EXAMPLE PROBLEM 9-5: Compute the chemical doses required to neutralize the softened water described by point 5 in Figure 9-3, using two-state recarbonation and one-stage recarbonation. Assume that the pH must be reduced to 8.7 during neutralization.

Solution:

A. *Two-Stage Recarbonation:*

1. Locate the point of minimum calcium concentration.

    (a) Starting at point 5, Figure 9-6, construct a vector vertically downward until the head of the vector intersects a horizontal line constructed through the ordinate value $C_1 = 0$. This is the point of minimum calcium concentration and is shown as point 7 in Figure 9-6. Note that if the vector had been extended beyond this point, it would have begun to intersect calcium lines of increasing concentration. First-stage recarbonation should be terminated at this point.

2. Determine the Ca, alkalinity, and pH for the settled water; i.e., determine the water characteristics at point 7 in Figure 9-6.

    (a) The Ca and pH values can be determined directly from the C-L diagram. However, in many cases it is very difficult to distinguish between the alkalinity lines in this region of the diagram. For those situations alkalinity can be computed from the C_2 value associated with the point of minimum calcium concentration.

$$C_2 = \text{Alk} - \text{Ca}$$

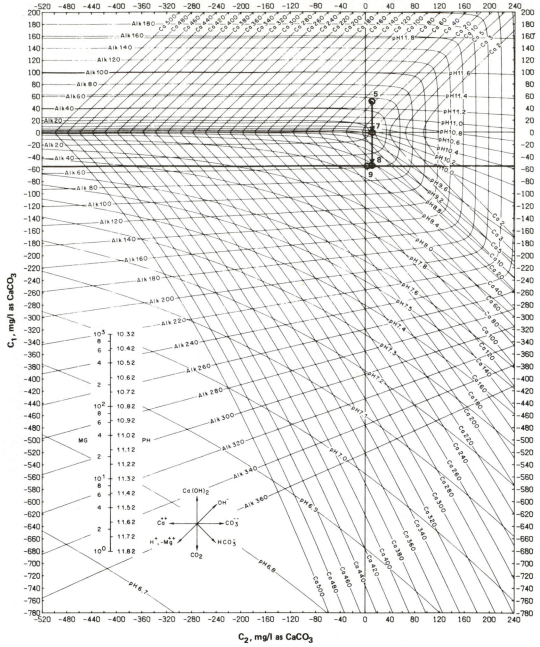

FIGURE 9-6 C-L diagram for a temperature of 15°C and an ionic strength of 0.01.

or

$$\text{Alk} = C_2 + \text{Ca}$$

For point 7, pH = 10.4, Ca = 10, and $C_2 = 18$. Then,

$$\text{alkalinity} = 18 + 10 = 28$$

3. Compute the CO_2 requirement for first-stage recarbonation.

    (a) The CO_2 requirement is given by the magnitude of the projection of the vector onto the C_1 axis.

    first-stage CO_2 dose = 50 mg/ℓ as $CaCO_3$
    (mg/ℓ as $CaCO_3$)

4. Determine the CO_2 requirement for second-stage recarbonation.

    (a) The direction format diagram cannot be applied beyond point 7 because the addition of CO_2, after the $CaCO_3$ particles precipitated during first-stage recarbonation have been removed, will produce an undersaturated condition. To circumvent this problem, compute the acidity of the final state at a pH of 8.7, based on the fact that CO_2 addition does not affect alkalinity as long as $CaCO_3$ does not precipitate. Equation 8-28 can be used for this calculation.

    acidity = 28 mg/ℓ as $CaCO_3$

    The CO_2 required to reduce the pH to 8.7 is given by the difference between the acidity of the final state (computed) and the acidity at point 7:

    second stage CO_2 dose = 28 − 0 = 28 mg/ℓ as $CaCO_3$
    (mg/ℓ as $CaCO_3$)

5. Calculate the total CO_2 requirement for first-stage and second-stage recarbonation:

    total CO_2 requirement = 50 + 28 = 78 mg/ℓ as $CaCO_3$
    (mg/ℓ as $CaCO_3$)

B. *One-Stage Recarbonation:*

1. Locate the point describing the limiting saturated state which can be attained by CO_2 addition.

    (a) Starting at point 5, Figure 9-6, construct a vector vertically downward until the head of the vector intersects calcium line 40 on the opposite side of the C_2 axis (shown as point 8 in Figure 9-6). The sequence of events occurring between points 5 and 8 can be visualized as $CaCO_3$ precipitation between points 5 and 7 and dissolution of the precipitated $CaCO_3$ between points 7 and 8. The direction format diagram cannot be used to show the reaction path past point 8 because any further CO_2 addition produces an undersaturated condition.

2. Evaluate the Ca, alkalinity, and pH for the saturated conditions at point 8. These values are read directly from the C-L diagram:

$$Ca = 40$$
$$Alk = 58$$
$$pH = 8.8$$

3. Compute the CO_2 dose requirement.

   (a) Remembering that alkalinity does not change with CO_2 as long as $CaCO_3$ does not precipitate, locate the final state at the intersection of the Alk = 58 and pH = 8.7 lines (shown as point 9 in Figure 9-6).

   (b) Construct a horizontal line through point 9 and determine the acidity of the final state.

$$C_1 = - \text{acidity}$$

   Thus,

$$\text{acidity} = -(-58) = 58 \text{ mg}/\ell \text{ as } CaCO_3$$

   (c) The CO_2 requirement is given by the change in acidity between the initial and final state (in this case between point 5 and point 9):

$$CO_2 \text{ requirement} = 50 + 58 = 108 \text{ mg}/\ell \text{ as } CaCO_3$$
$$(\text{mg}/\ell \text{ as } CaCO_3)$$

If a strong acid were used for neutralization, the first molecule would produce an undersaturated condition. Hence, the direction format diagram cannot be used in this case. To solve such a problem, remember that each equivalent of acid added will increase the acidity by one equivalent and decrease the alkalinity by one equivalent. A trial-and-error procedure should be followed to determine the acid dose required to reach the desired pH.

It should be understood that the use of C-L diagrams for computing the chemical requirements (either CO_2 or strong acid) for neutralizing does not recognize the chemical demand for unsettled $CaCO_3$ floc. This demand may, at times, be quite high.

Equilibrium Calculations

As noted earlier, the major disadvantage of using C-L diagrams is that unless the user fully understands their development, little insight of the mechanisms which control the water softening process is gained. An alternative to the use of C-L diagrams is the application of basic equilibrium concepts to elucidate the mechanisms of the water softening and neutralization processes as well as to calculate the required chemical dosages.

Before proceeding further, the reader should review the material contained in section 2-4. This material forms the basis of the equilibrium approach to chemical dose calculations.

Equilibrium expressions for calcium carbonate and magnesium hydroxide in aqueous solution are as follows:

$$CaCO_{3(s)} \rightleftharpoons Ca^{2+}_{(aq)} + CO^{2-}_{3(aq)} \qquad (9\text{-}18)$$

$$Mg(OH)_{2(s)} \rightleftharpoons Mg^{2+}_{(aq)} + 2OH^-_{(aq)} \qquad (9\text{-}19)$$

Drawing from fundamental thermodynamic considerations, the saturated equilibrium relationships for reactions 9-18 and 9-19 are

$$K_{SC} = (Ca^{2+})(CO_3^{2-}) \qquad (9\text{-}20)$$

$$K_{SM} = (Mg^{2+})(OH^-)^2 \qquad (9\text{-}21)$$

The effect of temperature on the solubility equilibrium constants for calcium carbonate and magnesium hydroxide is given by the following empirical equations (Loewenthal and Marais, 1976):

$$pK_{SC} = 0.01183t + 8.03 \qquad (9\text{-}22)$$

$$pK_{SM} = 0.0175t + 9.97 \qquad (9\text{-}23)$$

where t is the solution temperature in °C.

Equations 9-20 and 9-21 can be written in terms of molar concentrations by including the effect of ionic strength on the thermodynamic solubility equilibrium constant.

$$K'_{SC} = \frac{K_{SC}}{(\gamma_d)^2} = [Ca^{2+}][CO_3^{2-}] \qquad (9\text{-}24)$$

$$K'_{SM} = \frac{K_{SM}}{\gamma_d(\gamma_m)^2} = [Mg^{2+}][OH^-]^2 \qquad (9\text{-}25)$$

In reality, equations 9-24 and 9-25 are insufficient to calculate the solubility of either compound because there are competing reactions in solution which have not been considered. Complex ion formation is an important consideration for both calcium and magnesium ions and hence should be accounted for when modeling the softening process. Complex ion formation reactions which contribute to the soluble calcium and magnesium concentrations are listed in Table 9-3.

When writing a mass balance for soluble calcium and soluble magnesium all species which contribute to the soluble form must be considered. For example Figure 9-7 illustrates the effect the soluble species $CaOH^+$, $CaHCO_3^+$, and $CaSO_4^0$ have on the solubility of $CaCO_3$. For equilibrium to exist, the ion product $[Ca^{2+}][CO_3^{2-}]$ must equal K'_{SC}. Since a certain concentration of the Ca^{2+} ions formed from $CaCO_{3(s)}$ dissolution will be incorporated into the other soluble calcium species more $CaCO_{3(s)}$ must be dissolved if saturated equilibrium is to be established.

Water analyses generally provide information only on the total concentration of a particular ion and do not differentiate between the free ions and various complex ions. For samples containing HCO_3^-, CO_3^{2-}, SO_4^{2-}, and OH^- as the only complexing

TABLE 9-3 *Complex ion formation reactions of calcium and magnesium ions.**

Reaction	equilibrium constant	Temperature Correction (T in °K)
1. Calcium:		
(a) $Ca^{2+}_{(aq)} + OH^-_{(aq)} \rightleftharpoons CaOH^+_{(aq)}$	$K'_3 = \gamma_d K_3 = \dfrac{[CaOH^+]}{[Ca^{2+}][OH^-]}$	$pK_3 = -1.299 - 260.388\left[\dfrac{1}{T} - \dfrac{1}{298.15}\right]$
(b) $Ca^{2+}_{(aq)} + HCO^-_{3(aq)} \rightleftharpoons CaHCO^+_{3(aq)}$	$K'_4 = \gamma_d K_4 = \dfrac{[CaHCO^+_3]}{[Ca^{2+}][HCO^-_3]}$	$pK_4 = 2.95 - 0.0133(T)$
(c) $Ca^{2+}_{(aq)} + CO^{2-}_{3(aq)} \rightleftharpoons CaCO^0_{3(aq)}$	$K'_5 = \gamma_d^2 K_5 = \dfrac{[CaCO^0_3]}{[Ca^{2+}][CO^{2-}_3]}$	$pK_5 = 27.393 - \dfrac{4114}{T} - 0.05617(T)$
(d) $Ca^{2+}_{(aq)} + SO^{2-}_{4(aq)} \rightleftharpoons CaSO^0_{4(aq)}$	$K'_6 = \gamma_d^2 K_6 = \dfrac{[CaSO^0_4]}{[Ca^{2+}][SO^{2-}_4]}$	$pK_6 = \dfrac{691.70}{T}$
2. Magnesium:		
(a) $Mg^{2+}_{(aq)} + OH^-_{(aq)} \rightleftharpoons MgOH^+_{(aq)}$	$K'_7 = \gamma_d K_7 = \dfrac{[MgOH^+]}{[Mg^{2+}][OH^-]}$	$pK_7 = -0.684 - 0.0051(T)$
(b) $Mg^{2+}_{(aq)} + HCO^-_{3(aq)} \rightleftharpoons MgHCO^+_{3(aq)}$	$K'_8 = \gamma_d K_8 = \dfrac{[MgHCO^+_3]}{[Mg^{2+}][HCO^-_3]}$	$pK_8 = -2.319 + 0.011056(T) - 2.29812 \times 10^{-5}(T)^2$
(c) $Mg^{2+}_{(aq)} + CO^{2-}_{3(aq)} \rightleftharpoons MgCO^0_{3(aq)}$	$K'_9 = \gamma_d^2 K_9 = \dfrac{[MgCO^0_3]}{[Mg^{2+}][CO^{2-}_3]}$	$pK_9 = -0.991 - 0.00667(T)$
(d) $Mg^{2+}_{(aq)} + SO^{2-}_{4(aq)} \rightleftharpoons MgSO^0_{4(aq)}$	$K'_{10} = \gamma_d^2 K_{10} = \dfrac{[MgSO^0_4]}{[Mg^{2+}][SO^{2-}_4]}$	$pK_{10} = -\dfrac{707.07}{T}$

* Temperature corrections are from Truesdell and Jones (1973).

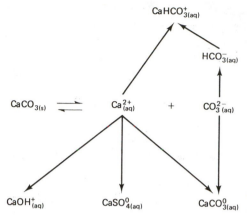

FIGURE 9-7 *Species affecting solubility of calcium carbonate.*

species, the total soluble calcium Ca_T, and total soluble magnesium Mg_T, are given by the following mass balance equations:

$$[Ca]_T = [Ca^{2+}] + [CaOH^+] + [CaHCO_3^+] + [CaCO_3^0] + [CaSO_4^0] \quad \textbf{(9-26)}$$

$$[Mg]_T = [Mg^{2+}] + [MgOH^+] + [MgHCO_3^+] + [MgCO_3^0] + [MgSO_4^0] \quad \textbf{(9-27)}$$

By solving the equilibrium expressions in Table 9-3 for each complex ion and then substituting the results into equations 9-26 and 9-27 the following relationships can be derived:

$$[Ca]_T = [Ca^{2+}](1 + K_3'[OH^-] + K_4'[HCO_3^-] + K_5'[CO_3^{2-}] + K_6'[SO_4^{2-}]) \quad \textbf{(9-28)}$$

$$[Mg]_T = [Mg^{2+}](1 + K_7'[OH^-] + K_8'[HCO_3^-] + K_9'[CO_3^{2-}] + K_{10}[SO_4^{2-}]) \quad \textbf{(9-29)}$$

At equilibrium the following relationships exist:

$$[Ca^{2+}] = \frac{K_{SC}'}{[CO_3^{2-}]} \qquad \text{(rearrangement of 9-24)}$$

$$[Mg^{2+}] = \frac{K_{SM}'}{[OH^-]^2} \qquad \text{(rearrangement of 9-25)}$$

$$[OH^-] = \frac{K_w'}{[H^+]} \qquad \text{(rearrangement of 2-91)}$$

$$[HCO_3^-] = \alpha_1 C_T \qquad \text{(rearrangement of 2-98)}$$

$$[CO_3^{2-}] = \alpha_2 C_T \qquad \text{(rearrangement of 2-99)}$$

Substitution of these values into equations 9-28 and 9-29 yields the following:

$$[Ca]_T = \frac{K_{SC}'}{\alpha_2 C_T}\left(1 + \frac{K_w' K_3}{[H^+]} + K_4'\alpha_1 C_T + K_5'\alpha_2 C_T + K_6[SO_4^{2-}]\right) \quad \textbf{(9-30)}$$

$$[Mg]_T = \frac{K_{SM}'[H^+]^2}{(K_w)^2}\left(1 + \frac{K_2' K_7'}{[H^+]} + K_8'\alpha_1 C_T + K_9'\alpha_2 C_T + K_{10}[SO_4^{2-}]\right) \quad \textbf{(9-31)}$$

In equations 9-30 and 9-31 C_T represents the equilibrium total carbonic species concentration and $[H^+]$ the equilibrium hydrogen ion concentration.

Figures 9-8 and 9-9 illustrate the relationship between total soluble species concentration, pH, and C_T for a solution with a temperature of 25°C. For convenience, both ionic strength and sulfate ion concentration have been assumed to be zero. The results show that the equilibrium carbonic species concentration has virtually no effect on the total soluble magnesium concentration but significantly affects the total soluble calcium concentration.

By applying the equilibrium relationships previously developed, it is possible to compute the chemical dosages required to achieve any desired degree of softening.

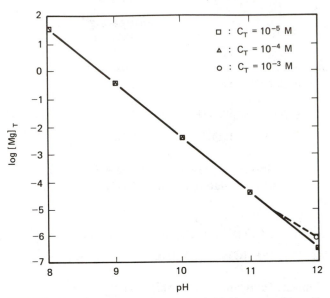

FIGURE 9-8 *Relationship between total soluble magnesium, pH and the final equilibrium total carbonic species concentration.*

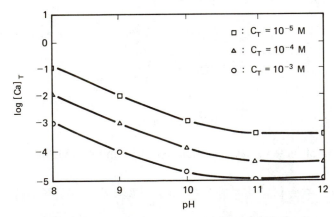

FIGURE 9-9 *Relationship between total soluble calcium, pH and equilibrium total carbonic species concentration.*

WATER SOFTENING AND NEUTRALIZATION   CHAPTER 9

The procedure to be used follows. Note that for waters low in dissolved solids it may be possible to ignore ionic strength effects and simplify the calculations.

A. Calculation of Required Lime Dose:

1. Assuming that both $Mg(OH)_2$ and $CaCO_3$ are infinitely soluble, find the chemical dose of lime (or NaOH) required to achieve the pH which results in the desired Mg^{2+} concentration.

    The required pH is most easily found by using a graph similar to Figure 9-8, which has been constructed for the water of interest and which accounts for ionic strength and temperature effects. Alternatively, equation 9-31 may be used in conjunction with equations 2-105 and 2-106, or equations 2-91 and 9-29 may be used, but such solutions can prove to be tedious.

2. The total concentration of carbonic species in the raw water (in moles/ℓ) can be calculated from the following equation:

$$[C_T] = \frac{1}{\alpha_0} \left[\frac{2\,[\text{Alk}] - \frac{K'_w}{[\text{H}^+]} + [\text{H}]^+}{\frac{2K'_1 K'_2}{[\text{H}^+]^2} + \frac{K'_1}{[\text{H}^+]}} \right] \qquad (2\text{-}113)$$

The α_0 and $[\text{H}^+]$ values are dependent on the pH of the raw water.

3. The addition of a strong base to increase the pH of the water will, of course, increase the OH^- concentration of the solution and hence the alkalinity by an equivalent amount (see equation 2-74) as long as $CaCO_3$ does not precipitate. Thus, the alkalinity at the pH achieved by lime addition can be computed from equation 2-114, since C_T does not change with lime addition as long as $CaCO_3$ is not precipitated nor CO_2 evolved (see equation 2-88).

$$[\text{Alk}] = \frac{1}{2}\left[[C_T]\alpha_1\left(\frac{2K'_2}{[\text{H}^+]} + 1\right) + \frac{K'_w}{[\text{H}^+]} - [\text{H}^+] \right] \qquad (2\text{-}114)$$

The C_T value (in moles/ℓ) used here is that obtained from the initial characterization of the raw water, and the hydrogen ion concentration and α_1 are to the pH to which the water must be adjusted for magnesium removal.

4. Since the change in alkalinity between the initial and final pH levels is directly related to the lime dose required to institute the pH change, the latter may be calculated as follows:

$$\text{Lime dose} = [\text{Alk}]_{\text{final}} - [\text{Alk}]_{\text{initial}} = \Delta[\text{Alk}] \qquad (9\text{-}32)$$
(moles/ℓ)

If NaOH is used for pH adjustment, only one-half mole of alkalinity as $CaCO_3$ is generated per mole of NaOH (see equation 2-108), so that the required NaOH dose will be twice that calculated for lime.

5. From simple mass balance considerations it is known that the amount of $Mg(OH)_2$ precipitated will be the same as the change in magnesium concentration between the initial and final solutions:

$$[Mg]_{\text{precipitated}} = [Mg^{2+}]_{\text{initial}} - [Mg^{2+}]_{\text{final}} = \Delta[Mg^{2+}] \qquad \text{(9-33)}$$

6. Since the loss of OH^- ions from solution during $Mg(OH)_2$ precipitation would cause both a decrease in alkalinity and pH, additional lime (or NaOH) is added to prevent this. The OH^- precipitated as $Mg(OH)_2$ is twice the $\Delta[Mg^{2+}]$, but lime supplies two OH^- ions per mole, so that additional lime required is equal to $\Delta[Mg^{2+}]$. The total lime requirement is, thus,

$$\text{Total lime dose} = \Delta[\text{Alk}] + \Delta[Mg^{2+}] \qquad \text{(9-34)}$$
$$(\text{moles}/\ell)$$

As before, if NaOH is used the additional requirement is twice as great as for lime.

B. Calculation of Required Soda Ash Dose: The pH required for magnesium precipitation (generally between 10 and 11) is also high enough to result in extremely low calcium levels, provided an appreciable carbonic species concentration remains after $CaCO_3$ precipitation (see Figure 9-9). In this region the total soluble calcium concentration may be controlled by the magnitude of the carbonic species concentration, and Figure 9-9 shows how the total soluble calcium concentration rapidly increases with a decrease in the equilibrium carbonic species concentration. As a result, in order to achieve the desired total soluble calcium concentration the equilibrium C_T value must be increased in many cases. This may be accomplished by the addition of soda ash, Na_2CO_3. In most cases, adequate calcium removal can be achieved by ensuring that a 1 mg/ℓ as $CaCO_3$ (10^{-5} mole/ℓ) C_T concentration is maintained at the pH required for magnesium removal. The necessary calculations are as follows.

1. Compute the total calcium content of the water:

$$[Ca^{2+}]_{\text{total}} = [Ca^{2+}]_{\text{initial}} + \text{total lime addition} \qquad \text{(9-35)}$$

If NaOH is used for pH adjustment, the total calcium concentration is equal to the initial calcium concentration.

2. Next, the amount of $CaCO_3$ precipitated must be calculated. This is accomplished based on mass balance considerations:

$$[Ca]_{\text{precipitated}} = [Ca]_{\text{total}} - [Ca]_{\text{residual}} \qquad \text{(9-36)}$$

The residual Ca is that which will remain dissolved in the water under the established equilibrium conditions. This value can be a specific selected value or can be assumed to be satisfactory, provided a C_T level of 1 mg/ℓ as $CaCO_3$ (10^{-5} M) is present. In the latter case, the $[Ca]_{\text{residual}}$ may be calculated from equation 9-30 (using the α and $[H^+]$ values at the pH required for magnesium removal, or chosen from a graph similar to Figure 9-9, but constructed for the particular water of interest).

3. Because $CaCO_3$ precipitation removes both calcium and carbonate ions from solution, there is an equivalent decrease in both calcium and C_T for each mole of $CaCO_{3(s)}$ formed. Hence, the decrease in C_T as a result of $CaCO_3$ precipitation is given by the expression

$$\Delta[C_T] = [Ca]_{precipitated} \qquad\qquad (9\text{-}37)$$

4. The calculation of the C_T residual which must exist to maintain the $[Ca]_{residual}$ calculated in step B-2 is dependent on the approach used in that step.
   (a) If a desired $[Ca]_{residual}$ was chosen, the required $C_{T\ residual}$ may be calculated based on equation 9-24. Assuming that $[Ca^{2+}] \simeq [Ca]_T$,

$$[C_T]_{residual} \simeq \frac{K'_{sc}}{[Ca]_T \alpha_2} \qquad\qquad (9\text{-}38)$$

   The α_2 value is that at the pH required for magnesium removal.
   (b) The second possibility is the assumption that a $C_{T\ residual}$ of $10^{-5}\ M$ would result in an acceptable $[Ca]_{residual}$, as stated in step B-2.

5. Knowing the C_T value for the solution and the $[C_T]$ residual desired allows the calculation of the C_T needed for $CaCO_3$ precipitation.

$$[C_T]_{available} = [C_T]_{initial} - [C_T]_{residual} \qquad\qquad (9\text{-}39)$$

6. If the $C_{T\ available} \geq \Delta C_T$, then there is a sufficient level for $CaCO_3$ precipitation, and no soda ash addition is required. In this situation there will be no noncarbonate hardness in the water.
   If however $C_{T\ available} < \Delta C_T$, then soda ash may be added to increase the CO_3^{2-} available. The soda ash requirement is equal to the difference in these values:

$$\text{soda ash requirement} = \Delta C_T - C_{T\ available} \qquad\qquad (9\text{-}40)$$
$$\text{(moles/}\ell\text{)}$$

This case would correspond to a water which contained noncarbonate hardness.

7. If no soda ash addition is required the residual soluble [Ca] can be found from the following relationship:

$$[C_T]_{initial} - [C_T]_{residual} = [Ca]_{initial} - [Ca]_{residual}$$

or rearranging

$$[C_T]_{initial} - [Ca]_{initial} = [C_T]_{residual} - [Ca]_{residual} \qquad\qquad (9\text{-}41)$$

Since both terms on the right side of the equation are unknown, another equation is needed. Equation 9-30 provides a second relationship between the residual C_T and Ca concentrations. Although there is no simple general solution to these equations, the substitution of one into the other will result in a quadratic equation with respect to $[Ca]_{residual}$, which can then be solved.

  C. *Calculation of CO_2 Dose:* In most water softening plants, after the chemical precipitates have been removed by settling, the pH of the softened water is generally adjusted to a value between 8 and 9. This is generally accomplished by the addition of carbon dioxide gas. The addition of CO_2 will not affect alkalinity as long as $CaCO_3$ does not precipitate because, for each ion of HCO_3^- or CO_3^{2-} formed, one or two H^+ ions, respectively, are generated. Thus, the net change in molar alkalinity is zero (equation 2-108). The addition of CO_2, however, does increase C_T, as can be seen from equation 2-88. The following procedure can be used to compute the CO_2 requirement for pH adjustment if the pH is lowered sufficiently to prevent any $CaCO_3$ from precipitating as a result of the increase in C_T.

1. Estimate the residual alkalinity concentration at the pH required for magnesium removal from the following mass balance equation:

$$[Alk]_{residual} = [Alk]_{initial} + [Alk]_{total\ lime} + [Alk]_{soda\ ash}$$
$$- [Alk]_{CaCO_3} - [Alk]_{Mg(OH)_2} \qquad (9\text{-}42)$$

All concentrations in equation 9-42 are expressed as moles/ℓ.

2. From equation 2-113 calculate the C_T value at the pH to be achieved from CO_2 addition.

$$(C_T)_{CO_2} = \frac{1}{\alpha_0}\left[\frac{\left(2\,[Alk]_{residual} - \dfrac{K'_w}{[H^+]} + [H^+]\right)}{\left(\dfrac{2K'_1 K'_2}{[H^+]^2} + \dfrac{K'_1}{[H^+]}\right)}\right] \qquad (2\text{-}113)$$

3. The difference in C_T between the initial state (see step B-4) and final state (equation 2-113) is equal to the CO_2 dose required.

$$[CO_2]_{required} = [C_T]_{CO_2\ initial} - [C_T]_{residual} \qquad (9\text{-}43)$$
$$(moles/\ell)$$

  Two important points concerning neutralization can be derived from the relationship between total soluble calcium concentration, pH, and the equilibrium carbonic species concentration (as shown in Figure 9-9):

1. There seems to be little necessity for the use of a two-stage recarbonation process to increase calcium removal. Increased calcium removal can be achieved by increasing the residual carbonic species concentration, which is accomplished by increasing the Na_2CO_3 dose (note that in the pH range

required for magnesium removal, Na_2CO_3 dose has little effect on the equilibrium pH value). This is basically how CO_2 addition induces additional $CaCO_3$ precipitation; it increases the C_T value, causing the total soluble calcium concentration to be depressed even though the pH is decreased. The obvious advantages of increasing C_T by Na_2CO_3 addition is that less chemical is required because the higher pH value is retained, and, more importantly, no additional clarifier is required.

2. In certain cases, afterprecipitation following recarbonation causes sand encrustation problems in the sand filter. Figure 9-9 suggests that a possible solution to this problem may be combined neutralization with strong acid and CO_2. This will produce the desired pH adjustment but with a smaller increase in the equilibrium C_T value than if CO_2 alone were used and will be cheaper than if only a strong acid were used. Figure 9-10 illustrates (in a relative sense) the benefits of using combined neutralization when afterprecipitation is a problem.

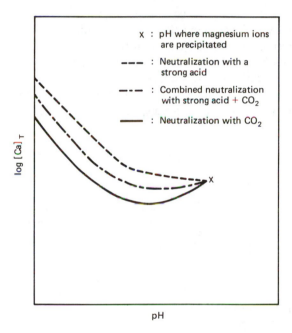

FIGURE 9-10 *Relationship between total soluble calcium concentration, pH and reagent used for neutralization.*

Split Treatment

Magnesium hydroxide scaling can be a problem in domestic hot-water heaters if a significant amount of magnesium is contained in the finished water. Merrill and Sanks (1977) developed Figure 8-7 to show the relationship between finished water pH finished water temperature, and the allowable magnesium concentration which will prevent $Mg(OH)_2$ fouling in water heaters. However, most workers follow the rule-of-thumb suggested by Larson, Lane, and Neff (1959) that the magnesium concentration in the finished water should not exceed 10 mg/ℓ as the ion (approximately 40 mg/ℓ as $CaCO_3$) if water heater fouling is to be prevented. This means that magnesium reduction will be required for raw waters containing more than about 40 mg/ℓ as $CaCO_3$ of magnesium hardness.

Acceptable calcium removal can generally be obtained at pH values near 10 (Figure 9-9 shows that with a residual C_T value of 10^{-4} M the total soluble calcium concentration should be approximately 13 mg/ℓ as $CaCO_3$ at pH 10). However, the Mg-pH nomograph in Figure 9-6 indicates that a pH greater than 11 is required to reduce the soluble magnesium concentration to a value less than 40 mg/ℓ as $CaCO_3$ when the temperature of the water is 15°C and the ionic strength is 0.01 M. This means that when a water contains more than 40 mg/ℓ as $CaCO_3$ of magnesium hardness a considerable amount of lime in excess of that required for calcium removal will have to be added to the water to achieve the pH required for magnesium hydroxide precipitation. Furthermore the elevated pH required for magnesium reduction will also increase the CO_2 requirement in the neutralization step.

To reduce the chemical requirements for treating water with more than 40 mg/ℓ as $CaCO_3$ of magnesium hardness, split treatment softening has been used. When split treatment softening is used the raw water is divided into two streams. Lime and soda ash are added to one stream to precipitate magnesium hydroxide and calcium carbonate. The treated water is settled and the clarified effluent is then mixed with the second stream which has bypassed treatment. Neutralization of the treated water can often be accomplished by the free carbon dioxide and alkalinity contained in the untreated water. When this is possible, recarbonation is not required.

The mixture of treated and bypassed water should have a magnesium concentration less than 40 mg/ℓ as $CaCO_3$. A typical flow schematic for a split treatment process is shown in Figure 9-11.

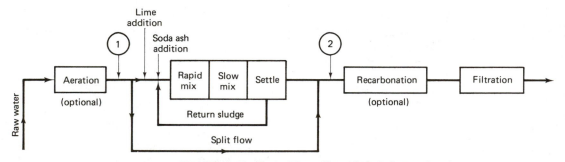

FIGURE 9-11 *Typical flow schematic for a lime-soda ash split treatment softening process.*

To calculate the chemical requirements for split treatment, it is necessary to know what fraction of the water will be bypassed. A material balance for magnesium between points 1 and 2 shown in Figure 9-11 gives

$$Q(Mg)_e = Q(1 - X)(Mg)_r - Q(1 - X)[(Mg)_r - (Mg)_t] + QX(Mg)_r \qquad \textbf{(9-44)}$$

or

$$(Mg)_e = X(Mg)_r + (1 - X)(Mg)_t \qquad \textbf{(9-45)}$$

where $(Mg)_e$ = magnesium concentration of the mixture of treated and bypassed water

$(Mg)_t$ = magnesium concentration of the treated water

$(Mg)_r$ = magnesium concentration of the raw water

X = fraction of the water bypassed

Q = raw water flow rate.

Equation 9-45 can be solved for the required magnesium concentration of the treated water for any fraction of the water bypassed.

$$(Mg)_t = \frac{(Mg)_e - (Mg)_r X}{1 - X} \qquad (9\text{-}46)$$

The upper limit for the fraction of water to bypass can be determined by noting that the lower limit for the magnesium concentration which can be achieved in the treated water is zero. Hence, setting $(Mg)_t = 0$ and solving for X gives

$$(X)_{max} = \frac{(Mg)_e}{(Mg)_r} \qquad (9\text{-}47)$$

EXAMPLE PROBLEM 9-6: A raw water has the following characteristics:

$$pH = 7.5$$
$$Ca^{2+} = 380 \text{ mg/}\ell \text{ as } CaCO_3$$
$$Mg^{2+} = 80 \text{ mg/}\ell \text{ as } CaCO_3$$
$$\text{alkalinity} = 100 \text{ mg/}\ell \text{ as } CaCO_3$$
$$\text{temperature} = 15°C$$
$$I = 0.01\ M$$

Determine the lime, soda ash, and CO_2 requirements for split treatment softening if the finished water pH is to be less than 9.0 and the magnesium concentration is not to exceed 40 mg/ℓ as $CaCO_3$.

Solution:

1. Determine the maximum fraction of water which can be bypassed from equation 9-47:

$$(X)_{max} = \frac{40}{80} = 0.5$$

Since this is the maximum fraction that can be bypassed and still reach the desired effluent Mg^{2+} concentration, 0.4 of the flow will actually be bypassed since Mg^{2+} is generally not removed to 0 mg/ℓ by chemical precipitation.

2. Compute the required magnesium concentration in the treated stream when 0.4 of the flow is bypassed:

$$(Mg)_t = \frac{40 - (80)(0.4)}{1 - 0.4} = 13.3 \text{ mg/}\ell \text{ as } CaCO_3$$

3. Evaluate the equilibrium state of the untreated water.

   (a) Locate the intersection of initial pH and initial alkalinity lines (shown as point 1 in Figure 9-12).

   (b) The calcium line which passes through point 1 is 440. Since this represents a concentration greater than 380, the water is unsaturated with respect to $CaCO_3$.

4. Compute the initial acidity of the untreated water.

   (a) Construct a horizontal line through point 1, and read the C_1 value at the point where this line intersects the ordinate:

   $$C_1 = -117$$

   Therefore,

   $$-117 = -(\text{acidity})$$

   or

   $$\text{acidity} = 117 \text{ mg/}\ell$$

5. Compute the C_2 value for the untreated water:

   $$C_2 = ([Alk] - [Ca]) = (100 - 380) = -280$$

6. Locate the system equilibrium point at the intersection of a horizontal line through $C_1 = -117$ and a vertical line through $C_2 = -280$ (shown as point 2 in Figure 9-12). The saturation condition is achieved by contacting the raw water with recycled $CaCO_3$ sludge.

7. Establish the pH required to produce the desired residual soluble magnesium concentration. This is obtained from the Mg-pH nomograph, which shows that a pH of 11.25 is required to reduce the soluble magnesium concentration to 13.3 mg/ℓ as $CaCO_3$.

8. Compute the change in the magnesium concentration as a result of $Mg(OH)_2$ precipitation:

   $$Mg = [Mg]_{initial} - [Mg]_{desired}$$
   $$= 80 - 13.3$$
   $$= 66.7 \text{ mg/}\ell \text{ as } CaCO_3$$

9. Construct a downward vector from point 2 which will account for the effects of $Mg(OH)_2$ precipitation.

   (a) Draw a horizontal line through the C_1 value of $-(117 + 66.7) = -183.7$.

   (b) Construct a vector, beginning at point 2, downward and to the left at 45° until it intersects the horizontal line through $C_1 = -183.7$ (shown as point 3 in Figure 9-12).

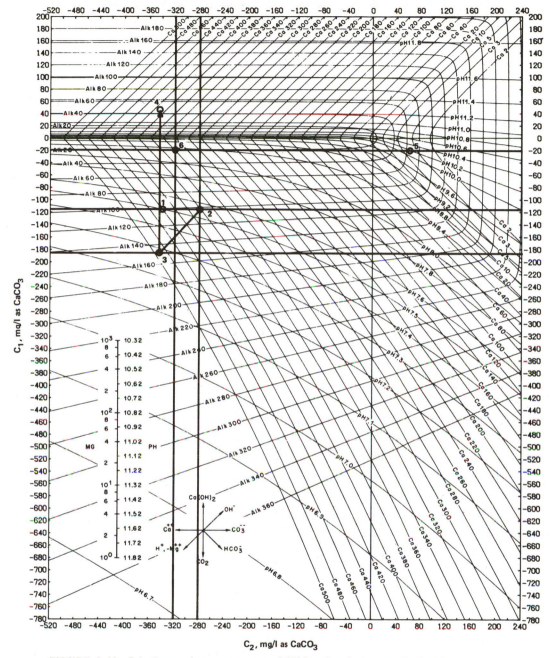

FIGURE 9-12 C-L *diagram for a temperature of 15°C and an ionic strength of 0.01.*

10. Construct a vertical vector, beginning at point 3, to intersect the pH = 11.25 line (shown as point 4 in Figure 9-12). The lime dose is equal to the magnitude of the projection of this vector onto the C_1 axis:

$$\text{lime dose} = 183.7 + 45 = 228.7 \text{ mg/}\ell \text{ as } CaCO_3$$
$$(\text{mg/}\ell \text{ as } CaCO_3)$$

11. Combine the treated and bypassed water, and evaluate the equilibrium state of the mixture.

    (a) Assume that the $CaCO_3$ is infinitely soluble and compute the Ca, alkalinity, and acidity of the mixed streams.

$$[Ca]_{mix} = [Ca]_{point\ 4}(0.6) + [Ca]_{raw}(0.4)$$
$$= (390)(0.6) + (380)(0.4)$$
$$= 386$$
$$[Alk]_{mix} = [Alk]_{point\ 4}(0.6) + [Alk]_{raw}(0.4)$$
$$= (45)(0.6) + (100)(0.4)$$
$$= 67$$
$$[Acd]_{mix} = [Acd]_{point\ 4}(0.6) + [Acd]_{raw}(0.4)$$
$$= (-45)(0.6) + (117)(0.4)$$
$$= 19.8$$

    (b) Construct a horizontal line through the ordinate value given by an acidity of 19.8, i.e., through $C_1 = -19.8$. Locate the intersection of this line and the alkalinity = 67 line (shown as point 5 in Figure 9-12).

    (c) The calcium line which passes through point 5 is 4. Since this represents a concentration less than 386, the mixture is oversaturated with respect to $CaCO_3$.

12. Determine the final equilibrium state of the system.

    (a) Remove the condition of infinite solubility and allow $CaCO_3$ to precipitate. Recall that (1) acidity remains constant during $CaCO_3$ precipitation, and (2) since equivalent amounts of alkalinity and Ca are removed, $([Alk] - [Ca]) = C_2$ will remain constant during $CaCO_3$ precipitation.

    (b) Construct a vertical line through the C_2 value of $([67] - [386]) = -319$. Locate the intersection of this line and the horizontal line through point 5 (shown as point 6 in Figure 9-12). The characteristics of the final equilibrium state at point 6 are pH = 8.4, alkalinity = 20, and Ca = 338. The high calcium concentration indicates that soda ash should be added to increase calcium removal. A good rule-of-thumb is to add enough soda ash to give a $[Ca]_{mix}$ value (calculated in step 11(a) of 100 mg/ℓ as $CaCO_3$.

Note: Steps 11 and 12 outline the procedure which should be followed in determining the composition of waters produced by blending.

WATER SOFTENING AND NEUTRALIZATION CHAPTER 9

Design Considerations

The reaction between free carbon dioxide and lime is shown by equation 9-3. Although this reaction does not contribute to hardness removal it must be considered because of the lime demand it represents. Because the amount of lime required to satisfy this reaction increases as the free CO_2 concentration increases, it may become uneconomical to remove CO_2 in this manner at high CO_2 concentrations. The State of Virginia guidelines suggest the possibility of CO_2 removal by aeration when the free carbon dioxide concentration exceeds 10 mg/ℓ.

In lime-soda ash or caustic soda-soda ash softening plants the softening process may be carried out by a unit sequence of rapid-mix, flocculation, and sedimentation (see Figure 9-11) or in a solids-contact softener where rapid-mix, flocculation, and sedimentation occur in a single unit. The rapid-mix step serves two functions: (1) to provide sufficient retention time for the feed chemical to dissolve and (2) to thoroughly mix the dissolved feed chemical with the water to be softened. Lime dissolves relatively slowly in water, and when used in the softening process, a retention time of 5 to 10 min in the rapid-mix unit is necessary for complete dissolution. However, caustic soda dissolves rapidly, and when used in the softening process a retention time of 30 sec is normally sufficient in the rapid-mix unit.

The purpose of the flocculation step is to provide the retention time required (generally 40 to 60 min) for the chemical precipitates to grow to a size large enough to be removed by gravity settling. Since the rate of crystal growth is increased in the presence of preformed precipitates, sludge recirculation is generally employed when solids-contact units are not used (see Figure 9-11).

Settling rates for chemical precipitates are a function of particle size and density. To ensure efficient removal of the precipitates formed in the softening process, a retention time of 2 to 4 hr is normally provided in the sedimentation basin, and polymers are often added to increase the particle size and thereby increase the settling rate of the particles.

Conventional softening is generally effective in reducing hardness, but it produces a very unstable water of high causticity. Such waters have objectionable tastes, cause filter-sand encrustation, and scale formation on pipes and valves. To eliminate these problems, the pH of the water is usually reduced by adding carbon dioxide, i.e., neutralization by recarbonation. Recarbonation basins should be about 8 ft deep and provide about 15 min of retention for completion of the chemical reactions.

9-2 ION EXCHANGE

Ion exchange is a reaction between a solid and a liquid resulting in replacement of an ion of the solid with an ion from the liquid which can be used to remove hardness from water. The selection of water softening by ion exchange over softening by chemical precipitation is largely determined by economic considerations. According to Weber (1972), the ability to produce water of essentially zero hardness by ion exchange softening is no real advantage, since a finished water hardness between 80 and 120 mg/ℓ is desirable for domestic use. He suggests that ion exchange is usually preferable to precipitation processes when:

1. Raw water contains low color and turbidity levels (does not require pretreatment).

2. Hardness is largely not associated with alkalinity; i.e. there are substantial amounts of noncarbonate hardness (soda ash is more expensive than lime).

3. There are variable hardness levels in raw water (requires constant adjustment of the chemical feed).

On the other hand, Weber (1972) indicated that chemical precipitation is usually preferable when:

1. Raw water requires clarification; i.e., the water contains high color and turbidity levels.

2. Hardness is largely associated with alkalinity; i.e., there is little noncarbonate hardness.

When ion exchange is used for softening, the resins employed in the process are generally cation exchange resins in the sodium form. For this type of resin a typical hardness exchange reaction is illustrated by equation 9-48:

$$2RNa + CaSO_4 \rightleftharpoons R_2Ca + Na_2SO_4 \qquad (9\text{-}48)$$

During the softening process the sodium ions are replaced by the hardness ions until the resin becomes exhausted and the feed water passes through the resin without a significant reduction in hardness. At that point, it becomes necessary to regenerate the resin. This is accomplished by washing with a strong salt solution. The high concentration of sodium ions in the salt solution drives the reaction to the left and regenerates the resin bed. The disposal of waste brine from the regeneration step is a problem which must be addressed when evaluating the feasibility of ion exchange softening.

The fundamentals of ion exchange are discussed in detail in Chapter 10, and an example problem (10-6) is presented in that chapter to illustrate the design of an ion exchange system for water softening. To avoid repetition, a discussion of ion exchange softening is postponed until Chapter 10.

PROBLEMS

9-1. Assume that a water has the following characteristics:

$$\text{calcium (Ca)} = 95 \text{ mg}/\ell$$
$$\text{magnesium (Mg)} = 33 \text{ mg}/\ell$$
$$\text{bicarbonate (HCO}_3) = 340 \text{ mg}/\ell$$
$$\text{pH} = 7.2$$

Compare the chemical dose requirements for softening when using lime and soda ash to the dose requirements when using caustic soda and soda ash.

9-2. Using a C-L diagram, determine the characteristics of the water described below when treated with various dosages of lime and 100 mg/ℓ of soda ash (all as $CaCO_3$). Plot the results, showing residual calcium, magnesium, alkalinity, and pH of the treated water as a function of lime dose.

$$\text{calcium} = 380 \text{ mg}/\ell \text{ as CaCO}_3$$
$$\text{magnesium} = 120 \text{ mg}/\ell \text{ as CaCO}_3$$
$$\text{alkalinity} = 260 \text{ mg}/\ell \text{ as CaCO}_3$$
$$\text{pH} = 7.0$$
$$\text{temperature} = 15°C$$
$$I = 0.01 \ M$$

9-3. Split treatment is to be used to soften a water described by the following analysis:

$$\text{calcium} = 3.5 \text{ meq.}/\ell$$
$$\text{magnesium} = 2.0 \text{ meq.}/\ell$$
$$\text{alkalinity} = 220 \text{ mg}/\ell \text{ as CaCO}_3$$
$$\text{pH} = 7.2$$

Criteria for the finished water are a maximum permissible magnesium hardness of 40 mg/ℓ as CaCO$_3$ and a final calcium hardness in the range of 40 to 50 mg/ℓ as CaCO$_3$. Determine the required lime and soda ash dose if the fraction of bypassed flow is to be 0.1 unit less than $(X)_{\max}$. Assume that the water temperature is 15°C and $I = 0.01 \ M$.

9-4. A municipality obtains raw water from two different sources. Characteristics of the two waters are as follows:
Source A:

$$\text{calcium} = 330 \text{ mg}/\ell \text{ as CaCO}_3$$
$$\text{magnesium} = 75 \text{ mg}/\ell \text{ as CaCO}_3$$
$$\text{alkalinity} = 200 \text{ mg}/\ell \text{ as CaCO}_3$$
$$\text{pH} = 7.2$$

Source B:

$$\text{calcium} = 180 \text{ mg}/\ell \text{ as CaCO}_3$$
$$\text{magnesium} = 25 \text{ mg}/\ell \text{ as CaCO}_3$$
$$\text{alkalinity} = 80 \text{ mg}/\ell \text{ as CaCO}_3$$
$$\text{pH} = 8.0$$

Determine the characteristics of the water to be treated if source A is pumped at the rate of 30 MGD, and source B is pumped at the rate of 20 MGD, and the sources are mixed prior to entering the treatment plant. Use a 15°C, 0.01I, C-L diagram for the calculations.

9-5. After settling, a lime-soda ash softened water has the following characteristics:

$$\text{pH} = 11.45$$
$$\text{calcium (Ca)} = 3 \text{ mg}/\ell \text{ as CaCO}_3$$
$$\text{alkalinity} = 120 \text{ mg}/\ell \text{ as CaCO}_3$$

Determine the dose of 12 N HCl required to reduce the pH to 8.0. Express the acid dose in terms of liters of acid per MG of water treated.

REFERENCES

CLARK, J.W., VIESSMAN, W., and HAMMER, M.J., *Water Supply and Pollution Control*, Harper & Row Publishers, Inc., New York (1977).

FAUST, S.D., and McWHORTER, J.G., "Water Chemistry," in *Handbook of Water Resources and Pollution Control*, [Eds.] H.W. GEHM and J.I. JACOB, Van Nostrand Reinhold Company, New York (1976).

KUNIN, R., "Water Softening by Ion Exchange," *Proc., Fourteenth Water Quality Conference,* University of Illinois, Urbana, IL (1972).

LARSON, T.E., LANE, R.W., and NEFF, C.H., "Stabilization of Magnesium Hydroxide in the Solids-Contact Process," *J. Am. Water Works Assoc.,* **51,** 1551 (1959).

LOEWENTHAL, R.E., and MARAIS, G.V.R., *Carbonate Chemistry of Aquatic Systems: Theory and Application,* Ann Arbor Science Publishers, Ann Arbor, MI (1976).

MERRILL, D.T., "Chemical Conditioning for Water Softening and Corrosion Control," paper Presented at the Fifth Environmental Engineers' Conference, Big Sky of Montana (June, 1976).

MERRILL, D.T. and SANKS, R.L., "Corrosion Control by Deposition of $CaCO_3$ Films: Part 1, A Practical Approach for Plant Operators," *J. Am. Water Works Assoc.,* **69,** 592 (1977).

SINGLEY, J.E., "Chemical Principles of Lime-Soda Softening," *Proc., Fourteenth Water Quality Conference,* University of Illinois, Urbana, IL (1972).

TRUESDELL, A.H., and JONES, B.F., "WATEQ, A Computer Program for Calculating Chemical Equilibria of Natural Waters," NTIS, U.S. Department of Commerce, PB 220464 (1973).

WEBER, W.J., Jr., *Physicochemical Processes for Water Quality Control,* Wiley-Interscience, New York (1972).

10

ION EXCHANGE

Common applications of ion exchange processes are: for water treatment to soften hard waters (removing polyvalent cations); for municipal wasterwater treatment to remove nitrogen and phosphorus and to demineralize wastewater for reuse; for industrial wastewater treatment to remove and/or recover many different ionic chemical species. Therefore, it is desirable that the environmental engineer have a basic understanding of the chemical principles governing the ion exchange process and be able to apply these principles to process evaluation and process design.

10-1 FUNDAMENTALS OF ION EXCHANGE

Ion exchange is defined as a process where an insoluble substance removes ions of positive or negative charge from an electrolytic solution and releases other ions of like charge into solution in a chemically equivalent amount. The process occurs with no structural changes in the resin. The ions in solution rapidly diffuse into the molecular network of the resin, where exchange occurs. The exchanged ions proceed by the same path into solution. At some point during the ion exchange process an ion exchange equilibrium is established.

The general reaction for the exchange of ions A and B on a cation exchange resin can be respresented as follows:

$$n\text{R}^-\text{A}^+ + \text{B}^{n+} \rightleftharpoons \text{R}_n^-\text{B}^{n+} + n\text{A}^+ \tag{10-1}$$

where R^- is an anionic group attached to the ion exchange resin, and A^+ and B^{n+} are ions in solution.

The equilibrium constant expression for this reaction is

$$K_{\text{A}^+}^{\text{B}^{n+}} = \frac{(\text{R}_n^-\text{B}^{n+})_\text{R}(\text{A}^+)_\text{S}^n}{(\text{R}^-\text{A}^+)_\text{R}^n(\text{B}^{n+})_\text{S}} \tag{10-2}$$

where $(R_n^- B^{n+})_R$ = activity of B^{n+} in the resin

$(B^{n+})_S$ = activity of B^{n+} in the solution

$(R^- A^+)_R$ = activity of A^+ in the resin

$(A^+)_S$ = activity of A^+ in the solution.

The term $K_{A^+}^{B^{n+}}$ is not actually a constant, since it is dependent on experimental conditions. Rather, it is referred to as a *selectivity coefficient*.

Equation 10-2 can be modified so that concentrations rather than activities are related:

$$K_{A^+}^{\backslash B^{n+}} = \frac{[R_n^- B^{n+}]_R [A^+]_S^n}{[R^- A^+]_R^n [B^{n+}]_S} \tag{10-3}$$

In the case of dilute aqueous solutions the activity coefficients for the solution phase can often be ignored. The internal ion concentrations in the resin, however, are often high, and the activity coefficients significant. As a result $K_{A^+}^{\backslash B^{n+}}$ values are generally valid only for narrow concentration ranges. Nevertheless, the relationship illustrated by equation 10-3 is useful for determining which ions will exchange in reasonable amounts and for estimating the amount of resin required to remove some quantity of an ion from solution.

The preference of an ion exchange material for one ion over another is often expressed in terms of the separation factor, or selectivity quotient, Q_S. For the reaction shown in equation 10-1,

$$Q_S = \frac{[R_n^- B^{n+}]_R [A^+]_S}{[R^- A^+]_R [B^{n+}]_S} \tag{10-4}$$

Note that the bracketed terms in equation 10-4 are the same as those in equation 10-3. In this case, however, they are each taken to the first power. The relationship between the selectivity coefficient and the separation factor is

$$K_{A^+}^{\backslash B^{n+}} = Q_S \left[\frac{[A^+]_S}{[R^- A^+]_R} \right]^{n-1} \tag{10-5}$$

In the instance of monovalent exchange, $n = 1$, and the separation factor and selectivity coefficient are identical.

The separation factor is useful because it indicates the preference of the ion exchanger for one ion relative to another. A Q_S value greater than 1 indicates the ion B^{n+} is preferred over the ion A^+. A separation factor less than unity would indicate the resin preferred ion A^+ over ion B^{n+}. When $Q_S = 1$ there is an equal preference for each ion. These relationships are often illustrated by isotherms such as those shown in Figure 10-1.

A number of different selectivity coefficients can be derived from Tables 10-1 and 10-2. Note, however, that separation factors are related to these values via equation 10-5. Note also that to obtain $K_{A^+}^{B^{3+}}$ from the K values given in Table 10-1 the expression should not be that indicated in the footnote to the table, but instead equation 10-6

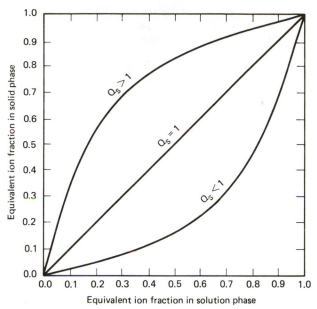

FIGURE 10-1 *Generalized ion exchange isotherms (after Environmental Protection Agency, 1975).*

should be used:

$$K_{A^+}^{B^{2+}} = \frac{K_{H^+}^{B^{2+}}}{(K_{H^+}^{A^+})^2} \tag{10-6}$$

The requirement for using this expression to obtain the selectivity coefficient for a univalent-divalent exchange is illustrated by the following reactions:

$$2R^-Na^+ + Ca^{2+} \rightleftharpoons R_2^-Ca^{2+} + 2Na^+ \tag{10-7}$$

where

$$K_{Na^+}^{\backslash Ca^{2+}} = \frac{[R_2^-Ca^{2+}][Na^+]^2}{[R^-Na^+]^2[Ca^{2+}]} \tag{10-8}$$

$$R^-Na^+ + K^+ \rightleftharpoons R^-K^+ + Na^+ \tag{10-9}$$

where

$$K_{Na^+}^{\backslash K} = \frac{[R^-K^+][Na^+]}{[R^-Na^+][K^+]} \tag{10-10}$$

$$2R^-K^+ + Ca^{2+} \rightleftharpoons R_2^-Ca^{2+} + 2K^+ \tag{10-11}$$

where

$$K_{K^+}^{\backslash Ca^{2+}} = \frac{[R_2^-Ca^{2+}][K^+]^2}{[R^-K^+]^2[Ca^{2+}]} \tag{10-12}$$

hence,

$$K_{K^+}^{\backslash Ca^{2+}} = \frac{K_{Na^+}^{\backslash Ca^{2+}}}{(K_{Na^+}^{\backslash K^+})^2} \tag{10-13}$$

TABLE 10-1 *Effect of degree of crosslinking on relative affinities of various cations for polystyrene cation exchange resins (after Abrams and Benezra, 1967).*

Ion	% Divinylbenzene		
	4	8	12
Monovalent Cations			
H	1.0	1.0	1.0
Li	0.9	0.85	0.81
Na	1.3	1.5	1.7
NH_4	1.6	1.95	2.3
K	1.75	2.5	3.05
Rb	1.9	2.6	3.1
Cs	2.0	2.7	3.2
Cu	3.2	5.3	9.5
Ag	6.0	7.6	12.0
Divalent Cations			
Mn	2.2	2.35	2.5
Mg	2.4	2.5	2.6
Fe	2.4	2.55	2.7
Zn	2.6	2.7	2.8
Co	2.65	2.8	2.9
Cu	2.7	2.9	3.1
Cd	2.8	2.95	3.3
Ni	2.85	3.0	3.1
Ca	3.4	3.9	4.6
Sr	3.85	4.95	6.25
Hg	5.1	7.2	9.7
Pb	5.4	7.5	10.1
Ba	6.15	8.7	11.6

Note: The constants given in this table are referred to H^+. To calculate $K_{A^+}^{B^+}$ from $K_{H^+}^{A^+}$ and $K_{H^+}^{B^+}$, use the formula $K_{A^+}^{B^+} = K_{H^+}^{B^+}/K_{H^+}^{A^+}$.

Although K_A^B and $K_A^{\prime B}$ are not numerically equivalent, these terms are used interchangeably in most situations.

On the basis of selectivity coefficients, relative affinities of ions for an ion exchanger can be quantitatively evaluated. This suggests that an order can be established for ions of the same valence on the basis of their selectivity coefficients. The affinity of an ion for a resin can be generalized by the following rules (Helms, 1973):

1. In general, ions of high valence are preferred over ions of low valence; i.e., the extent of the exchange reaction increases with increasing ion valence (e.g., $Fe^{3+} > Mg^{2+} > Na^+$; $PO_4^{3-} > SO_4^{2-} > NO_3^-$). This preference increases with a decrease in the total ionic concentration of the solution.

2. For ions of the same valence the extent of the exchange reaction increases with decreasing hydrated radius and increasing atomic number (e.g., $Ca^{2+} > Mg^{2+}$

TABLE 10-2 *Relative affinities of various anions for polystyrene-based strong-base anion exchange resins (after Abrams and Benezra, 1967).*

Ion	Relative Affinity	
	Type I*	Type II†
Hydroxide (reference)	1.0	1.0
Benzenesulfonate	500	75
Salicylate	450	65
Citrate	220	23
Iodide	175	17
Phenoxide	110	27
Bisulfate	85	15
Chlorate	74	12
Nitrate	65	8
Bromide	50	6
Bromate	27	3
Nitrite	24	3
Chloride	22	2.3
Bicarbonate	6.0	1.2
Iodate	5.5	0.5
Formate	4.6	0.5
Acetate	3.2	0.5
Fluoride	1.6	0.3

* Reactive group, $-CH_2N^{\oplus}(CH_3)_3$.
† Reactive group, $-CH_2N^{\oplus}(CH_3)_2C_2H_4OH$.
Note: The constants given in this table are refererred to OH^-. To calculate $K_{A^-}^{B^-}$ from $K_{OH^-}^{A^-}$ and $K_{OH^-}^{B^-}$, use the formula $K_{A^-}^{B^-} = K_{OH^-}^{B^-}/K_{OH^-}^{A^-}$.

$> Be^{2+}; K^+ > Na^+ > Li^+$). This type of response is a result of swelling pressure within the resin. Ions of larger hydrated radius increase the swelling pressure within the resin and decrease the affinity of the resin for such ions.

3. For a solution with a high total ionic concentration the extent of the exchange reaction follows no general rule and is often reversed. This type of response is the basis for the reversibility of regeneration.

4. The relationship between the degree of crosslinking and the size of the hydrated ion may affect the extent of the exchange reaction. If the resin has a high degree of crosslinking, the ion may be too large to penetrate into the matrix of the resin.

Weber (1972) gives the following advantages and disadvantages for choosing a resin with a high affinity for the ion to be exchanged:

1. *Advantages:*

   (a) Sharp breakthrough curve.

   (b) Shorter ion exchange column.

   (c) Greater flow rate applied to ion exchange column.

2. *Disadvantages:*

    (a) Higher regenerant concentration required.

10-2 TYPES OF ION EXCHANGE RESINS

In earlier times naturally occurring materials called *zeolites* were used for ion exchange resins. However most ion exchange resins in use today are synthetic materials made up of a polymer matrix (generally polystyrene chains held together by divinylbenzene crosslinks) with soluble ionic functional groups attached to the polymer chains. The total number and kind of functional groups in a resin determine the exchange capacity and ion selectivity while the polymer matrix provides insolubility and toughness to the resin. Resins are granular in nature and may have either a spherical or irregular shape. Although spherically shaped resins are generally used, the irregularly shaped form provides a larger surface area and higher void space in an ion exchange column. The larger void space results in less head loss across the column during operation.

    Ion exchange resins are usually classified in the following manner:

1. *Cation exchange resins (contain exchangeable cations):*

    (a) Strong-acid exchange resins (SAC).

    (b) Weak-acid exchange resins (WAC).

2. *Anion exchange resins (contain exchangeable anions);*

    (a) Strong-base exchange resins (SBA).

    (b) Weak-base exchange resins (WBA).

    Strong-acid exchange resins contain functional groups derived from a strong acid (normally sulfuric acid). Their degree of ionization is analogous to that of a strong acid (low pK_a) which permits the hydrogen to be dissociated and ready for exchange over a wide pH range. Weak-acid exchange resins, on the other hand, contain functional groups derived from a weak acid commonly of the carboxylic or phenolic form. Such resins are useful only within a fairly narrow pH range.

    Strong-base exchange resins contain functional groups which are either type I or type II quarternary ammonium groups, whereas weak-base exchange resins contain the primary, secondary, and/or tertiary amine as the functional group. The strong-base exchange resins are useful over a wide pH range, whereas the weak-base exchange resins are effective only within a fairly narrow pH range. The active groups associated with each of the different kinds of resins are listed in Table 10-3.

    An example of an exchange reaction for each type of resin is illustrated below (Helms, 1973):

1. *Strong-Acid Exchange Resins:* These resins split neutral salts and convert them to their corresponding acid.

$$R\text{—}SO_3^- : H^+ + NaCl \longrightarrow HCl + R\text{—}SO_3^- : Na^+ \qquad \textbf{(10-14)}$$

TABLE 10-3 *Chemical classification and dissociation constants of ion exchange resins, (after Panswad, 1975).*

Classification	Active Groups	Dissociation Constant pK_a	Typical configuration
Cation Exchange Resins			
Strong acid:	Sulfonic	1	$SO_3^- H^+$
	Metylene sulfonic	1	$CH_2SO_3^- H^+$
Weak acid:	Carboxylic	4–6	CH_2CHCH_2 $\mid$ $COO^- H^+$
	Phosphonic	2–3 7–8	$PO_3^{2+} H_2^+$
	Phenolic hydroxyl	9–10	$O^- H^+$
Anion Exchange Resins			
Strong base:	Quarternary ammonium (type I)	13	CH_2 $\mid$ $OH^-(N (CH_3)_3)^+$
	(type II)		CH_2 $\mid$ $((CH_3)_2 (C_2H_4OH) N)^+ OH^-$
Weak base:	Primary amine	6–9	CH_2NH_2
	Secondary amine	7–9	CH_2NRH
	Tertiary amine (aromatic matrix)	9–11	CH_2NR_2
	(aliphatic matrix)		$CHCH_2NCH_2^-$ $\mid$ \quad $\mid$ OH \quad CH_2

These resins can be regenerated with a strong acid such as H_2SO_4 or HCl.

$$2R\!\!-\!\!SO_3^- : Na^+ + H_2SO_4 \longrightarrow 2R\!\!-\!\!SO_3^- : H^+ + Na_2SO_4 \qquad \textbf{(10-15)}$$

The regeneration efficiency of these resins is 30 to 50%.

2. *Weak-Acid Exchange Resins:* These resins cannot split neutral salts but can remove cations associated with the water's alkalinity to form carbonic acid. In other words, acids weaker than the functional group must be formed if the exchange reaction is to proceed.

$$R\overset{\overset{\displaystyle O}{\|}}{-C}-O^-:H^+ + NaHCO_3 \longrightarrow R\overset{\overset{\displaystyle O}{\|}}{-C}-O^-:Na$$
$$+ \; (H_2CO_3 \; \rightleftharpoons \; CO_2\uparrow + H_2O) \qquad \text{(10-16)}$$

These resins can be regenerated with any acid stronger than the functional group.

$$R\overset{\overset{\displaystyle O}{\|}}{-C}-O^-:Na^+ + HCl \longrightarrow R\overset{\overset{\displaystyle O}{\|}}{-C}-O^-:H^+ + NaCl \qquad \text{(10-17)}$$

The regeneration efficiency of these resins is near 100%. However, their use is limited to waters having pH values greater than 7.0 and containing a high alkalinity.

3. *Strong-Base Exchange Resins:* These resins split neutral salts and convert them to their corresponding base.

$$R-NR_3^+:OH^- + NaCl \longrightarrow R-NR_3^+:Cl^- + NaOH \qquad \text{(10-18)}$$
$$R-NR_3^+:OH^- + HCl \longrightarrow R-NR_3^+:Cl^- + H_2O \qquad \text{(10-19)}$$

These resins can be regenerated with NaOH.

$$R-NR_3^+:Cl^- + NaOH \longrightarrow R-NR_3^+:OH^- + NaCl \qquad \text{(10-20)}$$

The regeneration efficiency of these resins is 30 to 50%.

4. *Weak-Base Exchange Resins:* These resins cannot split neutral salts but they can remove strong acids by adsorption.

$$R-NH_2 + HCl \longrightarrow R-NH_2 \cdot HCl \qquad \text{(10-22)}$$

These resins can be regenerated by eluting the adsorbed acid with bases such as NaOH, Na_2CO_3, and NH_4OH.

$$2R-NH_2 \cdot HCl + Na_2CO_3 \longrightarrow 2R-NH_2 + 2NaCl$$
$$+ \; (H_2CO_3 \; \rightleftharpoons \; CO_2\uparrow + H_2O) \qquad \text{(10-22)}$$
$$R-NH_2 \cdot HCl + NH_4OH \longrightarrow R-NH_2 + NH_4Cl + H_2O \qquad \text{(10-23)}$$

The regeneration efficiency of these resins is near 100%. However, their use is limited to waters having pH values less than 7.0.

Note: Salt-splitting is defined as the conversion of salts to their corresponding acids or bases by passage through strong-acid or strong-base ion exchange resins.

10-3 GENERAL CHARACTERISTICS OF ION EXCHANGE RESINS

To be effective as ion exchangers, resins must contain exchangeable ions within their structure, be insoluble in water, and provide enough space in their porous structure for ions to pass freely in and out of the polymer matrix.

Ion Exchange Capacity

In the design of an ion exchange system or the selection of an ion exchange resin, the capacity of the resin is important because of its effect on process efficiency and system cost. The capacity of an ion exchange resin is usually expressed as *total capacity* (theoretical or ultimate capacity) or *operating capacity* (breakthrough capacity). The total capacity is a measure of the total quantity of ions which theoretically can be exchanged (number of ionizable groups) per unit mass or per unit volume of resin (e.g., meq./ℓ, eq./ℓ, grains as $CaCO_3$/ft^3, kilograins as $CaCO_3$/ft^3, meq./g). Since resins are sold and measured by volume, volume capacities are normally preferred.

Operating capacity is a measure of the actual useful capacity of the resin for exchanging ions from a solution flowing through a fixed bed of resin particles under specified conditions. It may be defined as the capacity of a resin bed when the solute being removed from the feed solution exceeds some arbitrarily selected level. The operating capacity of a resin depends on the flow rate through the column, bed depth, selectivity coefficient, exchange ion size, amount of regenerant used, composition and concentration of the feed solutes, temperature, and desired quality of the product water. The operating capacity associated with any point along the breakthrough curve can be computed from equation 10-25.

$$X = \frac{\left[\begin{array}{c}\text{number of equivalents of the ion} \\ \text{of interest applied to the column}\end{array}\right]}{\text{resin volume}} - \frac{\left[\begin{array}{c}\text{number of equivalents of the ion of} \\ \text{interest passing through the column}\end{array}\right]}{\text{resin volume}} \quad \textbf{(10-24)}$$

or

$$X = \frac{C_o V_{op}}{V_r} - \frac{C_o}{V_r} \sum_{i=1}^{i=n} (Y_i)(V_i) \quad \textbf{(10-25)}$$

where
X = operating capacity (meq./mℓ)

C_o = influent concentration of ion of interest (meq./ℓ)

C_e = effluent concentration of ion of interest (meq./ℓ)

C_e/C_o = ratio of equivalents of anions or cations in the product water to the equivalents of anions or cations in the feed water. This term is sometimes referred to as *leakage*. Leakage is also defined, by many workers, as the appearance in the effluent of ions which are desired to be removed in the exchanger.

Leakage may be due either to unfavorable equilibrium or incomplete regeneration of the ion exchange resin.

$(C_e/C_o)_L$ = initial column leakage. Initial column leakage is due to incomplete regeneration of the resin at the bottom of the column. This means that the degree of initial leakage is dependent on the regeneration level and can be controlled by the amount of regenerant used.

$(C_e/C_o)_{op}$ = column leakage at termination of service run.

Y_i = leakage value used for a particular incremental summation. This is the C_e/C_o value associated with the midpoint of the appropriate V interval.

V_i = throughput volume increment associated with the interval of interest, $[(V)_{i+1} - (V)_i]$, (ℓ)

V_{op} = throughput volume when service run is terminated (ℓ)

V_r = volume of resin in column (mℓ). This is usually taken as the resin volume in the column after back-washing, settling, and draining.

EXAMPLE PROBLEM 10-1: Compute the operating capacity of the cation exchange resin which produced the breakthrough curve shown in Figure 10-2. The

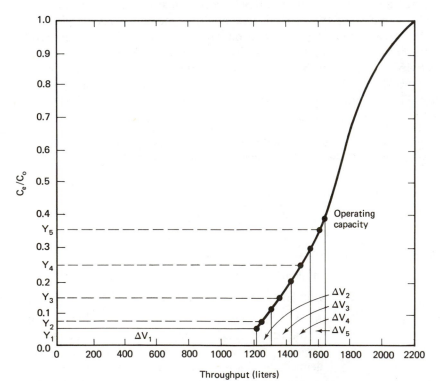

FIGURE 10-2 *Breakthrough curve for Example Problem 10-1.*

column was used to soften water which had an initial total hardness of 5 meq./ℓ. The volume of the resin in the column was 10,000 mℓ.

Solution:

1. Set the intervals to be used in the graphical integration, and tabulate the interval values for Y_i, V_i, and $(Y_i)(V_i)$. The intervals used in this problem are shown in Figure 10-2.

Sequence	Y_i	V_i		$(Y_i)(V_i)$
1	0.05	1210–0 =	1210 ℓ	60.5
2	0.075	1280–1210 =	70 ℓ	5.3
3	0.15	1440–1280 =	160 ℓ	24.0
4	0.25	1560–1440 =	120 ℓ	30.0
5	0.35	1660–1560 =	100 ℓ	35.0
				= 154.8

Note: In the first interval the upper limit rather than the midpoint for Y_i was used because the breakthrough curve was basically horizontal in this interval.

2. Compute the operating capacity of the resin from equation 10-25.

$$X = \frac{(5)(1660)}{10,000} - \frac{(5)(154.8)}{10,000} = 0.75 \text{ meq./m}\ell$$

The operating capacity is the capacity value used by engineers to design ion exchange columns. Abrams and Benezra (1967) note that capacity values furnished by manufacturers are generally for ideal conditions, and these values should be reduced by 15 to 20% to account for the nonideal behavior encountered during actual operation.

It is desirable that the environmental engineer be able to evaluate the effectiveness of a proposed ion exchange process by making only a small number of simple calculations. Anderson (1975) has developed a method which makes this possible for systems where strong ionic resins (SAC or SBA) are employed. This method is based on the fact that at 100% leakage ($C_e/C_o = 1.0$) the resin is in equilibrium with the influent solution. The resin composition at equilibrium can be interpreted in two different ways: (1) as the *limiting operating capacity* of the ion exchange column or (2) as the maximum regeneration level which can be attained with a regeneration solution of specified concentration.

When using this method to evaluate an ion exchange process, it must be understood that any equilibrium value computed is a limiting value which would not be achieved during actual column operation. Nevertheless, in many cases such numbers can provide useful information about anticipated process performance.

In his development Anderson (1975) rearranges equation 10-3 (assuming a monovalent-monovalent exchange reaction) so that the selectivity coefficient is expressed in terms of equivalent fractions. In converting concentration units to equivalent fractions, he introduces the following terms:

a) C = total anionic or cationic concentration of the solution (eq./ℓ)

b) $X_{A^+} = \dfrac{[A^+]_s}{C}$ (10-26)

where X_{A^+} is the equivalent fraction of the A^+ ion in solution.

c) $X_{B^+} = \dfrac{[B^+]_s}{C}$ (10-27)

where X_{B^+} is the equivalent fraction of the B^+ ion in solution.

d) $X_{A^+} + X_{B^+} = 1$ (10-28)

e) $\bar{X}_{A^+} = \dfrac{[R^-A^+]_R}{\bar{C}}$ (10-29)

where $\bar{X}_{A^+}$ is the equivalent fraction of the A^+ ion in the resin, and $\bar{C}$ is the total ionic concentration of the resin ($\bar{C}$ is taken as the resin capacity in eq./ℓ).

f) $\bar{X}_{B^+} = \dfrac{[R^-B^+]_R}{\bar{C}}$ (10-30)

where $\bar{X}_{B^+}$ is the equivalent fraction of the B^+ ion in the resin.

g) $\bar{X}_{A^+} + \bar{X}_{B^+} = 1$ (10-31)

Solving equations 10-26, 10-27, 10-29, and 10-30 for the appropriate concentration terms and substituting for these terms in equation 10-3 (assuming monovalent-monovalent exchange) gives

$$K_{A^+}^{\backslash B^+} = \frac{(\bar{C}\bar{X}_{B^+})(CX_{A^+})}{(\bar{C}\bar{X}_{A^+})(CX_{B^+})}$$ (10-32)

or

$$K_{A^+}^{\backslash B^+} = \frac{(\bar{X}_{B^+})(X_{A^+})}{(\bar{X}_A)(X_{B^+})}$$ (10-33)

Equations 10-28 and 10-31 can be rearranged into the following forms:

$$X_{A^+} = 1 - X_{B^+}$$ (10-34)
$$\bar{X}_{A^+} = 1 - \bar{X}_{B^+}$$ (10-35)

Substituting for X_{A^+} and $\bar{X}_{A^+}$ in equation 10-33 from equations 10-34 and 10-35 gives

$$K_{A^+}^{\backslash B^+} = \frac{(1 - X_{B^+})\bar{X}_{B^+}}{(1 - \bar{X}_{B^+})X_{B^+}}$$ (10-36)

or

$$\left[\frac{X_{\text{B}^+}}{1 - X_{\text{B}^+}}\right] K_{\text{A}^+}^{\text{B}^+} = \frac{\bar{X}_{\text{B}^+}}{1 - \bar{X}_{\text{B}^+}} \tag{10-37}$$

Equation 10-37 is valid for exchanges between monovalent cations or anions on fully ionized exchange resins. Anderson (1975) indicates that equation 10-37 has three important characteristics:

1. The term $\bar{X}_{\text{B}^+}/(1 - \bar{X}_{\text{B}^+})$ gives the state of the resin in an exchange column at the point when the influent and effluent ion concentrations are the same.

2. The term $\bar{X}_{\text{B}^+}$ defines the extent to which the resin can be converted to the B^+ form when the resin is in equilibrium with a solution of composition X_{B^+}.

3. The term $\bar{X}_{\text{B}^+}$ also shows the maximum degree of regeneration which can be achieved with a regenerant composition of X_{B^+}.

The use of equation 10-37 is illustrated in example problems 10-2 and 10-3, where the equilibrium resin composition is estimated from the influent solution composition and the selectivity coefficient for the exchange reaction.

EXAMPLE PROBLEM 10-2 (after Anderson, 1975): When nitrate ion removal is the primary treatment objective, what is the maximum volume of water that can be processed per liter of type I, strong-base, anion exchange resin if the total resin capacity is 1.3 eq./ℓ and $K_{\text{Cl}^-}^{\text{NO}_3^-} = 4$? Assume that the water has the following composition:

$$
\begin{array}{ll}
\text{Ca}^{2+} = 1.0 \text{ meq./}\ell & \text{Cl}^- = 3.0 \text{ meq./}\ell \\
\text{Mg}^{2+} = 1.0 \text{ meq./}\ell & \text{SO}_4^{2-} = 0.0 \text{ meq./}\ell \\
\underline{\text{Na}^+ = 2.5 \text{ meq./}\ell} & \underline{\text{NO}_3^- = 1.5 \text{ meq./}\ell} \\
\text{TDC} = 4.5 \text{ meq./}\ell & \text{TDA} = 4.5 \text{ meq./}\ell
\end{array}
$$

Solution:

1. For the equilibrium condition compute the nitrate equivalent fraction. This can be obtained from the influent solution nitrate concentration, since at equilibrium, leakage is 100% $(C_e/C_o = 1.0)$.

$$X_{\text{NO}_3^-} = \frac{1.5}{4.5} = 0.33$$

2. Compute the theoretical equilibrium resin composition with respect to the nitrate ion, using equation 10-37:

$$\frac{\bar{X}_{\text{NO}_3^-}}{1 - \bar{X}_{\text{NO}_3^-}} = 4\left[\frac{0.33}{1 - 0.33}\right]$$

$$\bar{X}_{\text{NO}_3^-} = 0.66$$

This means that 66% of the exchange sites will be utilized.

3. Determine the limiting operating capacity of the resin:

limiting operating
capacity (eq./ℓ) $= (1.3 \text{ eq./}\ell)(0.66) = 0.86 \text{ eq./}\ell$

4. Calculate the volume of water that can be treated during the service cycle:

$$\frac{0.86 \text{ (eq./}\ell \text{ of nitrate ion removed)}}{1.5 \times 10^{-3} \text{ (eq./}\ell \text{ of nitrate in solution)}} = 5.7 \times 10^2 \, \ell/\ell$$

Therefore,

$$\frac{570 \, \ell}{\ell} \times \frac{1 \text{ gal}}{3.78 \, \ell} \times \frac{28.3 \, \ell}{1 \text{ ft}^3} = 4267 \text{ gal/ft}^3$$

This would be a feasible process based on the volume of water treated per unit volume of resin. However, in actual practice regeneration of the resin is inefficient and requires a large volume of regenerant, making the cost of the process excessive.

EXAMPLE PROBLEM 10-3: A strong-acid cation exchange resin has been used for the removal of sodium ions. If the objective is to regenerate the resin to within 90% of its total exchange capacity, can a strong-acid solution having a total TDC of 1 eq./ℓ and a hydrogen ion concentration of 0.7 eq./ℓ be used for regeneration? Assume that $K_{\text{Na}^+}^{\text{H}^+} = 0.67$.

Solution:

1. For the equilibrium condition, compute the hydrogen equilibrium fraction:

$$X_{\text{H}^+} = \frac{0.7 \text{ eq./}\ell}{1 \text{ eq./}\ell} = 0.7$$

2. Compute the theoretical equilibrium resin composition with respect to the hydrogen ion:

$$\frac{\bar{X}_{\text{H}^+}}{1 + \bar{X}_{\text{H}^+}} = (0.67) \left[\frac{0.7}{1 - 0.7} \right]$$

$$\bar{X}_{\text{H}^+} = 0.61$$

This calculation indicates that a maximum of 62% percent of the exchange sites are all that can be converted using the acid solution described in this problem.

Anderson (1975) has also developed an equivalent fraction equation for a mono-valent-divalent ion exchange reaction. In this case, the exchange reaction may be represented as

$$2R^-A^+ + B^{2+} \rightleftharpoons 2A^+ + R^-B^{2+} \tag{10-38}$$

The selectivity coefficient equation for this reaction is

$$K_{\text{A}^+}^{\text{B}^{2+}} = \frac{[A^+]_S^2 [R^-B^{2+}]_R}{[R^-A^+]_R^2 [B^{2+}]_S} \tag{10-39}$$

Expressing equation 10-39 in the equivalent fraction form gives

$$\frac{\bar{X}_{\text{B}^{2+}}}{(1 - \bar{X}_{\text{B}^{2+}})^2} = K_{\text{A}^+}^{\text{B}^{2+}} \left[\frac{\bar{C}}{C} \right] \frac{X_{\text{B}^{2+}}}{(1 - X_{\text{B}^{2+}})^2} \tag{10-40}$$

This equation is valid for exchanges between a monovalent and a divalent ion on a fully ionized resin. In such an exchange the product $K_A^{\backprime B^{2+}}(\bar{C}/C)$ is called the *apparent-selectivity coefficient*. If this value is greater than 1, the exchange will be effective, thus, the solution concentration is an important consideration in evaluating the feasibility of any monovalent-divalent exchange reaction.

EXAMPLE PROBLEM 10-4 (after Anderson, 1975): Consider the nitrate removal problem outlined in example problem 10-2. If the ionic background of the water is changed so that the sulfate ion concentration is 3.0 meq./ℓ and the chloride concentration is 0.0 meq./ℓ, what is the maximum volume of water that can be processed per liter of resin? Assume that $K_{NO_3^-}^{\backprime SO_4^{2-}} = 0.04$ and that the resin capacity is 1.3 meq./ℓ.

Solution:

1. For the equilibrium condition compute the sulfate equilibrium fraction. This value can be obtained from the influent solution sulfate concentration, since, at equilibrium, leakage is 100% ($C_e/C_o = 1.0$):

$$X_{SO_4^{2-}} = \frac{3.0}{4.5} = 0.67$$

2. Compute the theoretical equilibrium resin composition with respect to the sulfate ion, using equation 10-40. In this problem, $C = TDS = 4.5$ meq./ℓ.

$$\frac{\bar{X}_{SO_4^{2-}}}{(1 - \bar{X}_{SO_4^{2-}})^2} = K_{NO_3^-}^{\backprime SO_4^{2-}}\left(\frac{\bar{C}}{C}\right)\left[\frac{X_{SO_4^{2-}}}{(1 - X_{SO_4^{2-}})^2}\right]$$

$$= (0.04)\left[\frac{1.3}{4.5 \times 10^{-3}}\right]\left[\frac{0.67}{(0.33)^2}\right]$$

$$= 213$$

Therefore,

$$\bar{X}_{SO_4^{2-}} = 0.93$$

3. Compute the equilibrium resin composition with respect to the nitrate ion:

$$\bar{X}_{NO_3^-} = 1 - \bar{X}_{SO_4^{2-}} = 0.07$$

4. Determine the limiting operating capacity of the resin with respect to the nitrate ion:

limiting operating
capacity (eq./ℓ) $= (1.3)(0.07) = 0.091$ eq./ℓ

5. Calculate the volume of water that can be treated during the service cycle:

$$\frac{0.091 \text{ (eq./}\ell \text{ of nitrate ion removed)}}{1.5 \times 10^{-3} \text{ (eq./}\ell \text{ of nitrate ion in solution)}} = 60.6 \text{ } \ell/\ell$$

Therefore,

$$60.6\frac{\ell}{\ell} \times \frac{1 \text{ gal}}{3.78 \text{ } \ell} \times \frac{28.3 \text{ } \ell}{1 \text{ ft}^3} = 453 \text{ gal/ft}^3$$

Even though the resin is selective for nitrate over sulfate ($K_{NO_3^-}^{\backslash SO_4^{2-}} < 1$), the presence of sulfate in the water significantly reduces the process effectiveness (see example problem 10-2). This implies that the ionic background of the water to be treated is an important consideration when evaluating the acceptability of ion exchange for a given treatment situation.

As noted by Anderson (1975), the equivalent fraction equations are useful for estimating the theoretical maximum operating capacity (not actual operating capacity) of a proposed ion exchange process. If a large volume of influent can be treated during the service cycle, the process may be considered feasible. However, it must be kept in mind that the application of these equations is restricted to strong-acid and strong-base exchange resins, since they are generally considered fully ionized over the entire pH range.

Approximate selectivity coefficients and exchange capacities for strong-acid cation exchange resins and strong-base anion exchange resins are shown in Tables 10-4 and 10-5.

TABLE 10-4 Approximate selectivity coefficients for strong-acid cation exchange resins (after Anderson, 1975).

$K_{H^+}^{\backslash Na^+}$	1.5	$K_{Na^+}^{\backslash Ca^{2+}}$	3 to 6
$K_{H^+}^{\backslash K^+}$, $K_{H^+}^{\backslash NH_4^+}$	2.5	$K_{Na^+}^{\backslash Mg^{2+}}$	1.0 to 1.5
$K_{Na^+}^{\backslash K^+}$	1.7	$\bar{C}$:	H^+ form, 1.8 eq./ℓ
$K_{Li^+}^{\backslash Na^+}$	2.0		Na^+ form, 2.0 eq./ℓ

TABLE 10-5 Approximate selectivity coefficients for strong-base anion exchange resins (after Anderson, 1975).

$K_{Cl^-}^{\backslash NO_3^-}$	4	$K_{Cl^-}^{\backslash SO_4^{2-}}$	0.15
$K_{Cl^-}^{\backslash Br^-}$	3	$K_{Cl^-}^{\backslash HSO_4^-}$	1.6
$K_{Cl^-}^{\backslash F^-}$	0.1	$K_{NO_3^-}^{\backslash SO_4^{2-}}$	0.04
$K_{Cl^-}^{\backslash HCO_3^-}$	0.5	$K_{OH^-}^{\backslash Cl^-}$	Type I, 15 to 20
			Type II, 1.5
$K_{Cl^-}^{\backslash CN^-}$	1.5	$\bar{C}$:	1.0 to 1.4 eq./ℓ

Moisture-Retention Capacity (Moisture Content)

When placed in water, ion exchange resins swell, and ions are able to diffuse into and out of their structure. The extent of swelling is a function of the degree of crosslinking. A low degree of crosslinking is associated with high moisture content (high moisture-retention capacity), a high level of resin swelling in an aqueous solution, and a low total exchange capacity. Because of the high level of resin swelling, resins which have a low degree of crosslinking undergo large contractions and expansions during regeneration and service. This must be considered when designing containing systems.

In contrast, resins which have higher degrees of crosslinking have higher total exchange capacities per unit volume, lower moisture contents, and lower levels of resin swelling.

Density

Ion exchange resin density may be expressed as bulk density (weight per unit total volume, i.e., the sum of the particle volume and the void volume) or as specific gravity (weight per unit particle volume). Bulk density values reported by resin manufacturers are usually for a fully hydrated resin after back-washing, settling, and draining. These values are used for shipping purposes, whereas specific gravity values are used for calculating acceptable resin back-wash rates and resin settling rates.

10-4 SYSTEM OPERATION

Ion exchange systems can be operated in one of four modes: batch, fixed-bed, fluidized-bed, and continuous. The fixed-bed system is by far the most common and is the only system considered in this text. A typical fixed-bed operating cycle consists of four steps: service, back-wash, regeneration, and rinse.

Service

The desired exchange reaction occurs during the service step. This step is characterized by an effluent concentration curve or breakthrough curve. The shape of that curve tells a lot about the effectiveness of the exchange reaction.

To illustrate the behavior of ion exchangers during the service cycle consider the case where a strong-acid cation exchange resin in the hydrogen form is used to remove sodium ions from solution. To begin the discussion, assume that the process is to be conducted in a batch reactor where a solution of NaCl is mixed with the ion exchange resin. The exchange reaction which occurs can be represented as

$$R^-H^+ + Na^+ \rightleftharpoons R^-Na^+ + H^+ \tag{10-41}$$

The equilibrium constant expression for this reaction is

$$K_{H^+}^{\backslash Na^+} = \frac{[R^-Na^+]_R [H^+]_s}{[R^-H^+]_R [Na^+]_s} \tag{10-42}$$

Equation 10-42 shows that the position of equilibrium for reaction 10-26 depends on the affinity of the hydrogen and sodium ions for the resin and the amounts of resin and sodium ions present. If the position of equilibrium is to be displaced far to the right, the relative amount of resin must far exceed the amount of sodium ions. However, experience has shown that the sodium ions can be removed from a specified volume of solution with much less resin than required in a single-stage process if only a small amount of resin is initially mixed with the solution and, after the system has equilibrated, the resin is removed and a small amount of fresh resin is added to the reactor. If this process is repeated until the sodium ions are removed, it will be found that the total amount of resin required for single-stage treatment is much greater than with multistage treatment.

In effect, what occurs in an ion exchange column is a kind of multistage treatment where the solution passing down through the column is repeatedly brought into contact with fresh resin. After a short period of operation the upper part of the resin bed becomes exhausted. The applied sodium chloride solution passes unchanged through this part of the bed, but farther into the bed it enters into the exchange zone, where the sodium ions displace the hydrogen ions from the resin. The displaced hydrogen ions pass through the lower part of the bed containing resin in the hydrogen form and exit the column at a concentration equivalent to the sodium concentration in the influent. As the column operation continues the exchange zone moves down the column until it reaches the bottom. At this point, sodium ions show in up the effluent; i.e., breakthrough occurs. The behavior of an ion exchange column is presented diagrammatically in Figure 10-3. The plots to the left and right of center represent Na^+ and H^+ ion concentrations vs. volume of solution passed through the column for the effluent and influent, respectively. The center plots illustrate the movement of the exchange zone through the depth of the column. Ideally, when a solution is passed through an ion

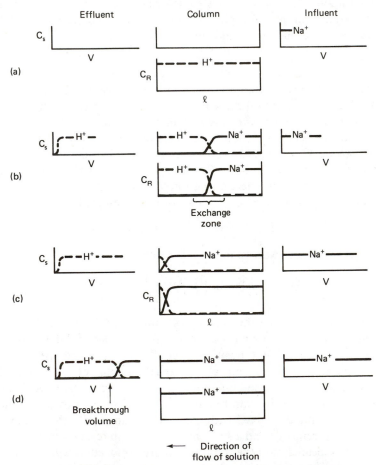

FIGURE 10-3 *Behavior of ion exchange columns. Displacement of H^+ ions by Na^+ ions in a cation exchange column (after Salmon and Hale, 1959).*

exchange column, the resin and solution should be in equilibrium at any point in the column; in practice, however, this is seldom achieved.

The shape of the front of the exchange zone as it moves through the column is very important in column operation as it determines the utilization efficiency of the resin. The closer the shape of the front approaches that of a vertical line, the more efficient the resin utilization. The shape of the front is influenced by flow rate and concentration of the feed solution. However, the controlling factor is the ion exchange isotherm (see Figure 10-4). If the ions initially attached to the resin (designated as A) are less strongly attracted to the resin than the ions in the feed solution (designated as B), the exchange isotherm is represented by Figure 10-4(a), and the shape of the front of the exchange zone will approach that of a vertical line (X_R^B represents the equivalent fraction of the B ion in the resin, and X_S^B represents the equivalent fraction of the B ion in solution). On the other hand, if the ions initially attached to the resin are more strongly attracted to the resin than the ions in the feed solution, the exchange isotherm is represented by Figure 10-4(b) and the shape of the front of the exchange zone is diffuse. In general, it can be said that the difference in total resin capacity and operating capacity increases as the value of the selectivity coefficient for the exchange reaction decreases; i.e., the breakthrough curve broadens as the selectivity coefficient for the exchange reaction decreases.

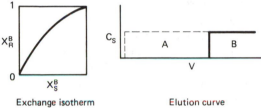

Exchange isotherm   Elution curve

(a) Ion B absorbed more strongly than ion A

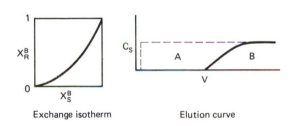

Exchange isotherm   Elution curve

(b) Ion B absorbed less strongly than ion A

FIGURE 10-4 *Replacement of ions in an ion exchange column under ideal conditions (after Salmon and Hale, 1959).*

Back-washing

Back-washing with product water is employed after the operating capacity of the ion exchanger has been reached. This is an upflow process used to prepare the resin for regeneration. Back-washing has several purposes: (1) break up resin clumps, (2) remove finely divided suspended material entrapped in the resin by filtration, (3) eliminate gas pockets, and (4) restratify the resin bed to ensure a uniform distribution of flow during downflow operations.

Regeneration

Regeneration displaces ions exchanged during the service run and returns the resin to its initial exchange capacity or to any other desired level, depending on the amount of regenerant used. In general, mineral acids are used to regenerate cation resins, and alkalies are used to regenerate anion resins. To minimize the regeneration time and the amount of regenerant used, the regenerant should provide a maximum peak elute concentration with minimum "tailing" of the elute. If the system is restored to its initial capacity, the number of equivalents of ions eluted from the resin during regeneration should equal the number of equivalents exchanged during the service cycle.

The *regeneration efficiency* is defined as the ratio of the total equivalents of ions removed from a resin to the total equivalents of ions present in the volume of regenerant used. Generally, the resin can be restored to full capacity by eluting all exchangeable ions. However, in many cases this may require that a large amount of regenerant be used, which can prove to be very costly. As a result, only a portion of the available exchange capacity is normally restored during the regeneration cycle. The extent of regeneration is referred to as the *regeneration level*.

The regeneration efficiency is higher for weak ionic resins than for strong ionic resins because the weak ionic resins' affinity is higher for the H^+ and OH^- ions. This means that regeneration is more favorable for weak ionic resins, with the result that less regenerant is required to achieve the same degree of exchange. This can be explained by considering that the value of the selectivity coefficient for the regeneration reaction is the reciprocal of the selectivity coefficient for the initial exchange reaction, and the value of the selectivity coefficient for the initial exchange reaction when strong ionic resins are employed is usually much larger than the value of the selectivity coefficient when weak ionic resins are employed.

Rinsing

After the regeneration step the ion exchange resin must be rinsed free of excess regenerant before being put back into operation. The rinsing procedure consists of two steps, using product water. The slow rinse of one bed volume displaces regenerant, and the waste from this rinse is combined with the regenerant brine for disposal. The fast rinse washes away excess ions, and the waste from that rinse is often collected and used for regenerant dilution water.

Operating Mode

The fixed-bed system can be operated in the downflow or upflow mode for service and in the downflow or upflow mode for regeneration. The operation is referred to as *countercurrent ion exchange* when service is in one direction and regeneration is in the opposite direction and *co-current ion exchange* when both service and regeneration are in the same direction (see Figure 10-5). According to Abrams (1973) countercurrent

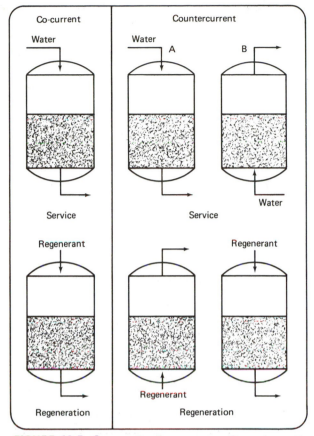

FIGURE 10-5 *Co-current and countercurrent ion exchange. By definition, countercurrent operation implies service in one direction and regeneration in the opposite direction. Service can be downflow as in (A) or upflow as in (B) (after Abrams, 1973).*

ion exchange is preferred for strong-acid cation or strong-base anion exchangers because of the relatively unfavorable equilibrium in the regeneration of these resins; i.e., the resin has a very high affinity for the exchanged ions and requires a considerable excess of regenerant to regenerate the resin bed. However, even with an excess of regenerant the resin at the effluent end of the column will not be completely regenerated. This will result in significant column leakage at the beginning of the next service run. Such a response is shown in Figure 10-6.

Abrams (1973) lists the following advantages of countercurrent ion exchange over co-current ion exchange for strong ionic resins:

1. Lower leakage of ions during service.

2. Lower consumption of regenerants.

3. Decrease in quantity of regenerant wastes.

4. Lower consumption of water for rinse and back-wash.

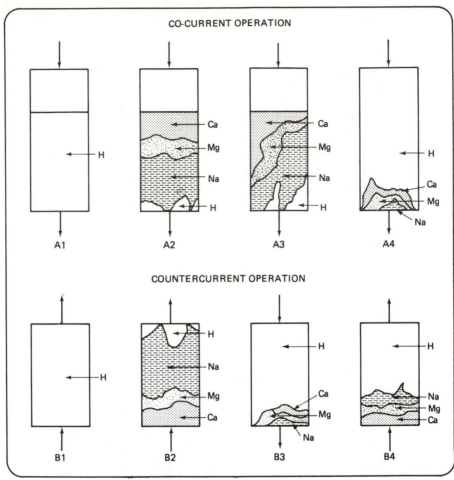

FIGURE 10-6 *Distribution of ions in resin bed. In (A1) the resin is completely converted to the hydrogen form. Column (A2) depicts the approximate distribution of calcium, magnesium, and sodium ions at the end of a downflow service run. Following exhaustion, the bed is backwashed during which the resin particles expand and mix freely. As a result, the distribution of ions is scattered even more randomly than indicated in (A3). During downflow regneration, the various cations are displaced downward by the excess of hydrogen ions leaving a residual of calcium, magnesium, and sodium at the bottom of the column (A4). The quantity of residual cations depends mainly on the acid dosage used during regeneration. In the following service run a few hydrogen ions are released in the upper portion of the bed and are exchanged for the residual ions at the bottom of the bed, thus causing some "leakage". This leakage is usually sodium, since this ion is most readily exchanged for hydrogen ions. If the free mineral acidity generated by ion exchange is high enough, magnesium and even calcium may also "leak" from the column. In (B1) the column is shown completely regenerated and ready for upflow service and downflow regeneration, i.e., countercurrent operation. In service the ion distribution (B2) is much the same as in (A2), but reversed. After downflow regeneration (B3), the distribution is similar to that in (A4). However, at the start of upflow service there is virtually no leakage of cations because all of the exchange occurs at the inlet rather than at the outlet end of the column (B4) (after Abrams, 1973).*

It should be understood that countercurrent operation is less advantageous with weak ionic resins.

10-5 THEORY OF ION EXCHANGE

Diffusion plays an important role in the ion exchange process. To reach an exchange site it is generally assumed that an ion from solution must first diffuse through a liquid film (Nernst film) surrounding the resin particle (film diffusion) and then through the resin particle to the exchange site (particle diffusion). The exchanged ion must follow the same process but in the reverse direction. The overall rate of the exchange process is controlled by whichever is the slowest process: the rate of diffusion through the liquid film or the rate of diffusion through the resin particle. For the general exchange reaction

$$RA + B \rightleftharpoons RB + A \qquad (10\text{-}43)$$

occurring in a *batch reactor*, Salmon and Hale (1959) suggest that an evaluation of the following relationship will indicate which diffusion process controls the rate of the exchange reaction:

$$\frac{\pi^2 D_r \delta [RB]_R}{3 D_o r [B]_S} \qquad (10\text{-}44)$$

where $[RB]_R$ = the final equilibrium concentration of the exchanging ion in the resin (meq./mℓ)

         $[B]_S$ = the final equilibrium concentration of the exchanging ion in the solution (meq./mℓ)

         D_r = the diffusion constant of the exchanging ion in the resin (cm^2/sec)

         D_o = the diffusion constant of the exchanging ion in the solution (cm^2/sec)

         δ = the thickness of the Nernst film (cm)

         r = the radius of the resin particles (cm).

If the value of equation 10-44 is less than 1, particle diffusion is the controlling process, whereas film diffusion is the controlling process when the value of equation 10-44 is greater than 1. If the value of equation 10-44 is near 1, then both processes exert about the same effect on the rate of the exchange process.

The rate of film diffusion can be increased by:

1. Increasing agitation to reduce the thickness of the Nernst film.

2. Increasing the concentration of the exchanging ions in the solution.

3. Increasing the total surface area of the resin by decreasing the size of the resin particle to be used in the exchanger.

4. Increasing the temperature, which increases the mobility of the ions.

Likewise, film diffusion will generally be controlling when:

1. The exchange capacity of the resin is high; i.e., the ionic concentration in the resin is high.
2. The resin has a low degree of crosslinking, resulting in a large D_r value.
3. The concentration of the exchanging ions in the solution is low.
4. The system receives little agitation.

In practical applications the overall rate of the exchange process is very important, and slow rates of exchange often impose a serious limitation on the usefulness of the process. The exchange quotient is a good way to evaluate the kinetics of the exchange process. The exchange quotient is defined as

$$\frac{(B)_t}{(B)_\infty} = \frac{\text{amount of ions exchanged in the resin until time } t}{\text{amount of exchangeable ions until the equilibrium state}} \qquad \textbf{(10-45)}$$

Figure 10-7 illustrates how the exchange quotient changes with time for exchange of hydrogen with sodium ions for different solution concentrations and different size resin particles. Figure 10-7(a) represents a film diffusion limiting case, and Figure 10-7(b) represents a particle diffusion limiting case.

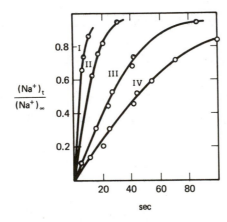
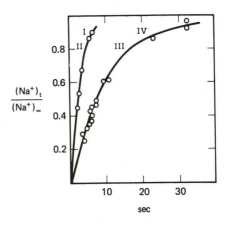

(a) Exchange of the hydrogen ions of a strongly acidic ion-exchange resin for sodium ions as a function of time at low solution concentrations (film diffusion):

I ~ particle size 0.05–0.1 mm;
    [Na⁺] = 0.048–0.050 M,
II ~ particle size 0.05–0.1 mm;
    [Na⁺] = 0.026–0.029 M,
III ~ particle size 0.3–0.4 mm;
    [Na⁺] = 0.045–0.05 M,
IV ~ particle size 0.3–0.4 mm;
    [Na⁺] = 0.023–0.028 M

(b) Exchange of the hydrogen ions of a strongly acidic ion exchange resin for solution ions as a function of time in the case of higher solution concentration (particle diffusion)

I ~ particle size 0.05–0.1 mm;
    [Na⁺] = 2.18 M,
II ~ particle size 0.05–0.1 mm;
    [Na⁺] = 1.09 M,
III ~ particle size 0.3–0.4 mm;
    [Na⁺] = 2.18 M,
IV ~ particle size 0.3–0.4 mm;
    [Na⁺] = 1.09 M

FIGURE 10-7 *Variation in exchange quotient with time (after Inczédy, et al., 1966).*

To treat the rate of the ion exchange process in a quantitative manner it is necessary that values of the diffusion constants D_r and D_o be known. Values for D_o are normally available in most chemical handbooks; however, values for D_r are not as common and must generally be evaluated in the laboratory. Values of D_r for some simple inorganic ions are presented in Table 10-6. Inczédy et al. (1966) have presented the following

TABLE 10-6 *Diffusion constants of some ions in ion exchange resins at 25°C. (after Inczédy et al., 1966).*

Ion	Ion-Exchange Resin	D_r. $(cm^2 \cdot sec^{-1})$
H$^+$	Dowex 50X4	9.14×10^{-6}
	8	5.40×10^{-6}
	16	2.20×10^{-6}
Na$^+$	Dowex 50X4	14.1×10^{-7}
	8	2.88×10^{-7}
	16	1.10×10^{-7}
	24	1.00×10^{-7}
K$^+$	Dowex 50X8.6	13.4×10^{-7}
Rb$^+$	Dowex 50X4	2.28×10^{-7}
	8.6	13.8×10^{-7}
Ag$^+$	Dowex 50X8.6	6.42×10^{-7}
	16	2.75×10^{-7}
	24	1.13×10^{-7}
Sr^{2+}	Dowex 50X8	3.38×10^{-8}
	16	2.98×10^{-9}
Zn^{2+}	Dowex 50X8	2.89×10^{-8}
	24	2.63×10^{-9}
Y^{3+}	Dowex 50X8	3.18×10^{-9}
	24	2.18×10^{-10}
Th^{3+}	Dowex 50X8	2.15×10^{-10}
HR + Na$^+ \longrightarrow$	Amberlite IRC 50	3.92×10^{-9}
BrO$_3^-$	Dowex 2X6	4.55×10^{-7}
Br$^-$	Dowex 2X6	3.87×10^{-7}
Cl$^-$	Dowex 2X6	3.54×10^{-7}
I$^-$	Dowex 2X6	1.33×10^{-7}
PO$_4^{3-}$	Dowex 2X6	0.57×10^{-7}

rule of thumb for estimating D_r values from D_o values obtained in aqueous solution:

1. D_r for monovalent cations: 1/3 to 1/10 of D_o.

2. D_r for bivalent cations: 1/5 to 1/100 of D_o.

3. D_r for trivalent cations: 1/10 to 1/1000 of D_o.

4. For common, medium, crosslinked ion exchange resins at room temperature, the average values of the diffusion constants for simple inorganic ions are $D_o = 10^{-5}$ cm^2/sec, $D_r = 3 \times 10^{-7}$ cm^2/sec.

Ion Exchange Column Operation

When a small volume of sodium chloride solution is passed through an ion exchange column containing a strongly acidic resin in the hydrogen form, the sodium ions will entirely replace the hydrogen ions in the upper part of the column. Below that part of the column where the sodium ions have completely replaced the hydrogen ions will exist a small part of the column in which the sodium ions have only partially replaced the hydrogen ions. Below that part of the column none of the hydrogen ions will have been replaced by sodium ions. This type of column response can be illustrated graphically by plotting the ratio of the sodium ion concentration in the resin to the total ion concentration in the resin along the X axis and the distance from the top of the column along the Y axis (see Figure 10-8). The distribution of the sodium ion

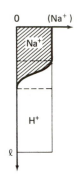

FIGURE 10-8 *Concentration of sodium and hydrogen ions as a function of length of the column at a given time (after Inczédy, et al., 1966).*

concentration in that part of the column containing both sodium and hydrogen ions can be represented by an S-shaped curve. As more sodium chloride solution is passed through the column the S-shaped curve moves down the column until it reaches the bottom. At that point, sodium ions begin to appear in the effluent and breakthrough is said to have occurred.

Two of the most common graphical representations of the changes which occur during the operation of an ion exchange column are as follows:

1. *Isochrone:* A plot of the ratio of the exchanging ion concentration in the resin to the total ion concentration in the resin at a given time against the length of the column.

2. *Isoplane:* A plot of the ratio of the exchanging ion concentration in the effluent to the total ion concentration in the effluent against the volume of solution passed through the column.

For the general exchange reaction

$$RA + B \rightleftharpoons RB + A \qquad \textbf{(10-43)}$$

it was shown earlier that the equilibrium constant expression for this reaction is

$$K_A^{\backslash B} = \frac{[RB]_R[A]_S}{[RA]_R[B]_S} \qquad \textbf{(10-46)}$$

If the value of the selectivity coefficient is greater than 1 and if the flow rate through the column is infinitely slow so that a state of equilibrium exists at every point in the column, the shape of the boundary separating the region of the resin containing the exchanging ions from the region containing the ions to be exchanged will approach that of a vertical line (see curve 1 in Figure 10-9). This boundary will move down the

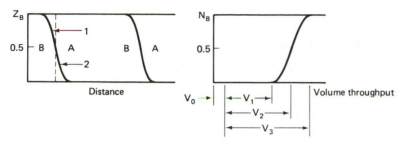

FIGURE 10-9 *Isochrone and isoplane for the exchange process when $K'^{B^+}_{A^+} > 1$ (after Inczédy, et al., 1966).*

Z_B = *the ratio of the exchanging ion concentration in the resin to the total ion concentration in the resin;*

N_B = *the ratio of the exchanging ion concentration in the effluent to the total ion concentration in the effluent.*

column in the direction of flow as more solution is passed through the exchanger. In practice, however, the rate of flow is significant and equilibrium is never totally attained in the column. Under these conditions the boundary separating the region of the resin containing the exchanging ions from the region containing the ions to be exchanged will be S-shaped (see curve 2 in Figure 10-9). This boundary will also move down the column in the direction of flow as more solution is passed through the exchanger. Both the isochrone and isoplane for equation 10-43, where $K'^B_A > 1$, are shown in Figure 10-9.

For the case where $K'^B_A < 1$, the S-shape of the boundary between the region of the resin containing the exchanging ions and the region containing the ions to be exchanged will flatten out as the front moves down the ion exchange column (see the isochrone representation shown in Figure 10-10). This occurs because the slope of any point along the front boundary is dependent on the concentration gradient

$$\frac{d[B]_S}{d[RB]_R} = \frac{d[A]_S}{d[RA]_R} \qquad (10\text{-}47)$$

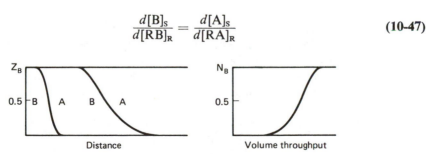

FIGURE 10-10 *Isochrone and isoplane for the exchange process when $K'^{B^+}_{A^+} < 1$ (after Inczédy, et al., 1966).*

This means that by the time the front reaches the end of the column (i.e., when break-through occurs) the front will be considerably flattened out, and also that the break-through curve will have a flattened A-form as shown by the isoplane representation of Figure 10-10.

Summarizing this discussion it can be stated that the shape of the exchange front at breakthrough and the breakthrough capacity of the resin bed depend on the following:

1. *Primary Consideration:*

   (a) The selectivity coefficient for the exchange reaction: If $K_A^{\backslash B} > 1$ the exchange front will be sharp. However, if $K_A^{\backslash B} < 1$ the front will flatten out as it moves toward the end of the column. In terms of breakthrough capacity, this value will be smaller when $K_A^{\backslash B} < 1$.

2. *Secondary Considerations:*

   (a) *Flow rate:* The front will be flatter and the breakthrough capacity less at higher flow rates.

   (b) *Degree of resin crosslinking:* The front will be flatter and the break-through capacity less the higher the degree of crosslinking.

   (c) *Size of resin particle:* The front will be flatter and the breakthrough capacity less the larger the size of the resin particle.

   (d) *Temperature:* The front will be flatter and the breakthrough capacity less at lower temperatures.

To characterize the operation of an ion exchange column Mayer and Tompkins (1947) define a parameter called a *theoretical plate*. In their development the exchange column was considered to consist of a number of theoretical plates within each of which equilibrium was attained between resin and solution. To relate this to column operation the height of a column slice *equivalent* of a theoretical plate is denoted by h. The efficiency of a column can then be expressed in terms of either the height of the theoretical plate or of the number of plates in the column, N. The more plates that a column contains (indicating smaller values of h), the better the column operates. The total bed depth, L, is then given by

$$N \times h = L \tag{10-48}$$

Inczédy et al. (1966) have presented the following mathematical relationships to describe the operation of an ion exchange column:

1. The throughput volume producing breakthrough of the exchanging ions in concentration C_B, for a column of height L, when film diffusion is rate controlling:

$$V = \left[\frac{X_T}{C} + \alpha \right] AL \left\{ 1 - \frac{h_f}{2L} \left[\left(\frac{K_A^{\backslash B}}{K_A^{\backslash B} - 1} \ln \frac{C}{C_B} \right) - \left(\frac{1}{K_A^{\backslash B} - 1} \ln \frac{C}{C - C_B} - 1 \right) \right] \right\} \tag{10-49}$$

where V = throughput volume (ml)

X_T = total exchange capacity of the resin (meq./ml of resin column)

C = total concentration of exchanging ions in influent (meq./ml)

C_B = concentration of exchanging ions in the effluent at breakthrough (meq./ml)

α = void fraction (free solution volume in the resin-filled column divided by the volume of the column without resin)

A = cross-sectional area of column (cm²)

L = depth of resin bed (cm)

$K_A^{\prime B}$ = selectivity coefficient for exchange reaction.

2. The throughput volume producing breakthrough of the exchanging ions in concentration C_B, for a column height L, when particle diffusion is the rate-controlling step:

$$V = \left[\frac{X_T}{C} + \alpha\right] AL \left\{1 - \frac{h_p + h_r}{2L}\left[\left(\frac{1}{K_A^{\prime B} - 1} \ln \frac{C}{C_B}\right) - \left[\left(\frac{K_A^{\prime B}}{K_A^{\prime B} - 1} \ln \frac{C}{C - C_B} - 1\right)\right]\right]\right\} \tag{10-50}$$

The value of h comprises three terms; i.e., h is a function of three different factors:

$$h = h_r + h_p + h_f \tag{10-51}$$

$$h_r = \text{the effect resin particle size has on } h \tag{10-52}$$

$$= 1.64r$$

where r represents the radius of the resin particle in centimeters.

$$h_p = \text{the effect particle diffusion has on } h$$

$$= \left[\frac{X_T/C}{(X_T/C) + \alpha}\right]^2 \frac{0.142 r^2 Q}{D_r(X_T/C)} \tag{10-53}$$

where D_r = diffusion constant of exchanging ion in resin (cm²/sec)

Q = linear flow rate through column (cm/sec).

$$h_f = \text{the effect film diffusion has on } h$$

$$= \left[\frac{X_T/C}{(X_T/C) + \alpha}\right]^2 \frac{0.266 r^2 Q}{D_o(1 + 70rQ)} \tag{10-54}$$

where D_o = diffusion constant of exchanging ion in solution (cm²/sec)

Important considerations in the design of an ion exchange column are the breakthrough capacity and the extent of column utilization. For a symmetrically shaped exchange zone, one-half the thickness of the zone at breakthrough is designated as ΔL, which is the length of the unutilized part of the column. The utilization fraction is, therefore, given by

$$\text{utilization fraction} = 1 - \frac{\Delta L}{L} \tag{10-55}$$

The breakthrough capacity of the column can be estimated from the relationship

$$\text{breakthrough capacity} = (L - \Delta L)AX_T \tag{10-56}$$

The value of ΔL, which is a function of the thickness of a theoretical plate is given by

$$\Delta L = \frac{h_f}{2}\left[\frac{K_A^{\text{'B}}}{K_A^{\text{'B}} - 1}\ln\frac{C}{C_B} - 1\right] + \frac{h_p + h_r}{2}\left[\frac{1}{K_A^{\text{'B}} - 1}\ln\frac{C}{C_B} + 1\right] \tag{10-57}$$

EXAMPLE PROBLEM 10-5: Determine the bed depth required to soften a water if it desired to achieve 95% utilization of the column. Assume that the following design criteria are applicable:

(a) Initial calcium concentration is 4×10^{-3} eq./ℓ.
(b) Allowable calcium concentration in the effluent, i.e., concentration at breakthrough is 4×10^{-4} eq./ℓ.
(c) Resin particles have a radius of 0.03 cm.
(d) Theoretical exchange capacity of resin is 1 eq./ℓ.
(e) Diffusion constant for calcium ion in resin particle is 3×10^{-7} cm²/sec.
(f) Diffusion constant for calcium ion in solution is 10^{-5} cm²/sec.
(g) Selectivity coefficient for $K_{\text{Na}}^{\text{'Ca}}$ is 5.
(h) $\alpha = 0.35$.
(i) Linear flow rate is 3 cm/sec.

Solution:

1. Compute the contribution of the resin particle to the theoretical plate height:

$$h_r = 1.64r$$
$$1.64(0.03)$$
$$= 0.049 \text{ cm}$$

2. Estimate the effect of particle diffusion on the theoretical plate height:

$$h_p = \left[\frac{X_T/C}{X_T/C + \alpha}\right]^2 \frac{0.142r^2Q}{D_r(X_T/C)}$$
$$= \left[\frac{\frac{1}{4} \times 10^3}{\frac{1}{4} \times 10^3 + 0.35}\right]^2 \frac{(0.142)(0.03)^2(3)}{(3 \times 10^{-7})(\frac{1}{4} \times 10^3)}$$
$$= 0.88 \text{ cm}$$

3. Estimate the effect of film diffusion on the theoretical plate height:

$$h_f = \left[\frac{X_T/C}{(X_T/C) + \alpha}\right]^2 \frac{0.266r^2Q}{D_o(1 + 70rQ)}$$
$$= \left[\frac{\frac{1}{4} \times 10^3}{\frac{1}{4} \times 10^3 + 0.35}\right]^2 \frac{(0.266)(0.03)^2(3)}{10^{-5}[1 + 70(0.03)(3)]}$$
$$= 1.71 \text{ cm}$$

4. Calculate the length of the unused portion of the bed:

$$\Delta L = \frac{h_f}{2}\left[\frac{K_{Na}^{\backslash Ca}}{K_{Na}^{\backslash Ca}-1}\ln\frac{C}{C_B}-1\right]+\frac{h_p+h_r}{2}\left[\frac{1}{K_{Na}^{\backslash Ca}-1}\ln\frac{C}{C_B}-1\right]$$

$$=\frac{1.71}{2}\left[\frac{5}{5-1}\ln\frac{4\times10^{-3}}{4\times10^{-4}}-1\right]+\frac{0.88+.049}{2}\left[\frac{1}{5-1}\ln\frac{4\times10^{-3}}{4\times10^{-4}}-1\right]$$

$$\Delta L = 2.6$$

5. Compute the total bed depth on the basis of 95% utilization of the bed:

$$\text{total bed depth} = \text{utilized portion} + \text{unused portion}$$

$$L = 0.95L + \Delta L = 0.95L + 2.6$$

$$L(1-0.95) = 2.6$$

$$L = 52 \text{ cm}$$

Calculating a required bed depth following the procedure outlined in example problem 10-5 requires the assumption that the column is restored to its theoretical exchange capacity during regenerating; in practice, this seldom occurs. Furthermore, the *degree of column utilization*, as used by most engineers, refers to the ratio of the operating exchange capacity to the theoretical exchange capacity, where the operating exchange capacity is established by the level to which the column is regenerated. In ion exchange terminology *efficiency* generally refers to *regeneration efficiency* which, as noted earlier, indicates the ratio of the operating exchange capacity to the exchange capacity which should theoretically be attained from the total volume of regenerant used, i.e., to the total equivalents of ions present in the volume of regenerant used.

For a column containing Amberlite IR-120 resin and processing water at a rate 2 gpm/ft³, which contains 500 mg/ℓ of hardness as $CaCO_3$, the relationships between column utilization, efficiency, and regeneration level are shown in Table 10-7. A 10% sodium chloride solution was used for regeneration in all cases.

TABLE 10-7 *Regeneration efficiency and column utilization as a function of regeneration level. (after Weber, 1972).*

Regeneration Level (lb NaCl/ft³ resin)	Hardness Removed (lb CaCO₃/ft³ resin)		Regeneration Efficiency (%)	Column Utilization† (%)
	Theoretical*	Actual		
1	0.85	0.83	98	14
2	1.70	1.32	78	22
3	2.55	1.83	72	30
5	4.25	2.85	67	47
10	8.50	3.90	46	64
15	12.75	4.65	36	76
20	17.00	5.00	29	82

* Based on quantity of NaCl used.
† Based on total exchange capacity of resin of 6.1 lb/ft³.

10-6 APPLICATIONS OF THE ION EXCHANGE PROCESS

Water Softening

According to Weber (1972) water softening by ion exchange is usually preferable to water softening by chemical precipitation when:

1. Raw water contains low color and turbidity levels (does not require pretreatment).
2. Hardness is largely not associated with alkalinity; i.e., there are substantial amounts of noncarbonate hardness (soda ash is more expensive than lime).
3. There are variable hardness levels in raw water (requires constant adjustment of the chemical feed).

On the other hand, water softening by chemical precipitation is usually preferable when:

1. Raw water requires clarification; i.e., the water contains high color and turbidity levels.
2. Hardness is largely associated with alkalinity; i.e., there is little noncarbonate hardness.

The most widely used ion exchange system for water softening is a continuous-flow, fixed-bed column employing a strong cationic resin in the sodium form. The sodium cation exchange softening process exchanges sodiums ions for all cations of two or more positive charges (polyvalent cations). Thus, when water containing hardness-causing ions passes through the exchanger, the sodium ions of the bed replace the hardness ions, with a resultant effluent water of close to zero hardness. The primary exchange reactions are

$$2R^-Na^+ + Ca^{2+} \rightleftharpoons R_2^-Ca^{2+} + 2Na^+ \tag{10-58}$$

$$2R^-Na^+ + Mg^{2+} \rightleftharpoons R_2^-Mg^{2+} + 2Na^+ \tag{10-59}$$

When the operating capacity of the bed has been reached the service cycle is terminated. The unit is then back-washed to clean the bed of filtered particulate matter and then regenerated. Since equations 10-58 and 10-59 represent equilibrium reactions, regeneration can be accomplished by a strong sodium chloride (salt) solution. The high sodium concentration reverses the exchange reactions, resulting in regeneration of the resin.

As noted earlier, countercurrent operation offers significant advantages over co-current operations for water softening. The principal advantage, however, is lower leakage for a given dosage of salt regenerant. Since countercurrent operation means that either service or regeneration is in the upflow mode, special precautions must be taken to ensure that the resin bed remains tightly packed during this mode of operation.

It has been found that the slightest fluidization will reduce either the service time or regeneration efficiency. One type of countercurrent operational scheme which has been used successfully is termed *countercurrent flow with air blockage* (see Figure 10-11). This type of operation applies 4 to 5 psi of air pressure above the resin bed during regeneration to prevent any upward movement of the particles. A collector is located just below the top of the resin bed to provide an exit for both air and regenerant. In certain cases, low density, inert, granular particles are placed at the top of the bed. Because of the low density of these particles they float at the top of the column during service and back-wash but are forced down by the air flow during regeneration to form a buffer zone around the collector. A number of different kinds of countercurrent processes are discussed by Abrams (1973).

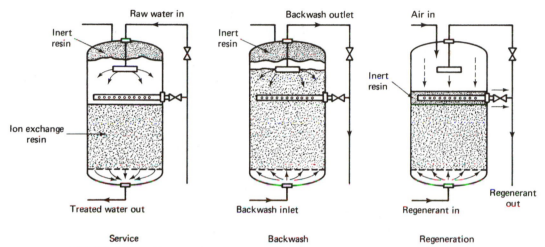

FIGURE 10-11 *Countercurrent flow with air blockage (after Abrams, 1973).*

Typical design criteria for an ion exchange, water softening system are as follows:

1. *Service Flow Rate:* Service flow is usually in the range of 5 to 8 gpm/ft² of bed cross-sectional area. The greater the flow rate, the lower the operating capacity of the bed.

2. *Pressure Drop:* The pressure drop (head loss) per unit depth of bed depends on the size of the resin and the service flow rate.

3. *Backwash Rate:* Backwash is normally applied at a rate to provide between 50 and 75% expansion of the resin bed. The rate used is dependent on the density of the resin and the temperature of the back-wash water.

4. *Regeneration:* For strong-acid and strong-base resins the regenerant solution concentration is normally between 2 to 10%, while a 1 to 5% solution is nor-

mally used for weak-base resins. The minimum contact time employed for regeneration is 30 min. This usually means a regeneration flow rate of 1 to 2 gpm/ft² of bed cross-sectional area. Normally the quantity of regenerant used will not completely regenerate the resin. Typical quantities of regenerant applied in regenerating certain Dowex resins are presented in Table 10-8. If the resin is to be completely regenerated much larger quantities of regenerant than the values shown in Table 10-8 will be required. The higher the regeneration level required, the lower the regeneration efficiency.

5. *Rinsing Requirement:* Rinsing the resin bed to remove excess regenerant normally requires 50 to 100 gal of water per ft³ of resin volume.

6. *Bed Depth:* A minimum bed depth of 30 in. is normally recommended; however, much greater depths are often used to obtain the desired operating capacity.

7. *Freeboard:* A freeboard length of 50 to 75% of the bed depth is normally provided.

TABLE 10-8 *Quantities of regenerant commonly used for certain Dowex resins (courtesy of Dow Chemical Company).*

	Desired Ionic Form	*Kind*	*Quantity*		*Concentration*
			lb/ft³	*mg/ml*	
Dowex 50W	H⁺	HCl	4–10	60–160	2–10%
Dowex MSC-1	H⁺	H₂SO₄	6–12	100–200	2–10%
	Na⁺	NaCl	5–10	80–160	6–25%
Dowex 1					
Dowex MSA-1	OH⁻	NaOH	4–8	60–120	2–10%
Dowex 21K					
Dowex 2	OH⁻	NaOH	3–6	50–100	2–10%
Dowex WGR	Free base	NaOH	2–4	30–60	2–4%
Dowex MWA-1	Free base	NH₃	1–2	15–30	1–2%

Regenerant concentration:
  Monovalent to polyvalent: low concentration
  Polyvalent to monovalent: high concentration

The theory of ion exchange is mathematically very complex, and as a result a precise description of the ion exchange process is of little value to the practicing engineer. It is fortunate, therefore, that resin manufacturers furnish design charts and tables for many different situations. Where no information is available, design data can be generated by setting up a pilot-scale ion exchange column and determining the resin performance empirically.

To design a simple ion exchange system such as that used at a municipal water treatment plant for water softening, it is convenient to use the graphs and tables supplied by resin manufacturers. One procedure which may be followed when designing a water softening system is outlined in example problem 10-6.

EXAMPLE PROBLEM 10-6: Design a fixed-bed ion exchange column to soften 20 MGD of water at a temperature of $10°C$. The raw water has a total hardness of 400 mg/ℓ $CaCO_3$, and it must be softened to 100 mg/ℓ $CaCO_3$. The maximum column diameter to be used is 10 ft.

Solution:

1. Determine the pretreatment requirements necessary to prevent resin blinding from suspended solids or resin oxidation because of the presence of oxidants in the influent. Typical pretreatment requirements are listed in Table 10-9.

TABLE 10-9 *Pretreatment requirements for ion exchange (after Sanks 1972).*

Contaminant	Effect	Removal
Suspended solids	Blinds resin particles	Coagulation and filtration
Organics	Large molecules (e.g., humic acids) foul strong base resins	Carbon sorption or use of weak-base resins only
Oxidants	Slowly oxidizes resins. Functional groups become labile	Avoid prechlorination or neutralize chlorine
Iron and manganese	Coats resin particles	Aeration

2. Select the resin to be used. For water softening a strong-acid cation exchange resin is recommended by resin manufacturers. In this problem the Duolite C-20 resin manufactured by Diamond Shamrock is selected. This resin removes the hardness ions, calcium and magnesium, by exchanging them for sodium ions. Sodium chloride is used to regenerate the exhausted resin.

3. Compile the manufacturer's data on the resin selected. For the Duolite C-20, Diamond Shamrock provides the following information:

   (a) *Softening capacity:*

Salt (lb/ft^3):	5	7.5	10	15	20
Resin capacity: (eq./ℓ)	0.82–1.05	1.1–1.4	1.35–1.55	1.5–1.8	1.7–1.9

   (b) *Leakage:* A maximum of 1% of the untreated water hardness

   (c) *Bed depth:* 30–36 in.

   (d) *Back-wash bed expansion:* 50% or more. The required back-wash flow rate varies with temperature (see Figure 10-12).

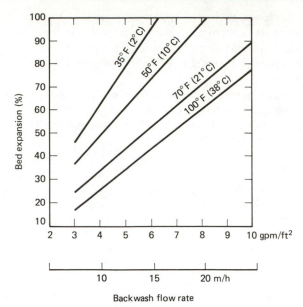

FIGURE 10-12 *Bed expansion vs backwash flowrate for duolite C-20 (Courtesy, Diamond Shamrock).*

(e) *Operating conditions:*

Operation	Rate	Solution	Time (min)
Service	2–5 gpm/ft^3	Water	—
Back-wash	6–8 gpm/ft^2 @50–75°F	Water	5–15
Regeneration	0.2–1 gpm/ft^3	10–20% NaCl	30–60
Rinse	1–5 gpm/ft^3	Water	10–40

4. Convert hardness concentration to meq./ℓ:

   (a) Influent hardness $= \dfrac{400 \text{ mg}/\ell}{50 \text{ mg/meq.}}$

   $= 8$ meq./ℓ

   (b) Effluent hardness $= \dfrac{100 \text{ mg}/\ell}{50 \text{ mg/meq.}}$

   $= 2$ meq./ℓ

5. Compute the hardness leakage in meq./ℓ, assuming leakage is 1%:

   hardness leakage $= (8 \text{ meq.}/\ell)(0.01)$

   $= 0.08$ meq./ℓ

6. Calculate the bypass fraction, using a hardness balance equation:

$$\begin{bmatrix} \text{hardness in} \\ \text{bypass stream} \end{bmatrix} + \begin{bmatrix} \text{hardness in} \\ \text{treated stream} \end{bmatrix} = \begin{bmatrix} \text{allowable hardness} \\ \text{in effluent} \end{bmatrix}$$

$$8f(Q) + 0.08(1 - f)Q = 2Q$$

or

$$8f + 0.08(1 - f) = 2$$

where $f =$ bypass fraction.

Therefore,

$$f = 0.242$$

7. Determine the quantity of the flow bypass:

$$\text{bypass flow} = (0.242)(20 \text{ MGD}) = 4.84 \text{ MGD}$$

8. Compute the quantity of water treated per day:

$$\text{water treated per day} = 20 - 4.84 = 15.16 \text{ MG}$$

9. Compute the total hardness to be removed per day:

total hardness
removed (meq./day)

$$= \left(8.0 \frac{\text{meq.}}{\ell} - 0.08 \frac{\text{meq.}}{\ell}\right)\left(15{,}160{,}000 \frac{\text{gal}}{\text{day}}\right)\left(3.78 \frac{\ell}{\text{gal}}\right)$$

$$= 453{,}854{,}016 \text{ meq./day}$$

10. Select the amount of regenerant to use and the associated resin capacity. From step 3(a) the following values are obtained:
    Salt: 10 lb/ft³
    Resin capacity: 1.4 eq./ℓ
    To account for nonideal operating conditions, the resin capacity will be reduced by 20%, and a value of 1.12 eq./ℓ will be used for design.

11. Design the exchange column, assuming a bed depth of 36 in. and a service rate of 3 gpm/ft³:

(a) Cross-sectional area per column

$$= \frac{\pi d^2}{4}$$

$$= \frac{\pi(10)^2}{4}$$

$$= 78.5 \text{ ft}^2$$

(b) Volume per column

$$= (\text{area})(\text{depth})$$

$$= (78.5 \text{ ft}^2)(\tfrac{36}{12})$$

$$= 235.5 \text{ ft}^3$$

(c) Total volume required

$$= \frac{\text{total flow}}{\text{loading rate}}$$

$$= \frac{15{,}160{,}000 \text{ gal/day}}{3 \frac{\text{gal}}{\text{min-ft}^3} \times 1440 \frac{\text{min}}{\text{day}}}$$

$$= 3509 \text{ ft}^3$$

(d) Number of
    columns required $= \dfrac{\text{total volume}}{\text{volume per column}}$

$$= \frac{3509 \text{ ft}^3}{235.5 \text{ ft}^3/\text{column}}$$

$$= 15 \text{ columns}$$

12. Compute the exchange capacity of each column:

$$\text{column capacity} = \left(1.12 \frac{\text{eq.}}{\ell}\right)(235.5 \text{ ft}^3)\left[28.3 \frac{\ell}{\text{ft}^3}\right]$$

$$= 7464 \text{ eq.}$$

$$= 7,464,000 \text{ meq.}$$

13. Compute the service time of column operation:

$$\frac{\text{service}}{\text{time}} = \frac{\text{column capacity}}{(\text{total hardness to be removed per day})/(\text{number of columns})}$$

$$= \frac{7,464,000 \text{ meq.}}{(453,854,016 \text{ meq./day})/(15)}$$

$$= 0.246 \text{ day}$$

$$= 6 \text{ hr}$$

14. Calculate the total weight of salt required per column:

$$\begin{array}{c}\text{total weight of salt} \\ \text{required for regeneration}\end{array} = \left[10 \frac{\text{lb}}{\text{ft}^3}\right](235.5 \text{ ft}^3)$$

$$= 2355 \text{ lb}$$

15. Determine the total weight of solution, assuming that a 10% salt solution is used:

$$\text{solution weight} = \frac{2355 \text{ lb}}{0.1}$$

$$= 23,550 \text{ lb}$$

16. Estimate the gallons of salt solution required for regeneration if the specific gravity of the salt solution is taken to be 1.07 (see Appendix IX):

$$\begin{array}{c}\text{gallons of salt} \\ \text{solution}\end{array} = \frac{23,550 \text{ lb}}{(1.07)(8.34 \text{ lb/gal})}$$

$$= 2642 \text{ gal}$$

17. Compute the regeneration time required, using a regeneration rate of 0.5 gpm/ft^3:

$$\begin{array}{c}\text{regeneration} \\ \text{time}\end{array} = \frac{2642 \text{ gal}}{\left[0.5 \frac{\text{gal}}{\text{min-ft}^3}\right](235.5 \text{ ft}^3)\left[60 \frac{\text{min}}{\text{hr}}\right]}$$

$$= 0.37 \text{ hr}$$

Use a minimum regeneration time of 30 min.

18. Assume that a 60% bed expansion is required for back-washing. Figure 10-12 gives the required back-wash rate as 5 gpm/ft^2.

19. Compute the total cycle time, assuming a 15 min back-wash and a 40 min rinse.

$$\text{cycle time} = 6.0 + 0.50 + \tfrac{15}{60} + \tfrac{40}{60}$$

$$= 7.4 \text{ hr}$$

Chromium Removal by Ion Exchange

Ion exchange can be used to remove either trivalent or hexavalent chromium from wastewater. Cation exchange resins can be used to remove trivalent chromium (Cr^{3+}), but anion resins must be used to remove hexavalent chromium, which will exist in solution as $HCrO_4^-$, CrO_4^{2-}, and $Cr_2O_7^{2-}$. Choice of resins is determined by treatment objectives.

Anion exchange resins are preferred if chromium is to be recovered for reuse. Such resins have a preference for chromate over dichromate and thus perform better in the pH range of 4.5 to 5.0, where the chromate ion is the dominant form. These resins are normally regenerated with NaOH, and the captured chromium is eluted as sodium chromate. Cation exchange resins can then be used to remove sodium from the eluate, leaving a chromic acid solution suitable for reuse. If reuse is not practiced, the eluted sodium chromate solution must be treated by chromium reduction and lime precipitation. The economics of ion exchange treatment are more favorable when the process is employed for recovery and reuse.

One method used for hexavalent chromium recovery by ion exchange is illustrated in Figure 10-13. In this operation the wastewater is initially passed through a column containing a strong-cationic resin in the hydrogen form. The hydrogen cation exchange process exchanges hydrogen ions for metal ions present in the wastewater. The effluent from the cation exchanger then passes through a column containing a strong-anionic resin in the hydroxyl form. The hydroxyl anion exchange process exchanges hydroxyl ions for the CrO_4^{2-} ions. To recover the chromium the anion exchanger is regenerated with a sodium hydroxide (NaOH) solution. The effluent from the regeneration process is a concentrated Na_2CrO_4 solution which is passed through a column containing a strong-cationic resin in the hydrogen form. This hydrogen cation exchange process

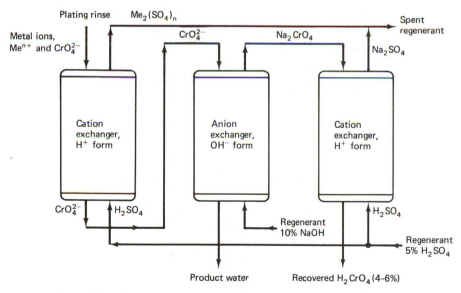

FIGURE 10-13 *Schematic of hexavalent chromium recovery by ion exchange (after Weber, 1972).*

exchanges hydrogen ions for sodium ions. The effluent from the exchanger is a chromic acid solution suitable for recovery.

If cation exchange resins are used, chromium must be in the trivalent state prior to removal by the resin. Ross (1968) suggests two situations in which cation exchange might be used for chromium removal. In one situation, trivalent chromium and other cationic contaminants would be removed from chromic acid solutions prior to reuse. In the other, cationic and anionic resins would be used in series to remove all species of chromium from dilute rinse waters. Wastewater treatment consisting of chromium reduction followed by ion exchange would normally not be economical compared to reduction and lime precipitation.

EXAMPLE PROBLEM 10-7: A plating operation generates 100,000 gal/day of waste at a temperature of 21°C. The primary components of the waste are

$$CrO_4^{2-} = 150 \text{ mg}/\ell$$
$$Fe^{3+} = 10 \text{ mg}/\ell$$
$$Cu^{2+} = 20 \text{ mg}/\ell$$
$$Zn^{2+} = 15 \text{ mg}/\ell$$
$$Ni^{2+} = 35 \text{ mg}/\ell$$
$$Na^{+} = 16 \text{ mg}/\ell$$

Design a chromic acid recovery process based on the schematic presented in Figure 10-13. The maximum column diameter to be used is 2 ft.

Solution:

1. Select the resins to be used in the process. For this problem the strong-cationic exchange resin selected is Diamond Shamrock's Duolite ES-26. The manufacturer's data on this resin are as follows:

Resin properties

Matrix:	Macroporous styrene-DVB copolymer
Functional group:	SO_3H
Physical form:	Gray, opaque beads
Specific gravity:	1.27 (H form)
Shipping weight:	50 lb/cu ft, net
Particle size:	20–50 mesh, U.S. Std Sieve
Uniformity coefficient:	1.8 maximum
Effective size:	0.45–0.62 mm
Moisture retention capacity:	48–52% (H form)
Total capacity:	1.9 eq./ℓ (41.6 Kgr/cu ft, as $CaCO_3$)

Suggested operating conditions

Minimum bed depth:	30 in.
Back-wash flow rate:	see figure below
Regen. concentration:	2–10% H_2SO_4
	5–10% HCl
Regen. flow rate:	0.4 to 1.0 gpm/cu ft
Regen. contact time:	minimum 40 min
Service flow rate:	2–5 gpm/cu ft
Pressure drop:	See figure below
Temperature:	Up to 250°F or 120°C (up to 300°F in sodium cycle)

Backwash Expansion vs. Flow Rate

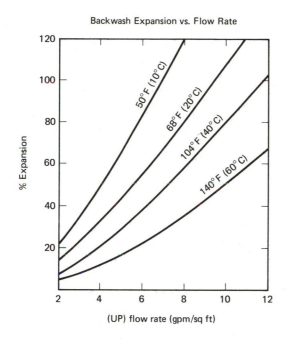

Pressure Drop vs. Flow Rate

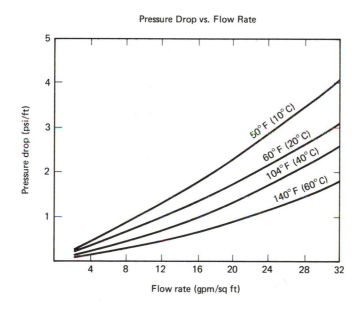

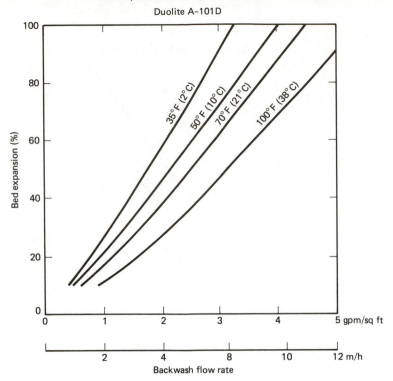

Bed Expansion vs. Backwash Flow Rate
Duolite A-101D

H₂SO₄ regeneration

	Capacity (Kgr/cu ft. as CaCO₃)		
lb/cu ft	25% Na	50% Na	75% Na
5	18.0	19.3	21.6
8	20.9	22.8	25.3
10	22.4	24.5	26.9
12	23.8	25.8	29.4
15	27.5	29.0	32.9

HCl regeneration

Dosage	Capacity (Kgr/cu ft. as CaCO₃)		
(lb/cu ft)	25% Na	50% Na	75% Na
3	17.3	18.6	19.8
5	23.7	24.7	26.0
8	28.4	28.7	30.5

Note: The percentage of Na values indicate the pro-
portion of sodium to total cations.

The strong-anionic exchange resin selected is Diamond Shamrock's Duolite A-101D. The manufacturer's data on this resin are given below:

Suggested operating conditions and parameters for Duolite A-101D

Minimum bed depth:	30 in. (75 cm)
Back-wash flow rate:	2–3 gpm/ft² (5–7 m/h*). See figure below
Regenerant concentration:	4% NaOH
Regenerant dosage:	5–10 lb NaOH/cu ft (80–160 g/ℓ)
Regeneration flow rate:	0.2 to 0.8 gpm/cu ft (1.5–6.5 bv/h†)
Regenerant displacement rate:	0.2 to 0.8 gpm/cu ft
Regeneration contact time:	40 to 60 min
Rinse flow rate:	2 gpm/cu ft (16 bv/h)
Rinse volume:	30–70 gal/cu ft (4–9 bv/h)
Service flow rate:	Up to 5 gpm/cu ft (40 bv/h)
Pressure drop:	See Figure below

* m/h: meters per hour
† bv/h: bed volumes per hour

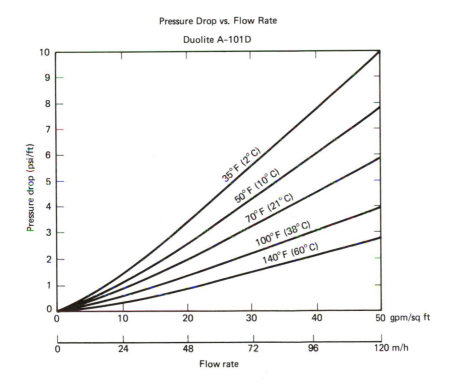

Pressure Drop vs. Flow Rate

Duolite A-101D

Matrix:	Crosslinked polystyrene
Functional groups:	Quaternary ammonium, $RC_6H_4CH_2N(CH_3)_3Cl$
Physical form:	Pale yellow, translucent beads
Specific gravity:	1.10 (Cl form, fully hydrated)
Apparent density:	670–720 g/ℓ (42–45 lb/cu ft) net
Void volume:	About 34% (2.5 gal/cu ft)
Particle size range:	0.3–1.1 mm (20–50 mesh (U.S. Standard Sieves)
Effective size:	0.50–0.55 mm
Uniformity coefficient:	1.3 to 1.6
Moisture retention capacity:	48–55% (Cl form)
Reversible swelling:	20% maximum (Cl to OH form)
pH limitations:	None
Temperature limitations in cyclic operations:	60°C (140°F) in OH form 100°C (212°F) in salt form
Chemical Resistance:	Unaffected by acids, alkalies, and solvents

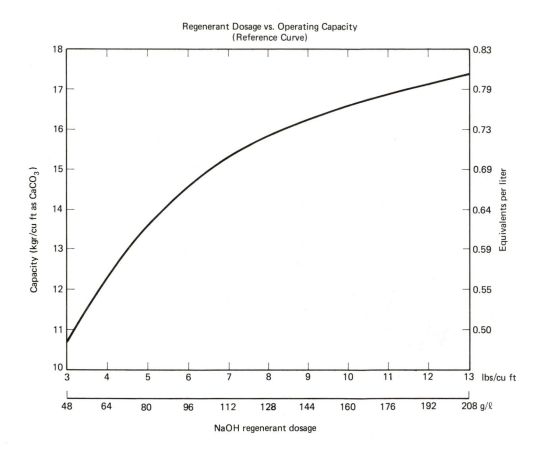

Regenerant Dosage vs. Operating Capacity
(Reference Curve)

Design of Primary Cation Exchanger:

1. Compute the total equivalents of metal ions to be removed per day.

   (a) Convert cation concentrations to equivalents per liter and compute total cation concentration:

   $$\text{eq. } Fe^{3+}/\ell = 10\,\frac{mg}{\ell} \times \frac{1\,mole}{55.8\,g} \times \frac{1g}{1000\,mg} \times \frac{3\,eq.}{1\,mole} = 0.00054\,\frac{eq.}{\ell}$$

   $$\text{eq. } Cu^{2+}/\ell = 20\,\frac{mg}{\ell} \times \frac{1\,mole}{63.5\,g} \times \frac{1\,g}{1000\,mg} \times \frac{2\,eq.}{1\,mole} = 0.00063\,\frac{eq.}{\ell}$$

   $$\text{eq. } Zn^{2+}/\ell = 15\,\frac{mg}{\ell} \times \frac{1\,mole}{65.4\,g} \times \frac{1\,g}{1000\,mg} \times \frac{2\,eq.}{1\,mole} = 0.00046\,\frac{eq.}{\ell}$$

   $$\text{eq. } Ni^{2+}/\ell = 35\,\frac{mg}{\ell} \times \frac{1\,mole}{58.7\,g} \times \frac{1g}{1000\,mg} \times \frac{2\,eq.}{1\,mole} = 0.00119\,\frac{eq.}{\ell}$$

   $$\text{eq. } Na^{+}/\ell = 16\,\frac{mg}{\ell} \times \frac{1\,mole}{23\,g} \times \frac{1\,g}{1000\,mg} \times \frac{1\,eq.}{1\,mole} = 0.00070\,\frac{eq.}{\ell}$$

   $$\text{total cation concentration} = 0.00054 + 0.00063 + 0.00046$$
   $$+ 0.00119 + 0.0007$$
   $$= 0.0035 \text{ eq.}/\ell$$

   (b) Calculate the total equivalents of metal ions removed per day:

   $$\text{total amount of ions removed (eq./day)} = 0.0035\,\frac{eq.}{\ell} \times 3.78\,\frac{gal}{\ell}$$
   $$\times 100,000\,\frac{gal}{day}$$
   $$= 1323 \text{ eq./day}$$

2. From the manufacturer's data select the amount of regenerant to use and the associated resin capacity.

   (a) Compute the percentage of sodium to total cations:

   $$\frac{0.0007}{0.0035} \times 100 = 20\%$$

   (b) Assume that a 10% HCl solution is to be used. The column for 25% (nearest to 20%) shows that for a regenerant dosage of 5 lb/ft³ of resin the exchange capacity of the bed is near 23.7 Kgr/ft³ as $CaCO_3$. To account for nonideal operating conditions, the resin capacity will be reduced by 20%, and a value of 19 Kgr/ft³ as $CaCO_3$ will be used for design.

   (c) Convert exchange capacity to eq./ℓ of resin:

   $$19\,\frac{Kgr}{ft^3} \times \frac{1\,g}{15.43\,gr} \times \frac{1000\,gr}{1\,Kgr} \times \frac{1\,ft^3}{28.32\ell} \times \frac{1\,mole}{100\,g} \times \frac{2\,eq.}{1\,mole}$$
   $$= 0.87 \text{ eq.}/\ell$$

3. Design the exchange column, assuming a bed depth of 36 in. and a service rate of 3 gpm/ft³.

(a) cross-sectional area per column $= \dfrac{\pi d^2}{4}$

$$= \dfrac{\pi(2)^2}{4}$$

$$= 3.14 \text{ ft}^2$$

(b) volume per column $=$ (area)(depth)

$$= (3.14 \text{ ft}^2)(\tfrac{36}{12})$$

$$= 9.4 \text{ ft}^3$$

(c) total volume required $= \dfrac{\text{total flow}}{\text{loading rate}}$

$$= \dfrac{100,000 \text{ gal/day}}{3 \dfrac{\text{gal}}{\text{min-ft}^3} \times 1440 \dfrac{\text{min}}{\text{day}}}$$

$$= 23.1 \text{ ft}^3$$

(d) number of columns required $= \dfrac{\text{total volume}}{\text{volume per column}}$

$$= \dfrac{23.1 \text{ ft}^3}{9.4 \text{ ft}^3}$$

$$= 2.46 \text{ Therefore use 3 columns}$$

4. Compute the exchange capacity of each column:

$$\text{column capacity} = \left(0.87 \dfrac{\text{eq.}}{\ell}\right)(9.4 \text{ ft}^3)\left[28.3 \dfrac{\ell}{\text{ft}^3}\right]$$

$$= 231 \text{ eq.}$$

5. Compute the service time of column operation:

$$\dfrac{\text{service}}{\text{time}} = \dfrac{\text{column capacity}}{(\text{total eq. of cations to be removed per day})/(\text{no. of columns})}$$

$$= \dfrac{231}{(1323 \text{ eq./day})/(3)}$$

$$= 0.524 \text{ days}$$

$$= 12.6 \text{ hr}$$

6. Calculate the amount of 10% HCl required per column. From Appendix IX, a 10% HCl solution contains 0.8741 lb of HCl per gallon.

$$\text{gallons} = \dfrac{1 \text{ gallon}}{0.8741 \text{ lb}} \times 5 \dfrac{\text{lb}}{\text{ft}^3} \times 9.4 \text{ ft}^3$$

$$= 54$$

7. Compute the regeneration time required, using a regeneration rate of 0.5 gpm/ft³:

$$\text{regeneration time} = \dfrac{54 \text{ gal}}{\dfrac{0.5 \text{ gal}}{\text{min-ft}^3} (9.4 \text{ ft}^3)\left(60 \dfrac{\text{min}}{\text{hr}}\right)}$$

$$= 0.19 \text{ hr}$$

Use a minimum regeneration time of 30 min, or 0.5 hr.

8. Assume that a 60% bed expansion is required for back-washing. The appropriate figure for the ES-26 resin shows that a backwash rate of 6.5 gpm/ft² is required.

9. Compute the total cycle time, assuming a 15-min back-wash and a 40 min rinse.

$$\text{cycle time} = 12.6 \text{ hr} + 0.5 \text{ hr} + \tfrac{15}{60} + \tfrac{40}{60}$$

$$= 14 \text{ hr}$$

Design of Anion Exchanger:

1. Compute the total equivalents of chromatic ion to be removed per day.

    (a) Convert chromatic ion concentration to equivalents per liter:

$$\text{eq. } CrO_4^{2-}/\ell = 150 \frac{mg}{\ell} \times \frac{1 \text{ mole}}{116 \text{ g}} \times \frac{1 \text{ g}}{1000 \text{ mg}} \times \frac{2 \text{ eq.}}{1 \text{ mole}} = 0.0026 \frac{\text{eq.}}{\ell}$$

    (b) Calculate the total equivalents of chromatic ions removed per day:

$$\begin{aligned}\text{total amount of ions} \\ \text{removed (eq./day)} \end{aligned} = 0.0026 \frac{\text{eq.}}{\ell} \times 3.78 \frac{\ell}{\text{gal}} \times 100{,}000 \frac{\text{gal}}{\text{day}}$$

$$= 983 \text{ eq./day}$$

2. From the manufacturer's data select the amount of regenerant to use and the associated resin capacity.

    (a) For a regenerant dosage of 10 lb/ft^3 of resin the exchange capacity of the bed is 0.76 eq./ℓ.

3. Design the exchange column, assuming a bed depth of 36 in. and a service rate of 3 gpm/ft^3.

    (a) cross-sectional area per column = 3.14 ft^2

    (b) volume per column = 9.4 ft^3

    (c) total volume required = 23.1 ft^3

    (d) number of columns required = 3 columns

4. Compute the exchange capacity of each column:

$$\text{column capacity} = \left(0.76 \frac{\text{eq.}}{\ell}\right)(9.4 \text{ ft}^3)\left(28.3 \frac{\ell}{\text{ft}^3}\right)$$

$$= 202 \text{ eq.}$$

5. Compute the service time of column operation:

$$\text{service time} = \frac{202}{(983 \text{ eq./day})/(3)}$$

$$= 0.616 \text{ days}$$

$$= 14.8 \text{ hr}$$

6. Calculate the amount of 4% NaOH required per column. A 4% solution of NaOH contains 0.3481 lb of NaOH per gallon (see Appendix IX).

$$\text{gallons} = \frac{1 \text{ gal}}{0.3481 \text{ lb}} \times 10 \frac{\text{lb}}{\text{ft}^3} \times 9.4 \text{ ft}^3$$

$$= 270 \text{ gal}$$

7. Compute the regeneration time required, using a regeneration rate of 0.5 gpm/ft³:

$$\text{regeneration time} = \frac{270 \text{ gal}}{0.5 \frac{\text{gal}}{\text{min-ft}^3} (9.4 \text{ ft}^3) \left(60 \frac{\text{min}}{\text{hr}}\right)}$$

$$= 0.96 \text{ hr}$$

8. Assume that a 60% bed expansion is required for back-washing. The appropriate figure for the A-101D resin shows that a back-wash rate near 5 gpm/ft² is required.

9. Compute the total cycle time, assuming a 15 min back-wash and a 40 min rinse.

$$\text{cycle time} = 14.8 \text{ hr} + 0.96 \text{ hr} + \tfrac{15}{60} + \tfrac{40}{60}$$

$$= 16.7 \text{ hr}$$

Design of Secondary Cation Exchanger:

1. Compute the equivalents of sodium per regeneration cycle to be removed. The amount of sodium to be removed is dependent on the amount of sodium hydroxide used during regeneration.

$$\begin{aligned}\text{sodium ions to be removed} \\ \text{(eq./regeneration cycle)}\end{aligned} = 10 \frac{\text{lb}}{\text{ft}^3} \times \frac{9.4 \text{ ft}^3}{1 \text{ column}} \times 3 \text{ columns}$$

$$\times 454 \frac{\text{g}}{\text{lb}} \times \frac{1 \text{ eq.}}{40 \text{ g}}$$

$$= 3201 \text{ eq./cycle}$$

2. Calculate the total flow to be handled. The flow going to the secondary cation exchanger is given by the total regenerant flow used:

$$\text{total flow} = 270 \frac{\text{gal}}{\text{column}} \times 3 \text{ columns}$$

$$= 810 \text{ gal}$$

3. From the manufacturer's data select the amount of regenerant to use and the associated resin capacity.

   (a) In this case the percentage of sodium to total cations is approximately 100%.

   (b) Assume that a 10% HCl solution is to be used and that a regenerant dose of 5 lb/ft³ will be applied. Since 100% Na is not given, a value of 26 Kgr/ft³ as $CaCO_3$ will be assumed and no adjustment for nonideal conditions will be made.

   (c) Convert exchange capacity to eq./ℓ of resin:

$$26 \frac{\text{Kgr}}{\text{ft}^3} \times \frac{1 \text{ g}}{15.43 \text{ gr}} \times \frac{1000 \text{ gr}}{1 \text{ Kgr}} \times \frac{1 \text{ ft}^3}{28.32 \, \ell} \times \frac{1 \text{ mole}}{100 \text{ g}} \times \frac{2 \text{ eq.}}{1 \text{ mole}} = 1.19 \text{ eq.}/\ell$$

4. Design the exchange column, assuming a bed depth of 60 in. and a service rate of 1 gpm/ft³.

   (a) cross-sectional area per column = 3.14 ft²

(b)   volume per column = (area)(depth)
$$= (3.14)(5)$$
$$= 15.7 \text{ ft}^3$$

(c)   total volume required $= \dfrac{\text{total flow}}{\text{loading rate}}$

$$= \dfrac{810 \text{ gal/cycle} \times \dfrac{24}{16.7} \dfrac{\text{cycle}}{\text{day}}}{1 \dfrac{\text{gal}}{\text{min-ft}^3} \times 1440 \dfrac{\text{min}}{\text{day}}}$$

$$= 0.81 \text{ ft}^3$$

(d)   number of columns required $= \dfrac{\text{total volume}}{\text{volume per column}}$

$$= \dfrac{0.81 \text{ ft}^3}{15.7}$$

$$\simeq 0.05 \text{ (Use 1 column)}$$

5.   Compute the exchange capacity of each column:

$$\text{column capacity} = \left(1.19 \dfrac{\text{eq.}}{\ell}\right)(15.7 \text{ ft}^3)\left(28.3 \dfrac{\ell}{\text{ft}^3}\right)$$

$$= 528.7 \text{ eq.}$$

Note: The column capacity of 528.7 eq. is less than the equivalents of sodium ions to be removed per regeneration cycle. It is desirable to be able to treat the entire regeneration flow from the anion exchanger before regeneration of the secondary cation exchanger is required. Hence, the number of columns required should be increased to achieve this objective.

6.   Calculate the number of columns required to treat the entire regeneration flow from the anion exchanger before regeneration of the secondary cation exchanger is required.

$$\text{No. of columns} = \dfrac{\text{total equivalents of Na to be exchanged}}{\text{exchange capacity per column}}$$

$$= \dfrac{3201}{528.7}$$

Number of columns = 6.05. Use 7 columns.

Nitrogen Removal by Ion Exchange

Ion exchange is often an attractive management process because it offers inherent process control, is amenable to automation, and is unaffected by temperature in the ranges generally encountered. Removal of nitrogen species, however, has until recently been impractical because conventional ion exchange resins prefer more common ions than the nitrogen species. This problem has been overcome by the discovery of a natural material which preferentially exchanges ammonium ions.

Clinoptilolite is a common constituent of bentonite deposits in the western United States. The largest such deposit is near Hector in southern California. Although clinoptilolite occurs widely, not all deposits produce a material strong enough to withstand the physical demands of columnar operation. Hector clinoptilolite has a relatively low attrition loss in use.

The total ion exchange capacity of clinoptilolite has been measured as 1.6–2.0 meq./g (EPA, 1975), which is slightly lower than average for zeolites. The operating capacity for ammonium ions in the presence of typical concentrations of other cations found in domestic wastewaters is about 0.4 meq./g.

The preference of an ion exchange material for one ion over another is often expressed in terms of the separation factor. Isotherms comparing the affinity of NH_4^+ for clinoptilolite and a strong acid, polystyrene resin (IR-120), are shown in Figure 10-14. Note that the separation factor is definitely greater than one for Hector clinoptilolite and less than one for the IR-120 resin. This characteristic of clinoptilolite makes an ion exchange approach to nitrogen removal feasible.

Clinoptilolite preferentially exchanges ammonium ions rather than other common cations in water, including Na^+, Mg^{2+}, and Ca^{2+} as illustrated in Figure 10-15. Note, however, that potassium ions are exchanged in preference to ammonium ions. Luckily, potassium concentrations in natural waters are generally less than the other cations listed.

Ames (1961) made two important observations during a study of ammonia removal by ion exchange. First, he found that the equilibrium for each cation pair is independent of others in a multicomponent system. Second, he noted a relationship between selectivity coefficients and the cation concentration ratios. In practice, these relationships (Figure 10-16 and 10-17) can be used to calculate the fraction of the clinoptilolite total ion exchange capacity occupied by ammonium ions under equilibrium conditions.

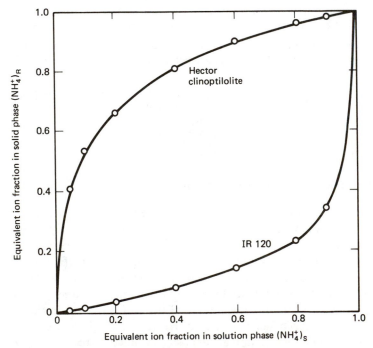

FIGURE 10-14 *The 23°C isotherm for reaction, $RCa^{2+} + 2 NH_4^+ = R(NH_4^+)_2 + Ca^{2+}$ with hector clinoptilolite and IR 120 (after EPA, 1975).*

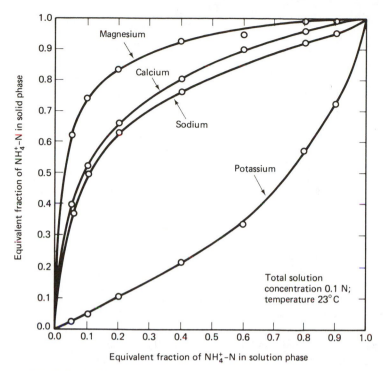

FIGURE 10-15 *Isotherm for exchange of NH_4^+ for K^+, Na^+, and Mg^{2+} on clinoptilolite (after EPA, 1975).*

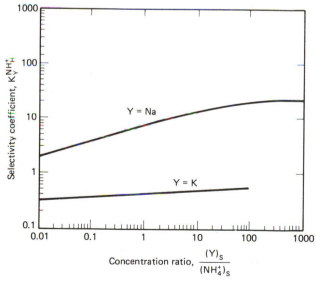

FIGURE 10-16 *Selectivity coefficients vs concentration ratios of sodium or potassium and ammonium in the equilibrium solution with hector clinoptilolite at 23°C for the reaction $(Y)_R + (NH_4^+)_S = 2(NH_4^+)_R + (Y)_S$ (after EPA, 1975).*

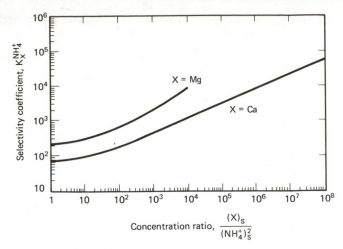

FIGURE 10-17 *Selectivity coefficients vs concentration ratios of calcium or magnesium and ammonium in the equilibrium solution with hector clinoptilolite at 23°C for the reaction $(X)_R + 2 (NH_4^+)_S = 2 (NH_4^+)_R + (X)_S$ (after EPA, 1971). Note: the experimental selectivity coefficients were determined in a reference solution of 0.1 N $CaCl_2$.*

EXAMPLE PROBLEM 10-8 (after EPA, 1971): Calculate the ammonium exchange capacity for clinoptilolite (i.e., the fraction of the clinoptilolite total ion exchange capacity occupied by ammonium ions under equilibrium conditions) for a wastewater containing 130 mg/ℓ of sodium ions, 15 mg/ℓ of potassium ions, 20 mg/ℓ of ammonium ions, 60 mg/ℓ of calcium ions, and 25 mg/ℓ of magnesium ions. Assume that the total ion exchange capacity of the clinoptilolite is 1.8 meq./g and the total ionic strength of the wastewater is 0.02 M at 25°C.

Solution:

1. To calculate the ammonium exchange capacity for clinoptilolite, use is made of the fact that the equivalent fractions of all the ions on the zeolite must sum to 1:

$$(NH_4)_R + (Ca)_R + (Mg)_R + (K)_R + (Na)_R = 1$$

2. Determine the molarity of the cations present.

$$[Na^+] = 130 \frac{mg}{\ell} \times \frac{1 \text{ g}}{1000 \text{ mg}} \times \frac{1 \text{ mole}}{23 \text{ g}} = 5.7 \times 10^{-3} \; M$$

$$[K^+] = 15 \frac{mg}{\ell} \times \frac{1 \text{ g}}{1000 \text{ mg}} \times \frac{1 \text{ mole}}{39.1 \text{ g}} = 3.8 \times 10^{-4} \; M$$

$$[NH_4^+] = 20 \frac{mg}{\ell} \times \frac{1 \text{ g}}{1000 \text{ mg}} \times \frac{1 \text{ mole}}{18 \text{ g}} = 1.2 \times 10^{-3} \; M$$

$$[Ca^{2+}] = 60 \frac{mg}{\ell} \times \frac{1 \text{ g}}{1000 \text{ mg}} \times \frac{1 \text{ mole}}{40 \text{ g}} = 3.0 \times 10^{-3} \; M$$

$$[Mg^{2+}] = 25 \frac{mg}{\ell} \times \frac{1 \text{ g}}{1000 \text{ mg}} \times \frac{1 \text{ mole}}{24.3 \text{ g}} = 2.1 \times 10^{-3} \; M$$

3. Compute the equilibrium solution concentration ratios for the various ions. Assume that at equilibrium the influent concentration equals the effluent concentration for all species.

$$\frac{[Na^+]_S}{[NH_4^+]_S} = 4.75$$

$$\frac{[K^+]_S}{[NH_4^+]_S} = 0.317$$

$$\frac{[Ca^{2+}]_S}{[NH_4^+]_S^2} = 2083$$

$$\frac{[Mg^{2+}]_S}{[NH_4^+]_S^2} = 1458$$

4. Determine the experimental selectivity coefficients $K_{K^+}^{NH_4^+}$ and $K_{Na^+}^{NH_4^+}$ from Figure 10-16.

   (a) For $\frac{[K^+]_S}{[NH_4^+]_S} = 0.317$, $K_{K^+}^{NH_4^+} = 0.33$

   (b) For $\frac{[Na^+]_S}{[NH_4^+]_S} = 4.75$, $K_{K^+}^{NH_4^+} = 11$

5. Determine the experimental selectivity coefficients $K_{Ca^{2+}}^{NH_4^+}$ and $K_{Mg^{2+}}^{NH_4^+}$ from Figure 10-17.

   (a) For $\frac{[Ca^{2+}]_S}{[NH_4^+]_S^2} = 2083$, $K_{Ca^{2+}}^{NH_4^+} = 760$

   (b) For $\frac{[Mg^{2+}]_S}{[NH_4^+]_S^2} = 1458$, $K_{Mg^{2+}}^{NH_4^+} = 2400$

6. Evaluate the $[Na^+]_R$ and $[K^+]_R$ concentrations in terms of $[NH_4^+]_R$.

   (a) For $[K^+]_R$, if it is assumed that activity coefficient corrections are not required for ions in the resin, equation 10-2 suggests that

$$K_{K^+}^{NH_4^+} = \frac{(\gamma_K)[K^+]_S(NH_4^+)_R}{(\gamma_{NH_4})[NH_4^+]_S(K^+)_R}$$

   Further, assume that activity coefficients may be neglected in univalent-univalent exchange but are necessary in univalent-divalent exchange. Then, the above expression becomes

$$0.33 = 0.317\frac{(NH_4^+)_R}{(K^+)_R}$$

   or

$$(K^+)_R = 0.96(NH_4^+)_R$$

   (b) Following the same procedure outlined in step 6(a) for $(Na^+)_R$ gives

$$(Na^+)_R = 0.43(NH_4^+)_R$$

7. Evaluate the $[Ca^{2+}]_R$ and $[Mg^{2+}]_R$ concentrations in terms of $[NH_4^+]_R$.

   (a) Compute the activity coefficient for divalent and monovalent ions from equation 1-47:

$$\gamma_D = 0.56, \qquad \gamma_M = 0.72$$

(b)   For $[Ca^{2+}]_R$, if it is assumed that activity coefficient corrections are not required for ions in the resin, equation 10-2 suggests that

$$K^{NH_4^+}_{Ca^{2+}} = \frac{(\gamma_{Ca})[Ca^{2+}]_S(NH_4^+)^2_R}{(\gamma_{NH_4})^2[NH_4^+]^2_S(Ca^{2+})_R}$$

or

$$760 = \frac{(0.56)(2083)(NH_4^+)^2_R}{(0.72)^2(Ca^{2+})_R}$$

$$(Ca^{2+})_R = 2.96(NH_4^+)^2_R$$

(b)   Following the same procedure outlined in step 7(a) for $(Mg^{2+})_R$ gives

$$(Mg^{2+})_R = 0.65(NH_4^+)^2_R$$

8.   Compute the ammonium exchange capacity of the clinoptilolite by substituting into the equation in step 1 from steps 6(a) and (b) and 7(a) and (b):

$$(NH_4^+)_R + 0.96(NH_4^+)_R + 0.43(NH_4^+)_R + 2.96(NH_4^+)^2_R + 0.6(NH_4^+)^2_R = 1$$

or

$$3.61(NH_4^+)^2 + 2.39(NH_4^+) - 1 = 0$$

Hence,

$$(NH_4^+) = \frac{-2.39 \pm \sqrt{(2.39)^2 - (4)(3.61)(-1)}}{2(3.61)}$$

$$= \frac{2.10}{7.22}$$

$$= 0.29$$

Therefore, the ammonium exchange capacity is

$$(0.29)(1.8 \text{ meq/g}) = 0.522 \text{ meq./g}$$

Koon and Kaufman (1971) empirically related resin ammonium concentration to cationic strength. Figure 10-18 shows this relationship, which was developed by using wastewaters having 16.4–19.0 mg/ℓ NH_4^+—N. Although this curve is empirical and represents a simplification of complex processes, it illustrates the effects of ionic competition for exchange sites and can be used to determine the ammonium exchange capacity of the clinoptilolite if the cationic strength of the wastewater is known.

A study of ammonia removal on clinoptilolite was made by McLaren and Farquhar (1973). They concluded that the greatest effect on the ammonia capacity was the initial ammonium ion concentration. They reported, for example, that capacity doubled with a five-fold increase in ammonia concentration. The effect of temperature variations within normal ambient ranges was shown to have no significant effect on the ammonia capacity of the clinoptilolite.

Regeneration of clinoptilolite can be accomplished in several ways. In one procedure a high-pH regenerant (such as a lime solution) is used to convert the ammonia to its unionized form. The regenerant can then be stripped of the ammonia. By another

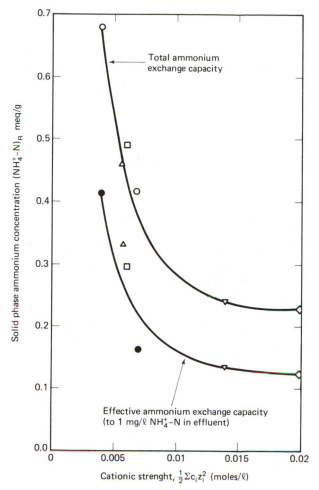

Where: C = concentration of the cation species i
        Z = valence of the cation species i

FIGURE 10-18 *Variation of ammonium exchange capacity with competing cation concentration for a three feet deep clinoptilolite bed (after Koon and Kaufman, 1971).*

method the regenerant passes through sulfuric acid, which then produces ammonium sulfate. The latter substance can be sold as fertilizer.

$$2NH_4^+ + H_2SO_4 \longrightarrow (NH_4)_2SO_4 + 2H^+ \qquad \textbf{(10-60)}$$

One important advantage of these processes, compared to the operation of typical ion exchangers, is that no large amounts of waste regenerant brine are produced.

Typical design criteria are presented below for an ion exchange column that uses clinoptilolite (Culp et al., 1978):

1. Feed pH: 4–8.

2. Hydraulic loading rate: 7.5 to 20 bed volumes per hour.

3. Clinoptilolite size: 20×50 mesh.

4. Bed depth: 3 to 6 ft.

5. Amount of regenerant (near neutral pH regeneration): 2% sodium chloride solution at near neutral pH requires about 25 to 30 bed volumes.

6. Regenerant flow rate: 10 bed volumes per hour.

PROBLEMS

10-1. Determine the concentration of a sulfuric acid solution theoretically required to regenerate a strong-acid cation exchange resin to 90% of its total capacity. Assume that $K_{Na^+}^{H^+} = 0.67$.

10-2. Using the breakthrough curve shown in Figure P10-2, compute the total capacity of the strong-acid cation exchange resin used to remove calcium ions from a water with an initial calcium concentration of 10 meq./ℓ. The ion exchange column used in the treatment process contained 8000 mℓ of resin.

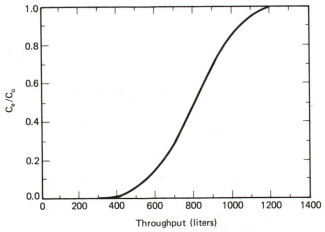

FIGURE P10-2 *Breakthrough curve for Problem 10-2.*

10-3. Estimate the void fraction of an unexpanded resin bed (uniform size particles), using the following equations:

(a) $V_s = \dfrac{g}{18\mu}(S_s - S)d^2$

where
V_s = terminal settling velocity of a spherical particle (cm/sec)
g = acceleration of gravity (cm/sec²)
μ = absolute viscosity of liquid (g/cm-sec)
S_s = mass density of particle (g/cm³)
S = mass density of liquid (g/cm³)
d = particle diameter (cm).

(b) $E_e = \left[\dfrac{V}{V_s}\right]^{0.2}$

where E_e = expanded bed porosity
           V = back-wash rate (cm/sec).

(c) $L_e = \dfrac{L(1 - E)}{(1 - E_e)}$

where L_e = depth of expanded bed (ft)
           L = depth of unexpanded bed (ft)
           E = porosity of unexpanded bed.

Assume that the back-wash rate is 6 gpm/ft², the water temperature is 10°C, the resin particle size is 0.5 mm, the mass density of the resin is 1.32 g/cm³, the unexpanded bed depth is 3 ft, and the relationship in Figure 10-12 is valid.

10-4. Estimate the concentration of the NaCl solution theoretically required to regenerate a strong-acid cation exchange resin to 80% of its total capacity of 2.0 eq./ℓ. Assume that $K_{Ca^{2+}}^{Na^+} = 0.17$ and that, at equilibrium, $C = X_{Na^+}$.

10-5. A wastewater is found to have the following composition

Na	120 mg/ℓ
Ca	50 mg/ℓ
NH$_4$	25 mg/ℓ
K	40 mg/ℓ
Mg	10 mg/ℓ
Cl	190 mg/ℓ
SO$_4$	90 mg/ℓ
HCO$_3$	220 mg/ℓ

If this solution is treated by using clinoptilolite resin, estimate the equilibrium ammonium loading in meq./g. Assume that the total capacity of the clinoptilolite is 1.80 meq./g.

10-6. The ammonia content of a wastewater is reduced from 12 mg/ℓ to 1.0 mg/ℓ by using clinoptilolite. If the average waste flow is 20 MGD, what amount of $(NH_4)_2SO_4$ could be produced annually?

REFERENCES

ABRAMS, I.M., and BENEZRA, L., "Ion-Exchange Polymers," in *Encyclopedia of Polymer Science and Technology*, John Wiley & Sons, Inc., New York (1967).

ABRAMS, I.M., "Countercurrent Ion Exchange with Fixed Beds," *Industrial Water Engineering*, **18** (January, 1973).

AMES, L.L., "Cation Sieve Properties of the Open Zeolites, Chabazite, Mordenite, Erionite and Clinoptilolite," *American Mineralogist*, **46**, 1120 (1961).

ANDERSON, R.E., "Estimation of Ion Exchange Process Limits by Selectivity Calculations," *AICHE Symposium Series*, **71**, 152, 236 (1975).

CULP, R.L., WESNER, G.M., and CULP, G.L., *Handbook of Advanced Wastewater Treatment*, Van Nostrand Reinhold Company, New York (1978).

ECKENFELDER and E.L. THACKSTON, Jenkins Publishing Co., Austin, TX (1972).

Environmental Protection Agency, "Wastewater Ammonia Removal by Ion Exchange, *Water Pollution Control Research Series*, 17010 ECZ 02/71 (1971).

Environmental Protection Agency, "Process Design Manual for Nitrogen Control," *EPA Technology Transfer* (1975).

HELMS, R.F., "Evaluation of Ion Exchange for Demineralization of Wastewater," Master's Thesis, University of Colorado, Boulder, CO (1973).

INCZÉDY, J., *et al.*, *Analytical Applications of Ion Exchangers*, Pergamon Press, Inc., New York (1966).

KOON, J.H., and KAUFMAN, W.J., "Optimization of Ammonia Removal by Ion Exchange Using Clinoptilolite," Environmental Protection Agency, Washington, DC (1971).

MAYER, S.W., and TOMPKINS, E.R., "Ion Exchange as a Separation Method", *J. Am. Chem. Soc.*, **69,** 2866 (1947).

MCLAREN, J.R., and FARQUHAR, G.J., "Factors Affecting Ammonia Removal by Clinoptilolite," *Journal Environmental Engineering Division*, ASCE, **99,** 429 (1973).

PANSWAD, T., "Ion Exchange Removal of Inorganic and Organic Wastewater Constituents," Ph.D. Dissertation, University of Colorado, Boulder, CO (1975).

ROSS, R.D. [Ed.], *Industrial Waste Disposal*, Van Nostrand Reinhold Company, New York (1968).

SALMON, J.E., and HALE, D.K., *Ion Exchange*, Academic Press, Inc., New York (1959).

SANKS, R.L., "Ion Exchange," in *Process Design in Water Quality Engineering*, [Eds.] W.W.

WEBER, W.J., Jr., *Physicochemical Processes for Water Quality Control*, Wiley-Interscience, New York (1972).

11

REMOVAL OF SOLUBLE ORGANIC MATERIALS
FROM WASTEWATER
BY CARBON ADSORPTION

The process of adsorption is a powerful tool for the environmental engineer and has many applications for treatment of municipal and industrial wastewater. Activated carbon is the most widely used commercial adsorbent and, in general, has a great capacity for the adsorption of organic molecules.

Applications of adsorption in municipal wastewater treatment fall into two categories: tertiary treatment and physical-chemical treatment (PCT). Tertiary treatment is the term used when adsorption is applied to remove residual organic material remaining in the effluent of biological treatment processes. Physical-chemical treatment processes are those in which raw wastewater is exposed only to physical or chemical treatment prior to carbon adsorption. In such cases adsorption alone, rather than biological treatment, is responsible for removal of dissolved organic impurities. Typical flow patterns for tertiary and physical-chemical treatment plants are shown in Figure 11-1. Adsorption is used in industrial wastewater treatment to remove organic materials such as color, phenols, detergents, cresols, or other toxic or nonbiogradable materials.

Activated carbon is marketed in both powdered and granular forms. Powdered activated carbons (PAC) are pulverized so that 95 to 100% will pass a 100-mesh sieve (149 microns nominal opening) and between 50 and 95% will pass a 325-mesh sieve (Fornwalt and Hutchins, 1966a). Granular activated carbon (GAC) is designated by sizes such as 12/20, 20/40, or 8/30. For example, a 20/40 carbon is comprised of particles that will pass through a U.S. Standard Mesh Size No. 20 screen (0.03 in.) but would be retained on a U.S. Standard Mesh Size No. 40 screen (0.017 in.). Granular carbon is slightly more expensive than powdered carbon but easier to handle, easier and cheaper to regenerate, and has better hydraulic properties when used in continuous-flow systems. Consequently, almost all continuous-flow adsorption systems employ granular carbon.

There are two general types of activated carbon: gas-adsorbent carbon, which is used to adsorb impurities from gases, and liquid-phase carbon, which is used to remove impurities from liquids and solutions (Fornwalt and Hutchins, 1966a). The primary

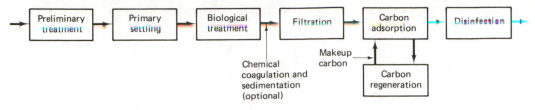

(a) Tertiary treatment plant.

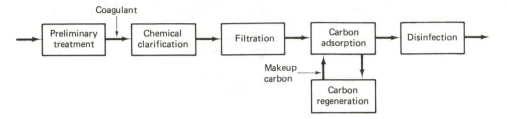

(b) Physical-chemical treatment plant.

FIGURE 11-1 *Typical flow patterns (after EPA, 1973).*

difference between these carbons is the distribution of pore sizes within the finished products. Gas-adsorbent carbons contain mostly micropores and thus exhibit a single peak in measurable pore-size distribution. This configuration permits rapid penetration of gas molecules but limits the penetration of liquids. Carbons produced from coconut shells frequently exhibit this property.

Liquid-phase carbons are characterized by a more uniform distribution of all three pore sizes. The macropores and transitional pores permit the entrance of large molecules and allow access of liquid to the micropores. Micropores usually account for at least 95% of the specific surface, and most adsorption occurs on the surfaces within these pores. Carbons produced from coal and cellulose generally exhibit this type of pore-size distribution and are commonly used for liquid-phase adsorption.

Commercially available carbons are usually described by a number of quality control parameters. These parameters cannot be used to predict how the carbon will perform on a given waste; that can only be done through laboratory and pilot plant studies. However, they do provide valuable information that can be used to select, for further study, those carbons processing certain desirable properties. Parameters usually reported by carbon suppliers are summarized in Table 11-1. Typical values for some of these parameters for three selected carbons are listed in Table 11-2.

11-1 ADSORPTION OF SOME COMMON ORGANIC SUBSTANCES

The adsorbability of complex mixtures such as wastewaters cannot be determined from adsorption data collected for the individual components of such mixtures. However, an understanding of the behavior of pure compounds can assist in predicting the behavior of complex mixtures and can thus prove valuable to the engineer.

TABLE 11-1 *Selected properties of activated carbon (after U.S. EPA, 1973).*

Property	Importance
Particle size distribution:	Rate of adsorption increases as particle size decreases. Head loss through packed column increases as particle size decreases. Size distribution expressed as percentage passing various size sieves.
Surface area:	A measure of the area available for adsorption. The larger the surface area, the greater the adsorptive capacity. Measured by determining the amount of nitrogen adsorbed by the carbon and reported as m^2/g.
Pore volume:	Measure of total macropore and micropore volume within the carbon particles. Measured in cc/g.
Iodine number:	Refers to milligrams of iodine adsorbed during standard test. Measures the volume present in pores from 10 to 28 Å in diameter. Carbons with a high percentage of pore sizes in this range would be suitable for adsorbing low molecular weight substances.
Molasses number:	Refers to milligrams of molasses adsorbed during standard test and measures the volume in pores greater than 28 Å in diameter. Carbons with a high percentage of this size pores would be suitable for adsorbing high molecular weight substances.
Abrasion number:	Measures ability of carbon to withstand handling and slurry transfer. This property is of limited value because measuring techniques are not reproducible.
Bulk density:	Useful in determining the volume occupied by a given weight of carbon.

TABLE 11-2 *Typical properties of several powdered and granular activated carbons (courtesy of Westvāco).*

Property	Granular Carbon Nuchar WV-W	Nuchar WV-G	Powdered Carbon Aqua (PAC)
Surface area (m^2/g)	900 (min)	1100 (min)	
Particle size	8×30	12×40	8–9 microns
Uniformity coefficient	1.8 (max)	1.8 (max)	—
Effective size (mm)	0.85–1.05	0.55–0.75	—
Apparent density (lb/ft³)	35–37	27–29	36 ± 4
Iodine number (mg/g)	850 (min)	1050 (min)	800 ± 100
Particle density (g/cm³)	1.45–1.55	1.30–1.40	1.4–1.5
Sieve analysis:			
through 100 mesh (%)			99–100
through 200 mesh (%)			97–99
through 325 mesh (%)			92–98

Guisti et al. (1974) studied the adsorption of various organic compounds onto activated carbon. Their results help to explain the effect of such characteristics as functionality, molecular weight, polarity, and solubility on adsorption. A summary of their results appears in Table 11-3.

Compound	Molecular Weight	Adsorbility $\left(\dfrac{g\ compound}{g\ carbon}\right)$	Remarks
Alcohols:			
Methanol	32.0	0.007	Compounds highly polar and highly soluble,
Ethanol	46.1	0.020	resulting in low amendability of com-
Propanol	60.1	0.038	pounds to adsorption. Polarity decreases
Butanol	74.1	0.107	as molecular weight increases, resulting in
n-Amyl alcohol	88.2	0.155	increased adsorbility with increasing mole-
n-Hexanol	102.2	0.191	cular weight.
Aldehydes:			
Formaldehyde	30.0	0.018	Aldehydes, like alcohols, are highly polar
Acetaldehyde	44.1	0.022	compounds. Polarity decreases as mole-
Propionaldehyde	58.1	0.057	cular weight increases, resulting in in-
Butyraldehyde	72.1	0.106	creased amenability to adsorption.
Amines:			
Di-N-Propylamine	101.2	0.174	Amenability to adsorption again limited by
Butylamine	73.1	0.103	polarity and solubility.
Di-N-Butylamine	129.3	0.174	
Allylamine	57.1	0.063	
Ethylenediamine	60.1	0.021	
Diethylenetriamine	103.2	0.062	
Aromatics:			
Benzene	78.1	0.080	Low polarity and low solubility of com-
Toluene	92.1	0.050	pounds makes them relatively easy to
Ethyl benzene	106.2	0.019	remove by adsorption onto carbon.
Phenol	94	0.161	Removal is also enhanced by π bonding
Hydroquinone	110.1	0.167	between compound and carbon surface.
Aniline	93.1	0.150	
Glycols:			
Ethylene glycol	62.1	0.0136	These compounds have multiple sites for
Diethylene glycol	106.1	0.053	hydrogen bonding, which gives them a hy-
Triethylene glycol	150.2	0.105	drophilic nature. Hydrophilic property
Tetraethylene glycol	194.2	0.116	makes compounds difficult to remove by
Propylene glycol	76.1	0.024	adsorption.

11-2 ADSORPTION STUDIES

The feasibility of using adsorption for a given waste treatment application must be determined by laboratory and pilot plant tests. In almost all cases the initial tests consist of laboratory-scale, constant temperature, batch adsorption studies. The procedure for conducting these tests has been presented by Fornwalt and Hutchins (1966a) as follows:

1. Select a carbon for use in the study. If more than one carbon is selected, the following procedure should be followed for each of the carbons.

2. Grind about 20 grams of the carbon until it all passes a 325-mesh screen. Grinding of the carbon will not significantly affect its adsorptive capacity but will increase the rate of adsorption.

3. Set up at least five flasks and add a selected amount of carbon to each flask. For example:

Flask No.	Carbon Added (g)
1	0
2	1
3	2
4	3
5	4

If these dosages do not provide the desired range of results, the test must be repeated with different dosages.

4. Analyze the wastewater to be treated, and determine the initial concentration of the impurity to be removed. Add a given amount (100–500 ml) of wastewater to each flask and place the flasks on a gyratory shaker table.

5. Shake the flasks until adsorption is complete. An agitation period of 1 to 2 hr is normally sufficient.

6. After shaking, filter the samples to remove the carbon and analyze the filterate for the impurity remaining in solution. The amount of impurity adsorbed by the carbon can then be determined.

The temperature should be held constant during the test, preferably at the value anticipated for the wastewater during treatment. Likewise, the pH of the sample should match the pH anticipated for the actual wastewater. In some cases, the carbon may affect the pH of the samples by preferentially adsorbing acids or bases or by supplying inorganics to solution (Fornwalt and Hutchins, 1966a). When this occurs, it may be desirable to adjust the treated samples to the original pH before analyzing for residual impurities.

Data collected during an adsorption study will describe the performance of the carbon and will yield valuable information if properly interpreted. The first step in interpreting the data is to plot an adsorption isotherm (Chapter 6) that will describe the equilibrium distribution of solute between the solid and liquid phases. Use of the Freundlich isotherm to analyze adsorption data is illustrated in the following example.

EXAMPLE PROBLEM 11-1: A wastewater contains 75 mg/ℓ of COD following biological treatment. Effluent guidelines require that the waste must contain no more than 15 mg/ℓ COD prior to discharge. An adsorption test was conducted with the results shown below. Plot the data according to the Freundlich isotherm and:

1. Determine if the desired effluent quality can be achieved by adsorption.
2. Determine the adsorptive capacity of the carbon at this level of effluent quality.
3. Determine the ultimate capacity of the carbon for this wastewater.
4. Calculate values of the constants K and n.

(1) Flask No.	(2) m Weight of Carbon (mg)	(3) Volume of Solution in Flask (ml)	(4) C Final COD Concentration (mg/ℓ)	(5) x Weight of Adsorbate Adsorbed (mg)	(6) x/m Adsorbate Adsorbed per Unit wt. of Carbon (mg/mg)
1	0	200	75	—	—
2	50	200	44	6.2	0.124
3	100	200	30	9.0	0.089
4	200	200	17.5	11.5	0.0575
5	500	200	6.75	13.65	0.0272
6	800	200	3.9	14.22	0.0177
7	1000	200	3.0	14.4	0.0144

Solution:

1. Determine if the desired effluent value can be achieved by adsorption. It is seen from inspection of the data that an effluent value of 15 mg/ℓ can be achieved by adsorption, since a value of 3 mg/ℓ was achieved in the test.
2. Determine the adsorptive capacity of the carbon at an effluent value of 15 mg/ℓ. In order to do this, it is necessary to plot the isotherm, which can be done as follows:

   (a) Calculate values of x and x/m from adsorption test data in columns 1–4.

   For flask #2,

   $$x = (75 \text{ mg/}\ell - 44 \text{ mg/}\ell) \times \frac{200 \text{ m}\ell \text{ in flask}}{1000 \text{ m}\ell/\text{liter}}$$

   $$= 31 \text{ mg/}\ell \times 0.2 \ \ell = 6.2 \text{ mg}$$

   $$\frac{x}{m} = \frac{6.2 \text{ mg}}{50 \text{ mg carbon}} = 0.124 \text{ mg/mg}$$

   (b) The Freundlich isotherm equation is

   $$\log \frac{x}{m} = \log K + \frac{1}{n} \log C$$

   Plot values of x/m vs. C on log-log paper to obtain the isotherm shown in Figure 11-2.

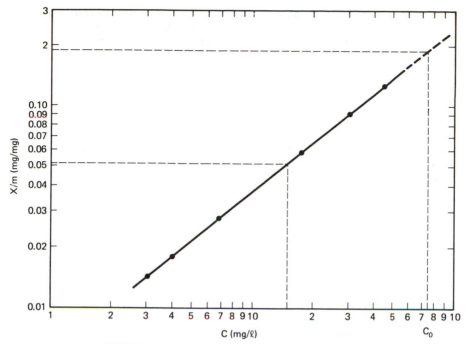

FIGURE 11-2 *Freundlich isotherm for Example 11-1.*

(c) Construct a vertical line from a C value of 15 mg/ℓ to intersect the isotherm line.

(d) The corresponding value of x/m represents the adsorptive capacity of the carbon for this equilibrium effluent concentration.
From Figure 11-2, read

$$\frac{x}{m} = 0.051 \text{ mg/mg}$$

3. Determine the ultimate capacity of the carbon for this waste. The ultimate capacity represents the amount of adsorbate adsorbed when the carbon is in equilibrium with the maximum (influent) solute concentration of 75 mg/ℓ.

(a) Construct a vertical line from the C value of 75 mg/ℓ to intersect the isotherm line.

(b) The corresponding x/m value represents the ultimate capacity of the carbon.
From Figure 11-2, read

$$\frac{x}{m} = 0.186 \text{ mg/mg}$$

4. Determine the value of the constants K and n.

(a) Calculate the value of the constant K which is equal to the intercept of the isotherm line at a value of C equal to 1.0. A log-log plot is used to linearize the Freundlich equation:

$$\frac{x}{m} = KC^{1/n} \qquad\qquad \text{(6-5)}$$

Since

$$\log\left(\frac{x}{m}\right) = \log K + \frac{1}{n}\log C$$

The intercept of such a line will occur at a value of $C = 1.0$ mg/ℓ such that $\log C = 0$. Then

$$\log\frac{x}{m} = \log K$$

or

$$K = \frac{x}{m}$$

From Figure 11-2,

$$K = 0.006$$

(b) Calculate the value of n by determining the slope of the line which is equal to $1/n$.

$$\frac{1}{n} = \frac{\log\dfrac{x}{m} - \log K}{\log C} = \frac{\log\left(\dfrac{x/m}{K}\right)}{\log C}$$

At $x/m = 0.10$, $C = 34.5$ mg/ℓ.

$$\frac{1}{n} = \frac{\log\left(\dfrac{0.10}{0.006}\right)}{\log 34.5} = \frac{1.22}{1.537} = 0.793$$

$$n = 1.26$$

11-3 INTERPRETATION OF ISOTHERMS

An isotherm plot can be analyzed to determine the performance of a given carbon or to compare the performance of two or more carbons; some typical plots of Freundlich isotherms will serve to illustrate how this might be done.

Assume that two carbons, A and B, were tested in the laboratory on the same wastewater. The results of this study are plotted as isotherms in Figure 11-3. This figure indicates that both carbons have some adsorptive power, but carbon A is obviously better than carbon B, since its isotherm lies above the isotherm for carbon B over the entire range of concentrations investigated. Thus, for any equilibrium solute concentration, C, the amount of solute adsorbed, x, per unit weight of carbon will be greater for carbon A than for carbon B. Furthermore, the fact that the isotherm for carbon A has a steep slope indicates that its adsorptive capacity increases at higher equilibrium solute concentrations over that at lower concentrations. Carbons that exhibit steep

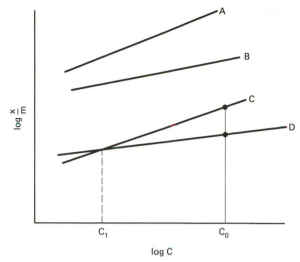

FIGURE 11-3 *Comparison of several carbons (after EPA, 1973 and Hassler, 1974).*

isotherms are, in general, more efficient for column application than are carbons with flat isotherms. However, it must be emphasized that the performance shown by the isotherm is indicative only of the performance of the carbon under the static test conditions of the isotherm test. Additional column tests would have to be conducted to evaluate the performance of the carbon in continuous-flow systems.

In some cases two carbons may yield isotherms as shown by lines C and D in Figure 11-3. It is observed that:

1. Both carbons have the same adsorptive capacity at an equilibrium concentration C_1.

2. Carbon C has more adsorptive capacity at higher equilibrium concentrations.

3. Carbon D has more adsorptive capacity at lower equilibrium concentrations.

If the wastewater being treated has an initial impurity concentration C_0, it is seen that carbon C has the highest adsorptive capacity when in equilibrium with this concentration. Since the objective in wastewater treatment is to exhaust the carbon as completely as possible, the carbon exhibiting the highest adsorptive capacity in equilibrium with the initial impurity concentration represents the preferred carbon (Fornwalt and Hutchins, 1966a).

Nonlinear Isotherms

Although linear isotherms are normally expected, some adsorption data may not produce a straight line for any of the isotherm equations. Nonlinear isotherms may indicate an error in data collection; however, in most cases, they result from some peculiarity of the adsorbent-adsorbate system.

A nonlinear isotherm exhibiting a well-defined breakpoint, such as isotherm *a* in Figure 11-4, may result if the waste being treated is composed of several compounds which are not equally adsorbable (Helbig, 1946). The upper linear portion of this isotherm indicates the removal of readily adsorbable material, whereas residual material remaining at the breakpoint is not as easily adsorbed by the carbon. If the slope below the breakpoint is favorable, additional removal by adsorption may be feasible. However, if the slope is steep, as for curve *a* in Figure 11-4, the breakpoint concentration may represent the minimum concentration that can be economically achieved by adsorption.

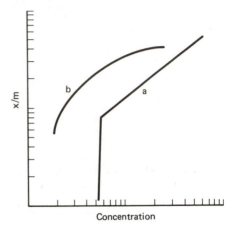

FIGURE 11-4 *Typical nonlinear isotherm.*

In some instances, when activated carbon is used for color removal, the resulting isotherms may exhibit gradual curvature throughout their entire length; see curve *b* in Figure 11-4. According to Helbig (1946), this situation is caused by components, such as iron, alkalies, and sulfides being extracted from the carbon, that impart color to the liquid. For example, iron is capable of causing increased color when extracted from carbon by liquids containing organic acids, and sulfides can increase color by reacting with traces of metal compounds present in the solutions being treated.

Nonlinear isotherms resulting from experimental errors are also frequently encountered in color removal systems (Helbig, 1946). The initial color of many solutions is usually rather dark and is difficult to measure accurately. If the initial color concentration is measured erroneously high, the x/m values will also be erroneously high. The error will be proportionately greater at the high equilibrium concentration values, resulting in an isotherm that is concave upward. If the initial color concentrations is measured erroneously low, the resulting isotherm will be concave downward. If nonlinear isotherms are suspected to be the result of experimental error, the isotherm tests should be repeated and the initial color measurement should be checked carefully.

11-4 TYPES OF ADSORPTION SYSTEMS

Carbon adsorption processes can be operated on either a batch or continuous-flow basis. In batch processes the carbon and wastewater are mixed together in a suitable reaction vessel until the concentration of solute has been reduced to the desired level.

Separation of the spent carbon leaves an effluent suitable for discharge; the carbon can be regenerated and reused.

Most continuous-flow adsorption systems are operated as fixed-bed adsorption columns. These systems are capable of treating large volumes of wastewater and are widely used for both municipal and industrial applications. Fixed-bed adsorbers may be operated as simple columns or as multiple columns in series. Furthermore, they may be operated in either the upflow or downflow mode. In downflow systems the carbon can serve for adsorption and for filtration of suspended solids; however, adsorption is more efficient and the beds are less expensive to operate if suspended solids are removed in advance.

Upflow columns may be operated either as packed or expanded beds. Packed beds require a high-clarity influent (less than 2.5 JTU turbidity) to prevent clogging, whereas expanded beds are capable of handling wastewater high in suspended solids, since the solids will move through the void spaces between the carbon particles and not clog the bed.

The use of upflow and downflow columns in series has been reported to optimize carbon usage and reduce operating costs (Strudgeon, 1976). The upflow bed is placed first in sequence and serves as a roughing contactor, while the downflow filter in the second position functions as a polishing unit.

11-5 DESIGN OF BATCH ADSORPTION SYSTEMS

The number and size of reaction vessels and the amount of carbon required to treat the wastewater must be determined for the design of batch adsorption systems. These factors are influenced by wastewater volume, flow rate, and the rate of the adsorption reaction.

Most batch adsorption systems are operated on a fill-and-draw basis and are limited to small wastewater volumes, since large volumes would require excessively large reaction vessels. If all the wastewater is generated over an 8–12 hr period, only one reaction vessel may be required. Once the reaction vessel is filled, carbon can be added and the mixture agitated until adsorption is complete. The vessel can then be drained and prepared to receive another quantity of wastewater. If wastewater is generated continuously, two tanks may be used and alternated in the fill-and-treat modes.

Reaction rates are important in fill-and-draw operations, since they influence turnaround times for the reaction vessel. Thorough mixing promotes good contact between the carbon and liquid and increases the reaction rate by reducing the thickness of the solvent film surrounding the particle, thus decreasing the resistance offered by film diffusion. In addition, reaction rates are more rapid for powdered carbon than for granular carbon, although the powdered carbon is more difficult to separate from the treated wastewater.

> EXAMPLE PROBLEM 11-2: A small industrial plant generates 50,000 gal of wastewater during an 8 hr operating day. The wastewater contains 100 mg/ℓ of phenol, which must be reduced to 1.0 mg/ℓ prior to discharge. An adsorption isotherm for this wastewater is shown in Figure 11-5. Design a batch adsorption system for treating this wastewater.

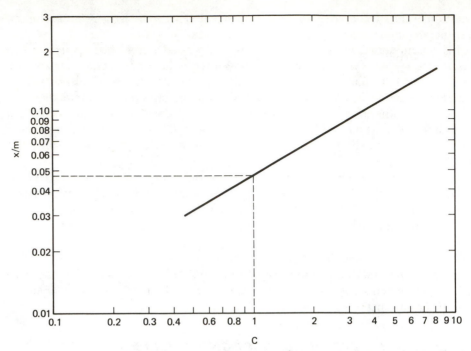

FIGURE 11-5 *Freundlich isotherm for Example 11-2.*

Solution:

1. Determine the reaction rate for the adsorption process. Reaction rates can be measured in the laboratory by mixing 0.5 g portions of carbon and 100 mℓ portions of wastewater in several flasks and agitating the mixtures for various time periods. The pH, temperature, and carbon size should be the same as those to be used in the full-scale system. The contents of one flask should be analyzed after 30 min, the second after 1 hr, etc., and the results should be plotted as shown in the graph. The shape of the graph curve indicates that the adsorption process requires about 4 hr to reach an equilibrium adsorbate concentration. This would be the minimum reaction time required for the full-scale system.

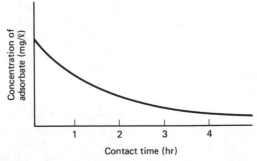

2. Determine the carbon dosage required to achieve the desired 1.0 mg/ℓ phenol concentration.

(a) Refer to Figure 11-5, which represents an isotherm for this wastewater. Construct a vertical line from the required effluent concentration of 1.0 mg/ℓ to intersect the isotherm line. Read the associated carbon capacity.

$$\frac{x}{m} = \frac{0.0475 \text{ mg phenol removed}}{\text{mg carbon}}$$

(b) Calculate the amount of phenol that must be adsorbed per day:

$$\text{phenol removed} = 0.050 \text{ MGD} \ (100 \text{ mg}/\ell - 1.0 \text{ mg}/\ell) \ 8.34 \frac{\text{lb}}{\text{gal}}$$

$$= 41.28 \text{ lb/day}$$

(c) Calculate the amount of carbon required per day to treat the wastewater:

$$\text{carbon requirement} = \frac{41.23 \text{ lb/day}}{\dfrac{0.0475 \text{ mg phenol removed}}{\text{mg carbon}}}$$

$$= 868 \text{ lb/day}$$

(d) Calculate the volume of this carbon if it has a bulk density of 24 lb/ft³:

$$\text{volume/day} = \frac{868 \text{ lb/day}}{24 \text{ lb/ft}^3} = 36.16 \text{ ft}^3\text{/day}$$

11-6 BEHAVIOR OF CARBON ADSORPTION COLUMNS

Adsorption columns behave in much the same way as ion exchange columns (Chapter 10) during operation. When wastewater is introduced at the top of a clean bed of activated carbon, most solute removal initially occurs in a rather narrow band at the top of the column, referred to as the *adsorption zone*. As operation continues, the upper layers of carbon become saturated with solute and the adsorption zone progresses downward through the bed. Eventually, the adsorption zone reaches the bottom of the column, and the solute concentration in the effluent begins to increase. A plot of effluent solute concentration vs. time usually yields an S-shaped curve, referred to as a *breakthrough curve*. The formation and movement of the adsorption zone and the resulting breakthrough curve are shown in Figure 11-6. The point on the S-shaped curve at which the solute concentration reaches its maximum allowable value is referred to as *breakthrough*. The point where the effluent solute concentration reaches 95 % of its influent value is usually called the *point of column exhaustion*.

The formation and movement of the adsorption zone has been described mathematically by Michaels (1952). The time required for the exchange zone to move the length of its own height down the column once it has become established is

$$t_z = \frac{V_s}{Q_w} \tag{11-1}$$

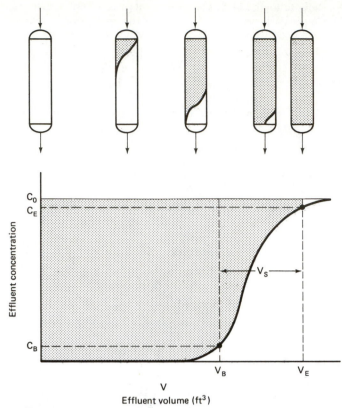

FIGURE 11-6 *Idealized breakthrough curve for carbon adsorption column.*

where V_s = total volume of wastewater treated between breakthrough and exhaustion (ft^3)

Q_w = wastewater flow rate (ft^3/sec).

The time required for the exchange zone to become established and move completely out of the bed is

$$t_E = \frac{V_E}{Q_w} \quad \text{(11-2)}$$

where V_E = total volume of wastewater treated to the point of exhaustion (ft^3).

The rate at which the adsorption zone is moving down through the bed is

$$U_z = \frac{h_z}{t_z} = \frac{h}{t_E - t_f} \quad \text{(11-3)}$$

where h_z = height of exchange zone (ft)

h = total column height (ft)

t_f = time required for the exchange zone to initially form (sec).

Rearranging equation 11-3 provides an expression for the height of the exchange zone, an important design parameter:

$$h_z = \frac{h(t_z)}{t_E - t_f} \tag{11-4}$$

All the terms in equation 11-4, with the exception of t_f, can be easily evaluated by conducting a laboratory column study. The value of t_f cannot be measured directly, but the limits for t_f can be established by further analysis of the adsorption zone.

If the carbon within the adsorption zone were completely exhausted, it would contain an amount of solute equal to

$$S_{\max} = C_0(V_E - V_B) \tag{11-5}$$

where C_0 = the initial solute concentration (lb/ft^3)

$(V_E - V_B)$ = volume of wastewater treated between breakthrough and exhaustion (ft^3).

However, only a fraction of carbon within the adsorption zone contains solute. The amount of solute that has been removed by the adsorption zone from breakthrough to exhaustion is the shaded area above the curve in Figure 11-6. This area represents an amount of solute, S_z, such that

$$S_z = \int_{V_B}^{V_E} (C_0 - C) \, dV \tag{11-6}$$

At breakthrough the fraction of carbon present in the adsorption zone still possessing ability to remove solute is

$$F = \frac{S_z}{S_{\max}} = \frac{\int_{V_B}^{V_E} (C_0 - C) \, dV}{C_0(V_E - V_B)} \tag{11-7}$$

$$F = \int_0^1 \left(1 - \frac{C}{C_0}\right) d\frac{(V - V_B)}{(V_E - V_B)} \tag{11-8}$$

If the adsorption zone is essentially saturated at breakthrough, the value of F will be very close to zero and the time required for the zone to form initially (t_f) will be approximately the same as the time required for the zone to move down a distance equal to its own height. If the adsorption zone is practically free of solute at the breakpoint, i.e., $F \simeq 1$, the time required for zone formation is very short. If the concentration front is characterized by a typical S-shaped curve, F is approximately 0.5.

The time for zone formation can then be written as a function of zone velocity:

$$t_f = (1 - F)t_z \tag{11-9}$$

Substituting equation 11-9 into equation 11-4 yields a ratio of zone height to total column height:

$$\frac{h_z}{h} = \frac{t_z}{t_E - (1 - F)t_z} \tag{11-10}$$

If all of the column were completely saturated, the total amount of solute adsorbed would be

$$S_T = \rho(h)\left[\frac{x}{m}\right]_s A_{cs} \tag{11-11}$$

where ρ = apparent packed density of carbon in the bed (lb/ft³)

            h = total height of carbon in bed (ft)

         $\left[\dfrac{x}{m}\right]_s$ = amount of solute adsorbed per unit weight of carbon at saturation. (lb/lb)

         A_{cs} = cross-sectional area of column (ft²).

At the breakpoint, an adsorption column will not be completely saturated but will be composed of the partially exhausted adsorption zone and the totally exhausted material located above the adsorption zone. The height of the exhausted zone is $(h - h_z)$, and the total amount of solute adsorbed in the column is

$$S_B = \left[\frac{x}{m}\right]_s \rho[(h - h_z) + Fh_z]A_{cs} \tag{11-12}$$

The percentage of the total column saturated at breakthrough is

$$\% \text{ saturation} = \frac{\left[\dfrac{x}{m}\right]_s \rho[(h - h_z) + Fh_z]A_{cs}}{\rho(h)\left[\dfrac{x}{m}\right]_s A_{cs}} \tag{11-13}$$

$$\% \text{ saturation} = \frac{h + (F - 1)h_z}{h} \times 100 \tag{11-14}$$

The shape of the breakthrough curve depends on the nature of the wastewater being treated. If there is only one adsorbable component in the wastewater, the adsorption zone will be short and the breakthrough curve will be steep like curve *a* in Figure 11-7 (Fornwalt and Hutchins, 1966a). If there is a mixture of components having different adsorbabilities, the adsorption zone will be deep and the breakthrough curve will be gradual like curve *b* in Figure 11-7. According to Fornwalt and Hutchins (1966a), if the breakthrough curve is steep such that breakthrough occurs near the point of column exhaustion (curve *a*) single column operation is feasible. However, if the breakthrough concentration is reached before the column is fully exhausted (curve *b*), a multiple column installation is preferable.

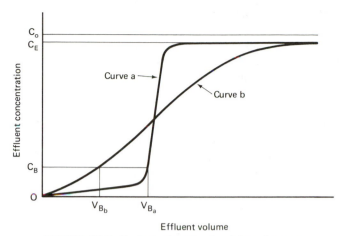

FIGURE 11-7 *Typical adsorption breakthrough curves.*

11-7 DESIGN OF CARBON ADSORPTION COLUMNS: LABORATORY PROCEDURE

The design of carbon adsorption columns must be preceded by laboratory tests in small-scale columns. The tests can be used to determine (1) the necessary column residence time, (2) the volume of wastewater treated before breakthrough occurs, (3) the head loss through the column, and (4) the shape of the column exhaustion curve (Fornwalt and Hutchins, 1966a). Results from the laboratory column study should be confirmed by pilot plant operation before final scale-up. Laboratory data should not be scaled up directly to the full-scale production unit.

When conducting a laboratory column study the engineer usually proceeds in the following manner:

1. Select one or more carbons with suitable physical and chemical characteristics.

2. Conduct a preliminary batch adsorption study (isotherm study) to determine the degree of treatment attainable by the various carbons. The results of this study can serve as a basis to compare the performance of the carbons and to select the carbon that is most effective.

3. Conduct a laboratory column test to obtain data for use in system design.

4. Analyze the laboratory data, using an appropriate design model.

Laboratory column tests should be conducted with the same granular carbon that is to be used in the production unit. Care must be taken when filling the test columns with carbon to avoid the entrapment of air that would produce channeling of flow and reduce contact between the carbon and liquid. Air entrapment can be avoided by slurrying the carbon with boiling water and feeding the hot slurry to the column (Fornwalt and Hutchins, 1966a).

Full-scale production columns may be 25 to 35 ft tall and 10 to 15 ft in diameter. Laboratory columns must be carefully designed and operated so as to produce results that will predict the performance of the full-scale columns. Important considerations include column height, column diameter, and liquid application rate.

Laboratory columns should be approximately the same height as the production unit in order to provide the most reliable data. Short columns may have to be arranged in series to achieve the desired depth. Furthermore, Conway and Ross (1980), recommend that the ratio of laboratory column diameter to carbon particle size be at least 25:1 so as to prevent wall effects from influencing test results. Fornwalt and Hutchins (1966a) suggest a column diameter of at least 4 in. to simulate the packing characteristics of a production column and to reduce wall effects.

The rate at which wastewater is applied to the laboratory column is an important test variable, since it determines the residence time in the column and thus the time of contact between the liquid and the carbon. Laboratory units should be operated at application rates which, based on column length, will yield empty-volume residence times of 25 to 50 min (Conway and Ross, 1980). Minimum application rates should not be less than 1 gpm/ft², corresponding to a linear velocity of 0.133 ft/min. Higher application rates, 2 to 8 gpm/ft², are typical of full-scale units and should be used in laboratory studies if column length is sufficient to provide the necessary residence time, t. Adsorption is actually a function of residence time rather than application rate and so residence times should be maintained constant during scale-up (Fornwalt and Hutchins, 1966a).

EXAMPLE PROBLEM 11-3: A laboratory adsorption column, 4 in. in diameter and 12 ft deep, is found to produce good results when operated at a flow of 10 gal/hr. Calculate the following:

1. The application rate in gpm/ft².
2. The residence time, t, in the column.
3. The volumetric flow rate, V_b, in bed volumes per hour, associated with this residence time.
4. The application rate that would yield the same residence time in a production column that was 10 ft in diameter and 30 ft tall.

Solution:

1. The application rate in gpm/ft²:

$$\text{column area} = \frac{\pi(4)^2}{4} = 12.57 \text{ in.}^2 = 0.087 \text{ ft}^2$$

$$\text{application rate} = \frac{10 \text{ gal/hr}}{60 \frac{\text{min}}{\text{hr}} \times 0.087 \text{ ft}^2} = 1.915 \text{ gpm/ft}^2$$

2. The residence time, t, in the column:

$$\text{linear velocity} = \frac{1.915 \text{ gpm/ft}^2}{7.48 \text{ gal/ft}^3} = 0.256 \text{ ft/min}$$

$$t = \frac{12 \text{ ft}}{0.256 \text{ ft/min}} = 46.87 \text{ min}$$

3. The volumetric flow rate, V_b, in bed volumes per hour:

    The residence time can be converted to a volumetric flow rate if the fractional void volume in the column is known. Fractional void volumes for granular carbon columns normally range from 0.40 to 0.55 and can be assumed to average 0.50 (Fornwalt and Hutchins, 1966a). Thus,

$$V_b = \frac{0.50}{t}$$

and

$$V_b = \frac{0.50}{46.87 \text{ min/60 min/hr}} = 0.64 \frac{\text{bed volumes}}{\text{hr}}$$

4. The application rate for a 10 ft diameter by 30 ft deep column to give the same residence time as in the laboratory column:

$$\text{required linear velocity} = \frac{30 \text{ ft}}{46.87 \text{ min}} = 0.640 \text{ ft/min}$$

$$\text{application rate} = 0.640 \frac{\text{ft}}{\text{min}} \times 7.48 \frac{\text{gal}}{\text{ft}^3} = 4.707 \frac{\text{gpm}}{\text{ft}^2}$$

11-8 DESIGN OF CARBON ADSORPTION COLUMNS: MATHEMATICAL MODELS

Data collected during laboratory and pilot plant tests serve as a basis for the design of full-scale adsorption columns. A number of mathematical models have been developed for use in design. Several of these design models, along with examples for their use, are explained in subsequent sections.

Design Based on Mass Transfer Model

Weber (1972) has presented a design approach which consists of using data from batch laboratory studies to design continuous-flow adsorption columns. The batch data are first transformed into a theoretical breakthrough curve which serves as a basis for design, using the concepts presented by Michaels (1952).

If data from a batch adsorption study are plotted on arithmetic paper (as shown in Figure 11-8), an experimental equilibrium curve will result. An *operating line* can be constructed by considering an adsorbate materials balance over the column (Humenick, 1977). The amount of adsorbate on the carbon is related to the amount of adsorbate in solution by

$$\left[\frac{x}{m}\right] = \frac{C}{C_0}\left[\left[\frac{x}{m}\right]_0 - \left[\frac{x}{m}\right]_r\right] + \left[\frac{x}{m}\right]_r \qquad \text{(11-15)}$$

where $\left[\dfrac{x}{m}\right]$ = amount of adsorbate on the carbon (lb/lb)

$\left[\dfrac{x}{m}\right]_0$ = value of $\left[\dfrac{x}{m}\right]$ in equilibrium with initial influent concentration C_0

$\left[\dfrac{x}{m}\right]_r$ = residual amount of adsorbate on the carbon after regeneration.

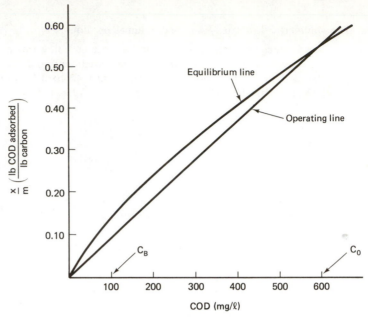

FIGURE 11-8 *Plot of equilibrium and operating lines.*

Assuming the fresh carbon is being used, $\left[\dfrac{x}{m}\right]_r$ is initially equal to zero and equation 11-15 becomes

$$\frac{x}{m} = \frac{C}{C_0}\left[\frac{x}{m}\right]_0 \tag{11-16}$$

Thus, when $C = C_0$, $\left[\dfrac{x}{m}\right]$ is equal to $\left[\dfrac{x}{m}\right]_0$, and the coordinate C_0, $\left[\dfrac{x}{m}\right]_0$ represents a point on the operating line. Since the operating line must also pass through the origin, it can be constructed as shown in Figure 11-8.

According to Weber (1972), the rate of transfer of solute from solution over a differential depth of column, dh, is given by

$$(F_w)\,dC = Ka(C - C^*)\,dh \tag{11-17}$$

where F_w = wastewater flow rate (lb/min-ft²)

Ka = overall mass-transfer coefficient, which includes the resistances offered by film diffusion and pore diffusion

C^* = equilibrium concentration of solute in solution corresponding to an adsorbed concentration, $\left[\dfrac{x}{m}\right]$.

The term $(C - C^*)$ is the driving force for adsorption and is equal to the distance between the operating line and equilibrium curve at any given value of $\left[\dfrac{x}{m}\right]$. Integrating equation 11-17 and solving for the height of the adsorption zone,

$$h_z = \frac{F_w}{Ka} \int_{C_B}^{C_E} \frac{dC}{(C - C^*)} \qquad \text{(11-18)}$$

where C_B = concentration of solute in effluent at breakthrough

C_E = concentration of solute in effluent at exhaustion.

For any value of h less than h_z, corresponding to a concentration C between C_B and C_E, equation 11-18 can be written as

$$h = \frac{F_w}{Ka} \int_{C_B}^{C} \frac{dC}{(C - C^*)} \qquad \text{(11-19)}$$

Dividing equation 11-19 by 11-18 results in

$$\frac{h}{h_z} = \frac{\int_{C_B}^{C} \dfrac{dC}{C - C^*}}{\int_{C_B}^{C_E} \dfrac{dC}{C - C^*}} = \frac{V - V_B}{V_E - V_B} \qquad \text{(11-20)}$$

Graphical integration of equation 11-20 makes it possible to plot a theoretical breakthrough curve and determine the operating characteristics necessary to design an adsorption column. The procedure is illustrated by the following example.

EXAMPLE PROBLEM 11-4:   A batch adsorption study was conducted to investigate the removal of COD from a refinery wastewater by using activated carbon. Results of the study indicate that the Freundlich isotherm can be described by

$$\frac{x}{m} = 0.004C^{0.77}$$

Determine the design of an 18 ft deep by 4 ft diameter fixed-bed, continuous-flow adsorption column to treat a flow of 5.0 gpm/ft². The wastewater has an initial COD of 600 mg/ℓ, which must be reduced to 100 mg/ℓ prior to discharge. The carbon selected for use has an apparent density of 35 lb/ft³ and a packed density (bed density) of 30 lb/ft³. The value of Ka has been found to be 1100 lb/min-ft³ at a temperature of 20°C.

Solution:

1. Assume various values of C, and calculate the corresponding values of $\dfrac{x}{m}$

    using the Freundlich equation presented above.
        If $C = 150$ mg/ℓ,

$$\frac{x}{m} = 0.004(150)^{0.77} = 0.190 \; \frac{\text{lb COD}}{\text{lb carbon}}$$

Plot the resulting equilibrium line shown in Figure 11-9.

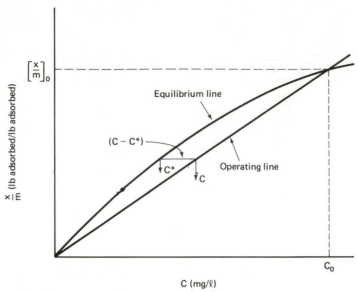

FIGURE 11-9   Equilibrium and operating lines for Example 11-3.

2. Construct an operating line passing through the origin and through the point given by the following coordinates:

$$C_0 = 600 \text{ mg}/\ell$$

$$\left[\frac{x}{m}\right]_0 = 0.55 \frac{\text{lb COD}}{\text{lb carbon}}$$

3. Determine values of $C - C^*$ from Figure 11-9 for various values of $\frac{x}{m}$, and calculate the corresponding values of $(C - C^*)^{-1}$.

TABLE 11-4

(1)	(2)	(3)	(4)	(5)	(6)
$\frac{x}{m}$	C	$(C - C^*)$	$(C - C^*)^{-1}$	$\int_{C_B}^{C_E} (C - C^*)^{-1} dC$	$\frac{(V - V_B)}{(V_E - V_B)}$
0.09	100 (C_B)	41	0.0244	0	0
0.15	164	52	0.0192	1.43	0.113
0.20	220	59	0.0169	2.425	0.191
0.25	274	59	0.0169	3.345	0.264
0.30	327	53	0.0189	4.21	0.332
0.35	381	48	0.0208	5.25	0.414
0.40	435	43	0.0233	6.465	0.510
0.45	490	28	0.0357	8.065	0.636
0.50	545	17	0.0588	10.62	0.838
0.524	570 (C_E)	9	0.111	12.68	1.000

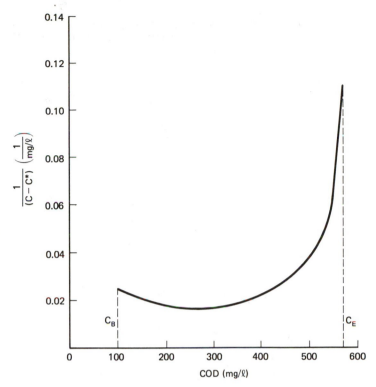

FIGURE 11-10 Curve to evaluate $\int_{CB}^{CE} \frac{dc}{c - c^*}$.

4. Construct a plot of $(C - C^*)^{-1}$ vs. C (see Figure 11-10), and determine the area under the curve by graphical integration to evaluate the term

$$\int_{C_B}^{C_E} \frac{dC}{C - C^*}$$

Since $C - C^*$ approaches infinity as C approaches C_0, it is necessary to terminate the plot at a C value somewhat less than C_0. A value of $C_E = 570$ mg/ℓ was selected. This corresponds to $0.95C_0$, which is frequently considered to be the point of exhaustion for adsorption columns. The area under the curve is determined to be

$$\int_{C_B}^{C_E} \frac{dC}{(C - C^*)} = 12.68$$

5. The depth of the adsorption zone can be calculated by equation 11-18

$$h_z = \frac{F_w}{Ka} \int_{C_B}^{C_E} \frac{dC}{(C - C^*)}$$

$$= \frac{(5 \text{ gpm/ft}^2) \times (8.34 \text{ lb/gal})}{1100 \text{ lb/min-ft}^3} \times 12.68 \qquad \textbf{(11-18)}$$

$$= \frac{507.9 \text{ lb/min-ft}^2}{1100 \text{ lb/min-ft}^3} = 0.46 \text{ ft}$$

6. Determine the fractional capacity of the adsorption zone at breakthrough. This can be determined by graphical integration of a plot of $(V - V_B)/(V_E - V_B)$ vs. C/C_0. Values of $(V - V_B)/V_E - V_B)$ are shown in Table 11-4. These values were calculated by dividing the values in column 5 by the last number in column 5, which corresponds to the value of C_E. (Refer to equation 11-20.) The shaded area in Figure 11-11 represents the fractional capacity, F.

$$F = 0.320 \text{ (by graphical integration)}$$

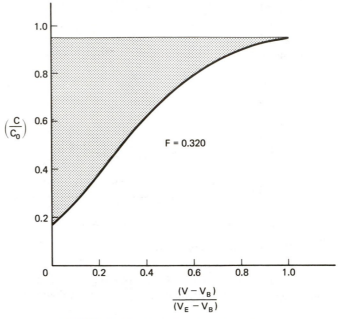

FIGURE 11-11 *Theoretical breakthrough curve.*

7. Calculate the percent saturation for the column at breakthrough:

$$\% \text{ saturation} = \frac{18 \text{ ft} + (0.321 - 1)0.46 \text{ ft}}{18 \text{ ft}} \times 100$$

$$= \frac{18 \text{ ft} - 0.31 \text{ ft}}{18 \text{ ft}} \times 100 = 98.27\%$$

(11-14)

8. Calculate the capacity of the column to remove COD.
   (a) The weight of carbon in the column is determined as follows:

$$\text{weight of carbon} = 18 \text{ ft} \times \frac{(4)^2 \pi}{4} \times 30 \text{ lb/ft}^3 = 6785.7 \text{ lb}$$

   (b) At exhaustion, the $\frac{x}{m}$ value (see Figure 11-9) is 0.55 (lb/COD)/(lb carbon). Thus, the amount of COD adsorbed at exhaustion, based on 98.27% saturation, is

lb COD
adsorbed = 6785.7 lb carbon $\times$ 0.55 lb COD/lb carbon $\times$ 0.9827

= 3667.57 lb

9. Calculate the time required to exhaust the column, based on a wastewater flow of 5 gpm/ft², with an initial COD of 600 mg/ℓ:

COD
concentration = $(600 \text{ mg}/\ell) \times \left(10^{-3} \frac{\text{g}}{\text{mg}}\right) \times \left(2.205 \times 10^{-3} \frac{\text{lb}}{\text{g}}\right) \times 28.32 \frac{1}{\text{ft}^3}$

= 0.0375 lb/ft³

wastewater
application
rate $= (5.0 \text{ gpm/ft}^2) \times \left(\frac{\pi(4)^2}{4} \text{ ft}^2\right) \times \left(1440 \frac{\text{min}}{\text{day}}\right)$

= 90,478 gal/day

COD appli-
cation rate = $(90{,}478 \text{ gal/day})(0.0375 \text{ lb/ft}^3)\left(\frac{1}{7.48} \times \frac{\text{ft}^3}{\text{gal}}\right)$

= 454 lb/day

run time $= \dfrac{3667.57 \text{ lb COD/run}}{454 \text{ lb COD/day}} = 8.08$ days/run

The reported advantage of the mass-transfer design approach is that the necessary data can be collected by batch laboratory studies, which are less time-consuming and less expensive to conduct than the continuous-flow studies required by other methods. A serious limitation of this approach is the difficulty of evaluating the mass-transfer coefficient, Ka, that will exist in the actual continuous-flow column. Film diffusion generally controls mass transfer in continuous-flow systems; however, there is no practicable method for determining film diffusion coefficients from the rapidly mixed batch studies utilized in the mass-transfer approach. A correlation technique for determining Ka has been presented by Keinath and Weber (1968) and demonstrated by Humenick (1977). While the approach is straightforward mathematically, it requires that the diffusivity of the solute be determined. Although this parameter can be calculated with some degree of accuracy for single component solutions, it cannot be determined accurately for the complex wastewaters which must frequently be treated by adsorption. Consequently, other approaches, although more expensive and time-consuming, may often prove better for design.

Design Based on Bohart-Adams Equation

Bohart and Adams (1920) proposed an equation for the design of carbon adsorption columns that has been widely used. Their equation, which is based on surface reaction rate theory, can be represented as follows:

$$\ln \left(\frac{C_0}{C_B} - 1\right) = \ln \left(e^{KN_0 \frac{x}{V}} - 1\right) - KC_0 t \qquad \textbf{(11-21)}$$

where C_0 = initial concentration of solute (lb/ft³)

C_B = desired concentration of solute at breakthrough (lb/ft³)

K = rate constant (ft³ liquid/lb carbon-hr)

N_0 = adsorptive capacity of carbon (lb/ft³)

x = depth of carbon bed (ft)

V = linear flow velocity of feed to bed (ft/hr, gpm/ft²)

t = service time of column under above conditions (hr).

The equation can be rearranged to yield an expression for service time, t. Realizing that

$$e^{KN_0\frac{x}{V}} \gg 1 \tag{11-22}$$

the equation simplifies to

$$\ln\left(\frac{C_0}{C_B} - 1\right) = \ln e^{KN_0\frac{x}{V}} - KC_0 t \tag{11-23}$$

Since

$$\ln e^{KN_0\frac{x}{V}} = KN_0\frac{x}{V} \tag{11-24}$$

we have

$$(V)\ln\left(\frac{C_0}{C_B} - 1\right) = KN_0\frac{x}{V} - KC_0 t \tag{11-25}$$

which is the same as

$$\ln\left(\frac{C_0}{C_B} - 1\right) = KN_0 x - KC_0 t(V) \tag{11-26}$$

dividing both sides by KN_0 yields

$$\frac{V}{KN_0}\ln\left(\frac{C_0}{C_B} - 1\right) = x - \frac{VC_0 t}{N_0} \tag{11-27}$$

Solving for t,

$$t = \frac{N_0}{C_0 V}x - \frac{1}{C_0 K}\ln\left(\frac{C_0}{C_B} - 1\right) \tag{11-28}$$

The form of the Bohart-Adams equation, shown as equation 11-28, can be used to determine the service time, t, of a column of bed depth, x, given the values of N_0, C_0, and K which must be determined for laboratory columns operated over a range of velocity values, V.

Setting $t = 0$ and solving equation 11-28 for x yields

$$x_0 = \frac{V}{KN_0}\ln\left(\frac{C_0}{C_B} - 1\right) \tag{11-29}$$

where x_0 is the minimum column height necessary to produce an effluent concentration C_B. This value may be taken as being the height of the adsorption zone referred to in section 11-6.

Evaluation of the constants in the Bohart-Adams equation is illustrated by the following example.

EXAMPLE PROBLEM 11-5: A wastewater contains 12 mg/ℓ of phenol, which must be reduced to 0.5 mg/ℓ by carbon adsorption prior to discharge. A series of column studies is conducted in the laboratory, using 1 in. diameter Plexiglass columns with bed depths and flow rates as shown in the table. The throughput volume and time associated with a breakthrough concentration of 0.50 mg/ℓ are also given.

Determine values for the Bohart-Adams constants K and N_0, and find the value of x_0 for each flow rate.

Flow rate (gpm/ft^2)	Bed Depth (ft)	Throughput Volume (gal)	Time (hr)
	3.0	818	1000
2.5	5.0	1809	2213
	7.0	2789	3412
	3.0	589	400
4.5	5.0	1452	987
	9.0	3181	2162
	5.0	1146	438
8.0	9.0	2773	1060
	12.0	3989	1525

Solution:

1. The Bohart-Adams equation, 11-28, is

$$t = \frac{N_0}{C_0 V} x - \frac{1}{C_0 K} \ln \left(\frac{C_0}{C_B} - 1 \right) \qquad \textbf{(11-28)}$$

This equation is of the form $y = mx + b$, which plots as a straight line on arithmetic paper. A plot of service time (t) to breakthrough vs. bed depth (ft) should yield a line with a slope equal to $N_0/C_0 V$ and an intercept of

$$-\frac{1}{C_0 K} \ln \left(\frac{C_0}{C_B} - 1 \right)$$

Values of N_0 and K can be determined from the plot.

2. Plot the data as shown in Figure 11-12, and determine the slope and intercept of each line.

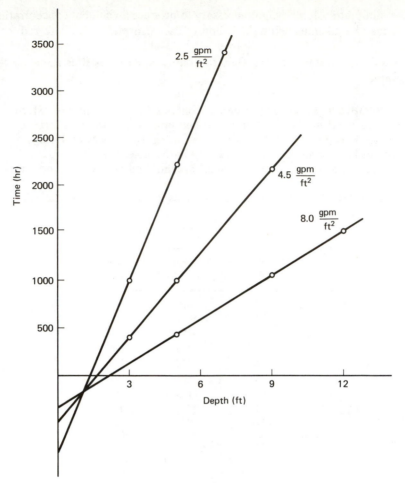

FIGURE 11-12 *Plot to determine slope and intercept for Bohart-Adams equation.*

3. Calculate the values of N_0, using the slopes of the lines.

$$\text{slope} = \frac{N_0}{C_0 V}$$

where V = linear velocity of flow

              C_0 = initial phenol concentration (lb/ft³).

For 2.5 gpm/ft²:

$$V = 2.5 \frac{\text{gpm}}{\text{ft}^2} \times \frac{1}{7.48} \frac{\text{ft}^3}{\text{gal}} \times 60 \frac{\text{min}}{\text{hr}} = 19.23 \text{ ft/hr}$$

$$C_0 = (12 \text{ mg}/\ell) \times \left(10^{-3} \frac{\text{g}}{\text{mg}}\right) \times \left(2.205 \times 10^{-3} \frac{\text{lb}}{\text{g}}\right) \times 28.32 \frac{1}{\text{ft}^3}$$

$$= 7.49 \times 10^{-4} \text{ lb/ft}^3$$

$$\text{slope} = 603 \text{ hr/ft} = \frac{N_0}{C_0(V)}$$

$$N_0 = 603\, \frac{\text{hr}}{\text{ft}}(C_0)(V)$$

$$= 603\, \frac{\text{hr}}{\text{ft}}\left(7.49 \times 10^{-4}\, \frac{\text{lb}}{\text{ft}^3}\right)\left(19.23\, \frac{\text{ft}}{\text{hr}}\right)$$

$$= 8.68 \text{ lb/ft}^3$$

Values of N_0 for other flow rates can be calculated in a similar manner.

4. Calculate the K values, using the intercepts of the lines:

$$\text{intercept} = -\frac{1}{C_0 K} \ln\left(\frac{C_0}{C_B} - 1\right)$$

For 2.5 gpm/ft^2:

$$\text{intercept} = -800 \text{ hr} = -\frac{1}{7.49 \times 10^{-4} \text{ lb/ft}^3(K)} \ln\left(\frac{12 \text{ mg}/\ell}{0.5 \text{ mg}/\ell} - 1\right)$$

$$K = -\frac{1}{(7.49 \times 10^{-4} \text{ lb/ft}^3)(-800 \text{ hr})} \ln(24 - 1)$$

$$= 5.23\, \frac{\text{ft}^3}{\text{lb-hr}}$$

Other K values can be calculated in a similar manner.

5. Calculate the x_0 values for the various flow rates. x_0 may be considered to be the height of the exchange zone.

$$x_0 = \frac{V}{KN_0} \ln\left(\frac{C_0}{C_B} - 1\right)$$

For 2.5 gpm/ft^2:

$$x_0 = \frac{19.23 \text{ ft/hr}}{\left(5.23\, \frac{\text{ft}^3}{\text{lb-hr}}\right)(8.68 \text{ lb/ft}^3)} \ln\left(\frac{12}{0.5} - 1\right)$$

$$= 1.33 \text{ ft}$$

6. Summarize the Bohart-Adams constants in tabular form:

Flow rate (gpm/ft^2)	V (ft/hr)	Slope (hr/ft)	Intercept (hr)	N_0 (lb/ft^3)	K $\left(\frac{\text{ft}^3}{\text{lb-hr}}\right)$	x_0 (ft)
2.5	19.23	603	−800	8.68	5.23	1.33
4.5	36.09	294	−488	7.95	8.58	1.66
8.0	64.17	155	−330	7.45	12.38	2.18

EXAMPLE PROBLEM 11-6: Using the data from example Problem 11-5, design an adsorption column to treat a waste flow of 10,000 gal/day containing 12 mg/ℓ phenol. The required effluent concentration is 0.05 mg/ℓ, and it is desired to have the column operate for 60 days before reaching exhaustion. Operation is to be 8 hr/day, 5 days week.

Solution:

1. Plot the N_0 and K values from example problem 11-5, as shown in Figure 11-13, to show the variation in these parameters with flow rate.

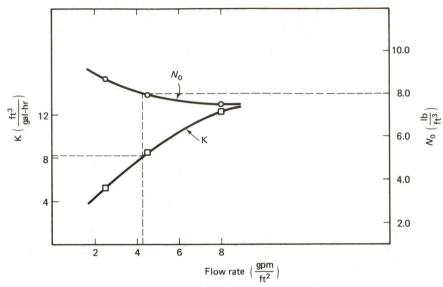

FIGURE 11-13 *Variation of K and N_0 with flow rate.*

2. Calculate the wastewater flow rate in gallons per minute, assuming operation will be 8 hr/day.

$$Q \text{ (gpm)} = \frac{10{,}000 \text{ gal/day}}{8 \dfrac{\text{hr}}{\text{day}} \times 60 \dfrac{\text{min}}{\text{hr}}} = 20.83 \text{ gpm}$$

3. Select a column diameter of 2.5 ft, and calculate the corresponding hydraulic loading rate:

$$\text{area} = \frac{\pi(2.5)^2}{4} = 4.90 \text{ ft}^2$$

$$\text{hydraulic loading} = \frac{20.83 \text{ gpm}}{4.90 \text{ ft}^2} = 4.25 \text{ gpm/ft}^2$$

4. Enter Figure 11-13 at this hydraulic loading rate and find the proper values of N_0 and K:

$$N_0 = 8.05 \text{ lb/ft}^3$$

$$K = 8.20 \frac{\text{ft}^3}{\text{gal-hr}}$$

5. Calculate the depth of bed required for 60 day operation. Solving equation 11-28 for bed depth (x) yields

$$x = \frac{C_0 V}{N_0}\left[t + \frac{1}{C_0 K}\ln\left(\frac{C_0}{C_B} - 1\right)\right]$$

$$= \frac{t(C_0)V}{N_0} + \frac{V}{N_0 K}\ln\left(\frac{C_0}{C_B} - 1\right)$$

where $C_0 = 12$ mg/ℓ $= 7.49 \times 10^{-4}$ lb/ft^3

$C_B = 0.5$ mg/ℓ

$$V = 4.25 \text{ gpm/ft}^2 \times \frac{1}{7.48}\frac{\text{ft}^3}{\text{gal}} \times 60 \frac{\text{min}}{\text{hr}} = 34.09 \frac{\text{ft}}{\text{hr}}$$

$t = 60$ days $\times$ 8 hr/day $= 480$ hour

$$x = \frac{(480 \text{ hr})\left(7.49 \times 10^{-4}\frac{\text{lb}}{\text{ft}^3}\right)\left(34.09\frac{\text{ft}}{\text{hr}}\right)}{8.05\frac{\text{lb}}{\text{ft}^3}}$$

$$+ \frac{\left(34.09\frac{\text{ft}}{\text{hr}}\right)}{\left(8.05\frac{\text{lb}}{\text{ft}^3}\right)\left(8.20\frac{\text{ft}^3}{\text{gal-hr}}\right)}\ln\left(\frac{12}{0.5} - 1\right)$$

$$= 1.52 + 1.62 = 3.14 \text{ ft}$$

6. Calculate the amount of carbon required to fill the bed:

volume of carbon $= 4.90$ ft^2 $\times$ 3.14 ft $= 15.39$ ft^3

7. Calculate the amount of carbon required on an annual basis, assuming no regeneration:

$$\frac{\text{number of carbon}}{\text{changes per year}} = \frac{52 \text{ weeks} \times 5\frac{\text{days}}{\text{week}}}{60 \text{ days/change}} = 4.33$$

annual volume $= 4.33$ changes $\times$ $15.39\frac{\text{ft}^3}{\text{change}} = 66.64$ ft^3

Design Based on Bed Depth/Service Time (BDST) Approach

At least nine individual column tests must be conducted to collect the laboratory data required for the Bohart-Adams approach, an expensive and time-consuming task. A technique has been presented by Hutchins (1973) which requires only three column tests to collect the necessary data. In this technique, called the *bed depth/service time* (BDST) *approach*, the Bohart-Adams equation is expressed as

$$t = ax + b \qquad\qquad \textbf{(11-30)}$$

where $\quad a = \text{slope} = \dfrac{N_0}{C_0(V)}$

$\quad\quad\quad b = \text{intercept} = \dfrac{1}{K(C_0)} \ln \left(\dfrac{C_0}{C_B} - 1 \right).$

According to BDST, if a value of a is determined for one flow rate, values for other flow rates can be calculated by multiplying the original slope a by the ratio of the original and new flow rates. It is not necessary to adjust the b value, since this term is assumed to be insignificantly affected by changing flow rates.

It is also proposed that data collected at one influent solute concentration be adjusted by the BDST technique and used to design systems for treating other influent solute concentrations. If a laboratory test is conducted at solute concentration C_1, yielding an equation of the form

$$t = a_1 x + b_1 \tag{11-31}$$

it is possible to predict the equation for concentration C_2 as follows:

$$a_2 = a_1 \frac{C_1}{C_2} \tag{11-32}$$

$$b_2 = b_1 \left(\frac{C_1}{C_2} \right) \frac{\ln (C_2/C_F - 1)}{\ln (C_1/C_B - 1)} \tag{11-33}$$

where $\quad a_1 = $ slope at concentration C_1

$\quad\quad\quad a_2 = $ slope at concentration C_2

$\quad\quad\quad b_1 = $ intercept at concentration C_1

$\quad\quad\quad b_2 = $ intercept at concentration C_2

$\quad\quad\quad C_F = $ effluent concentration at influent concentration C_2

$\quad\quad\quad C_B = $ effluent concentration at influent concentration C_1.

This method of estimating the effect of solute concentration on adsorption is only approximate. Additional laboratory tests should be conducted at the concentration of interest to obtain data that are more reliable.

EXAMPLE PROBLEM 11-7: A laboratory adsorption column operating at a linear flow rate of 1.0 gpm/ft² was used to study the removal of phenol from a wastewater. The phenol concentration was reduced from 100 mg/ℓ to 2 mg/ℓ, with the bed depths and service times as tabulated below.

Use the BDST approach to predict the operating time for a 4 ft deep bed operating at a flow rate of 1.5 gpm/ft². Also, predict the operating time for this column if the influent concentration of phenol increased from 100 mg/ℓ to 140 mg/ℓ at a flow rate of 1.5 gpm/ft².

Bed Depth (ft)	Time to Breakthrough (hr)
2	85
4	95
6	400
8	545

Solution:

1. Plot the data as shown in Figure 11-14, and determine the equation for the resulting line:

$$t = ax + b \qquad \textbf{(11-30)}$$

$$a = \text{slope of the line} = \frac{400 \text{ hr} - 250 \text{ hr}}{6 \text{ ft} - 4.1 \text{ ft}}$$

$$= \frac{150 \text{ hr}}{1.9 \text{ ft}} = 78.95 \frac{\text{hr}}{\text{ft}}$$

$b = \text{intercept of the line} = -70$ (from plot)

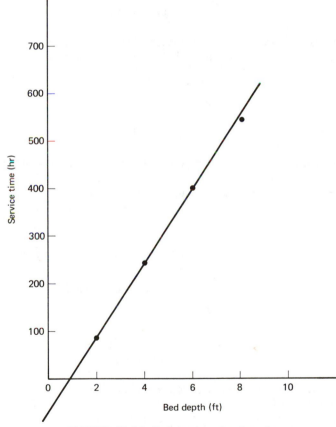

FIGURE 11-14 Bed depth service time plot.

Thus, for a linear flow rate of 1.0 gpm/ft² the equation for the line is

$$t = 78.95x - 70$$

2. Adjust the slope term a to the desired flow rate of 1.5 gpm/ft². It is not necessary to adjust the b term.

$$a_1 = \frac{a(V)}{V_1}$$

where a = slope of the line at the original flow rate (V)

   V = original flow rate (gpm/ft²)

   V_1 = revised flow rate (gpm/ft²)

   a_1 = revised slope for new flow rate V_1

$$a_1 = 78.95\frac{(1.0 \text{ gpm/ft}^2)}{(1.5 \text{ gpm/ft}^2)}$$

$$= 52.63$$

The predicted equation at $V_1 = 1.5$ gpm/ft² is

$$t = 52.63x - 70$$

3. Calculate the predicted service time for a 4 ft deep bed operating at the linear flow rate of 1.5 gpm/ft², with an influent phenol concentration of 100 mg/ℓ:

$$t = 52.63(4) - 70 = 210.52 - 70$$

$$= 140.52 \text{ hr}$$

4. If the influent phenol concentration changes from 100 to 140 mg/ℓ at a flow rate of 1.50 gpm/ft², both the a and b terms must be corrected.

   (a) The slope term from the 1.0 gpm/ft² test has already been corrected to the new flow rate of 1.5 gpm/ft². This value must now be corrected for the new influent concentration of 140 mg/ℓ.

$$a_2 = a_1\frac{C_1}{C_2}$$

where a_1 = slope value for 1.0 gpm/ft² corrected to a flow rate of 1.5 gpm/ft²

   C_1 = old influent concentration (mg/ℓ)

   C_2 = new influent concentration (mg/ℓ)

thus,

$$a_2 = 52.63\frac{100 \text{ mg/}\ell}{140 \text{ mg/}\ell} = 37.59.$$

   (b) The intercept term must be corrected for an influent concentration of 140 mg/ℓ:

$$b_2 = b_1\frac{C_1}{C_2}\frac{\ln (C_2/C_F - 1)}{\ln (C_1/C_B - 1)}$$

where b_1 = intercept at influent concentration C_1

C_1 = influent concentration of 100 mg/ℓ

C_2 = influent concentration of 140 mg/ℓ

C_B = effluent concentration at influent concentration C_1

C_F = effluent concentration at influent concentration C_2

$C_B = C_F = 2.0$ mg/ℓ

Thus,

b_2 = slope corrected for change in influent solute concentration

$$= -70\left(\frac{100}{140}\right)\frac{\ln\left(\frac{140}{2.0} - 1\right)}{\ln\left(\frac{100}{2} - 1\right)}$$

$$= -70(0.714)\frac{\ln 69}{\ln 49}$$

$$= -49.98\left(\frac{4.234}{3.891}\right) = -54.39.$$

(c) Thus, the new equation corrected for flow rate and influent concentration is

$$t = 37.59x - 54.39$$

(d) The operating time for a 4 ft deep bed under these conditions is

$$t = 37.59(4) - 54.39 = 95.97 \text{ hr}$$

Design Based on Mathematical/Graphical Approach

Fornwalt and Hutchins (1966b) have presented a mathematical/graphical approach for scaling up adsorption columns from laboratory or pilot plant data. This approach makes it possible to calculate the total weight of carbon needed to treat a wastewater and also predicts the number of full-scale columns required in series to yield the desired results.

To use this approach, sufficient small-scale column tests must be conducted to establish the shape of the exhaustion curve and to determine the volumetric flow rate, V_b, in bed volumes per hour, required to achieve the necessary treatment. The volumetric flow rate is defined as

$$V_b = \frac{Q_w}{h \times A_{cs}} \tag{11-34}$$

where Q_w = wastewater flow (gal/hr)

h = column height (ft)

A_{cs} = column cross-sectional area (ft²).

Once the volumetric flow rate is determined, the volume of carbon B required in a production column can be calculated:

$$B = \frac{Q_w}{\left[7.48\,\frac{\text{gal}}{\text{ft}^3}\right] \times V_b} \tag{11-35}$$

where $B =$ required volume of carbon (ft³).

The pounds, w, of carbon required in the column will be equal to the volume, B, multiplied by the bulk density, D, of the carbon in lb/ft³.

$$W = B(D) \tag{11-36}$$

The operating life of the column will be the time required at a given flow rate to reach the maximum allowable breakthrough concentration, C_B, of impurity in the effluent. This operating life can be calculated as

$$t = \frac{W}{Q_w/V_a} \tag{11-37}$$

where $t =$ operating life of the carbon column (hr.)

$V_a =$ gallons of wastewater treated per pound of carbon at the maximum allowable effluent concentration (C_B).

Rewriting equation 11-37 yields

$$t = \frac{B(D)}{Q_w/V_a} = \frac{Q_w(D)V_a}{7.48(V_b)(Q_w)} = \frac{V_a(D)}{7.48V_b} \tag{11-38}$$

The required weight of carbon as calculated by equation 11-36 can be installed in a single column or in a number of columns arranged in series or in parallel. The number of columns required in series installations can be determined graphically from a plot of the column exhaustion curve.

The graphical analysis for columns in series is based on a plot of an exhaustion curve for a laboratory column as shown in Figure 11-15. This curve is similar to the curve of Figure 11-6 except that the abscissa is in units of gallons of wastewater treated per pound of carbon rather than just gallons treated. The shape of the exhaustion curve for the laboratory column would be representative of the shape of the curve for a lead column in a series installation. If it is assumed that all columns in the series will have similarly shaped exhaustion curves, the curves for any number of other columns can be constructed similar to that for the laboratory column. However, since the influent to subsequent columns will be partly purified, the actual curves will be more favorable than that for the laboratory column, and the assumption of similar break-through curves will add a factor of safety in the design process (Fornwalt and Hutchins, 1966b).

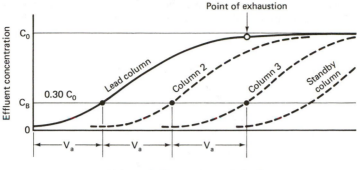

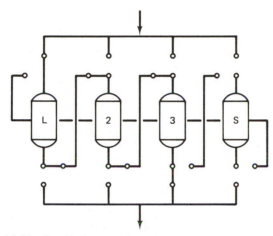

FIGURE 11-15 *Graphical approach to the design of adsorption columns in series (after Fornwalt and Hutchins, 1966).*

The graphical analysis is started by constructing exhaustion curves for other columns, identical to the original curve but shifted to the right a distance V_a. Additional curves are constructed until the carbon in the lead column is shown to be exhausted when the carbon in the last column has an effluent concentration equal to the allowable concentrations C_B. An additional column is then added to serve as a standby in case one of the other columns is out of service for maintenance or reloading. The use of these techniques is illustrated by the following example.

EXAMPLE PROBLEM 11-8: An industrial plant discharges 6000 gal/hr of a wastewater containing 150 mg/ℓ of COD. The plant is required to reduce the COD to 25 mg/ℓ prior to discharge. A laboratory study indicates that the desired effluent concentration can be achieved in an adsorption column operating at 0.25 bed volumes per hours. Determine the following:

1. The volume of carbon required to treat the wastewater.

2. The weight of carbon required if the carbon has a bulk density of 23.0 lb/ft³.

3. The operating life of a single carbon column under the above conditions. The exhaustion curve determined for a laboratory column is shown in Figure 11-16.

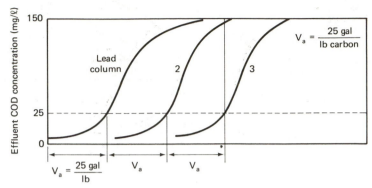

FIGURE 11-16 Column breakthrough curves for Example 11-8.

4. The number of columns that must be arranged in series to yield the desired results.

Solution:

1. The volume of carbon required to treat the wastewater can be calculated from equation 11-35:

$$B = \frac{Q_w}{7.48 V_b}$$

$$= \frac{6000 \text{ gal/hr}}{7.48 \frac{\text{gal}}{\text{ft}^3} \times \left(0.25 \frac{\text{bed volumes}}{\text{hour}}\right)} \qquad \textbf{(11-35)}$$

$$= 3208.5 \text{ ft}^3$$

2. The weight of carbon can be determined from equation 11-36:

$$W = B(D)$$

$$= (3208.5 \text{ ft}^3)(23.0 \text{ lb/ft}^3) \qquad \textbf{(11-36)}$$

$$= 73,795.5 \text{ lb}$$

3. The operating life of a carbon column under the above conditions is, using equation 11-38,

$$t = \frac{V_a(D)}{7.48 V_b} \qquad \textbf{(11-38)}$$

$$V_a = 25 \frac{\text{gal}}{\text{lb carbon}} \quad \text{(from Figure 11-15)}$$

$$t = \frac{25 \frac{\text{gal}}{\text{lb carbon}} \times 23 \text{ lb/ft}^3}{7.48 \times 0.25 \frac{\text{bed volumes}}{\text{hr}}} = 307.48 \text{ hr}$$

4. The number of columns that must be arranged in series can be determined from Figure 11-16. If the exhaustion curve is duplicated at intervals V_a apart, it is seen that *three* columns will be required. The size of these columns can be calculated from the results of parts 1, 2, and 3 above.

PROBLEMS

11-1. For the data of Problem 6-1, determine the daily amount of carbon required to treat 250,000 gal of wastewater in a batch reactor. A final TOC of 15 mg/ℓ is required.

11-2. A laboratory study was conducted to collect design data for an adsorption column to treat a wastewater containing 15 mg/ℓ of ABS. The study was conducted in $1\frac{1}{2}$ in. diameter columns and the time required to reduce the ABS concentration from 15 mg/ℓ to 1.0 mg/ℓ was determined.

Flow rate (gpm/ft²)	Bed Depth (ft)	Throughput Volume (gal)	Time (hr)
	2.5	1399	950
2.0	4.5	3277	2225
	6.0	4749	3225
	2.5	1178	400
4.0	4.5	3057	1038
	6.0	4418	1500
	3.0	1031	200
7.0	6.0	3608	700
	9.0	6185	1200

Use the Bohart-Adams approach to determine the time to exhaustion for a full-scale, 6 ft deep by 4 ft diameter bed of carbon treating a flow of 30,000 gal/day. Assume 8 hr/day operation of the column.

11-3. A laboratory column study was conducted to ascertain the removal of color from an industrial wastewater. The following data were collected at a flow rate of 1.5 gpm/ft². Use the BDST approach to predict the depth of a 5 ft diameter bed of carbon that must operate 200 hr between regenerations when treating a flow of 4,000 gal/hr.

Bed Depth (ft)	Time to Breakthrough (hr)
2	110
4	317
6	525
8	735

11-4. Wastewater from a chemical factory contains 100 mg/ℓ COD, which must be reduced to 10 mg/ℓ prior to discharge. An adsorption study indicates that the desired concentration can be achieved in an adsorption column operating at 0.20 bed volumes per hour. Using the method proposed by Fornwalt and Hutchins, determine the following:

(a) The volume of carbon required to treat the wastewater if the flow rate is 5000 gal/hr.

(b) The weight of carbon required if the carbon has a bulk density of 24.0 lb/ft³.

(c) The operating life of a single column if the gallons of wastewater treated per pound of carbon (V_a) is 23.0.

REFERENCES

BOHART, G.S., and ADAMS, E.Q., "Some Aspects of the Behavior of Charcoal with Respect to Chlorine," *J. Am. Chem. Soc.*, **42,** 523 (1920).

CONWAY, R.A., and ROSS, R.D., "*Handbook of Industrial Waste Disposal,*" Van Nostrand Reinhold Company, New York (1980).

FORNWALT, H.J., and HUTCHINS, R.A., "Purifying Liquids with Activated Carbon," *Chemical Engineering*, **73,** 8, 179 (April 11, 1966a).

FORNWALT, H.J., and HUTCHINS, R.A., "Purifying Liquids with Activated Carbon," *Chemical Engineering*, **73,** 10, 155 (May 9, 1966b).

GIUSTI, D.M., CONWAY, R.A., and LAWSON, C.T., "Activated Carbon Adsorption of Petrochemicals," *Journal Water Pollution Control Federation*, **46,** 5, 947 (1974).

HASSLER, J.W., *Purification with Activated Carbon*, Chemical Publishing Company, New York (1974).

HELBIG, W.A., "Colloid Chemistry," Vol. 6, [Ed.] J. ALEXANDER, Van Nostrand Reinhold Company, New York (1946).

HUMENICK, M.J., "*Water and Wastewater Treatment: Calculations for Chemical and Physical Processes*", Marcel Dekker, Inc., New York (1977).

HUTCHINS, R.A., "New Method Simplifies Design of Activated-Carbon Systems," *Chemical Engineering*, **80,** 19, 133 (August 20, 1973).

KEINATH, T.M., and WEBER, W.J., Jr., "A Predictive Model for the Design of Fluid-Bed Adsorbers," *Journal Water Pollution Control Federation*, **40,** 5 (Part 1), 741 (1968).

MICHAELS, A.S., "Simplified Method of Interpreting Kinetic Data in Fixed-Bed Ion Exchange," *Industrial and Engineering Chemistry* **44,** 1922 (1952).

STRUDGEON, G.E., "Upflow-Downflow Carbon Adsorption," *AICHE Symposium Series—Water*, 1976, **73,** 166, pp. 43–53.

U.S. Environmental Protection Agency, *Process Design Manual for Carbon Adsorption*, EPA 625/1–71–002a (October, 1973).

WEBER, W.J., Jr., *Physicochemical Processes for Water Quality Control*, Wiley-Interscience, New York (1972).

12

FLUORIDE REMOVAL

12-1 INTRODUCTION

Fluoride has beneficial and detrimental effects on the environment and is of great significance to the environmental engineer. The beneficial effect of fluoride is its capability to help prevent dental cavities when an optimum amount is present in drinking water. However, on the negative side, long-term consumption of water containing excessive amounts of fluoride can lead to fluorosis of the teeth and bones. Furthermore, high concentrations of fluoride in streams can be toxic to aquatic organisms, and the presence in the atmosphere of fluoride dusts and gases can be detrimental to both plant and animal life.

Fluoride commonly occurs in the earth's crust as fluorspar (CaF_2), cryolite (Na_3AlF_6), and fluorapatite ($Ca_{10}F_2(PO_4)_6$) and ranks thirteenth among the elements in order of abundance (AWWA, 1971). Fluoride is present in sea water at a concentration of approximately $1.4\ mg/\ell$, and concentrations as high as 9–$10\ mg/\ell$ are not uncommon in some ground waters. Such waters should be treated for fluoride removal if they are to be used as a source of drinking water.

In addition to its natural occurrence, fluoride is relocated and concentrated in the environment as a result of industrial activities. It is widely used in many manufacturing processes and is present in large quantities in effluents from glass manufacture, electroplating operations, aluminum and steel production, and the manufacture of electronic parts (Zabban and Jewett, 1967). Furthermore, large quantities of fluorides are released when fertilizer is manufactured from phosphate rock containing fluorapatite. Effluents from such industrial operations must be properly treated and carefully controlled to prevent excessive levels of fluoride in the environment.

Relatively little research has been done on the subject of fluoride removal and little effort has been made to develop theoretical design models that embody the theories and principles discussed earlier in this text. Consequently, the design of fluoride removal processes is, in many cases, largely empirical and based on past experience.

Information that is available on fluoride removal is widely scattered throughout the literature. The purpose of this chapter is to present the available information in a useful form. The first part of the chapter is devoted to a discussion of the processes used to defluoridate drinking water supplies and industrial waste discharges. While some of the processes are applicable to both water and wastewater, the chemical complexity of wastewater, compared to water, results in certain differences in process design and application. Thus, the latter portion of this chapter, which is devoted to treatment applications, will be separated into water and wastewater for ease of presentation.

12-2 PRECIPITATION OF FLUORIDE WITH CALCIUM

Calcium salts, including $Ca(OH)_2$, $CaSO_4$, and $CaCl_2$, can be used to precipitate fluoride as insoluble CaF_2. The reaction using lime is as follows:

$$Ca(OH)_2 + 2HF \longrightarrow CaF_2\downarrow + 2H_2O \qquad (12\text{-}1)$$

Theoretically, precipitation should affect the reduction of fluoride levels to the limit of solubility of fluoride in the CaF_2 system. The equilibria that affect the solubility of fluoride include not only the CaF_2 itself but also the protonated species HF and HF_2^- as well as the ion-pair CaF^+. Appropriate squilibrium constants are as follows (Butler, 1964):

$$CaF_{2(s)} \rightleftharpoons Ca^{2+} + 2F^-, \qquad (K_s)_{CaF_2} = 4.0 \times 10^{-11}$$
$$Ca^{2+} + F^- \rightleftharpoons CaF^+, \qquad (K)_{CaF^+} = 10$$
$$H^+ + F^- \rightleftharpoons HF, \qquad (K)_{HF} = 1.5 \times 10^3$$
$$HF + F^- \rightleftharpoons HF_2^-, \qquad (K)_{HF_2^-} = 3.9$$

The solubility of fluoride in an ideal solution can be calculated by the procedures explained in Chapter 3.

> **EXAMPLE PROBLEM 12-1:** Calculate the theoretical minimum concentration of fluoride that can be achieved by precipitating the fluoride as CaF_2. Consider residual calcium concentrations of 40 mg/ℓ and 400 mg/ℓ.

> **Solution:**
>
> 1. The equilibria of interest are as follows:
>   - (a) $CaF_{2(s)} \rightleftharpoons Ca^{2+} + 2F^-,$ $\qquad K_{CaF_2} = 4.0 \times 10^{-11}$
>   - (b) $Ca^{2+} + F^- \rightleftharpoons CaF^+,$ $\qquad K_{CaF^+} = 10$
>   - (c) $H^+ + F^- \rightleftharpoons HF,$ $\qquad K_{HF} = 1.5 \times 10^3$
>   - (d) $HF + F^- \rightleftharpoons HF_2^-,$ $\qquad K_{HF_2^-} = 3.9$

2. Develop the equation that describes the total soluble fluoride concentration.

(a) From the above equilibria expressions,

$$K_{CaF_2} = [Ca^{2+}][F^-]^2$$

$$[F^-] = \left[\frac{K_{CaF_2}}{Ca^{2+}}\right]^{1/2} = \left[\frac{10^{-10.4}}{[Ca^{2+}]}\right]^{1/2}$$

$$K_{CaF^+} = \frac{[CaF^+]}{[Ca^{2+}][F^-]}$$

$$[CaF^+] = (K_{CaF^+})[Ca^{2+}][F^-] = 10[Ca^{2+}][F^-]$$

$$K_{HF} = \frac{[HF]}{[H^+][F^-]}$$

$$[HF] = (K_{HF})[H^+][F^-] = 1.5 \times 10^{+3}[H^+][F^-]$$

$$K_{HF_2^-} = \frac{[HF_2^-]}{[HF][F^-]}$$

$$[HF_2^-] = (K_{HF_2^-})[HF][F^-] = (K_{HF_2^-})(K_{HF})[H^+][F^-]^2$$
$$= (3.9)(1.5 \times 10^{+3})[H^+][F^-]^2$$

(b) The expression for total soluble fluoride is

$$[F]_T = [F^-] + [CaF^+] + [HF] + [HF_2^-]$$

Substituting equilibrium expressions:

$$[F]_T = [F^-] + (K_{CaF^+})[Ca^{2+}][F^-] + K_{HF}[H^+][F^-]$$
$$+ (K_{HF_2^-})(K_{HF})[H^+][F^-]^2$$

Factoring out an $[F^-]$:

$$[F]_T = [F^-]\{1 + K_{CaF^+}[Ca^{2+}] + (K_{HF})[H^+] + (K_{HF_2^-})(K_{HF})[H^+][F^-]\}$$

Substituting for $[F^-]$:

$$[F]_T = \left[\frac{K_{CaF_2}}{[Ca^{2+}]}\right]^{1/2}\left\{1 + (K_{CaF^+})[Ca^{2+}] + (K_{HF})[H^+]\right.$$
$$\left. + (K_{HF_2^-})(K_{HF})[H^+]\left[\frac{K_{CaF_2}}{[Ca^{2+}]}\right]^{1/2}\right\}$$

or

$$[F]_T = \left[\frac{10^{-10.4}}{[Ca^{2+}]}\right]^{1/2}\left\{1 + (10)[Ca^{2+}] + (1.5 \times 10^3)[H^+]\right.$$
$$\left. + (3.9)(1.5 \times 10^3)[H^+]\left[\frac{10^{-10.4}}{[Ca^{2+}]}\right]^{1/2}\right\}$$

Thus, the total soluble fluoride concentration is seen to be a function of pH and residual Ca^{2+} concentration.

(c) Calculate the soluble fluoride concentration over a range of pH values for calcium concentrations of (a) 40 mg/ℓ (10^{-3} M), (b) 400 mg/ℓ (10^{-2} M). Tabulate or plot the results.

Example: $[Ca^{2+}] = 10^{-3}\ M$, pH 4.

$$[F]_T = \left[\frac{10^{-10.4}}{[10^{-3}]}\right]^{1/2} \left\{1 + 10(10^{-3}) + 1.5 \times 10^3[10^{-4}]\right.$$

$$\left. + (3.9)(1.5 \times 10^3)[10^{-4}]\left[\frac{10^{-10.4}}{[10^{-3}]}\right]^{1/2}\right\}$$

$$= 1.995 \times 10^{-4}\{1 + 0.01 + 0.15 + 0.0001167\}$$

$$= 2.314 \times 10^{-4}\ M$$

If $Ca(OH)_2$ is used as a source of lime, the pH will increase with increasing calcium dosage. Consequently, if alkalinity is present, precipitation of $CaCO_3$ will become a factor in the equilibrium at high pH values.

In addition to the factors considered above, the solubility of fluoride compounds is influenced by temperature, ionic strength, and the common ion effect. Furthermore, fluoride is capable of forming soluble complexes with many ions, including S_i^{2+}, Al^{3+}, Fe^{3+}, and B^{3+}, which are frequently found in waste solutions. Consequently, residual fluoride concentrations achieved in practice are seldom as low as those determined from theoretical solubility calculations.

12-3 REMOVAL OF FLUORIDE WITH ALUM

Alum was one of the first chemicals investigated for use in removing fluoride from drinking water supplies. It is still being used, alone and in combination with other chemicals, as a defluoridating agent.

When added to water, alum reacts with the alkalinity in the water to produce insoluble $Al(OH)_3$.

$$Al_2(SO_4)_3 \cdot 14.3H_2O + 3Ca(HCO_3)_2 \longrightarrow Al(OH)_3$$
$$+ 3CaSO_4 + 14.3H_2O + 6CO_2 \tag{12-2}$$

Rabosky and Miller (1974) suggest that fluoride ions are removed from solution by adsorption onto the $Al(OH)_3$ particles. The $Al(OH)_3$ and adsorbed fluorides can then be separated from the water by sedimentation.

12-4 FLUORIDE REMOVAL BY ION EXCHANGE/SORPTION

Fluoride can be removed from water or wastewater by a number of processes that involve either ion exchange or adsorption. These processes can be applied to either concentrated or dilute solutions, and they are capable of providing complete removal of the fluoride ion under proper conditions. The economics are generally more favorable when the processes are used to treat dilute solutions to achieve low residual fluoride concentrations. Thus, they are commonly used to defluoridate drinking water supplies or dilute fluoride-bearing wastewaters. Media used for defluoridation include bone char, synthetic ion exchange materials, and activated alumina.

Bone Char

The uptake of fluoride onto the surface of bone was one of the early methods suggested for defluoridation of water supplies (Smith and Smith, 1937). The process was reportedly one of ion exchange, in which the carbonate radical of the apatite comprising bone, $Ca(PO_4)_6 \cdot CaCO_3$, was replaced by fluoride to form an insoluble fluorapatite (Smith and Smith, 1937).

$$Ca(PO_4)_6 \cdot CaCO_3 + 2F^- \longrightarrow Ca(PO_4)_6 \cdot CaF_2 + CO_3^{2-} \qquad (12\text{-}3)$$

It was soon recognized that bone char, produced by carbonizing bone at temperatures of 1100–1600°C, had qualities that were superior to those of unprocessed bone. Thus, bone char replaced bone as a defluoridating agent. Bone char in sizes between 28 and 48 mesh has been used with success in many full-scale installations for defluoridation of drinking water (Harman and Kalichman, 1965). When exhausted, the column is regenerated by application of a 1.0 % solution of caustic soda, which converts the fluorapatite to hydroxyapatite. The fluoride is removed from the column as soluble sodium fluoride. Regeneration is followed by a caustic rinse and an acid wash to restore the pH to a favorable level. In the regenerated form, the hydroxyl radical becomes the exchange material in the defluoridation reaction (Maier, 1953). Typical properties of bone char are listed in Table 12-1.

TABLE 12-1 Typical properties of bone char (after Mantell, 1945).

Constituents and Properties	Content
$Ca_3(PO_4)_2$	73.50%
$CaCO_3$	8.50%
Iron and aluminum oxide	0.40%
Magnesia	0.20%
Acid insoluble ash	0.30%
Total volatile	16.50%
Apparent density (lb/ft^3)	40–46

Synthetic Ion Exchange Media: Anionic and cationic ion exchangers have both been used for fluoride removal. Benson et al. (1940) investigated a two-stage ion exchange process in which sodium ions were removed from solution by a cationic material during the first stage and replaced with hydrogen ions in accord with the following equation:

$$2NaF + H_2Z \rightleftharpoons H_2F_2 + Na_2Z \qquad (12\text{-}4)$$

The hydrogen fluoride was removed from solution by an anionic exchange material during the second stage:

$$2R_3N + H_2F_2 \rightleftharpoons 2R_3NHF \qquad (12\text{-}5)$$

A 10 mg/ℓ fluoride solution was reduced to a concentration below 1.0 mg/ℓ when two pairs of beds were used in series.

Amberlite IRA 400 (manufactured by Rohm and Haas), an ion-exchange material consisting of an 8% crosslinked polystyrene matrix of the trimethylbenzylammonium type, has been used to remove fluoride from process rinse waters (Staebler, 1974). The strongly basic nature of this material caused the pH of the solution to increase from 6.4 to approximately 11.6 as a result of passage through the column.

TABLE 12-2 *Characteristics of Amberlite IRA-400 (courtesy of Rohm and Haas).*

Type	Strongly Basic
Bulk density	44 lb/ft^3
Mesh	20–50
Void volume	40–45%
Moisture content (drained)	46% (by weight)

Activated Alumina: Activated alumnia is a granular, highly porous material consisting essentially of aluminum trihydrate, Al_2O_3. It is widely used as a commercial desiccant and in many gas-drying processes.

Activated alumina has been used successfully for defluoridation of drinking water supplies. Some researchers have concluded that removal was the result of ion exchange (Savinelli and Black, 1958), but Wu and Nitya (1979) have shown that the process is one of adsorption and follows the Langmuir isotherm.

Activated alumina can be regenerated with HCl, H_2SO_4, alum, or NaOH. The use of NaOH, followed by a neutralization step to remove residual NaOH from the bed, appears to be the most practical approach to regeneration. Typical properties of activated alumina are shown in Table 12-3.

TABLE 12-3 *Typical properties of and specifications for activated alumina, F-1 type (courtesy of Alcoa).*

Constituents and Properties	Content
Al_2O_3	92.00%
Na_2O	0.90%
Fe_2O_3	0.08%
SiO_2	0.09%
Loss on ignition (1100°C)	6.50%
Form	Granular
Surface area (sq. m/g)	210
Size	1/4″–14 mesh
Bulk density, loose (g/cu cm)	0.83
Bulk density, packed (g/cu cm)	0.88
Specific gravity	3.3

12-5 DEFLUORIDATION OF WATER SUPPLIES

The 1962 USPHS Drinking Water Standards established recommended optimum as well as maximum fluoride concentrations as shown in Table 12-4. The 1975 EPA Interim Primary Drinking Water Regulations, established under the provisions of the Safe Drinking Water Act (PL 93-523), established maximum contaminant levels (MCL) for 10 inorganic chemicals, including fluoride. The MCL for fluoride was the same as the maximum concentration previously established by the USPHS. Since water consumption, and thus fluoride intake, increase with increasing air temperature, the standards are established as a function of annual average maximum daily air temperature.

TABLE 12-4   1962 USPHS fluoride limitations for drinking water

Annual Average Maximum Daily Air Temperature (°F)	Fluoride Concentration (mg/ℓ)	
	Recommended	Maximum*
50.0–53.7	1.7	2.4
53.8–58.3	1.5	2.2
58.4–63.8	1.3	2.0
63.9–70.6	1.2	1.8
70.7–79.2	1.0	1.6
79.3–90.5	0.8	1.4

* Corresponds to EPA 1975 Interim Primary Drinking Water Regulations.

According to the Safe Drinking Water Act, all waters that contain more than the allowable maximum fluoride concentration must be defluoridated prior to use as public water supplies. Surface waters are generally low in fluorides, but ground waters frequently contain high fluoride concentrations. It has been estimated that more than 1100 public water supplies, serving approximately 4.2 million people, utilize water that exceeds the recommended maximum level for fluoride (Rubel and Woosley, 1978). At least 138 communities reportedly use water supplies with 4.0 mg/ℓ or more of fluorides (AWWA, 1971). Most of these communities are located in Arizona, Colorado, Illinois, Iowa, New Mexico, Ohio, Oklahoma, California, South Dakota, and Texas (Rubel and Woosley, 1978).

Very few defluoridation plants have ever been constructed, as suggested by the large number of public water supplies that exceed the recommended maximum level for fluoride. If the provisions of PL 93-523 are strictly enforced, many municipalities must either install defluoridation facilities or switch to low-fluoride sources of supply. Since so few treatment facilities have been built, no single technique has emerged as the best approach for defluoridation.

Bone char was used as a defluoridating agent at the Britton, South Dakota, plant which operated from 1948 until the city switched to a new source of supply in 1971. Fluoride removal was reported to be pH dependent, and it was found that lowering the pH of the raw water would increase cycle lengths by a factor of more than two

(Maier, 1971). Bone char, however, is acid-soluble, so an operating pH of 7.1 was selected to provide reasonably good fluoride removal while minimizing the loss of media. The system produced a treated water with an average fluoride content of 1.5 mg/ℓ from a raw water containing 6.7 mg/ℓ of fluoride (Maier, 1971).

Alum was investigated for use at a municipal plant in La Crosse, Kansas, to treat a soft, highly mineralized water supply containing 3.6 mg/ℓ of fluoride (Culp and Stoltenberg, 1958). Good fluoride removal was achieved in the pH range of 5.5 to 7.5, which is the pH range corresponding to minimum solubility of $Al(OH)_3$, and at an alum dosage of 225 mg/ℓ. An alum dose of 315 mg/ℓ was required to reduce the fluoride to 1.0 mg/ℓ.

The most promising approach to defluoridation of municipal water supplies appears to be removal with packed beds of granular activated alumina. Relatively large defluoridation plants employing activated alumina have been in operation for a number of years in Desert Center, California; Vail, Arizona; and Gila Bend, Arizona (Rubel and Woosley, 1978). Raw water characteristics for these plants are presented in Table 12-5.

Operation of an activated alumina system involves four separate modes: treatment, back-wash, regeneration, and neutralization. To describe system operation, it is convenient to start by assuming that the bed is in an exhausted state. The exhausted bed can be regenerated with HCl, H_2SO_4, alum, or NaOH. Regeneration with H_2SO_4 has been reported to result in the formation of cementitious compounds which plug and solidify the bed, and the use of HCl leads to high losses of alumina per regeneration cycle (Zabban and Helwick, 1975). Furthermore, the chloride ion contained in the HCl is not precipitated when the spent regenerant is neutralized with lime, which results in an increased total dissolved solids concentration in the waste stream from the process (Zabban and Helwick, 1975). Sodium hydroxide has been widely used in practice, and this appears to be the best choice of regenerant. The following sequence of operations

TABLE 12-5 *Water analyses at existing operations (after Rubel and Woosley, 1978).*

Facility		Lake Tamarisk	Ricon Water Co.	Town of Gila Bend
Ca	(mg/ℓ)	11	51	54
Mg	(mg/ℓ)	0.5	5.8	2.5
Na	(mg/ℓ)	58	151	402
SO$_4$	(mg/ℓ)	40	261	144
Cl	(mg/ℓ)	67	22	582
Hardness	(mg/ℓ)	30	152	146
M Alkalinity	(mg/ℓ)	77	171	52
P Alkalinity	(mg/ℓ)	0	0	0
Fe	(mg/ℓ)	0.2	< 0.05	0.2
SiO$_2$	(mg/ℓ)	22	55	21
F	(mg/ℓ)	7.5	4.5	5.0
TDS	(mg/ℓ)	409	650	1210
pH		7.9	7.5	8.0

describes the procedure involved in regenerating an exhausted bed of activated alumina (Rubel and Woosley, 1978):

1. An upflow back-wash at 8–9 gpm/ft² with raw water. This back-wash expands the bed and removes any suspended solids that might have been trapped in the bed. Normal back-wash time is approximately 10 min.

2. An upflow regeneration step employing a 1 % (by weight) NaOH solution at 2.5 gpm/ft² for approximately 35 min.

3. An upflow rinse at 5.0 gpm/ft² for approximately 30 min.

4. A final regeneration step in the downflow direction, using 1 % NaOH at $2\frac{1}{2}$ gpm/ft² for approximately 35 min.

This series of operations strips fluoride from the bed and restores the removal capacity of the activated alumina. However, following these operations the entire bed is in the pH range of 12.5 to 13.0 as a result of the caustic solution used for regeneration. Fluoride removal by activated alumina is strongly pH dependent. Batch adsorption data presented by Wu and Nitya (1979) show that very little removal occurs at pH 11.0, and optimum removal occurs at pH 5.0 (see Figure 12-1). Consequently, the regenerated bed will not be effective in removing fluoride unless its pH is adjusted.

Bed neutralization is accomplished by adjusting the raw water pH with sulfuric acid. The procedure used at the Gila Bend, Arizona, defluoridation plant has been described as follows (Rubel and Woosley, 1978). After regeneration the raw water pH is initially adjusted to 2.5, and the water is fed to the bed at the normal treatment rate of 5–6 gpm/ft². Neutralization of the bed is evidenced by a drop in the pH of effluent water as shown in Figure 12-2. Fluoride will not be removed from the raw water during the early stages of neutralization because of the caustic condition of the bed. The first water through the regenerated bed must be discharged to waste.

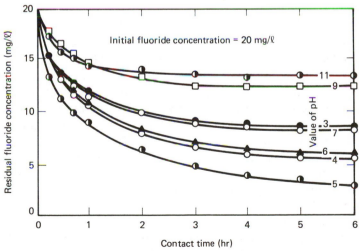

FIGURE 12-1 *Effect of* pH *and time of contact on fluoride removal of activated alumina adsorption. Fluoride/alumina ratio = 0.004 mg/mg (after Wu and Nitya, 1979).*

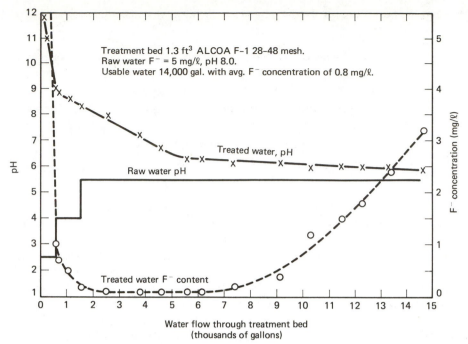

Treatment bed 1.3 ft³ ALCOA F-1 28-48 mesh.
Raw water F⁻ = 5 mg/ℓ, pH 8.0.
Usable water 14,000 gal. with avg. F⁻ concentration of 0.8 mg/ℓ.

FIGURE 12-2 *Typical pilot-plant run Gila Bend, Arizona (after Rubel and Woosley, 1978).*

When the pH of the effluent water drops to 9.0–9.5, the bed will begin to remove some fluoride as shown in Figure 12-2. At this point the raw water pH is adjusted to 4.0 and the neutralization process continues. When the effluent stream shows a pH of 8.5, the raw water is adjusted to pH 5.5, and it is maintained at that pH throughout the remainder of the run.

Fluoride removal will be nearly 100% during the early part of a run but will decrease toward the end of the run as the bed becomes exhausted. Low fluoride water (0.1 mg/ℓ) produced during the early part of the run can be blended with high fluoride water (3.0 mg/ℓ) produced during the latter stages of the run to yield a water with an acceptable level of fluoride. If a suitable reservoir is not available for blending, staggered regeneration of the treatment beds may be employed to maximize system performance. During staggered regeneration, low fluoride water from the fresh bed is blended with water containing a higher concentration of fluoride from the nearly exhausted second bed, to produce a finished water with an acceptable fluoride concentration. The maximum fluoride concentration at which a run must be terminated will depend on the fluoride concentration achievable by blending. Blending also serves to smooth out pH variations; however, supplemental pH control is usually required to produce a stabilized water.

Waste, amounting to about 4% of the total plant throughout, is produced during back-washing, regeneration, and the early part of neutralization. The back-wash water is composed only of raw water and may be discharged to surface waters or storm sewers. The neutralization waste has a high pH, and the regeneration waste has a high pH and a high concentration of fluoride ions. Several existing plants concentrate the regeneration waste in lined evaporation ponds with eventual fluoride recovery or disposal. Disposal techniques must be chosen to conform to local water pollution control guidelines. Operating costs for activated alumina systems, including chemicals, electricity, bed replacement, replacement parts, and labor, are estimated to be in the range of 8–20¢ per thousand gallons (Rubel and Woosley, 1979).

The ability of activated alumina to remove fluoride depends on the chemistry of the water being treated. Such factors as hardness, silica, and boron, if present in the water, will interfere with fluoride removal and reduce the efficiency of the system. Information necessary for the design of activated alumina systems must be collected through laboratory and pilot plant tests of the water to be treated.

Batch adsorption studies (Chapter 11) can be used to evaluate the effects of pH and time of contact on fluoride removal. Results of such studies can also be used to evaluate reaction rate constants and to determine the fluoride removal capacity of activated alumina for a given water. Continuous-flow column studies should be used to establish final design information.

Most activated alumina systems employ beds that contain a 5 ft depth of media, with at least two beds per installation. Design flow rates are in the range of 5.0–7.5 gpm/ft² (1.0–1.5 gpm/ft³ of media) (Rubel, 1979). A 5 min superficial residence time (flow time through the bed neglecting the volume of bed material) appears to be the minimum time needed to achieve maximum removal efficiency. Rubel and Woosley (1979) indicate removal capacities in the range of 2000–4000 grains/ft³ when the raw water pH is approximately 5.5. Capacities are reported to drop to 500 grains/ft³ at a raw water pH of 7.0.

The following examples will serve to illustrate how laboratory and plot plant data can be analyzed.

EXAMPLE PROBLEM 12-2: A batch adsorption study was conducted by placing varying amounts of activated alumina in five beakers containing 1.0 ℓ of water having an initial fluoride concentration of 10.0 mg/ℓ. The test was conducted at pH 5.0, and a steady-state fluoride concentration was achieved after 1 hr of contact. Determine the adsorptive capacity of the activated alumina at an effluent fluoride concentration of 0.9 mg/ℓ, based on the following data. Assume that Languir's isotherm applies.

Beaker No.	Weight of Alumina (g) (m)	Volume of Sample (mℓ)	Initial F⁻ (mg/ℓ)	Final F⁻ (mg/ℓ) (C)	Wt. of F⁻ Absorbed (mg) (x)	F⁻ Absorbed per Unit Wt. of Alumin (mg/g) (x/m)
1	0	1000	10	10	0	—
2	1.750	1000	10	2.00	8.00	4.57
3	2.800	1000	10	1.25	8.75	3.125
4	3.600	1000	10	1.0	9.00	2.50
5	4.600	1000	10	0.8	9.20	2.00

Solution:

1. The linear form of the Langmuir isotherm is

$$\frac{1}{(x/m)} = \frac{1}{b} + \frac{1}{abc}$$

where x = amount of material absorbed (mg)

m = weight of absorbent (g)

c = concentration of material remaining in solution after adsorption (mg/ℓ)

a and b = constants.

2. Plot test data values of $1/(x/m)$ vs. $1/c$ to obtain a Langmuir isotherm:

Beaker No.	$\dfrac{1}{(x/m)}$ (g/mg)	$\dfrac{1}{c}$ (ℓ/mg)
1	—	0.10
2	0.219	0.500
3	0.320	0.800
4	0.400	1.000
5	0.500	1.250

3. Construct a vertical line at a value of

$$\frac{1}{c} = \frac{1}{0.9} = 1.11$$

4. Read the value of $1/(x/m)$ from Figure 12-3.

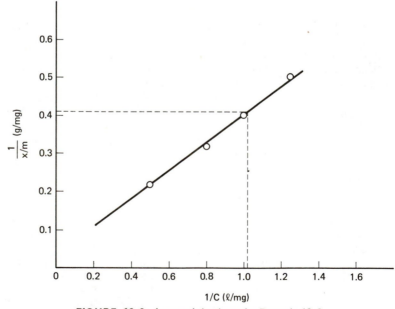

FIGURE 12-3 *Langmuir isotherm for Example 12-2.*

$$\frac{1}{x/m} = 0.41$$

This corresponds to an adsorptive capacity of 2.439 mg of F^- per gram of activated alumina.

5. Assuming a density for the activated alumina of 0.88 g/cm³ (Table 12-3), the adsorptive capacity can be expressed as

$$\frac{2.439 \text{ mg} \times 15.432 \text{ grains/g}}{1.0 \text{ g} \times 0.88 \text{ cm}^3/\text{g} \times 1000 \text{ mg/g}} = 0.04277 \text{ grain/cm}^3$$

or

$$= 0.04277 \text{ grain/cm}^3 \times \frac{1}{3.531 \times 10^{-5} \text{ ft}^3/\text{cm}^3}$$

Thus, capacity = 1211 grains/cft³

12-6 DEFLUORIDATION OF WASTEWATERS

The amount of fluoride allowed in treated effluents has been established by the EPA in the form of effluent limitations. The guidelines for several industries are summarized in Table 12-6.

TABLE 12-6   Summary of EPA effluent limitations for fluoride.

Industrial Category	Fluoride Limitation (30 day Average)
Glass manufacturing:	
Television tubes	0.07 lb/1000 lb furnace pull
Incandescent lamp envelope	0.115 lb/1000 lb furnace pull
Phosphate manufacturing:	
Phosphorous production	0.05 lb/1000 lb product
Na_3PO_4 manufacturing	0.15 lb/1000 lb product
Nonferrous metals:	
Aluminum smelting (Hall-Heroult process)	1.0 lb/1000 lb product
Fertilizer manufacturing:	
Phosphate	15 mg/ℓ
Inorganic chemicals:	
Hydrogen fluoride	15 mg/ℓ
Aluminum fluoride	0.34 lb/1000 lb product
Sodium silicofluoride	0.003 lb/1000 lb product
Plastics and synthetics:	
Polytetrafluorethylene	0.60 lb/1000 lb product

Many industrial wastewaters contain fluoride ion concentrations ranging from 40 to 100,000 mg/ℓ. Obviously, fluoride must be removed from these wastes prior to discharge so as to comply with existing regulatory requirements. The most common approach to treatment of fluoride-laden wastewaters is chemical precipitation with lime. Ion exchange or coagulation, using alum, polylectrolytes, or iron salts, have frequently been employed following precipitation to maximize fluoride removal.

Removal by Chemical Precipitation

In many cases lime alone has been used to precipitate fluoride from wastewater. Miller (1974) described a treatment system used by Western Electric to remove fluoride from a metal-finishing waste. A flow schematic for the treatment system is shown in Figure 12-4. The wastewater, containing approximately 40 mg/ℓ fluoride and having

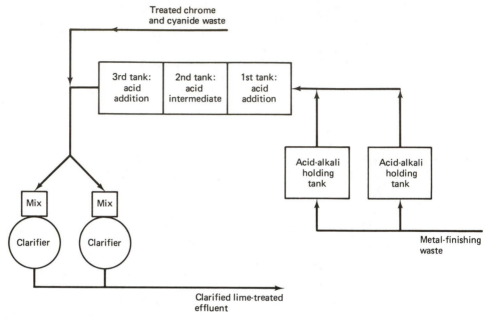

FIGURE 12-4 *Fluoride removal flow diagram (after Miller, 1974).*

a pH of 2–3, flows into two holding tanks and then into a neutralization tank, where slaked lime is added to raise the pH to 11–12 and promote precipitation of fluoride according to the following reaction:

$$Ca(OH)_2 + 2HF \longrightarrow \underline{CaF_2} \downarrow + 2H_2O \qquad (12\text{-}6)$$

Mixing and reaction time is provided in the second tank, followed by H_2SO_4 addition in the third tank to achieve a final pH of 7–9. The reaction is

$$Ca(OH)_2(\text{excess}) + H_2SO_4 \longrightarrow CaSO_4 + 2H_2O \qquad (12\text{-}7)$$

Theoretical solubility calculations suggest that an effluent fluoride concentration of approximately 7.8 mg/ℓ could be achieved by lime precipitation. Typically, effluent fluoride concentrations in the range of 12–18 mg/ℓ were achieved in practice. Greatly increased chemical dosages were required to consistently produce effluent fluoride concentrations of 12 mg/ℓ.

At one time it was believed that pH values of 11.0 or above were necessary to achieve good fluoride removal using lime (Zabban and Jewett, 1967). Later it was realized that the excess calcium contributed by the high lime dosages, rather than the high pH, was responsible for precipitation of CaF_2. In fact, Rohrer (1974) has shown that good fluoride removal can be achieved in the pH range of 5.7 to 8.0 by using a mixture of lime and $CaCl_2$. The highly soluble $CaCl_2$ provides excess calcium ions without causing the pH increase associated with the use of lime. Fluoride levels have been reduced from 725 mg/ℓ to 15–20 mg/ℓ by using a 2:1 mixture of lime and $CaCl_2$ (Rohrer, 1974). Furthermore, sludge volumes are reportedly less than those resulting from hydrated lime addition, and, since a pH of 11.0 or greater is not reached, the reneutralization step normally required in a lime precipitation process is eliminated.

Some researchers have reported that CaF_2 precipitates very slowly from solution and that detention times of 1 to 6 days are required to allow for post-precipitation (Rohrer, 1974; Zabban and Jewett 1967). Solids-contact clarifiers have been used for some lime treated effluents (Zabban and Jewett, 1967), and these units should help to control post-precipitation through the effects of crystal seeding. However, Miller (1974) has shown that the solubility of fluoride increases (and thus fluoride removal efficiency decreases) in the presence of excess solid CaF_2. The increased solubility was thought to result from CaF_2 forming a soluble complex with hydroxyl radicals. This observation suggests that if solids-contact units are used the solids level in these units should be controlled very carefully to minimize post-precipitation without increasing fluoride solubility.

Removal by Precipitation/Coagulation

Alum, polyelectrolytes, and iron salts have all been used as coagulants following chemical precipitation to remove fluoride from wastewaters. Final fluoride concentrations as low as 1–2 mg/ℓ have been reported for this process.

Rabosky and Miller (1974) investigated fluoride removal from a sodium fluoride solution using a two-stage process of lime precipitation followed by alum and poly-electrolyte coagulation. The initial fluoride concentration of approximately 200 mg/ℓ was reduced to 4–6 mg/ℓ following lime precipitation. This concentration was further reduced to approximately 2.0 mg/ℓ by coagulation with 25 mg/ℓ Al^{3+} and 2 mg/ℓ polyelectrolyte. The optimum pH for coagulation was in the range of 6.0 to 7.0.

Link and Rabosky (1976) reported on the calcium precipitation and iron salt coagulation for removing fluoride from a wastewater having an initial fluoride concentration of 200 mg/ℓ. Fluoride concentrations could be reduced from approximately 200 mg/ℓ to 6 mg/ℓ, using 4000 mg/ℓ of lime.

Laboratory studies indicated that very low fluoride levels could be achieved if the clarified effluent from lime precipitation was further treated with a 4:1 mixture of alum and sodium hexametaphosphate. For example, the effluent fluoride concentration could be reduced from 17 mg/ℓ to 5 mg/ℓ by using a combined chemical dose of 200 mg/ℓ, and a 2.0 mg/ℓ fluoride concentration could be achieved by using a combined dose of 600 mg/ℓ. The reduction in fluorides was attributed to the formation of a highly insoluble reaction product whose chemical formula was estimated to be $NaPO_3 \cdot AlF_3$.

REFERENCES

AWWA, *Water, Quality and Treatment* 3rd ed., McGraw-Hill Book Company, New York (1971).

BENSON, D.L., PORTH, D.L., and Sweeney, O.R., "The Removal of Fluorides from Water by Ionic Exchange," *Proc. Iowa Academy of Science*, **47,** 221 (1940).

BUTLER, J.N., "*Ionic Equilibrium: A Mathematical Approach*," Addison-Wesley Publishing Co., Inc. (1964).

CULP, R.L., and STOLTENBERG, H.A., "Flouride Removal at LaCrosse, Kansas," *J. Am. Water Works Assoc.*, **50,** 3, 423 (March, 1958).

HARMON, J.A., and KALICHMAN, S.G., "Defluoridation of Drinking Water in Southern California," *J. Am. Water Works Assoc.*, **57,** 2, 245, (February, 1965).

LINK, W.E., and RABOSKY, J.G., "Fluoride Removal from Wastewater, Employing Calcium Precipitation and Iron Salt Coagulation," *Proc. Thirty-first Purdue Industrial Waste Conference*, pp. 485–500 (May 1976).

MAIER, F.J., "Defluoridation of Municipal Water Supplies," *J. Am. Water Works Assoc.*, **45,** 8, 879, (August, 1953).

MAIER, F.J., "Water Defluoridation at Britton: End of an Era," *Public Works*, **6,** 70 (June, 1971).

MANTELL, C.L., *Adsorption*, McGraw Hill Book Company, New York (1945).

MILLER, D.G., "Fluoride Precipitation in Metal Finishing Waste Effluent," *AICHE Symposium Series, Water*, 1974, pp. 39–46.

"National Interim Primary Drinking Water Regulations," U.S. Environmental Protection Agency EPA-570/9–76–003. (1974)

RABOSKY, J.G., and MILLER, J.P., "Fluoride Removal by Lime Precipitation and Alum and Polyelectrolyte Coagulation," *Proc. Twenty-ninth Purdue Industrial Waste Conference*, pp. 669–676 (1974).

ROHRER, K.L., "Lime, $CaCl_2$, Beat Fluoride Wastewater," *Water and Wastes Engineering*, **11,** 11, 66, November, 1974).

RUBEL, F., and WOOSLEY, R.D., "Removal of Excess Fluoride from Drinking Water," *EPA Technical Report*, EPA 570/9–78–001 (January, 1978).

RUBEL, F., and WOOSLEY, R.D., "The Removal of Excess Fluoride from Drinking Water by Activated Alumina," *J. Am. Water Works Assoc.*, **71,** 1, 45, (January, 1979).

RUBEL, F., personal communication, Tucson, AZ (July, 1979).

SAVINELLI, E.A., and BLACK, A.P., "Defluoridation of Water with Activated Alumina," *J. Am. Water Works Assoc.*, **50,** 1, 33 (January, 1958).

SMITH, H.R., and SMITH, L.C., "Bone Contact Removes Fluoride," *Water Works Engineering*, **90,** 5, 600 (May, 1937).

STAEBLER, C.J., Jr., "Treatment and Recovery of Fluoride Industrial Wastes," EPA-660/2–73–024, U.S. Environmental Protection Agency (March, 1974).

Wu, Y.C., and Nitya, A., "Water Defluoridation with Activated Alumina," *Journal of the Environmental Engineering Division*, ASCE, **105**, EE2, 357 (April, 1979).

Zabban, W, and Jewett, H.W., "The Treatment of Fluoride Wastes," *Proc., Twenty-third Purdue Industrial Waste Conference*, pp. 706–715 (May 2–4, 1967).

Zabban, W., and Helwick, R., "Defluoridation of Wastewater," *Proc. Thirtieth Purdue Industrial Waste Conference*, pp. 479–492 (May 6–8, 1975).

13

APPLICATIONS OF REDOX CHEMISTRY

13-1 INTRODUCTION

Certain chemicals deserve special consideration for wastewater treatment because of the threat that they pose to the environment. The Environmental Protection Agency has compiled a list of 297 specific chemicals and compounds, referred to as hazardous substances, that fall into this category. The list includes organic compounds, heavy metals, cyanide, and asbestos.

Many of the hazardous chemicals are widely used in industrial processes and consequently are often present in objectionable concentrations in industrial wastewater. Some of these materials can be removed from a wastewater by conventional treatment processes; however, others require the application of specific treatment technology.

The chemical treatment of many wastewaters relies on oxidation-reduction reactions to convert harmful materials to a harmless or less harmful waste product. This type of treatment process is controlled by measuring the electrical potential of the chemical system with respect to a known reference and maintaining the electrical potential at a value that ensures the absence of the undesirable chemical species by the continued addition of a suitable oxidizing or reducing agent. In this chapter the following applications of the principles of oxidation-reduction are discussed: (1) treatment of metal-plating wastewaters for the removal of cyanide and chromium and (2) treatment of wastewater effluents for ammonia removal by breakpoint chlorination.

13-2 TREATMENT OF METAL-PLATING WASTEWATER

In the electroplating process the item to be plated functions as a cathode in a plating bath containing an electrolyte solution of the plating metal. As the item being plated is moved through the bath, some of the plating solution is carried along with it. This solution, referred to as *dragout*, is removed in subsequent rinse baths as the item progresses through the process.

Metal-plating wastes can be divided into two categories: (1) concentrated wastes resulting from the discharge of spent plating baths and (2) dilute wastes from rinsing and cleaning operations. The plating baths are normally discharged intermittently, whereas rinse waters may be discharged on either a batch (intermittent) or continuous-flow basis. Typical concentrations of contaminants in plating baths and rinse waters have been reported as shown in Table 13-1.

TABLE 13-1 *Typical concentrations of contaminants in plating baths and rinse waters (after Burford and Masselli in Rudolfs, 1953).*

Plating Process	Concentration in Bath (mg/ℓ)	Concentration in Rinse water, mg/ℓ*	
		0.5 gph Dragout	2.5 gph Dragout
Chromium	207,000 Cr	431 Cr	2155 Cr
Copper-cyanide	12,400 Cu	2.8 Cu	14 Cu
	28,000 CN	58 CN	290 CN
Cadmium	23,000 Cd	48 Cd	240 Cd
	57,700 CN	120 CN	600 CN
Zinc	33,800 Zn	70 Zn	350 Zn
	48,900 CN	102 CN	510 CN
Brass	21,000 Cu	44 Cu	220 Cu
	5,250 Zn	11 Zn	55 Zn
	47,500 CN	99 CN	495 CN

* Concentrations estimated based on a 4 gpm rinse rate, assuming "dragouts" of 0.5 and 2.5 gallons per hour (gph).

Approaches to Treatment

Each industrial plant is responsible for ensuring that its wastewater receives proper treatment as prescribed by law. To meet this responsibility a plant may discharge its wastewater into a municipal treatment plant or it may construct and operate its own treatment facilities. If it discharges into a municipal treatment facility, an industrial plant must still bear ultimate responsibility for the effects of its wastewater on the environment. Consequently, a plant must pretreat its wastewater to remove those things that would interfere with or would not normally be removed by conventional treatment processes. The Environmental Protection Agency has established maximum metal concentrations that are permitted to be discharged into municipal sewer systems. Concentrations applicable to electroplating shops discharging more than 10,000 gal day are shown in Table 13-2. Pretreatment is required to ensure that these concentrations are not exceeded.

Effluent guidelines will also be established by the EPA for those industrial plants that elect to treat their own wastewater prior to discharge. Currently, regulatory agencies are enforcing limits similar to those shown in Table 13-2.

TABLE 13-2 *Discharge limitations for selected metal contaminants.*

	Pretreatment	Effluent Guidelines
Parameter	Max. Concentration Allowed for Discharge to Sanitary Sewer (Avg. of 4 Consecutive days)*	Approximate Conc. (mg/ℓ) (daily avg.)†
Cadmium	0.7 mg/ℓ	0.1 mg/ℓ
Copper	2.7 mg/ℓ	0.5 mg/ℓ
Chromium (hexavalent)	—	0.1 mg/ℓ
Chromium (total)	4.0 mg/ℓ	0.5 mg/ℓ
Cyanide (total)	1.0 mg/ℓ	0.1 mg/ℓ
Zinc	2.6 mg/ℓ	0.8 mg/ℓ
Nickel	2.6 mg/ℓ	0.5 mg/ℓ
Lead	0.4 mg/ℓ	

* *Federal Register*, Vol. 46, No. 18, January 28, 1981.

† Values obtained from Alabama Water Improvement Commission, Montgomery, AL.

Approaches to treating a particular wastewater depend on the composition of the wastewater. Wastewater that contains cyanide must be separated from other wastewaters and given special treatment to destroy or detoxify the cyanide. Likewise, wastewater that contains chromium must receive special treatment to destroy the toxicity of this chemical. Wastewaters that contain no chromium or cyanide may require only treatment for removal of objectionable heavy metals prior to discharge.

Treatment of Cyanide Wastewater

Cyanide is highly toxic for fish but somewhat less toxic for man and microorganisms. Concentrations of 0.02 mg/ℓ have been reported to be lethal for certain species of fish, whereas a concentration of 0.20 mg/ℓ is allowable for drinking water supplies. Natural body mechanisms are capable of detoxifying small amounts of cyanide as they are ingested, which not only offers some protection against cyanide poisoning but also prevents cyanides from accumulating in the human body. Cyanide ingestion does not become lethal unless the capacity of this natural detoxifying mechanism is exceeded. The minimum lethal dose for an average size man has been estimated to be 180–200 mg of 95–100% sodium cyanide (American Electroplaters' Society, 1969). Moreover, many microorganisms can acclimate to relatively high concentrations of cyanides, and concentrations up to 30 mg/ℓ have been successfully treated in biological treatment processes (Echenfelder, 1966). Thus, the discharge of cyanide into the environment is limited by its extreme toxicity for fish rather than its effects on man and microorganisms.

Most cyanide used in industrial processes is added in the form of sodium cyanide (NaCN) or hydrogen cyanide (HCN). Sodium cyanide hydrolyzes to form hydrocyanic acid and sodium hydroxide:

$$NaCN + H_2O \rightleftharpoons HCN + NaOH \tag{13-1}$$

In wastewater, cyanide usually exists as CN^-, and HCN, or in the form of a complex ion. The ionization of HCN is pH dependent as shown below.

$$HCN \rightleftharpoons H^+ + CN^- \qquad (K_a = 4.8 \times 10^{-10} \text{ at } 25°C) \qquad \textbf{(13-2)}$$

A distribution diagram for HCN, Figure 13-1, shows that practically all the cyanide is present as HCN at pH 7.0 or below, and exists as CN^- at pH 11.5 or above.

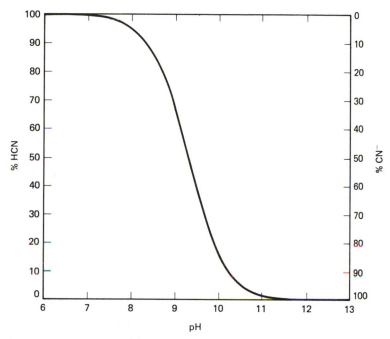

FIGURE 13-1 *Relationship between HCN and CN at various pH values (after Chamberlain and Synder, 1955).*

Cyanide forms complexes with a number of metals, including zinc, iron, nickel, and cadmium, which are frequently found, along with cyanide, in industrial wastewater. Zinc and cadmium complexes are unstable and highly toxic, whereas iron and nickel complexes are relatively stable and much less toxic. It is necessary that these complex ions be destroyed to ensure complete wastewater treatment.

Treatment of wastewater containing cyanide is directed at detoxifying the wastewater by destroying the cyanide with chemical or electrolytic oxidation or by concentrating the cyanide for recovery or reuse through evaporation or ion exchange. The form of cyanide, whether free or complexed, and its concentration must be considered when selecting a treatment process.

Under proper conditions, wastewaters containing highly toxic cyanide can be rendered safe for discharge by the two-stage process of alkali-chlorination. This is the most popular method for treatment of cyanide wastewater and is adaptable to either batch or continuous-flow applications.

Step 1. *Chlorine oxidation of cyanide to cyanate:*

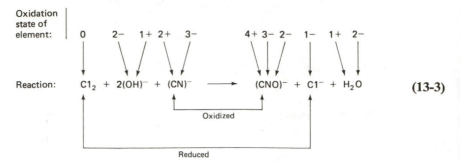

$$Cl_2 + 2(OH)^- + (CN)^- \longrightarrow (CNO)^- + Cl^- + H_2O \qquad (13\text{-}3)$$

If it is assumed that the reaction goes to completion, molar ratios suggest the following chlorine and caustic dose per mg/ℓ of cyanide oxidized to cyanate:

Chlorine:

$$\frac{2 \times 35.45}{26} = 2.7 \text{ mg/}\ell \text{ of chlorine required per mg/}\ell \text{ of } CN^- \text{ present}$$

Caustic:

$$\frac{2 \times 17}{26} = 1.3 \text{ mg/}\ell \text{ of hydroxide required per mg/}\ell \text{ of } CN^- \text{ present. If}$$
hydroxide is added in the form of sodium hydroxide, the dose will be 3.1 mg/ℓ of sodium hydroxide per mg/ℓ of CN^-.

The cyanate ion is almost 1000 times less toxic than cyanide, and as a result in many treatment situations the oxidation process is only taken to this point.

Step 2. *Chlorine oxidation of cyanate to bicarbonate and nitrogen gas:*

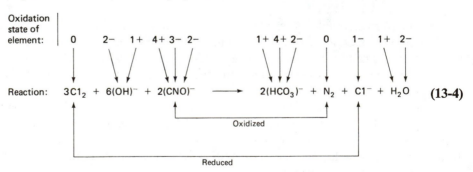

$$3Cl_2 + 6(OH)^- + 2(CNO)^- \longrightarrow 2(HCO_3)^- + N_2 + Cl^- + H_2O \qquad (13\text{-}4)$$

If it is assumed that the reaction goes to completion, molar ratios suggest the following chlorine and caustic dose per mg/ℓ of cyanate oxidized to bicarbonate and nitrogen gas:

Chlorine:

$$\frac{3 \times 35.45}{42} = 2.5 \ mg/\ell \text{ of chlorine required per } mg/\ell \text{ of } CNO^- \text{ present}$$

Caustic:

$$\frac{6 \times 17}{42} = 2.4 \ mg/\ell \text{ of hydroxide required per } mg/\ell \text{ of } CNO^- \text{ present. If}$$
hydroxide is added in the form of sodium hydroxide, the dose will be 5.7 mg/ℓ of sodium hydroxide per mg/ℓ of CNO^-.

It should be understood that a portion of the total soluble cyanide concentration may be in the form of complex ions. Since these ions are in equilibrium with the CN^- ion, the cyanide associated with the complexes will be removed in the oxidation process because as free CN^- ions are oxidized the complex ion will dissociate in an effort to reestablish equilibrium. The CN^- ions which are freed will then be oxidized. Thus, chemical doses should be based on the total soluble cyanide concentration and not just the CN^- concentration. Generally, this value can be computed from the amount of cyanide used in the plating process.

Although the stoichiometric method is commonly employed to compute chemical dose requirements for cyanide removal, the process is much more complicated than is suggested by equations 13-3 and 13-4. Chamberlain and Snyder (1955) propose that the oxidation of cyanide to cyanate progresses through the intermediate compound cyanogen chloride according to the following reaction sequence:

$$CN^- + OCl^- + H_2O \longrightarrow CNCl + 2OH^- \qquad \textbf{(13-5)}$$

$$CNCl + 2OH^- \longrightarrow CNO^- + Cl^- + H_2O \qquad \textbf{(13-6)}$$

overall reaction: $\quad CN^- + OCl^- \longrightarrow CNO^- + Cl^- \qquad \textbf{(13-7)}$

The oxidation of cyanides to cyanogen chloride (CNCl), equation 13-5, occurs instantaneously at all pH levels. However, the hydrolysis of CNCl to cyanate, equation 13-6, occurs very slowly at pH values around 7.0 but increases rapidly in the pH range of 8.5–9.0 as shown Figure 13-2. Cyanogen chloride is a highly toxic gas, and it is essential that it hydrolyze quickly to avoid a buildup of dangerous concentrations in solution. Consequently, first-stage chlorination should be conducted at pH 9–10 in order to ensure rapid completion of reaction 13-7. The overall reaction for the chlorine oxidation of cyanide to cyanate shows that the hypochlorite ion (OCl^-) is the active chlorine group in the oxidation process.

Cyanate is only about one-thousandth as toxic as cyanide, and treatment can be stopped at this point if regulatory guidelines permit the discharge of cyanate. If the discharge of cyanate is unacceptable, complete destruction of cyanide can be accomplished by continued chlorination in second-stage treatment.

According to Chamberlain and Snyder (1955), hypochlorous acid is the active chlorine species that accomplishes oxidation of cyanate to bicarbonate and nitrogen gas:

$$2CNO^- + 3HOCl \longrightarrow 2HCO_3^- + N_2 + 3Cl^- + H^+ \qquad \textbf{(13-8)}$$

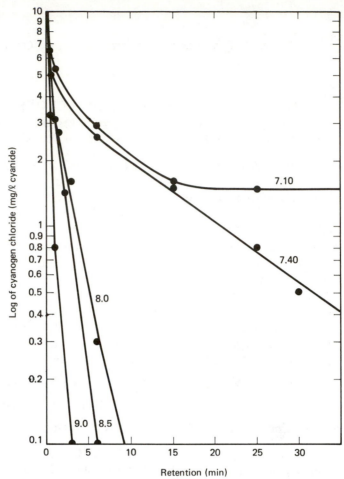

FIGURE 13-2 *Effect of pH on hydrolysis of cyanogen chloride to cyanates in presence of a chloromine residual (after Chamberlain and Synder, 1955).*

Thus, for optimum results the oxidation of the cyanate ion should be conducted at a pH low enough to ensure that HOCl is the predominant active chlorine species.

Example problem 13-1 illustrates the procedures required for computing the chlorine dose required for oxidation and the operating redox potential required for process control.

EXAMPLE PROBLEM 13-1: A metal-plating wastewater contains $100 \text{ mg}/\ell$ of CN^- ions. It is desired to reduce the CN^- ion concentration to $0.1 \text{ mg}/\ell$ by chlorine oxidation before the wastewater is discharged from the plant. Compute the required chlorine dose at a pH of 8 for the oxidation of cyanide to cyanate. Also, compute the operating redox potentials required for process control. Assume that the operating temperature is 25°C and the initial chloride concentration in wastewater is zero.

1. Calculate the equilibrium constant for the chlorine oxidation of cyanide from equation 1-7 and 1-55:

$$\Delta G^{\circ}_{rxm} = [(-23.3)+(-31.37)]-[(41.2)+(-8.8)]$$
$$= -87.07 \text{ KCal}$$

Thus,

$$\Delta G^{\circ}_{rxm} = -1.36\log(Ka)_{eq.}$$

or

$$\log(Ka)_{eq.} = 63.8$$
$$(Ka)_{eq.} = 10^{63.8}$$

The large $(Ka)_{eq}$ value indicates that for all practical purposes, the reaction goes to completion

2. Compute the cyanate concentration associated with the cyanide reduction assuming the reaction goes to completion.

$$[CNO^-] = \frac{0.1 \text{ g}/\ell}{26 \text{ g/mole}} = 3.8 \times 10^{-3} \text{ moles}/\ell$$

3. Compute the chloride concentration associated with the cyanide reduction assuming chlorine added using Cl_2.

The chemical equation representing the oxidation of cyanide with chlorine has the form

$$CN^- + OCl^- \rightarrow CNO^- + Cl^-$$

This equation suggests that for every mole$/\ell$ of cyanide oxidized to cyanate, 1 mole$/\ell$ of chloride will be produced. Thus,

$$[Cl^-] = 1(\times 3.8 \times 10^{-3}) = 3.8 \times 10^{-3} \text{ moles}/\ell$$

Also, an equal amount of chloride will be produced from the reaction

$$Cl_2 + H_2O \rightleftharpoons HOCl + H^+ + Cl^-$$

Thus,

$$[Cl^-]_T = 2 \times 3.8 \times 10^{-3} = 7.6 \times 10^{-3} \text{ moles}/\ell$$

4. Determine the equilibrium concentrations of $[CN^-]$ and $[OCl^-]$

$$10^{63.18} = \frac{[CNO^-][Cl^-]}{[CN^-][OCl^-]}$$

Since stoichiometric amounts of OCl^- to CN^- are being added, assume x to be the amounts of CN^- and OCl^- remaining. Then

$$x^2 = \frac{(3.8 \times 10^{-3})(7.6 \times 10^{-3})}{10^{63.18}}$$

$$x = 1.37 \times 10^{-34}$$

TREATMENT OF METAL-PLATING WASTEWATER

5. Compute E°_{cell} for the reaction.

$$CNO^- + H_2O + 2e^- \rightleftharpoons CN^- + 2OH^- : E^\circ = -0.97 \text{ V}$$

$$OCl^- + H_2O + 2e^- \rightleftharpoons Cl^- + 2OH^- : E^\circ = 0.89 \text{ V}$$

Therefore,

oxidation: $CN^- + 2OH^- \rightleftharpoons CNO^- + H_2O + 2e^- : E^\circ = 0.97 \text{ V}$

reduction: $OCl^- + H_2O + 2e^- \rightleftharpoons Cl^- + 2OH^- \quad : E^\circ = 0.89 \text{ V}$

overall: $CN^- + OCl^- \rightleftharpoons CNO^- + Cl^- \qquad : E^\circ_{cell} = 1.86 \text{ V}$

6. Compute the *equilibrium* cell potential for the cyanide oxidation, based on the hydrogen electrode.

$$E_{cell\text{-}h} = E^\circ_{cell} - \frac{RT}{mF} \ln Q$$

$$= 1.86 - \frac{(1.98)(298)(2.3)}{(2)(23{,}061)} \log \left[\frac{(3.8 \times 10^{-3})(7.6 \times 10^{-3})}{(1.37 \times 10^{-34})(1.37 \times 10^{-34})} \right]$$

$$E_{cell\text{-}h} = 0.0$$

At equilibrium $E_{cell\text{-}h}$ should be zero.

7. Calculate the total chlorine dose.

$$\text{chlorine dose } (\text{mg } Cl_2/\ell) = (3.8 \times 10^{-3})(2)(35.45)(10^3)$$

$$= 269 \text{ mg}/\ell$$

8. Compute α_1 at pH 8 for HOCl.

$$\alpha_1 = \frac{1}{1 + \dfrac{[H^+]}{Ka}}$$

$$= \frac{1}{1 + \dfrac{10^{-8}}{3.2 \times 10^{-8}}}$$

$$\alpha_1 = 0.7639$$

9. Compute E_{meter} for the process. It will be assumed that the rate of the reaction is sufficiently slow at pH 8.0 that by providing the proper retention time a cynide concentration of 0.1 mg/ℓ will be established at the time the wastewater leaves the reactor even though a chlorine dose of 3.8×10^{-3} moles/ℓ has been applied. Of course, at equilibrium the cyanide concentration would be reduced to 1.32×10^{-34} moles/ℓ.

(a) Compute the percentage completion of the reaction.

$$\%\text{completion} = \left[\frac{0.1}{100}\right] 100 = 0.1\%$$

(b) Compute the cell potential based on the hydrogen electrode.

$$E_{cell\text{-}h} = 1.86 + \frac{(1.98)(298)(2.3)}{(2)(23,061)}$$

$$\times \log\left[\frac{(3.8 \times 10^{-3})(7.6 \times 10^{-3})}{[(0.001)(3.8 \times 10^{-3})(0.764)(0.001)(3.8 \times 10^{-3})]}\right]$$

$$= 1.49$$

(c) Calculate the theoretical meter reading, assuming that an inert platinum indicator electrode and a calomel reference electrode are used:

$$E_{meter} = E_{cell\text{-}h} - E_{reference}$$
$$= 1.49 - 0.244$$
$$E_{meter} = 1.246 \text{ V}$$

Electrolytic Treatment of Cyanide Wastewater

Cyanides contained in wastewater can be destroyed by electrolytic oxidation under proper conditions. In this process cathodes and anodes are immersed in a tank containing the cyanide solution, and an appropriate current is imposed on the system. Destruction of cyanide has been reported to occur as follows (Drogon and Pasek, 1965):

1. Cyanogen complexes are destroyed, and the cyanide is liberated or converted to cyanate:

$$[Cu(CN)_3]^{2-} \longrightarrow Cu^+ + 3CN^- \qquad (13\text{-}9)$$

$$[Cu(CN)_3]^{2-} + 6OH^- - 7\,e^- \longrightarrow Cu^{2+} + 3CNO^- + 3H_2O \qquad (13\text{-}10)$$

2. Free cyanide ions in solution are oxidized to cyanate:

$$CN^- + 2OH^- - 2\,e^- \longrightarrow CNO^- + H_2O \qquad (13\text{-}11)$$

3. The cyanate ion in solution then hydrolyzes:

$$CNO^- + 2H_2O \longrightarrow NH_4^+ + CO_3^{2-} \qquad (13\text{-}12)$$

4. Simultaneous oxidation of cyanate occurs at the anode:

$$2CNO^- + 4OH^- - 6\,e^- \longrightarrow 2CO_2 + N_2 + 2H_2O \qquad (13\text{-}13)$$

5. The copper liberated by reaction 13-9 is plated out at the cathode and can be recovered and sold:

$$Cu^+ + e^- \longrightarrow Cu \qquad (13\text{-}14)$$

Optimum operating conditions have been reported to be an anode current density of 37 amp/ft² at 9 to 10 V.

Electrolytic oxidation is seemingly economical and efficient for use with concentrated wastes such as spent plating baths, cyanide strip solutions, and pre-plating cyanide dips (American Electroplaters' Society, 1969). The results of a typical production run as reported by Easton (1967) are shown in Table 13-3. In these runs the cyanide concentration was reduced from 75,000 mg/ℓ to 0.2 mg/ℓ after 18 days of electrolyzing, and all copper was essentially plated out after 11 days. If time is not available for such a prolonged electrolyzing period, it is possible to stop the process at some intermediate point and complete destruction of the residual cyanide with chlorination (Easton, 1967).

TABLE 13-3 Changes in wastewater characteristics during electrolytic decomposition (after Easton, 1967).

Days of Decomposition	Cyanide (mg/ℓ)	Copper (g/ℓ)	Sodium Carbonate (g/ℓ)	pH
Start	75,000	26	60	12.2
1	50,000	23	—	11.7
4	12,500	11	—	10.4
6	5,980	6	81	10.0
8	2,200	2	—	9.8
11	750	0	—	9.7
18	0.2		95	9.3

Electrolytic oxidation is somewhat limited for treating dilute solutions such as rinse waters because the low conductivity of these solutions results in poor current efficiency. Drogon and Pasek (1965) have shown that process efficiency can be improved for dilute solutions if NaCl is added to the wastewater. The chloride behaves as follows during electrolysis:

$$2Cl^- - 2\,e^- \longrightarrow Cl_2 \tag{13-15}$$

$$Cl_2 + OH^- \longrightarrow Cl^- + HOCl \tag{13-16}$$

$$HOCl + OH^- \longrightarrow OCl^- + H_2O \tag{13-17}$$

The hypochlorite formed in reaction 13-17 converts the CN^- to nontoxic end products.

$$2CN^- + 5OCl^- + 2OH^- \longrightarrow 5Cl^- + N_2 + 2CO_3^{2-} + H_2O \tag{13-18}$$

Wastewaters containing an average cyanide concentration of 200 mg/ℓ required a 4 oz/gal (29,960 mg/ℓ) addition of NaCl (Drogon and Pasek, 1965). Such large chemical requirements appear to make the process impractical for dilute wastewaters in most cases.

Treatment of Chromium Wastewater

Chromium can exist as chromium(II), chromium(III), or chromium(VI). Chromium(VI), or hexavalent chrome, is toxic to man when ingested and produces lung tumors when inhaled. The exact level of chromium that can be continuously consumed by humans without adverse effects is not known; however, a limit of 0.05 mg/ℓ, which appears to be very conservative, has been adopted by the Environmental Protection Agency as the maximum allowable concentration in public water supplies (EPA, 1976).

Plating baths are prepared by adding hexavalent chromium in the form of sodium dichromate ($Na_2Cr_2O_7 \cdot H_2O$) or chromium trioxide (CrO_3). When sodium dichromate is used, it dissociates to produce the divalent dichromate ion ($Cr_2O_7^{2-}$). When chromium trioxide is used, it immediately dissolves in water to form chromic acid according to the following reaction:

$$CrO_3 + H_2O \longrightarrow H_2CrO_4 \tag{13-19}$$

Chromic acid is considered a strong acid, although it never completely ionizes. Its ionization has been described as follows (Chamberlain and Day, 1956):

$$H_2CrO_4 \longrightarrow H^+ + HCrO_4^- \qquad K_a = 0.83 \text{ at } 25°C \tag{13-20}$$
$$\text{acid chromate ion}$$

$$HCrO_4^- \longrightarrow H^+ + CrO_4^{2-} \qquad K_a = 3.2 \times 10^{-7} \text{ at } 25°C \tag{13-21}$$
$$\text{chromate ion}$$

Moreover, the dichromate ion ($Cr_2O_7^{2-}$) will exist in equilibrium with the acid chromate ion as follows:

$$Cr_2O_7^{2-} + H_2O \rightleftharpoons 2HCrO_4^- \qquad K_a = 0.0302 \text{ at } 25°C \tag{13-22}$$

The logarithmic concentration diagram of Figure 13-3 indicates that $HCrO_4^-$ is the predominant species between pH 1.5 and 4.0, that $HCrO_4^-$ and CrO_4^{2-} exist in equal amounts at pH 6.5, and that CrO_4^{2-} predominates at higher pH values. Chrome plating wastewater are generally somewhat acid, and the acid chromate ion, $HCrO_4^-$, is predominant in this wastewater.

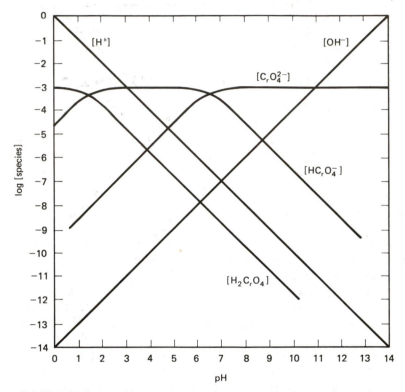

FIGURE 13-3 *Logarithmic concentration diagram for $10^{-3}M\ H_2CrO_4$ solution.*

Chemical treatment of chromium wastewater is usually conducted in two steps. In the first step hexavalent chromium is reduced to trivalent chromium by use of a chemical reducing agent. The trivalent chromium is precipitated during the second stage of treatment.

Sulfur dioxide (SO_2), sodium bisulfite ($NaHSO_3$), and sodium metabisulfite ($Na_2S_2O_5$) are commonly used as reducing agents. All these compounds react to produce sulfurous acid when added to water, according to the following reactions (Chamberlain and Day, 1956):

$$SO_2 + H_2O \rightleftharpoons H_2SO_3 \qquad \text{(13-23)}$$

$$Na_2S_2O_5 + H_2O \rightleftharpoons 2NaHSO_3 \qquad \text{(13-24)}$$

$$NaHSO_3 + H_2O \rightleftharpoons H_2SO_3 + NaOH \qquad \text{(13-25)}$$

It is the sulfurous acid produced from these reactions that is responsible for reduction of hexavalent chromium. The reaction is shown in equation 13-26:

$$2H_2CrO_4 + 3H_2SO_3 \longrightarrow Cr_2(SO_4)_3 + 5H_2O \qquad \textbf{(13-26)}$$

The typical amber color of the hexavalent chromium solution will turn to a pale green once the chromium has been reduced to the trivalent state. Although this color change is a good indicator, redox control is usually employed.

The theoretical amount of sulfurous acid required to reduce a given amount of chromium can be calculated from equation 13-26. The actual amount of sulfurous acid required to treat a wastewater will be greater than this because other compounds and ions present in the wastewater will consume acid. Primary among these is dissolved oxygen, which oxidizes sulfurous acid to sulfuric acid according to the following:

$$H_2SO_3 + \tfrac{1}{2}O_2 \rightleftharpoons H_2SO_4 \qquad \textbf{(13-27)}$$

Each part of dissolved oxygen initially present in the wastewater produces 6.1 parts of sulfuric acid.

Undissociated sulfurous acid is responsible for the reduction of hexavalent chromium. Consequently, the reduction reaction is strongly pH dependent because of the effect of pH on acid dissociation.

$$H_2SO_3 \rightleftharpoons H^+ + HSO_3^- \qquad K_a = 1.72 \times 10^{-2} \text{ at } 25°C \qquad \textbf{(13-28)}$$
$$HSO_3^- \rightleftharpoons H^+ + SO_3^{2-} \qquad K_a = 1.0 \times 10^{-7} \text{ at } 25°C \qquad \textbf{(13-29)}$$

The dissociation as a function of pH and the effect of pH on reaction rate is shown in Figures 13-4(a) and (b), respectively. Obviously, the reaction proceeds much faster at low pH values, where the concentration of undissociated sulfurous acid is highest. As a result, chromium reduction processes are generally conducted at pH values of 2–3 to maximize reaction rates and minimize the volume of reaction vessels. Sulfuric acid is generally added to reduce the pH of the wastewater to the desired level and to maintain it at that level throughout treatment. If the pH is not maintained at the desired level but is allowed to increase during treatment, the reaction may not go to completion in the retention time available, and unreduced hexavalent chromium may exist in the effluent. The amount of acid required to depress the pH to the level selected for chrome reduction will depend on the alkalinity of the wastewater being treated. This acid requirement can be determined by titrating a sample of wastewater with sulfuric acid to the desired pH in the absence of a reducing agent.

In addition to the sulfuric acid required for pH adjustment, some amount of acid is consumed by the reduction reaction (equation 13-26). If sulfur dioxide is used as the reducing agent, it will provide all the acid consumed by this reaction, and additional acid will not be required. However, if sodium bisulfite or sodium metabisulfite are used, additional acid must be supplied to satisfy the acid demand. This acid requirement is stoichiometric and can be calculated from equations 13-32 to 13-35. The cal-

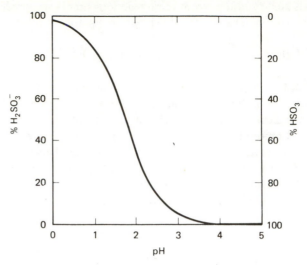

(a) Relationship between H_2SO_3 and HSO_3^- at various pH values

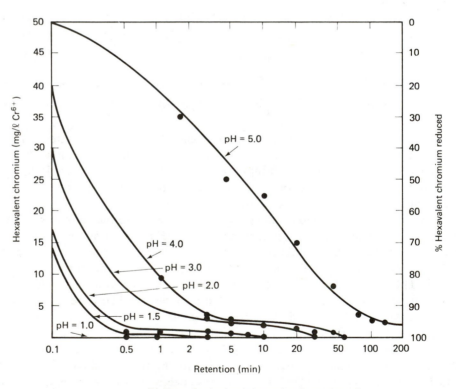

(b) Rate of reduction fo hexavalent in presence of
excess sulphur dioxide at various pH levels

FIGURE 13-4 *Effects of* pH *on reduction of hexavalent chromium
(after Chamberlain and Day, 1956).*

culations are complicated by the fact that the acid demand will depend not only on the reducing agent used but on the pH at which the reduction is carried out (Chamberlain and Day, 1956). The pH affects the distribution of sulfate and bisulfate ions in solution, since it affects the dissociation of sulfuric acid. Sulfuric acid is generally considered a strong acid but actually dissociates as follows (Butler, 1964):

$$H_2SO_4 \rightleftharpoons H^+ + HSO_4^- \tag{13-30}$$

$$HSO_4^- \rightleftharpoons H^+ + SO_4^{2-} \qquad K_a = 1.02 \times 10^{-2} \text{ at } 25°C \tag{13-31}$$

The first hydrogen is completely dissociated in solutions more dilute than 1 molar. Regarding the second dissociation, Figure 13-5 indicates that only the sulfate ion, SO_4^{2-}, is present at pH levels above 3.0 and that the sulfate and bisulfate ions exist in about equal amounts at pH 2.0. The bisulfate ion predominates at pH 1.0.

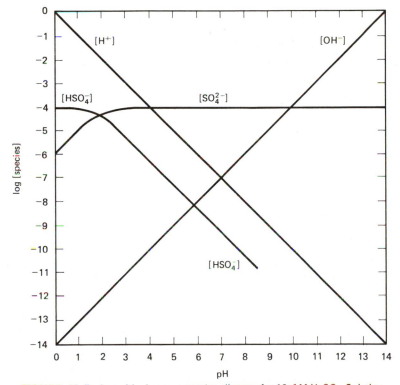

FIGURE 13-5 *Logarithmic concentration diagram for $10^{-3} M\, H_2SO_4$ Solution.*

The acid requirements associated with sodium bisulfite or sodium metabisulfite at pH 3.0 and above (SO_4^{2-} only) and at pH 2.0 ($SO_4^{2-} \simeq HSO_4^-$) can be calculated from the following equations presented by Chamberlain and Day (1956). (Similar equations could be developed for pH values between 2 and 3 as a function of the SO_4^{2-} and HSO_4^- distribution.)

At pH 3.0 to 4.0:

$$3NaHSO_3 + 1.5H_2SO_4 + 2H_2CrO_4 \rightleftharpoons$$
$$Cr_2(SO_4)_3 + 1.5NaSO_4 + 5H_2O \tag{13-32}$$

$$1.5Na_2S_2O_5 + 1.5H_2SO_4 + 2H_2CrO_4 \rightleftharpoons$$
$$Cr_2(SO_4)_3 + 1.5Na_2SO_4 + 3.5H_2O \tag{13-33}$$

At pH 2.0:

$$3NaHSO_3 + 2H_2SO_4 + 2H_2CrO_4 \rightleftharpoons$$
$$Cr_2(SO_4)_3 + Na_2SO_4 + NaHSO_4 + 5H_2O \tag{13-34}$$

$$1.5Na_2S_2O_5 + 2H_2SO_4 + 2H_2CrO_4 \rightleftharpoons$$
$$Cr_2(SO_4)_3 + Na_2SO_4 + NaHSO_4 + 3.5H_2O \tag{13-35}$$

Precipitation of Trivalent Chromium Following Reduction

Following reduction of hexavalent chromium, sodium hydroxide or lime is added to the wastewater to neutralize the pH and precipitate the trivalent chromium. This lime will also react with heavy metals such as iron, copper, and zinc and with any residual sodium sulfate, sulfurous acid, or sodium bisulfite. Using chromium salt as an example of all heavy metals, the following reactions apply (Chamberlain and Day, 1956):

$$Cr_2(SO_4)_3 + 3Ca(OH)_2 \longrightarrow 2Cr(OH)_3 + 3CaSO_4 \tag{13-36}$$
$$H_2SO_4 + Ca(OH)_2 \longrightarrow CaSO_4 + 2H_2O \tag{13-37}$$
$$2NaHSO_4 + Ca(OH)_2 \longrightarrow CaSO_4 + Na_2SO_4 + 2H_2O \tag{13-38}$$
$$H_2SO_3 + Ca(OH)_2 \longrightarrow CaSO_3 + 2H_2O \tag{13-39}$$
$$2NaHSO_3 + Ca(OH)_2 \longrightarrow CaSO_3 + Na_2SO_3 + 2H_2O \tag{13-40}$$

Chromium hydroxide is an amphoteric compound (see Figure 13-6) and exhibits minimum solubility in the pH range of 7.5 to 10.0. Effluents from chromium reduction processes should be neutralized to the range of zero solubility (pH 8.5–9.0) to minimize the amount of soluble chromium remaining in solution.

Base requirements cannot be calculated precisely because the exact composition of the wastewater is usually not known. The best method for determining these requirements is to titrate a sample of treated wastewater to an endpoint of pH 8.5 to 9.0, using the selected base, and to determine the chemical dosage from the titration curve. Such a titration will include not only the base required to neutralize the wastewater, but also the base required to react with all residual chemicals in the treated wastewater. If a sample of treated wastewater is not available, the base requirement can be estimated by taking a sample of raw wastewater through the following procedure (Chamberlain and Day, 1956):

1. Add H_2SO_4 to the wastewater to reduce the pH to the level chosen for treatment and then back titrate to pH 8.5–9.0, using the selected base.

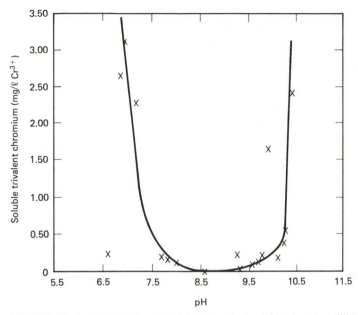

FIGURE 13-6 *pH range for complete removal of soluble chromium (III) as chromic hydroxide (after Chamberlain and Day, 1956).*

2. Calculate the amount of base required to precipitate the chromium, using equation 13-36.

3. Calculate the amount of base required to react with the excess sulfite as shown in equations 13-39 and 13-40. This requirement will be a function of pH.

4. Calculate the amount of base required to neutralize the H_2SO_4 produced from the dissolved oxygen initially present in the wastewater. Reactions are shown in equations 13-37 and 13-38.

The sum of these quantities will represent the base requirement. Use of this procedure is illustrated by the following example.

EXAMPLE PROBLEM 13-2: Rinse water from a chromium-plating operation has the following characteristics:

pH	6.5
Cr^{6+}	20 mg/ℓ
Total Cr	102 mg/ℓ
Total iron	34 mg/ℓ
Total zinc	35 mg/ℓ
Suspended solids	25 mg/ℓ
Dissolved oxygen	3.0 mg/ℓ

Calculate the amount of sulfur dioxide and acid required to treat this wastewater at pH 3.0, and determine the amount of base required to precipitate the chromium and neutralize the treated wastewater.

A sample of raw wastewater was titrated with H_2SO_4 and neutralized with $Ca(OH)_2$, yielding the results shown in Figures 13-7 (a) and 13-7 (b).

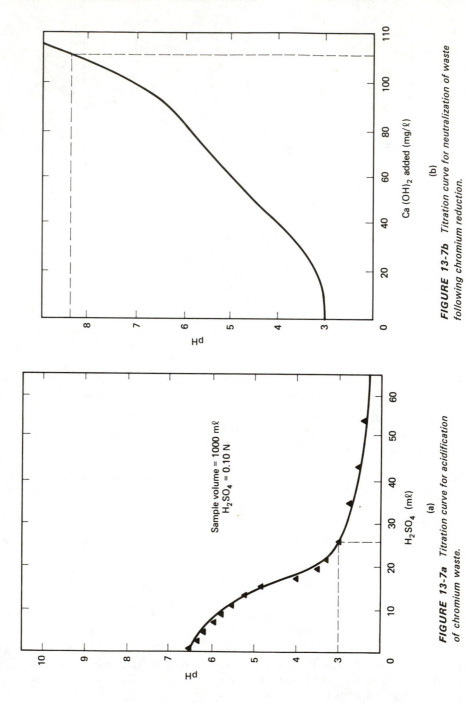

FIGURE 13-7b Titration curve for neutralization of waste following chromium reduction.

(b)

FIGURE 13-7a Titration curve for acidification of chromium waste.

(a)

Sample volume = 1000 ml
H_2SO_4 = 0.10 N

Solution: If SO_2 is used as the reducing agent, the chemical requirements consist of the amount of acid required to reduce the pH to the level selected for treatment and the amount of SO_2 required for reduction of hexavalent chromium. Additional SO_2 would be required for other reacting species such as dissolved oxygen present in the wastewater.

1. Referring to Figure 13-7(a), 26 mℓ of 0.1N H_2SO_4 is required to acidify 1 ℓ of wastewater to pH 3.0. This is equal to an acid dose of

$$26 \text{ m}\ell/\ell \times 0.1N = 2.6 \frac{\text{meq. } H_2SO_4}{\ell}$$

The acid feed rate required to achieve this dose will depend on the normality of acid used for treatment.

2. The amount of SO_2 required to reduce Cr^{6+} and react with dissolved oxygen can be calculated from equations 13-23, 13-26, and 13-27.

$$SO_2 + H_2O \longrightarrow H_2SO_3 \qquad \text{(13-23)}$$

$$2H_2CrO_4 + 3H_2SO_3 \longrightarrow Cr_2(SO_4)_3 + 5H_2O \qquad \text{(13-26)}$$

$$H_2SO_3 + \tfrac{1}{2}O_2 \rightleftharpoons H_2SO_4 \qquad \text{(13-27)}$$

Molar ratios can be used to calculate SO_2 requirements, assuming the reactions go to completion.

For Cr^{6+} from Equation 13-23 (since 3 moles of SO_2 yield 3 moles H_2SO_3):

$$\frac{3 \times (32 + 2(16))}{2 \times 51.99} = 1.85 \text{ mg}/\ell \text{ } SO_2 \text{ per mg}/\ell \text{ } Cr^{6+}$$

For O_2 from Equation 13-27

$$\frac{32 + (2(16))}{\tfrac{1}{2} \times 2(16)} = 4 \text{ mg}/\ell \text{ } SO_2 \text{ per mg}/\ell \text{ } O_2$$

Therefore,

$$SO_2 \text{ required (mg}/\ell) = (1.85)20 \text{ mg}/\ell \text{ } Cr^{6+} + (4) \text{ mg}/\ell \text{ } O_2$$

$$= 49 \text{ mg}/\ell$$

3. Calculate the amount of $Ca(OH)_2$ required to neutralize the wastewater and precipitate the chromium. Assume $Ca(OH)_2$ to be 90% pure.

(a) The amount of $Ca(OH)_2$ required to neutralize the wastewater to pH 8.5, as indicated by Figure 13-7(b), is found to be *105 mg/ℓ* (neglecting volume change resulting from acid addition).

(b) The amount of $Ca(OH)_2$ required to precipitate the chromium can be calculated from equation 13-36:

$$Cr_2(SO_4)_3 + 3Ca(OH)_2 \longrightarrow 2Cr(OH)_3 + 3CaSO_4 \qquad \text{(13-36)}$$

$$\frac{3(40 + 2(16) + 2(1))}{2(51.99)} = 2.14 \text{ mg}/\ell \text{ of } Ca(OH)_2 \text{ per mg}/\ell \text{ of } Cr^{3+}$$

The total lime requirement must be based on the combined total of trivalent and hexavalent chromium present in the raw wastewater.

$$Ca(OH)_2 \text{ requirement (assume 90\% pure)} = \frac{2.14 \times 102 \text{ mg/}\ell}{0.90}$$

$$= 242.53 \text{ mg/}\ell$$

(c) The amount of lime required to react with excess sulfite or bisulfite is as follows: the sulfite or bisulfite in the treated wastewater will result from the dissociation of H_2SO_3 produced by addition of sulfur dioxide. Since the stoichiometric amount of SO_2 was added, it can be assumed that all the H_2SO_3 reacted with the Cr^{6+} and that no excess H_2SO_3 was present to dissociate into sulfite or bisulfite.

   If excess SO_2 had been added, its distribution between H_2SO_3 and HSO_3^- would have to be determined from Figure 13-4(a). The appropriate amount of lime could then be calculated, using equations 13-39 and 13-40.

(d) The amount of lime required to neutralize the H_2SO_4 produced from dissolved oxygen initially present in the wastewater can be calculated from equations 13-27, 13-37, and 13-38:

$$H_2SO_4 + Ca(OH)_2 \longrightarrow CaSO_4 + 2H_2O \qquad \textbf{(13-37)}$$

$$2NaHSO_4 + Ca(OH)_2 \longrightarrow CaSO_4 + NaSO_4 + 2H_2O \qquad \textbf{(13-38)}$$

Equation 13-27 can be used to calculate the amount of H_2SO_4 produced:

$$\frac{(2(1) + 32 + 16(4))}{\frac{1}{2}(2)(16)} = 6.12 \text{ mg/}\ell \; H_2SO_4 \text{ produced}$$
$$\text{per mg/}\ell \; O_2 \text{ present}$$

$$H_2SO_4 \text{ produced} = 6.12 \times 3.0 \text{ mg/}\ell \; O_2 = 18.36 \text{ mg/}\ell$$

This H_2SO_4 will dissociate as shown in equations 13-30 and 13-31. At pH 3.0 only the sulfate ion will exist, and equation 13-38 does not apply. The lime requirement can be calculated from equation 13-37.

$$\frac{(40 + 2(16) + 2(1))}{(2(1) + 32 + 4(16))} = 0.76 \text{ mg/}\ell \text{ lime per mg/}\ell \; H_2SO_4$$

$$Ca(OH)_2 \text{ required (assuming 90\% pure)} = \frac{0.76 \times 18.36 \text{ mg/}\ell}{0.90}$$

$$= 15.5 \text{ mg/}\ell$$

$$\text{Total amount of 90\% } Ca(OH)_2 = 105 + 242.53 + 15.5$$

$$= 363.03 \text{ mg/}\ell$$

EXAMPLE PROBLEM 13-3: Rework example problem 13-2, using sodium bisulfite as the reducing agent at pH 3.0.

Solution: The amount of acid required to reduce the pH to 3.0 would be the same as determined previously. Since $NaHSO_3$ is used, additional acid would be required to satisfy the reduction reaction.

The amount of acid and sodium bisulfite required for reduction at pH 3.0 can be calculated from equation 13-32, using molar ratios.

$$3NaHSO_3 + 1.5H_2SO_4 + 2H_2CrO_4 \rightleftharpoons$$
$$Cr_2(SO_4)_3 + 1.5NaSO_4 + 5H_2O \qquad \text{(13-32)}$$

Sodium Bisulfite Requirement;

$$\frac{3 \times (23 + 1(1) + 32 + 16(3))}{(2 \times 51.99)} = 3.00 \text{ mg/} \ell \text{ } NaHSO_3 \text{ per mg/} \ell \text{ } Cr^{6+}$$

$$NaHSO_3 \text{ required} = 3.00 \times 20 \text{ mg/} \ell = 60 \text{ mg/} \ell$$

Sulfuric Acid Requirement;

$$\frac{1.5(2(1) + 32 + 4(16))}{2(51.99)} = 1.42 \text{ mg/} \ell \text{ } H_2SO_4 \text{ per mg/} \ell \text{ } Cr^{6+}$$

$$H_2SO_4 \text{ required} = \frac{1.42 \times 20 \text{ mg/} \ell}{49 \text{ mg/meq.}} = 0.58 \text{ } \frac{\text{meq. } H_2SO_4}{\ell}$$

This amount of acid would be in addition to the acid required to acidify the wastewater to pH 3.0, as determined in Example 13-2.

$$\text{total } H_2SO_4 = 2.6 \text{ meq./} \ell + 0.58 \text{ meq./} \ell = 3.18 \text{ meq./} \ell$$

13-3 AMMONIA REMOVAL BY BREAKPOINT CHLORINATION

Nitrogen is one of the elements necessary for all life. In its natural elemental state it is a diatomic gas which comprises 78% of the atmosphere. From this large reservoir nitrogen is chemically incorporated into both organic and inorganic compounds, which in turn react or decompose to reform the element. The entire process, much of which is biologically mediated, is referred to as the *nitrogen cycle*. The importance of nitrogen to life itself can easily be recognized by its presence in all amino acids, which in turn are the building blocks from which proteins are made.

In aquatic systems excessive concentrations of nitrogen compounds can cause problems. The primary adverse effects are as follows:

1. Ammonia, a common nitrogen compound in domestic wastewater, is toxic to aquatic life in low concentrations.

2. Organic nitrogen compounds and ammonia are both oxidized in aquatic systems, with a concomitant loss of dissolved oxygen in the system.

3. In instances where nitrogen is limiting to growth in a particular aquatic ecosystem, discharge of nitrogen compounds can promote the growth of nuisance algae. This picture is complicated, however, by the fact that certain species of algae can satisfy their nitrogen requirements directly from the atmosphere, using a process known as *fixation*.

4. High nitrate levels in drinking water can produce a human health hazard.

Because of these problems, considerable attention has been focused in recent years on the removal of nitrogen from wastewaters.

The Chemical Forms of Nitrogen

Nitrogen chemistry is complicated by the multiplicity of oxidation states the element assumes in its compounds. Those of greatest importance to environmental engineers, however, are the -3, $+3$, and $+5$ oxidation states, of which common examples are ammonia (NH_3), nitrite ion (NO_2^-), and nitrate ion (NO_3^-).

Animal wastes contain appreciable amounts of nitrogen incorporated into organic compounds, primarily unassimilated proteins. Urine contains nitrogen principally in the form of urea, $(NH_2)_2CO$. Bacterial action on such organic matter results in its degradation and the release of ammonia. The ammonia so produced may then be further oxidized to nitrite by bacteria such as *Nitrosomonas*, and the nitrite produced from this reaction can be oxidized to nitrate by other bacteria such as *Nitrobacter*. These biologically mediated reactions—collectively referred to as *nitrification*—may be represented as follows:

$$2NH_3 + 3O_2 \longrightarrow 2NO_2^- + 2H^+ + 2H_2O \qquad \text{(13-41)}$$

$$2NO_2^- + O_2 \longrightarrow 2NO_3^- \qquad \text{(13-42)}$$

Early investigators in the public health field were sometimes able to use the above relationships to estimate how recently environmental contamination of water had occurred. Since raw wastewater contains a great deal of organic nitrogen and ammonia, if these compounds comprised a large portion of the total nitrogen found in a water sample, it was inferred that the pollution was probably of recent origin and, therefore, of greater potential danger. Over a period of time these forms would be transformed to nitrate nitrogen in an aerobic environment. The presence of primarily nitrate nitrogen in a water sample was taken to indicate older, and thus less dangerous, pollution. Today, bacteriological tests are used for more reliable evaluation of potential health hazards in water.

To determine the total nitrogen content of a water sample, it is necessary to sum the various species. To accomplish such addition, similar units are necessary, and concentrations of nitrogen species are therefore often expressed in terms of their nitrogen content. For example, NH_3 has a molecular mass of 17, of which 14/17 is due to the element nitrogen. Therefore, $17 \, mg/\ell \, NH_3$ is equivalent to $14 \, mg/\ell \, NH_3$-N (read

NH_3-as-N, or ammonia nitrogen). Since the measurement of organic nitrogen does not identify any particular compound, this value is always in terms of the element.

EXAMPLE PROBLEM 13-4: A water sample contains 10 mg/ℓ ammonia, 2 mg/ℓ nitrite, and 25 mg/ℓ nitrate. Calculate the total inorganic nitrogen concentration.

Solution: Total inorganic nitrogen = NH_3-N + NO_2^--N + NO_3^--N.

$$NH_3\text{-N} = 10\,\frac{mg}{\ell} \times \frac{14}{17} = 8.2\ mg/\ell$$

$$NO_2^-\text{-N} = 2\,\frac{mg}{\ell} \times \frac{14}{46} = 0.6\ mg/\ell$$

$$NO_3^-\text{-N} = 25\,\frac{mg}{\ell} \times \frac{14}{62} = 5.6\ mg/\ell$$

$$\text{total inorganic nitrogen} = 14.4\ mg/\ell$$

EXAMPLE PROBLEM 13-5: How much nitrate would be produced from the total oxidation of (1) 5 mg/ℓ NH_3? (2) 5 mg/ℓ NH_3-N?

Solution:

1. Equations 13-41 and 13-42 indicate that 1 mole of NH_3 can be transformed into 1 mole of NO_3^-. Therefore,

$$\frac{mg}{\ell}\,NO_3^- = 5\,\frac{mg}{\ell}\,NH_3 \times \frac{62}{17} = 18.2$$

2. By relating species concentration to nitrogen, calculations can be simplified. Since nitrogen is conserved in the reaction,

$$5\,\frac{mg}{\ell}\,NH_3\text{-N must produce } 5\,\frac{mg}{\ell}\,NO_3^-\text{-N}$$

Ammonia is a gas at room temperature, with a pungent odor familiar to most people. The typical concentration of ammonia in raw domestic wastewater is 25 mg/ℓ NH_3-N. Ammonia is extremely soluble in water (about 90 g/ℓ in cold water), wherein it reacts to form the ammonium ion

$$NH_3 + H_2O \rightleftharpoons NH_4^+ + OH^- \tag{13-43}$$

From this equation it can be seen that ammonia produces a basic solution when it dissolves in water. The dominant ammonia form in neutral or acidic solutions is the ammonium ion, however. Many times, NH_4^+ and NH_3 are used interchangeably when total concentrations are discussed. There are, however, important differences.

One very important difference in these species lies in the area of toxicity. It has been determined that unionized NH_3 is the toxic form, and that ammonium ion is of itself not a threat to biota. One common water quality criterion recommends a limit of 0.02 mg/ℓ unionized NH_3 in recovery waters (EPA, 1976). The amount of total ammonia (both NH_4^+ and NH_3 species) corresponding to this criterion is of course dependent on pH (equation 13-43). At low pH values a relatively large concentration of total ammonia might not be toxic, since only a small portion of the total would be present in the unionized form. At higher pH levels nearly all the ammonia would be present as NH_3. The relationship is illustrated in Figure 13-8, which also shows the effect of temperature on the relationship.

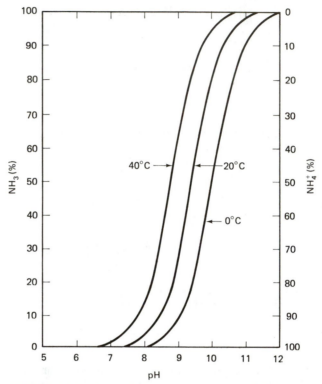

FIGURE 13-8 *Effects of pH and temperature on distribution of ammonia and ammonium ion in water (after EPA, 1975).*

EXAMPLE PROBLEM 13-6: Determine whether a receiving water containing 1.5 mg/ℓ total NH_3 at a pH of 7.5 would exceed the water quality criterion for ammonia toxicity. The constant for the acid dissociation of the ammonium ion at 25°C is as follows:

$$NH_4^+ \rightleftharpoons NH_3 + H^+ \qquad pK_a = 9.245 \qquad \textbf{(13-44)}$$

Solution:

1. Set up the equilibrium constant expression

$$\frac{[NH_3][H^+]}{[NH_4^+]} = 10^{-9.245}$$

2. Calculate $[NH_3]_{total}$ and $[H^+]$:

$$\text{total } NH_3 = 1.5 \frac{mg}{\ell} \times \frac{1 \text{ mole}}{17{,}000 \text{ mg}} = 8.82 \times 10^{-5} M = 10^{-4.05} M$$

$$[H^+] = 10^{-pH} = 10^{-7.5} M$$

3. Recognizing that $[NH_3] + [NH_4^+] = [NH_3]_{total}$,

$$\frac{[NH_3](10^{-7.5})}{10^{-4.054} - [NH_3]} = 10^{-9.245}$$

$$10^{-7.5}[NH_3] = 10^{-13.299} - 10^{-9.245}[NH_3]$$

$$[NH_3] = 10^{-5.799} M$$

4. Convert $[NH_3]$ to mg/ℓ:

$$mg/\ell \ NH_3 = \left(10^{-5.799} \frac{mole}{\ell}\right)\left(\frac{17{,}000 \text{ mg } NH_3}{\text{moles } NH_3}\right)$$

$$= 0.0270 \frac{mg}{\ell} \text{ unionized } NH_3$$

5. Since the criterion is $0.02/\ell$ unionized NH_3, it is exceeded by the above conditions.

Emerson et al. (1975) established the temperature dependence of the pK value for ammonium ion dissociation:

$$pK_a = 0.09018 + \frac{2729.92}{T} \tag{13-45}$$

where T represents temperature in °K. Using this relationship, temperature effects can be included in the evaluation of ammonia toxicity.

The influence of ammonia on dissolved oxygen concentration in streams is related to its oxidation by bacteria. The overall reaction may be represented by the summation of equations 13-41 and 13-42.

$$2NH_3 + 4O_2 \longrightarrow 2NO_3^- + 2H^+ + 2H_2O \tag{13-46}$$

The stoichiometric molar ratio $NH_3 : O_2$ is $1:2$, which is a mass ratio of $0.27:1$. Thus, about 3.7 mg of oxygen is utilized in the oxidation of every mg of ammonia via nitrification. These numbers of course do not indicate the rate at which oxygen would be utilized but only reflect the ultimate result. In effect, the above numbers produce a theoretical, ultimate, nitrogenous oxygen demand.

Organic nitrogen is measured in the laboratory by means of the Kjeldahl procedure, and the result is termed *total Kjeldahl nitrogen* (TKN). The TKN value is indicative of both organic nitrogen and ammonia in a sample. Thus, the concentration of organic nitrogen itself is the difference between the TKN value and the ammonia nitrogen concentration.

The organic nitrogen content of raw domestic wastewater is normally about 15 mg/ℓ as nitrogen. Most of this is removed or converted to ammonia in the course of secondary treatment.

EXAMPLE PROBLEM 13-7: Using the following data from a secondary wastewater treatment plant, determine the percent removal of total nitrogen and of organic nitrogen:

	Influent (mg/ℓ)	Effluent (mg/ℓ)
TKN	40	8.2
NH_3	30	9
NO_2^-	0	4
NO_3^-	0	20

Solution:

1. Convert all values to terms of nitrogen. The conversion factors are as follows:

   (a) for NH_3, $14/17 = 0.824$

   (b) for NO_2^-, $14/46 = 0.304$

   (c) for NO_3^-, $14/62 = 0.226$

   This results in the following:

	Influent (mg/ℓ)	Effluent (mg/ℓ)
TKN	40	8.2
NH_3-N	24.7	7.4
NO_2^-	0	1.2
NO_3^-	0	4.6

2. Since the TKN value includes ammonia, total nitrogen is the sum of the TKN, NO_2^-, and NO_3^- values.

$$\text{total influent N} = 40 \text{ mg/}\ell$$
$$\text{total effluent N} = 14 \text{ mg/}\ell$$
$$\% \text{ removal} = \frac{40 - 14}{40} \times 100 = 65\%$$

3. Organic nitrogen is the difference between TKN and NH_3-N.

$$\text{influent organic N} = 40 - 24.7 = 15.3 \text{ mg}/\ell$$
$$\text{effluent organic N} = 8.2 - 7.4 = 0.8 \text{ mg}/\ell$$
$$\% \text{ removal} = \frac{15.3 - 0.8}{15.3} \times 100 = 95\%$$

Nitrite and Nitrate

Nitrite (NO_2^-) is usually present in only small amounts in natural waters, or even wastewater treatment plant effluents, since it is an intermediate product in the oxidation of ammonia. It may also be produced by the reduction of nitrate, which occurs during a biologically mediated process known as *denitrification.*

Nitrate is a common ion found in most natural waters. It is the end product of the oxidation of ammonia (Equation 13-45). Because it is an algal nutrient, it can promote nuisance algal blooms under certain circumstances. Nitrate has also been identified as a potential health hazard to young infants. This is due to the potential reduction of nitrate to nitrite in the stomachs of infants, which can then bond with the hemoglobin of affected babies, thus diminishing the transfer of oxygen to the body's cells and resulting in a bluish skin color (National Academy of Sciences, 1978). The medical problem just described is termed *methemoglobinemia*; it is not a problem in adults because their stomachs are more acidic, thus preventing the growth of the organisms which reduce nitrate. To minimize the threat to infants, the drinking water criterion for nitrates has been set at 10 mg/ℓ NO_3^-—N (EPA, 1976).

It is noted here that a common technique used in nitrate analysis—the cadmium reduction method—measures nitrate and nitrite species simultaneously. Thus, it is common to see values of (NO_3^- + NO_2^-)-N reported. To obtain true nitrate concentrations, the nitrite would have to be measured separately and subtracted. However, since nitrite levels are usually low, they are often ignored.

The nitrate ion is the most oxidized form of nitrogen and chemically unreactive in dilute aqueous solution. All common metal nitrate salts are soluble; nitric acid (HNO_3) is a strong acid. Nitrate has little tendency to form complexes with metal ions in dilute aqueous solutions.

Nitrogen Removal

The annual nitrogen loading in domestic wastewater is 8–12 lb per capita. There are primarily four processes utilized in advanced wastewater treatment to remove this nitrogen. These processes are nitrification-denitrification, ammonia stripping, breakpoint chlorination, and ion exchange. Of these processes only breakpoint chlorination employs the principles of oxidation-reduction and, as such, is the only process discussed in this chapter.

Early experience with the use of chlorine for disinfection showed that the "chlorine demand" of the water had to be met before residual chlorine could be measured in the water. Some of this demand was found to be due to inorganic reducing agents, such as sulfide or ferrous ions, in the water. But ammonia nitrogen was often found to play a dominant role because of the formation of a group of compounds known as *chloramines*. The following sequence of chemical reactions is now known to occur:

$$NH_4^+ + HOCl \rightleftharpoons \underset{\text{monochloramine}}{NH_2Cl} + H_2O + H^+ \qquad \text{(13-47)}$$

$$NH_2Cl + HOCl \rightleftharpoons \underset{\text{dichloramine}}{NHCl_2} + H_2O \qquad \text{(13-48)}$$

$$NHCl_2 + HOCl \rightleftharpoons \underset{\text{nitrogen trichloride}}{NCl_3} + H_2O \qquad \text{(13-49)}$$

$$2NH_2Cl + HOCl \longrightarrow N_2 + 3HCl + H_2O \qquad \text{(13-50)}$$

Both HOCl and OCl$^-$ are commonly called *free chlorine* species, whereas the chloramines are referred to as *combined chlorine* forms.

The chlorination of water containing ammonia characteristically results in an initial increase in chlorine residuals, followed by a decline and then, finally, another increase. A typical breakpoint curve is shown in Figure 13-9. Note that after the initial,

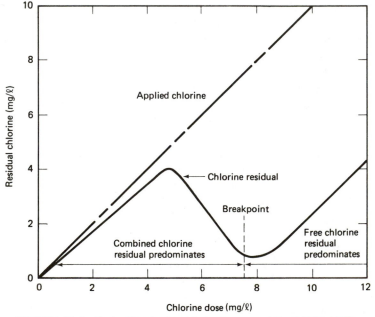

FIGURE 13-9 *Typical breakpoint chlorination curve (after WPCF, 1977).*

immediate chlorine demand is satisfied, monochloramines and dichloramines increase (equations 13-47 and 13-48), but then they decrease as the chlorine dose is increased (equations 13-48 and 13-49). The second upturn in the curve is the breakpoint, and beyond that point any added chlorine produces free chlorine residuals.

The primary interest in these reactions stemmed from the fact that the free chlorine residuals are better disinfecting agents than the combined forms. Therefore, by practicing breakpoint chlorination, better disinfection and shorter contact times could be achieved.

The usefulness of breakpoint chlorination in nitrogen removal can readily be seen by the summation of equations 13-47 and 13-50:

$$2NH_4^+ + 3HOCl \longrightarrow N_2 + 3H_2O + 3HCl + 2H^+ \qquad \text{(13-51)}$$

If this equation is, in turn, combined with the equation in Ex. Prob. 1–10, page 30, a different form results:

$$2NH_4^+ + 3Cl_2 \longrightarrow N_2 + 6HCl + 2H^+ \qquad \text{(13-52)}$$

Clearly, if sufficient chlorine is reacted with the ammonia in water, the nitrogen will be removed from the solution in the form of nitrogen gas. Theoretically, the atomic ratio of $Cl:N$ to achieve breakpoint is $3:1$, which by mass is $7.6:1$.

Some studies have shown evidence for the production of nitrous oxide (N_2O) or nitrate during the breakpoint process. A study by Pressley et al. (1972), however, found that the principal product was nitrogen gas. The specific product is dependent on process variables such as pH, contact time, and initial $Cl:N$ ratios. It is known, for example, that the formation of NCl_3 is favored under low pH conditions after the breakpoint is reached. This compound is irritating, and its production is to be avoided.

The $Cl:N$ ratio required to remove nitrogen from wastewaters is always greater than the stoichiometric ratio of $7.6:1$. However, pretreatment of wastewater does lower the $Cl:N$ ratio required to near stoichiometric values. For example, a $Cl:N$ mass ratio of less than $8:1$ was needed for a lime-clarified and filtered secondary effluent. But ratios as high as $10:1$ are required in practice.

The hydrolysis of chlorine in water produces a strong acid (HCl) in addition to HOCl. Moreover, the reaction of HOCl with ammonia also produces acid. The net result (equation 13-52) is that alkalinity is destroyed, or the pH of the solution is lowered if the buffer capacity is insufficient. From the stoichiometric relationship, it can be seen that 2 moles of ammonium ion will produce 8 moles of hydrogen ion. Since each mole of hydrogen would react with 0.5 mole of alkalinity (as $CaCO_3$), it can be calculated that 14.3 mg/ℓ of alkalinity as $CaCO_3$ would be destroyed for each mg/ℓ of ammonia nitrogen consumed. In practice, about 15 mg/ℓ of alkalinity are required.

Breakpoint chlorination has been found to be a practical means of nitrogen removal from wastewater, particularly when the ammonia concentrations are relatively low. The disadvantages of the process include the cost of chlorine and also the introduction of dissolved solids to the water, which renders the water less desirable for reuse.

The increase in dissolved solids is dependent in part on the form in which the chlorine is added. For example, when chlorine gas is used in breakpoint chlorination, 6 moles of HCl and 2 additional moles of hydrogen ions result where only 2 moles of the ammonium ion had been before. This is a net gain of 185 g of dissolved solids per

2 moles of ammonium ion reacted, or a mass ratio of 6.6:1. The actual increase is larger than this because larger than stoichiometric ratios are generally required.

If hypochlorite salts are used in the process, the associated cation will also be added to the treated water. And if the acid production is offset by the addition of a base such as CaO or NaOH, even greater amounts of dissolved solids are added. The contribution of dissolved solids by several chemicals that may be used in breakpoint chlorination is summarized in Table 13-4.

TABLE 13-4 *Effects of chemical addition on total dissolved solids in breakpoint chlorination (after EPA, 1975).*

Chemical Addition	TDS Increase: NH_4^+-N Consumed
Breakpoint with chlorine gas	6.2:1
Breakpoint with sodium hypochlorite	7.1:1
Breakpoint with chlorine gas—Neutralization of all acidity with lime (CaO)	12.2:1
Breakpoint with chlorine gas—neutralization of all acidity with sodium hydroxide (NaOH)	14.8:1

EXAMPLE PROBLEM 13-8: A clarified wastewater containing 5 mg/ℓ NH_3-N is to be breakpoint chlorinated. The water has an alkalinity of 100 mg/ℓ as CaCO. Determine

1. The stoichiometric requirement for Cl_2.

2. If the water is sufficiently buffered.

3. The increase in dissolved solids to be expected.

Solution:

1. The stoichiometric requirement is for a Cl:N mass ratio of 7.6:1.

$$5 \frac{mg}{\ell} NH_3\text{-N} \times \frac{7.6}{1} = 38 \frac{mg}{\ell} Cl_2$$

2. Alkalinity consumed is at least 14.3 mg/ℓ as $CaCO_3$ per mg/ℓ N consumed. Use 15 mg/ℓ alkalinity as a precaution.

$$5 \frac{mg}{\ell} NH_3\text{-N} \times \frac{15}{1} = 75 \frac{mg}{\ell} \text{ alkalinity to } CaCO_3$$

Therefore, there appears to be sufficient alkalinity to prevent a significant change in pH.

3. A minimum of 6.6 mg/ℓ of added dissolved solids can be expected for each mg/ℓ N reacted.

$$5 \frac{mg}{\ell} \, NH_3\text{-}N \times 6.6 = 33 \, mg/\ell \text{ dissolved solids}$$

Since chlorine can lead to a toxicity problem in receiving waters, wastewater treated by breakpoint chlorination is sometimes dechlorinated. This process is commonly achieved by one of three reagents:

1. Sulfur dioxide:

$$SO_2 + Cl_2 + H_2O \longrightarrow H_2SO_4 + 2HCl \qquad \textbf{(13-53)}$$

2. Sodium metabisulfite:

$$Na_2S_2O_5 + Cl_2 + 3H_2O \longrightarrow 2NaHSO_4 + 2HCl \qquad \textbf{(13-54)}$$

3. Activated carbon:

$$2Cl_2 + C + 2H_2O \longrightarrow 4HCl + CO_2 \qquad \textbf{(13-55)}$$

PROBLEMS

13-1. Review the article by Chamberlain and Snyder (1955). Using the stoichiometric approach outlined by these authors, calculate the amount of chlorine and NaOH required to treat the wastewater from a metal-plating process by two-stage alkali-chlorination. The wastewater has the following composition:

$$pH = 10.5$$
$$Cu^{2+} = 6 \, mg/\ell$$
$$Fe^{3+} = 2 \, mg/\ell$$
$$Ni^{2+} = 25 \, mg/\ell$$
$$Zn^{2+} = 39 \, mg/\ell$$
$$CN^- = 10 \, mg/\ell$$

13-2. Determine the amount of $NaHSO_3$ and H_2SO_4 required to treat a chromium-plating wastewater at pH = 4.0. Composition of the wastewater and titration curves for acidification and wastewater neutralization are shown in Figures P13-2(a) and P13-2(b), respectively. Assume wastewater neutralization is accomplished with $Ca(OH)_2$ that is 70% pure.

$$pH = 6.0$$
$$Cr^{6+} = 25 \, mg/\ell$$
$$\text{total iron} = 30 \, mg/\ell$$
$$\text{dissolved oxygen} = 5.0 \, mg/\ell$$

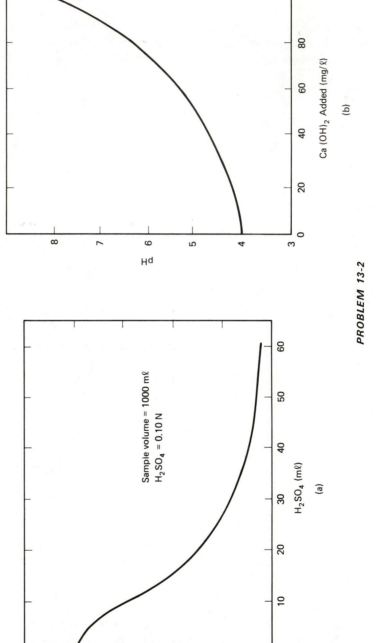

Sample volume = 1000 mℓ
H_2SO_4 = 0.10 N

H_2SO_4 (mℓ)

(a)

Ca (OH)$_2$ Added (mg/ℓ)

(b)

PROBLEM 13-2

13-3. A water sample is analyzed and found to contain 2.5 mg/ℓ ammonium ion, 0.5 mg/ℓ nitrite, and 15 mg/ℓ nitrate. The Kjeldahl nitrogen in this sample is found to be 3.0 mg/ℓ. Calculate:

    (a)   The organic nitrogen concentration in the sample.

    (b)   The total inorganic nitrogen concentrations in the sample.

    (c)   The potential nitrate concentration in the sample if all nitrogen were oxidized to the nitrate form.

    (d)   The potential oxygen consumed if the reactions in part (c) proceeded.

13-4. A wastewater treatment plant discharges 15 MGD of effluent into a well-buffered stream which has a 7 day, 10 year low flow of 150 cu ft/sec. If the summer temperature of the stream reaches 30°C and if the measured pH is 7.30, calculate the *total ammonia* which can be allowed in the effluent without exceeding the toxicity criterion of 0.02 mg/ℓ unionized ammonia in the stream. The influence of the effluent on stream temperature may be ignored.

13-5. Assume that HOCl is 50 times more effective a disinfectant than OCl$^-$. Then, using the same total dose of chlorine, what would be the relative gain or loss in disinfecting action if the pH of the solution changed from 7.5 to 7.0? Assume a temperature of 20°C.

13-6. A wastewater containing 12 mg/ℓ NH$_3$ is to be breakpoint chlorinated. The water has an alkalinity of 120 mg/ℓ as CaCO$_3$. Find:

    (a)   The stoichiometric requirement for chlorine.

    (b)   The amount of lime which is required, if any, to provide buffering capacity.

    (c)   The increase in dissolved solids to be anticipated.

REFERENCES

American Electroplaters' Society, "A Report on the Control of Cyanides in Plating Shop Effluents," *Plating*, 1107 October, (1969).

BUTLER, J.H., *Ionic Equilibrium, A Mathematical Approach*, Addison-Wesley Publishing Co., Inc., Reading, MA (1964).

CHAMBERLAIN, N.S., and DAY, R.V., "Technology of Chrome Reduction with Sulphur Dioxide," *Proc. Eleventh Purdue Industrial Waste Conference*, 129 (1956).

CHAMBERLAIN, N.S., and SYNDER, H.B., Jr., "Technology of Treating Plating Wastes," *Proc. Tenth Purdue Industrial Waste Conference*, 277 (1955).

DROGON, J., and PASEK, L., "Continuous Electrolytic Destruction of Cyanide Waste," *Electroplating and Metal Finishing*, p. 310 (September, 1965).

EASTON, J.K., "Electrolytic Decomposition of Concentrated Cyanide Plating Wastes," *Journal Water Pollution Control Federation*, **39**, 1621, (1967).

ECKENFELDER, W.W., *Industrial Water Pollution Control*, McGraw-Hill Book Company, New York (1966).

EMERSON, K., RUSSO, R.C., LUND, R.E., and THURSTON, R.V., "Aqueous Ammonia Equilibrium Calculations: Effect of pH and Temperature," *J. Fish. Res. Bd. Can.*, **32** (12), 2379 (1975).

Environmental Protection Agency, *Process Design Manual for Nitrogen Control*, U.S. Government Printing Office, Washington, DC (1975).

Environmental Protection Agency, *Quality Criteria for Water*, U.S. Government Printing Office, Washington, DC (1976).

Environmental Protection Agency, "National Interim Primary Drinking Water Regulations," Office of Water Supply, EPA-570/9-76-003 (1976).

National Academy of Sciences, *Nitrates: An Environmental Assessment*, U.S. Government Printing Office, Washington, DC (1978).

PRESSLEY, T.A., BISHOP, D.F., and ROAN, S.G., "Ammonia-Nitrogen Removal by Breakpoint Chlorination," *Env. Sci. Tech.*, **6** (7): 622 (1972).

ROSS, R.D., [Ed.] *Industrial Waste Disposal*, Van Nostrand Reinhold Company, New York (1968).

ROUSE, J.V., "Removal of Heavy Metals from Industrial Effluents", *Jour, EED ASCE*, **102**, EE5, (October 1976).

RUDOLFS, W., *Industrial Wastes—Their Disposal and Treatment*, Van Nostrand Reinhold Company, New York (1953).

SAWYER, C.N., and MCCARTY, P.L., *Chemistry for Sanitary Engineers*, 2nd ed., McGraw-Hill Book Company New York (1967).

SHINSKEY, F.G., *pH and pIon Control in Process and Waste Streams*, John Wiley & Sons, New York (1973).

Water Pollution Control Federation and American Society for Civil Engineers, *Wastewater Treatment Plant Design*, Lancaster Press, Inc., Lancaster, PA (1977).

14

IRON AND MANGANESE REMOVAL

Iron and manganese are natural constituents of soil and rocks. Iron, one of the most abundant elements, is found much more frequently than manganese. This variation probably accounts for the fact that waters containing soluble iron and manganese generally have an iron content which is greater than the manganese content. Furthermore, waters are seldom found which have iron levels greater than 10 mg/ℓ or manganese levels greater than 2 mg/ℓ.

Iron exists in the $+2$ or $+3$ oxidation states, whereas manganese exists in either the $+2, +3, +4, +6,$ or $+7$ oxidation states. However, Morgan and Stumm (1964) have shown that in waters containing dissolved oxygen, iron(III), and manganese(IV) are the only stable oxidation states for iron and manganese. The chemical forms associated with these two species are highly insoluble, indicating that waters containing dissolved oxygen will contain very little soluble iron or manganese. On the other hand, appreciable levels of iron and manganese may be found in ground water and in the water at the bottom levels of lakes and reservoirs, where anaerobic conditions favor the reduction of iron(III) and manganese(IV) to the soluble iron(II) and manganese(II) forms.

The presence of significant amounts of either or both of these metals in a water supply can create several problems for the consumer. According to Breland and Robinson (1967), the following problems can be attributed to the presence of iron:

1. Large concentrations of iron impart a metallic taste to the water.

2. Industrial products such as paper, textiles, or leather may be discolored.

3. Household fixtures such as porcelain basins, bathtubs, glassware, and dishes are stained.

4. Clothes may stain a yellow or brown-yellow color.

5. Iron precipitates clog pipes and promote the growth of gelatinous masses of iron bacteria. These bacteria slough off and create "red water."

6. Iron bacteria may cause odor and taste problems due to low flow conditions.

The problems associated with manganese are very similar to those for iron:

1. Manganese may produce a taste problem at high concentrations.
2. Manganese may cause a discoloration of industrial products similar to that caused by iron.
3. Household fixtures may be stained a brown or black color.
4. Clothes may become dingy or grayish.

Because of the nuisance condition created by the presence of these metals, the EPA recommends that the iron and manganese concentration in drinking water not exceed 0.3 mg/ℓ and 0.05 mg/ℓ, respectively. Any water which contains these substances above the recommended levels should be treated for their removal. The basic method of removing both of these metals depends on oxidation to their higher insoluble oxidation state [iron(III) and manganese(IV)] and separation of the solid phase from the liquid phase.

14-1 EQUILIBRIA GOVERNING IRON AND MANGANESE SOLUBILITY

In most groundwaters and in the hypolimnetic waters of reservoirs, iron and manganese exist principally in the reduced form. In the absence of carbonate and sulfide species, the solubility of the two metals is controlled by the metal hydroxide solid phase.

$$Fe(OH)_{2\,(s)} \rightleftharpoons Fe^{2+}_{(aq)} + 2OH^-_{(aq)} \qquad (14\text{-}1)$$

$$Mn(OH)_{2\,(s)} \rightleftharpoons Mn^{2+}_{(aq)} + 2OH^-_{(aq)} \qquad (14\text{-}2)$$

Assuming an ideal solution, the saturated equilibrium relationships for reactions 14-1 and 14-2 are

$$K_{S\,Fe} = [Fe^{2+}][OH^-]^2 \qquad (14\text{-}3)$$

where

$$K_{S\,Fe} = 10^{-15.1} \text{ at } 25°C$$

and

$$K_{S\,Mn} = [Mn^{2+}][OH^-]^2 \qquad (14\text{-}4)$$

where

$$K_{S\,Mn} = 10^{-12.9} \text{ at } 25°C$$

In reality, equations 14-3 and 14-4 do not accurately describe the solubility of either compound because there are competing reactions in solution which have not been considered. Complex ion formation is an important consideration for both iron(II) and manganese(II) ions and hence should be accounted for. Complex ion formation reactions which increase the solubility of iron(II) hydroxide and manganese(II) hydroxide are listed in Table 14-1.

	Reaction	K at $25°C$	Reference
1.	Iron:		
	(a) $Fe^{2+}_{(aq)} + OH^-_{(aq)} \rightleftharpoons FeOH^-_{(aq)}$	$K_1 = 10^{5.56}$	4
	(b) $Fe^{2+}_{(aq)} + 3OH^-_{(aq)} \rightleftharpoons Fe(OH)^-_{3(aq)}$	$K_2 = 10^{9.67}$	4
2.	Manganese:		
	(a) $Mn^{2+}_{(aq)} + OH^-_{(aq)} \rightleftharpoons MnOH^+_{(aq)}$	$K_3 = 10^{3.9}$	4
	(b) $Mn^{2+}_{(aq)} + 3OH^-_{(aq)} \rightleftharpoons Mn(OH)^-_{3(aq)}$	$K_4 = 10^{8.3}$	4

When writing a mass balance for soluble iron and soluble manganese, all species which contribute to the soluble form must be considered. For waters containing OH^- as the only complexing species, the total soluble iron(II), $[Fe]_T$, and total soluble manganese(II), $[Mn]_T$, are indicated by the following mass balance equations:

$$[Fe]_T = [Fe^{2+}] + [FeOH^+] + [Fe(OH)^-_3] \tag{14-5}$$

$$[Mn]_T = [Mn^{2+}] + [MnOH^+] + [Mn(OH)^-_3] \tag{14-6}$$

By solving the equilibrium expressions in Table 14-1 for each complex ion and then substituting the results into equations 14-5 and 14-6, the following relationships can be derived:

$$[Fe]_T = [Fe^{2+}] + K_1[Fe^{2+}][OH^-] + K_2[Fe^{2+}][OH^-]^3 \tag{14-7}$$

$$[Mn]_T = [Mn^{2+}] + K_3[Mn^{2+}][OH^-] + K_4[Mn^{2+}][OH^-]^3 \tag{14-8}$$

Since

$$[OH^-] = \frac{K_w}{[H^+]} \tag{2-9}$$

equations 14-7 and 14-8 may be written as

$$[Fe]_T = [Fe^{2+}]\left[1 + \frac{K_1 K_w}{[H^+]} + \frac{K_2(K_w)^3}{[H^+]^3}\right] \tag{14-9}$$

$$[Mn]_T = [Mn^{2+}]\left[1 + \frac{K_3 K_w}{[H^+]} + \frac{K_4(K_w)^3}{[H^+]^3}\right] \tag{14-10}$$

or

$$[Fe]_T = \frac{K_{S\,Fe}}{[OH^-]^2}\left[1 + \frac{K_1 K_w}{[H^+]} + \frac{K_2(K_w)^3}{[H^+]^3}\right] \tag{14-11}$$

$$[Mn]_T = \frac{K_{S\,Mn}}{[OH^-]^2}\left[1 + \frac{K_3 K_w}{[H^+]} + \frac{K_4(K_w)^3}{[H^+]^3}\right] \tag{14-12}$$

The equilibrium soluble iron and manganese concentrations at any pH may be obtained directly from equations 14-11 and 14-12, or they may be estimated from log [species] vs. pH diagrams. The plotting equations which are needed to construct equilibrium diagrams for the $Fe(OH)_2$-H_2O system and the $Mn(OH)_2$-H_2O system are presented in Table 14-2; the appropriate diagrams are shown by Figures 14-1 and 14-2. According

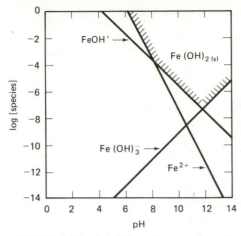

FIGURE 14-1 Solubility diagram for the $Fe(OH)_2 - H_2O$ system.

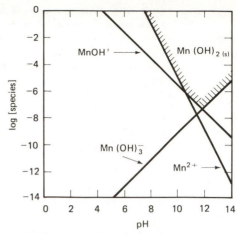

FIGURE 14-2 Solubility diagram for the $Mn(OH)_2 - H_2O$ system.

TABLE 14-2 Plotting equations for $Fe(OH)_2$-H_2O and $Mn(OH)_2$-H_2O systems

Reaction	Plotting Equation
1. Iron:	
(a) $Fe(OH)_{2(s)} \rightleftharpoons Fe^{2+}_{(aq)} + 2OH^-_{(aq)}$	$\log [Fe^{2+}] = 12.9 - 2\ pH$
(b) $Fe^{2+}_{(aq)} + OH^-_{(aq)} \rightleftharpoons FeOH^+_{(aq)}$	$\log [FeOH^+] = 4.5 - pH$
(c) $Fe^{2+}_{(aq)} + 3OH^-_{(aq)} \rightleftharpoons Fe(OH)^-_{3(aq)}$	$\log [Fe(OH)^-_3] = pH - 19.4$
2. Manganese:	
(a) $Mn(OH)_{2(s)} \rightleftharpoons Mn^{2+}_{(aq)} + 2OH^-_{(aq)}$	$\log [Mn^{2+}] = 15.1 - 2\ pH$
(b) $Mn^{2+}_{(aq)} + OH^-_{(aq)} \rightleftharpoons MnOH^+_{(aq)}$	$\log [MnOH^+] = 5.0 - pH$
(c) $Mn^{2+}_{(aq)} + 3OH^-_{(aq)} \rightleftharpoons Mn(OH)^-_{3(aq)}$	$\log [Mn(OH)^-_3] = pH - 18.6$

to these figures, the minimum solubility of both $Fe(OH)_2$ and $Mn(OH)_2$ occurs near pH 12. Since the pH of most natural waters is generally within 1 pH unit of neutral, it is possible to have fairly high levels of iron and manganese in waters where the soluble concentrations are controlled by $Fe(OH)_2$ and $Mn(OH)_2$ precipitation. However, most natural waters have a significant amount of alkalinity present (i.e., carbonates are present) and, as a result, within the pH range of 6.5 to 9.5 the soluble iron and manganese content of these waters are often controlled by the solubility of their carbonates.

$$FeCO_{3(aq)} \rightleftharpoons Fe^{2+}_{(aq)} + CO^{2-}_{3(aq)} \qquad \textbf{(14-13)}$$

$$MnCO_{3(s)} \rightleftharpoons Mn^{2+}_{(aq)} + CO^{2-}_{3(aq)} \qquad \textbf{(14-14)}$$

Assuming an ideal solution, the saturated equilibrium relationships for reactions 14-11 and 14-12 are

$$K_{SC\ Fe} = [Fe^{2+}][CO^{2-}_3] \qquad \textbf{(14-15)}$$

where $\quad K_{SC\ Fe} = 10^{-10.7}$ at 25°C.

and

$$K_{SC\ Mn} = [Mn^{2+}][CO_3^{2-}] \qquad\qquad (14\text{-}16)$$

where $\quad K_{SC\ Mn} = 10^{-10.4}$ at 25°C.

For the situation where iron(II) carbonate and manganese(II) carbonate control the soluble metal concentration and OH$^-$ ions are considered to be the only complexing specie present, the mass balance expressions for total soluble metal concentration are

$$[Fe]_T = \frac{K_{SC\ Fe}}{[CO_3^{2-}]}\left[1 + \frac{K_1 K_w}{[H^+]} + \frac{K_2 (K_w)^3}{[H^+]^3}\right] \qquad\qquad (14\text{-}17)$$

$$[Mn]_T = \frac{K_{SC\ Mn}}{[CO_3^{2-}]}\left[1 + \frac{K_3 K_w}{[H^+]} + \frac{K_4 (K_w)^3}{[H^+]^3}\right] \qquad\qquad (14\text{-}18)$$

Since

$$[CO_3^{2-}] = \alpha_2 C_T \qquad\qquad (2\text{-}99)$$

where $\quad \alpha_2 =$ ionization fraction

$\quad\quad\quad C_T =$ total equilibrium carbonic specie concentration,

these equations may be expressed as

$$[Fe]_T = \frac{K_{SC\ Fe}}{\alpha_2 C_T}\left[1 + \frac{K_1 K_w}{[H^+]} + \frac{K_2 (K_w)^3}{[H^+]^3}\right] \qquad\qquad (14\text{-}19)$$

$$[Mn]_T = \frac{K_{SC\ Mn}}{\alpha_2 C_T}\left[1 + \frac{K_3 K_w}{[H^+]} + \frac{K_4 (K_w)^3}{[H^+]^3}\right] \qquad\qquad (14\text{-}20)$$

When the carbonate solid phase controls the equilibrium soluble iron and manganese concentrations, the soluble concentration of these species at any pH may be estimated from equations 14-19 and 14-20, or they may be obtained from log [species] vs. pH diagrams. The plotting equations which are needed to construct equilibrium diagrams for the FeCO$_3$-H$_2$O system and the MnCO$_3$-H$_2$O system are presented in Table 14-3,

TABLE 14-3 Plotting equations for FeCO$_3$-H$_2$O and MnCO$_3$-H$_2$O systems.

Reaction	Plotting Equation
1. Iron:	
(a) $FeCO_{3(s)} \rightleftharpoons Fe^{2+}_{(aq)} + CO^{2-}_{3(aq)}$	$\log[Fe^{2+}] = -10.7 - \log \alpha_2 - \log C_T$
(b) $Fe^{2+}_{(aq)} + OH^-_{(aq)} \rightleftharpoons FeOH^+_{(aq)}$	$\log[FeOH^+] = pH - 19.1 - \log \alpha_2 - \log C_T$
(c) $Fe^{2+}_{(aq)} + 3OH^-_{(aq)} \rightleftharpoons Fe(OH)^-_{3(aq)}$	$\log[Fe(OH)^-_3] = 3pH - 43 - \log \alpha_2 - \log C_T$
2. Manganese:	
(a) $MnCO_{3(s)} \rightleftharpoons Mn^{2+}_{(aq)} + CO^{2-}_{3(aq)}$	$\log[Mn^{2+}] = -10.4 - \log \alpha_2 - \log C_T$
(b) $Mn^{2+}_{(aq)} + OH^-_{(aq)} \rightleftharpoons MnOH^+_{(aq)}$	$\log[MnOH^+] = pH - 20.5 - \log \alpha_2 - \log C_T$
(c) $Mn^{2+}_{(aq)} + 3OH^-_{(aq)} \rightleftharpoons Mn(OH)^-_3$	$\log[Mn(OH)^-_3] = 3pH - 44.1 - \log \alpha_2 - \log C_T$

and the appropriate diagrams are presented in Figures 14-3 through 14-6. From these figures it can be seen that the minimum solubility of $FeCO_3$ occurs near pH 8, and the minimum solubility of $MnCO_3$ occurs near pH 10. Furthermore, a comparison of Figures 14-3 and 14-4 and Figures 14-5 and 14-6 indicates that, for a given pH, the solubility of iron(II) carbonate and manganese(II) carbonate is inversely proportional to the equilibrium total carbonic species concentration; i.e., the greater the total carbonic species concentration, the lower the soluble metal concentration. Since alkalinity in most natural waters is due to the carbonic acid system, the inverse relationship also exists between alkalinity and the soluble metal concentration.

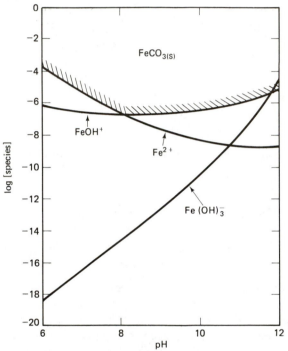

FIGURE 14-3 *Solubility diagram for the $FeCO_3$-H_2O system for $C_T = 10^{-2}$ M.*

In Chapter 3 it is noted that where two or more precipitation reactions are competing, the ion-ratio method may be used to predict the order of precipitation of solids. For the $FeCO_3$-$Fe(OH)_2$-H_2O and $MnCO_3$-$Mn(OH)_2$-H_2O systems the total soluble metal concentrations are given by

$$[Fe]_T = [Fe^{2+}]\left[1 + \frac{K_1 K_w}{[H^+]} + \frac{K_2 (K_w)^3}{[H^+]^3}\right] \tag{14-9}$$

$$[Mn]_T = [Mn^{2+}]\left[1 + \frac{K_3 K_w}{[H^+]} + \frac{K_4 (K_w)^3}{[H^+]^3}\right] \tag{14-10}$$

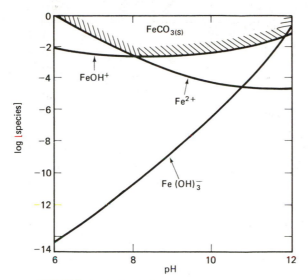

FIGURE 14-4 *Solubility diagram for the $FeCO_3$-H_2O system for $C_T = 10^{-6}$ M.*

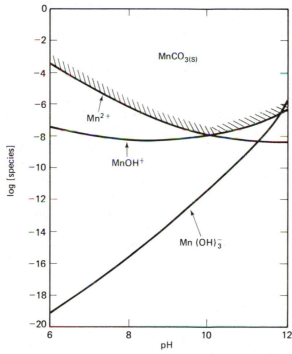

FIGURE 14-5 *Solubility diagram for the $MnCO_3$-H_2O system for $C_T = 10^{-2}$ M.*

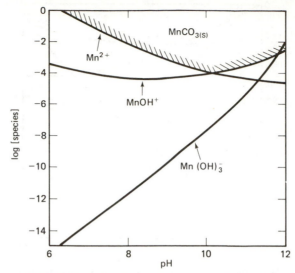

FIGURE 14-6 *Solubility diagram for the $MnCO_3$-H_2O system for $C_T = 10^{-6}$ M.*

Equation 3-111 can be used to determine the pH, where the solid phase change between the carbonate and hydroxide system occurs. When the pH of the water is less than this value, carbonate will control the solubility, and equations 14-19 and 14-20 would be used to compute the soluble metal concentration. When the pH of the water is greater than the computed pH, hydroxide will control the solubility, and equations 14-11 and 14-12 would be used to compute the soluble metal concentrations.

To remove iron and manganese from water, the engineer takes advantage of the fact that iron(III) hydroxide and manganese(IV) dioxide are much less soluble than either iron(II) hydroxide and manganese(II) hydroxide or iron(II) carbonate and manganese(II) carbonate. The solubility diagram for iron(III) hydroxide, developed in Chapter 3 (see Figure 3-2), indicates that the minimum $Fe(OH)_3$ solubility occurs over the pH range of 7 to 10, where the equilibrium soluble iron concentration should be near 10^{-9} mole/ℓ. There is no significant aqueous chemistry for MnO_2, since this chemical species is extremely insoluble. Because of this, no solubility equilibrium diagram will be constructed for this compound.

In situations where only iron removal is required (the initial manganese(II) concentration of the water is less than 0.05 mg/ℓ), the reaction pH is generally between 7 and 8. If the water contains a large initial alkalinity, iron(II) removal may result from two different mechanisms: precipitation of iron(II) in the form of $FeCO_3$ and oxidation of iron(II) to iron(III) and precipitation of $Fe(OH)_3$. Both mechanisms may occur simultaneously, resulting in very rapid iron removal.

To simplify the discussion in this section, it has been assumed that OH^- ions are the only complexing species in solutions. In most situations this is a reasonable assumption; however, there may be cases where other inorganic complexing species such as HCO_3^-, SO_4^{2-}, and PO_4^{3-}, and organic complexing species such as humic and fulvic acids may become significant. The formation of metal complexes with these species will tend to increase the soluble concentration of iron and manganese beyond the values predicted by the mass balance expressions which were developed in this section.

14-2 OXIDATION OF IRON(II) AND MANGANESE(II)

Most of the treatment methods in use today for the removal of iron and manganese from water supplies depend on the oxidation of soluble iron(II) and manganese(II) to the insoluble iron(III) and manganese(IV) forms. The most commonly used oxidizing agents are oxygen, chlorine, and potassium permanganate.

In the presence of oxygen both iron(II) and manganese(II) are oxidized. The oxidations may be presented as follows:

$$4Fe^{2+}_{(aq)} + O_{2(g)} + 10H_2O_{(\ell)} \rightleftharpoons 4Fe(OH)_{3(s)} + 8H^+_{(aq)} \tag{14-21}$$

$$2Mn^{2+}_{(aq)} + O_{2(g)} + 2H_2O_{(\ell)} \rightleftharpoons 2MnO_{2(s)} + 4H^+_{(aq)} \tag{14-22}$$

Stoichiometrically, $(2 \times 16)/(4)(55.8) = 0.14$ mg/ℓ of oxygen will oxidize 1 mg/ℓ of iron(II), and $(2 \times 16)/(2)(54.9) = 0.29$ mg/ℓ of oxygen will oxidize 1 mg/ℓ of manganese(II). Also, $(8 \times 1)/(4)(55.8)$ or 0.036 mg/ℓ of hydrogen ions will be produced for each 1 mg/ℓ of iron(II) oxidized and $(4 \times 1)/(2)(54.9)$ or 0.036 mg/ℓ of hydrogen ions will also be produced for each 1 mg/ℓ of manganese oxidized. Since the production of 1 meq./ℓ of hydrogen ions will destroy 1 meq./ℓ of alkalinity, the amount of alkalinity destroyed during the oxidation process is $(0.036)(50/1)$ or 1.8 mg/ℓ expressed as $CaCO_3$ per mg/ℓ of iron(II) or manganese(II) oxidized. If insufficient alkalinity is present, the H^+ ion concentration will increase during the course of the reaction. This will result in a depression of the pH and a concurrent decrease in the rate of the reaction. The effect of pH on the rate of these two reactions (equations 14-21 and 14-22) is discussed in the following section.

EXAMPLE PROBLEM 14-1: Consider the reaction represented by equation 14-21:

$$4Fe^{2+}_{(aq)} + O_{2(g)} + 10H_2O_{(\ell)} \rightleftharpoons 4Fe(OH)_{3(s)} + 8H^+_{(aq)}$$

Compute the equilibrium constant from free energy of formation data.

Solution:

1. From Appendix I, obtain $\Delta G^\circ_{formation}$ for each chemical species in the balanced chemical equation.

Species	$\Delta G^\circ_{formation}$ (KCal/mole)
$Fe^{2+}_{(aq)}$	-18.85
$O_{2(g)}$	0
$H_2O_{(\ell)}$	-56.69
$Fe(OH)_{3(s)}$	-166.50
$H^+_{(aq)}$	0

2. Compute the $\Delta G^\circ_{reaction}$ value from equation 1-37:

$$\Delta G^\circ_{reaction} = [4(-166.50) + 8(0)] - [4(-18.85) + (0) + 10(-56.69)]$$
$$= -23.7 \, KCal$$

3. Calculate $(K_a)_{eq.}$ from equation 1-55:

$$\log (K_a)_{eq} = \frac{-23.7}{-1.364} = 17.37$$

or

$$(K_a)_{eq} = 10^{17.37}$$

The value of $10^{17.37}$ indicates that the position of equilibrium lies far to the right; i.e., the large equilbrium constant value indicates that the oxygenation reaction proceeds virtually to stoichiometric completion under standard state conditions.

Both chlorine and hypochlorites are effective oxidizing agents and have been used to oxidize iron(II) and manganese(II) to their insoluble forms.

$$2Fe^{2+}_{(aq)} + Cl_{2(g)} + 6H_2O_{(\ell)} \rightleftharpoons 2Fe(OH)_{3(s)} + 2Cl^-_{(aq)} + 6H^+_{(aq)} \quad \textbf{(14-23)}$$
$$Mn^{2+}_{(aq)} + Cl_{2(g)} + 2H_2O_{(\ell)} \rightleftharpoons MnO_{2(s)} + 2Cl^-_{(aq)} + 4H^+_{(aq)} \quad \textbf{(14-24)}$$

Theoretically, 0.64 mg/ℓ of Cl_2 will oxidize 1 mg/ℓ of iron(II), while 1.29 mg/ℓ of Cl_2 will oxidize 1 mg/ℓ of manganese(II). During the reactions 2.70 mg/ℓ of alkalinity as $CaCO_3$ will be destroyed for each mg/ℓ of iron(II) oxidized, whereas 3.64 mg/ℓ of alkalinity as $CaCO_3$ will be destroyed for each mg/ℓ of manganese(II) oxidized. In actual practice, there are side reactions which consume chlorine, and as a result the required chlorine dose is usually greater than the dose predicted by stoichiometry.

Oxidation of iron(II) and manganese(II) may also be accomplished with potassium permanganate.

$$3Fe^{2+}_{(aq)} + KMnO_{4(aq)} + 7H_2O_{(\ell)} \rightleftharpoons 3Fe(OH)_{3(s)} + MnO_{2(s)}$$
$$+ K^+_{(aq)} + 5H^+_{(aq)} \quad \textbf{(14-25)}$$
$$3Mn^{2+}_{(aq)} + 2KMnO_{4(aq)} + 2H_2O_{(\ell)} \rightleftharpoons 5MnO_{2(s)} + 2K^+_{(aq)} + 4H^+_{(aq)} \quad \textbf{(14-26)}$$

Stoichiometrically, 0.94 mg/ℓ of $KMnO_4$ will oxidize 1 mg/ℓ of iron(II), and 1.92 mg/ℓ of $KMnO_4$ is required to oxidize 1 mg/ℓ of manganese(II). 1.50 mg/ℓ of alkalinity as $CaCO_3$ will be destroyed per mg/ℓ iron(II) oxidized, and 1.21 mg/ℓ of alkalinity as $CaCO_3$ will be destroyed per mg/ℓ of manganese(II) oxidized. In actual practice, the required potassium permanganate dose would normally be less than the dose predicted by stoichiometry. Acording to Engelbrecht et al. (1965), this is due to the formation of MnO_2, which catalyses the following secondary reactions:

$$2Fe^{2+}_{(aq)} + 2MnO_{2(s)} + 5H_2O_{(\ell)} \rightleftharpoons 2Fe(OH)_{3(s)} + Mn_2O_{3(s)} + 4H^+_{(aq)} \quad \textbf{(14-27)}$$
$$3Mn^{2+}_{(aq)} + MnO_{2(s)} + 4H_2O_{(\ell)} \rightleftharpoons 2Mn_2O_{3(s)} + 8H^+_{(aq)} \quad \textbf{(14-28)}$$

Note that in all cases (equations 14-21, 14-22, 14-23, 14-24, 14-25, and 14-26) manganese consumes about twice as much oxidant as iron; this is because manganese involves a two unit change in oxidation state as opposed to a one unit change for iron.

It should be understood that although stoichiometry will provide the oxidant dose required for a particular treatment situation, it will not predict the reactor retention time required to allow complete oxidation of the iron and manganese to occur. Reactor retention time is controlled by the kinetics of the reactions.

14-3 KINETICS OF IRON(II) AND MANGANESE(II) OXIDATION

Various investigators have studied the rate of iron(II) and manganese(II) oxidation, using oxygen as the oxidizing agent, and a number of these studies are summarized in this section. On the other hand, very little kinetic information is available where oxidants other than oxygen were employed in rate studies. However, some general comments may be found in the literature, which may be useful in the design of reaction basins where chlorine or permanganate is to be used. For example, White (1972) stated that the chlorine oxidation of iron(II) proceeds over a wide pH range but that a value of 7 is satisfactory. He also suggested a pH range of 7 to 10 (normally a pH greater than 8.5 is required for effective Mn removal) for the chlorine oxidation of manganese(II). According to White (1972), manganese oxidation with chlorine normally requires 2 to 4 hr to reach completion. Sundstrom and Klei (1979) indicated that the rate of iron(II) and manganese(II) oxidation is several times faster with chlorine than with oxygen, and in the pH range of 6 to 9 permanganate oxidation is even faster than chlorine oxidation. They also noted that regardless of the oxidant used, the rate of manganese(II) oxidation is always slower than the rate of iron(II) oxidation and that a reaction pH near 10 is required if manganese(II) oxidation is to be completed within a reasonable reactor retention time. O'Conner (1971), on the other hand, stated that $KMnO_4$ will oxidize manganese(II) to MnO_2 within 5 min over a broad pH range.

Kinetics of Iron(II) Oxidation with Oxygen

Stumm and Lee (1961) studied the oxidation of iron(II) by oxygen in bicarbonate solutions in the neutral pH range. They determined a rate law of the following form:

$$\frac{d[Fe^{2+}]}{dt} = -k[P_{O_2}][OH^-]^2[Fe^{2+}] \tag{14-29}$$

where $\dfrac{d[Fe^{2+}]}{dt}$ = rate of iron(II) oxidation, moles liter^{-1} min^{-1}

k = reaction rate constant. This constant was found to have a value of 8×10^{13} liter^2 mole^{-2} atm^{-1} min^{-1} at 20.5°C.

P_{O_2} = partial pressure of oxygen in the gas phase (atm)

$[OH^-]$ = hydroxyl ion concentration (moles/ℓ)

$[Fe^{2+}]$ = iron(II) concentration at any time (moles/ℓ).

Since

$$[OH^-] = \frac{K_w}{[H^+]} \qquad \text{(2-9)}$$

an apparent reaction rate constant expression can be introduced which has the form

$$K_{app} = \frac{k[P_{O_2}](K_w)^2}{[H^+]^2} \qquad \text{(14-30)}$$

For most situations in process chemistry a value of 0.21 atm can be assumed for P_{O_2}, and a value of 10^{-14} can be assumed for K_w (the ionization constant for water does vary with temperature, and at 20.5°C a value of $10^{-14.16}$ is more precise). In this case equation 14-30 reduces to

$$K_{app} = \frac{1.68 \times 10^{-15}}{[H^+]^2} \qquad \text{(14-31)}$$

Equation 14-31 indicates an inverse relationship between the rate of iron(II) oxidation and the hydrogen ion concentration. Since an increase in hydrogen ion concentration results in a decrease in pH, equation 14-31 may also be interpreted as showing an increase in the rate of oxidation with an increase in solution pH. Data illustrating this effect are presented in Figure 14-7.

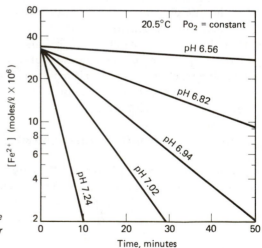

FIGURE 14-7 Relationship between the rate of iron (II) oxidation and solution pH (after Stumm and Lee, 1961).

When K_{app} is substituted into equation 14-29, the resulting equation has the form

$$\frac{d[Fe^{2+}]}{dt} = -K_{app}[Fe^{2+}] \qquad \text{(14-32)}$$

indicating that the oxidation of iron(II) with oxygen is a pseudo first-order reaction. The validity of this rate law is supported by the fact that for a specified pH a semi-log plot of $[Fe^{2+}]$ vs. time will give a linear trace (see Figure 14-7).

Ghosh (1976) has observed that high alkalinity levels can have an affect on the rate of iron(II) oxidation with oxygen. In his work he found that the rate of iron oxidation was adequately described by equation 14-32 for waters that have buffer intensity or β values between 1.0×10^{-4} and 4.0×10^{-3} eq./ℓ-pH (recall β varies directly with alkalinity). However, for waters with buffer intensities greater than 4.0×10^{-3} eq./ℓ-pH, he proposes a modification of equation 14-29 to include a β term.

$$\frac{d[Fe^{2+}]}{dt} = -k_1[\beta]^{1/2}[P_{O_2}][OH^-]^2[Fe^{2+}] \qquad \textbf{(14-33)}$$

where β = buffer intensity (eq./ℓ-pH)

k_1 = reaction rate constant. The value of k_1 was found to be 1.3×10^{15} at 25°C for values of β between 4.0×10^{-3} to 1.5×10^{-2} eq./ℓ-pH and pH ranging from 5.5 to 7.2.

In this situation the K_{app} term became

$$K_{app} = \frac{2.73 \times 10^{-12}[\beta]^{1/2}}{[H^+]^2} \qquad \textbf{(14-34)}$$

It is not unusual to encounter situations where the actual oxidation rates are less than the rates predicted by either equation 14-29 or 14-33. This may be the case for low alkalinity waters, where the buffer system is slow to respond to localized acidity changes induced by the oxidation reaction (Stumm and Lee, 1961). The presence of organic material in water may also retard the rate of iron(II) oxidation (Theis and Singer, 1974).

EXAMPLE PROBLEM 14-2: Determine the plug-flow reactor volume required to reduce the iron(II) concentration from 5 mg/ℓ to 0.1 mg/ℓ by oxygenation. Assume that the pH is to be maintained at 7 and the buffer intensity of the water is 1.0×10^{-4} eq./ℓ-pH.

Solution:

1. Compute K_{app} from equation 14-31:

$$K_{app} = \frac{1.68 \times 10^{-15}}{(10^{-7})^2} = 0.168 \text{ min}^{-1}$$

2. Calculate the required reaction time:

$$t_{pF} = \frac{1}{K}\left[\ln\left(\frac{C_o}{C_e}\right)\right]$$

$$t_{pF} = \frac{1}{0.168} \ln\left[\left(\frac{5}{0.1}\right)\right] = 23.3 \text{ min}$$

3. Determine the volume requirement per MGD of water treated.

$$V = Qt_{pF}$$

$$= 1 \frac{MG}{day} \times 23.3 \text{ min} \times \frac{1 \text{ day}}{1440 \text{ min}}$$

$$= 0.016 \text{ MG/MGD of water treated}$$

EXAMPLE PROBLEM 14-3: According to Ghosh (1976), the energy of activation for reaction 14-21 is 23 KCal/mole. Using this value for E_a, evaluate the reaction rate constant at 15°C for the rate law presented by equation 14-29.

Solution:

Compute k from equation 5-68:

$$\log \left[\frac{k}{8 \times 10^{13}} \right] = \frac{23,000}{2.3(1.98)} \left[\frac{288 - 293.5}{(288)(293.5)} \right]$$

$$\log k = 13.574$$

$$k = 3.75 \times 10^{13} \text{ liter}^2 \text{ mole}^{-2} \text{ atm}^{-1} \text{ min}^{-1}$$

Kinetics of Manganese(II) Oxidation with Oxygen

The kinetic behavior of $Mn(II)$-O_2-H_2O systems has been studied extensively by Morgan (1967a, 1967b). He observed that the rate of oxidation of manganese(II) by oxygen was relatively slow at pH values below 9.5 (see Figure 14-8) and did not follow the pseudo first-order rate law which described the oxidation of iron(II) by oxygen. If the oxidation rate followed first-order kinetics, a plot of $[Mn^{2+}]$ vs. time on semi-log would be expected to yield a linear trace. However, Figure 14-9 shows that this is not the case for the oxidation of manganese (the $[Mn^{2+}]_0$ term shown in this figure represents the initial manganese(II) content of the water. Since this is a constant value, the slope of this line obtained by $\log [Mn^{2+}]$ vs. t will be the same as the slope obtained by plotting $\log ([Mn^{2+}]/\text{constant})$ vs. t. Thus, the plot presented in Figure 14-9 does provide a valid test for first-order kinetics). Morgan (1967b) found that the kinetic data

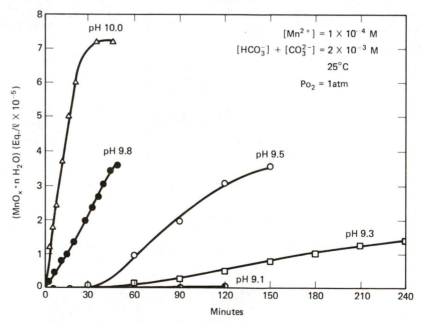

FIGURE 14-8 Oxygenation of manganese (II) (after Morgan, 1967a).

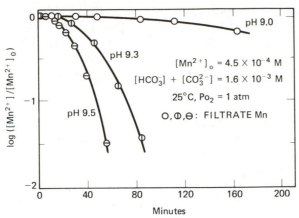

FIGURE 14-9 *Removal of manganese (II) by oxygenation (after Morgan, 1967b).*

for the manganese(II) oxidation and removal process were consistent with an auto-catalytic kinetic model.

In Chapter 5 an integrated rate law for an autocatalytic reaction is presented, which has the following form:

$$\ln\left[\frac{C_{A_0}(C_T - C_A)}{C_A(C_T - C_{A_0})}\right] = (C_{A_0} + C_{P_0})k_a t \qquad \text{(5-63)}$$

where C_{A_0} = initial reactant concentration (moles/ℓ)

C_{P_0} = initial product concentration (moles/ℓ)

C_T = moles/ℓ of reactant plus moles/ℓ of product at any time during the course of the reaction (moles/ℓ)

C_A = reactant concentration at any time during the course of the reaction (moles/ℓ)

k_a = autocatalytic reaction rate constant.

Since C_{A_0}, C_T, and C_{P_0} are constants, equation 5-63 can be written as

$$\log\left[A\left(\frac{C_T - C_A}{C_A}\right)\right] = K_1 t \qquad \text{(14-35)}$$

or

$$\log\left[A\left(\frac{C_T}{C_A} - 1\right)\right] = K_1 t \qquad \text{(14-36)}$$

where

$$A = \frac{C_{A_0}}{(C_T - C_{A_0})}$$

$$K_1 = \frac{(C_{A_0} + C_{P_0})k_a}{2.3}$$

Since A is constant, a plot of $\log [(C_T/C_A) - 1]$ vs. t will give a linear trace having the same slope as a plot of $\log [A(C_T/C_A - 1]$ vs. t if the reaction is autocatalytic. Using the data from Figure 14-9, Morgan (1967b) constructed such a plot (see Figure 14-10) and found that the removal of manganese(II) followed an autocatalytic reaction model. In this plot Morgan assumed that the initial product concentration was much smaller than the reactant concentration, so that its value could be neglected in the C_T sum term $C_T = C_{A_0} + C_{P_0}$.

Morgan (1967a) also compared the pH dependence of iron(II) and manganese(II) oxidation (see Figure 14-11). He found that, in both cases, the oxidation rate is second-order with respect to the OH^- ion concentration, even though the pH range for rapid manganese oxidation is higher than the pH range for iron(II) oxidation.

Stumm and Morgan (1972) indicated that both manganese(II) oxidation and removal rates follow the autocatalytic reaction model, and that the rate dependence on oxygen partial pressure is the same as that noted for iron(II). Thus, an apparent

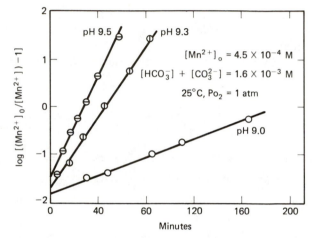

FIGURE 14-10 *Data of figure 14-9 plotted according to an autocatalytic model (after Morgan, 1967b).*

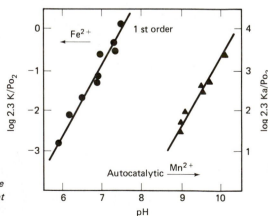

FIGURE 14-11 *Effect of pH on manganese (II) and iron (II) oxygenation rate constants at 25°C (after Morgan, 1967a).*

IRON AND MANGANESE REMOVAL

rate constant expression can be defined for manganese(II), which has the same form as equation 14-30.

$$K_{app} = \frac{(C_{A_0} + C_{P_0})k_a[P_{O_2}](K_w)^2}{2.3[H^+]^2} \qquad \text{(14-37)}$$

The integrated rate law relationships for manganese(II) oxidation or removal is obtained by expressing equation 14-36 in the following form:

$$\log\left[A\left(\frac{[Mn^{2+}]_0}{[Mn^{2+}]} - 1\right)\right] = K_{app}t$$

where $[Mn^{2+}]$ = manganese(II) concentration at any time (moles/ℓ).

To explain the autocatalytic nature of manganese(II) removal, Morgan (1967b) proposes that Mn^{2+} ions are rapidly adsorbed by MnO_2 particles which are slowly formed during the reaction. The adsorbed ions are then oxidized at a slow rate. This proposed reaction scheme is described by the following sequence of reactions:

$$Mn^{2+}_{(aq)} + \tfrac{1}{2}O_{2(g)} \xrightarrow{\text{slow}} MnO_{2(s)} \qquad \text{(14-38)}$$

$$Mn^{2+}_{(aq)} + MnO_{2(s)} \xrightarrow{\text{fast}} Mn^{2+}MnO_{2(s)} \qquad \text{(14-39)}$$

$$Mn^{2+}MnO_{2(s)} + \tfrac{1}{2}O_{2(g)} \xrightarrow{\text{slow}} 2MnO_{2(s)} \qquad \text{(14-40)}$$

EXAMPLE PROBLEM 14-4: Determine the rate constant, k_a, for the removal of manganese(II) at 25°C. Use the data associated with the pH 9.3 plot presented in Figure 14-10. Neglect the initial concentration of MnO_2.

Solution:

1. Compute the value of K_{app} at pH = 9.3:

$$K_{app} = \text{slope} = \frac{9.30 - (-0.20)}{60 - 40} = 0.075$$

2. Calculate k from equation 14-37:

$$0.075 = \frac{(4.5 \times 10^{-4})k_a(1)(10^{-28})}{2.3(10^{-9.3})^2}$$

$$k_a = 9.63 \times 10^{11}$$

EXAMPLE PROBLEM 14-5: A water contains 10 mg/ℓ of iron(II) and 2 mg/ℓ of manganese (II) and has a measured alkalinity of 220 mg/ℓ as $CaCO_3$. If the initial pH of the water is 7.2, determine the lime dose required to maintain a reaction pH of 10. The initial calcium concentration is 400 mg/ℓ as $CaCO_3$, and the water temperature is 15°C. Assume that $I = 0.01\ M$ and that chlorine is to be used as the oxidant.

Solution:

1. Evaluate the equilibrium state of the untreated water.

   (a) Locate the intersection of the initial pH and initial alkalinity line (shown as point 1 on Figure 14-12).

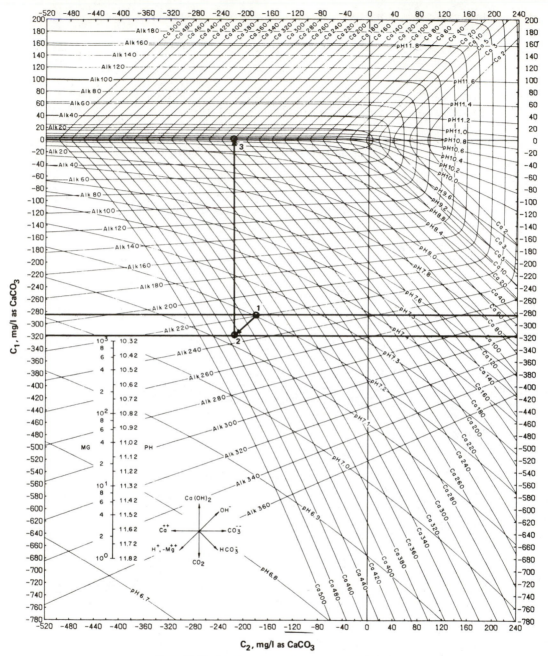

Figure 14-12: C-L Diagram for a Temperature of 15°C and
an Ionic Strength of 0.01

FIGURE 14-12 *C-L Diagram for a temperature of 15°C
and an ionic strength of 0.01.*

(b) The calcium line which passes through point 1 is 400. Since this is the same as the initial calcium concentration, the water is saturated with respect to $CaCO_3$. Thus, the directional format diagram can be applied at point 1.

2. Compute the amount of H^+ ions produced in the oxidation reactions.

(a) The amount of H^+ ions produced as a result of iron oxidation is computed from equation 14-23:

$$10 \left[\frac{6 \times 1}{2 \times 55.8} \right] = 0.54 \text{ mg}/\ell \text{ as } H^+ \text{ ions}$$

or

$$0.54 \left[\frac{50}{1} \right] = 27 \text{ mg}/\ell \text{ as } CaCO_3$$

(b) The amount of H^+ ions produced as a result of manganese oxidation is computed from equation 14-24:

$$2 \left[\frac{4 \times 1}{1 \times 54.9} \right] = 0.15 \text{ mg}/\ell \text{ as } H^+ \text{ ions}$$

or

$$0.15 \left[\frac{50}{1} \right] = 7 \text{ mg}/\ell \text{ as } H^+ \text{ ions}$$

(c) The total amount of H^+ ions produced is

$$H^+ \text{ mg}/\ell \text{ as } CaCO_3 = 27 + 7 = 34$$

3. Compute the initial acidity of the untreated water.

(a) Construct a horizontal line through point 1 and read the C_1 value at the point where this line intersects the ordinate:

$$C_1 = -285$$

Therefore,

$$-285 = - \text{ (acidity)}$$

or

$$\text{acidity} = 285 \text{ mg}/\ell \text{ as } CaCO_3$$

4. Construct a downward vector from point 1, which will account for the effects of H^+ ion production during the oxidation reactions.

(a) Draw a horizontal line through C_1 at a value of $-(285 + 34) = -319$.

(b) Beginning at point 1, construct a vector downward and to the left at $45°$ until it intersects the horizontal line through $C_1 = -319$ (shown as point 2 on Figure 14-12).

5. Construct a vertical vector, beginning at point 2, to intersect the pH $= 10$ line (shown as point 3 on Figure 14-12). The required lime dose is equal to the magnitude of the projection of this vector onto the C_1 axis.

$$\text{lime dose} = 319 + 3 = 322$$
$$(\text{mg}/\ell \text{ as } CaCO_3)$$

Reh (1972) lists the following processes which are often used either separately or in combination to remove iron and manganese from water:

1. Oxidation with oxygen, chlorine, or potassium permanganate followed by sedimentation and filtration. Lime or NaOH addition may be required for pH adjustment.

2. Lime softening, which removes iron and manganese along with other hardness-causing metals.

3. Ion exchange operating on the sodium cycle. In this case, the water should not be oxygenated before passing through the ion exchanger if fouling of the bed by precipitated $Fe(OH)_3$ and MnO_2 is to be avoided.

4. Manganese zeolite process. In this process sodium greensand zeolite is converted to manganese zeolite by treating it with solutions of potassium permanganate and manganese(II) salt. The manganese zeolite has an exchange capacity of about 0.09 lb/ft³ for iron or manganese. The service flow rate is approximately 3 gpm/ft². Normally, about 0.18 lb/ft³ of potassium permanganate is required for regeneration.

By far the most common method used for iron and manganese removal is aeration followed by sedimentation and filtration. Aeration raises the pH of the water slightly by stripping CO_2 and introduces oxygen for the oxidation or iron(II) and manganese-(II). A coke tray aeration system is normally used for aeration, although other types of aeration devices have been employed. O'Connor (1971) states that the sedimentation unit is normally not included if the total iron and manganese concentration is less than 10 mg/ℓ. He indicates that when provided in iron and manganese removal plants, most sedimentation tanks function as quiescent reaction basins rather than solid-liquid separation units. He therefore suggests that consideration be given to stirring the contents of these units to promote agglomeration and adsorption prior to filtration.

According to Reh (1972), so-called organic iron and manganese is not effectively removed by oxygen, chlorine, or permanganate oxidation. When this material is present, he recommends coagulation with alum, followed by sedimentation and filtration.

PROBLEMS

14-1. Compute the equilibrium constant from free energy of formation data for the following reactions:

$$2Fe^{2+}_{(aq)} + Cl_{2(g)} + 6H_2O_{(\ell)} \rightleftharpoons 2Fe(OH)_{3(s)} + 2Cl^-_{(aq)} + 6H^+_{(aq)}$$

$$Mn^{2+}_{(aq)} + Cl_{2(g)} + 2H_2O_{(\ell)} \rightleftharpoons MnO_{2(s)} + 2Cl^-_{(aq)} + 4H^+_{(aq)}$$

14-2. If a water contains 10 mg/ℓ of iron(II) and 2 mg/ℓ of manganese(II), what is the theoretical potassium permanganate dose required for iron and manganese removal?

14-3. Determine the reaction time required in a plug-flow reactor to reduce the manganese-(II) concentration of a water from 2 mg/ℓ to 0.05 mg/ℓ. Assume that the initial MnO_2 content is to be maintained at 100 mg/ℓ by sludge recycle. The reaction temperature is 25°C, the oxygen partial pressure is 1 atm, and the reaction pH is 10.

14-4. A water contains 10 mg/ℓ of iron(II) and 2 mg/ℓ of manganese(II). Determine the reaction time required in a plug-flow reactor to reduce the iron(II) concentration to 0.1 mg/ℓ and the manganese(II) concentration to 0.05 mg/ℓ if a pH of 10 is to be maintained during the reaction. Use the following rate constant values:

$$\text{iron:} \quad k = 8 \times 10^{13}$$

$$\text{manganese:} \quad k = 9.63 \times 10^{11}$$

Assume that the initial MnO_2 content is to be maintained at 50 mg/ℓ by sludge recycle.

REFERENCES

BRELAND, E.D., and ROBINSON, L.R., JR., "Iron and Manganese Removal from Low Alkalinity Groundwaters," Report to the Water Resources Research Institute, Mississippi State University, MS (1967).

ENGELBRECHT, R.S., O'CONNOR, J.T., and GHOSH, M., "Significance and Removal of Iron in Water Supplies," *Fourth Annual Environmental Engineering and Water Resources Conference*, Vanderbilt University, Nashville, TN (1965).

GHOSH, M.M., "Oxygenation of Ferrous Iron(II) in Highly Buffered Waters," in *Aqueous-Environmental Chemistry of Metals*, [Ed.] A.J. RUBIN, Ann Arbor Science Publishers, Ann Arbor, MI (1976).

Lange's Handbook of Chemistry, [Ed.] J.A. DEAN, McGraw-Hill Book Company, New York (1974).

MORGAN, J.J. and STUMM, W., "The Role of Multivalent Metal Oxides in Limnological Transformation, as Exemplified by Iron and Manganese," *Proc. Second International Water Pollution Research Conference*, New York (1964).

MORGAN, J.J., "Applications and Limitations of Chemical Thermodynamics in Water Systems," in *Equilibrium Concepts in Natural Water Systems*, [Ed.] R.F. GOULD, American Chemical Society, Washington, DC (1967a).

MORGAN, J.J., "Chemical Equilibria and Kinetic Properties of Manganese in Natural Waters," in *Principles and Applications in Water Chemistry*, [Eds.] S.D. FAUST and J. HUNTER, John Wiley & Sons, Inc., New York (1967b).

O'CONNOR, J.T., "Iron and Manganese," in *Water Quality and Treatment*, American Water Works Association, Inc., Denver, CO (1971).

REH, C.W., "Water Supply Engineering for the Removal of Iron and Manganese," *Proc. Fourteenth Water Quality Conference*, University of Illinois, Urbana, IL (1972).

STUMM, W., and LEE, G.F., "Oxygenation of Ferrous Iron," *Industrial and Engineering Chemistry*, **53**, 143 (1961).

STUMM, W., and MORGAN, J.J., *Aquatic Chemistry*, Wiley-Interscience, New York (1972).

Sundstrom, D.W., and Klei, H.E., *Wastewater Treatment*, Prentice-Hall, Inc., Englewood Cliffs, NJ (1979).

Theis, T.L., and Singer, P.C., "The Stabilization of Ferrous Iron by Organic Compounds in Natural Waters," in *Trace Metals and Metal-Organic Interactions in Natural Waters*, [Ed.] P.C. Singer, Ann Arbor Science Publishers, Ann Arbor, MI (1974).

White, G.C., *Handbook of Chlorination*, Van Nostrand Reinhold Company, New York (1972).

APPENDICES

APPENDIX I

Table of ΔH_f° and ΔG_f° values at 25 °C.

The values of ΔH_f° and ΔG_f° in the table are expressed as kilocalories per mole. Although not listed, the enthalpy and free energy of formation of any element (e.g. Al, Pb, S) in its standard state is taken as zero.

Formula	State	ΔH_f°	ΔG_f°
Aluminum			
Al^{3+}	aq	-127	-116
$AlPO_4$ berlinite	c	-404.4	-382.7
AlO_2^-	aq	-219.6	-196.8
$Al(OH)^{2+}$	aq		-165.9
$Al(OH)_3$	amorphous	-305	
$Al(OH)_4^-$	aq	-356.2	-310.2
$Al_2(SO_4)_3 \cdot 6H_2O$	c	-1269.53	-1104.82
Ammonium			
NH_3	g	-11.02	-3.94
NH_4^+	aq	-31.67	-18.97
NH_4OH	aq	-87.50	-63.04
NH_4HCO_3	c	-203.0	-159.2
NH_4Cl	c	-75.15	-48.51
NH_4NO_3	c	-87.37	-43.98
Cadmium			
Cd^{2+}	aq	-17.3	-18.58
$CdCl_2$	aq	-98.04	-81.28
$CdCl_3^-$	aq	-134.1	-116.4
$Cd(CN)_2$	aq	53.9	63.9
$Cd(CN)_4^{2-}$	aq	102.3	121.3
$CdCO_3$	c	-179.4	-160.0
$Cd(NH_3)_4^{2+}$	aq	-107.6	-54.1
$CdOH^+$	aq		-62.4
$Cd(OH)_2$	c	-134	-113.2
	aq	-128.08	-93.73
$Cd(OH)_3^-$	aq		-143.6
$Cd(OH)_4^{2-}$	aq		-181.3
Calcium			
Ca^{2+}	aq	-129.77	-132.18
$CaCO_3$	c	-288.45	-269.78
	aq		-262.76
CaO	c	-151.9	-144.4
$Ca(OH)_2$	c	-234.80	-214.22
$CaSO_4$	c	-340.27	-313.52
	aq		-312.67
$CaHPO_4$	c	-435.2	-401.5

Formula	State	ΔH_f°	ΔG_f°
Calcium(Cont'd)			
$Ca(H_2PO_4)_2$	c	−744.4	−672
$Ca_3(PO_4)_2$	c	−986.2	−929.7
$CaOH^+$	aq		−171.55
$CaHCO_3^+$	aq		−273.67
$Ca(OCl)_2$	c	−178.6	
Carbon			
CN^-	aq	36.0	41.2
HCN	g	32.3	29.8
CO_2	g	−94.05	−94.25
CO_3^{2-}	aq	−161.84	−126.17
HCO_3^-	aq	−165.39	−140.26
H_2CO_3	aq	−167.22	−148.9
HCNO	aq	−34.9	−23.3
CNO^-	aq	−34.9	−23.3
Chlorine			
Cl_2	g	0	0
HCl	aq	−39.95	−31.37
Cl^-	aq	−39.95	−31.37
HOCl	aq	−27.83	−19.11
OCl^-	aq	−25.6	−8.8
Chromium			
Cr^{2+}	aq	−34.3	−42.1
Cr^{3+}	aq	−61.2	−51.5
$Cr_2O_7^{2-}$	aq	−364.0	−315.4
$Cr(OH)_2$	c		−140.5
$Cr(OH)_3$	c	−247.1	−215.3
$Cr(OH)_2^+$	aq		−151.2
Copper			
Cu^+	aq	12.4	12.0
Cu^{2+}	aq	15.39	15.53
$Cu(CN)_2^-$	aq		61.6
$Cu(CN)_3^{2-}$	aq		96.5
$Cu(NH_3)^{2+}$	aq	−9.3	3.72
$Cu(NH_3)_2^{2+}$	aq	−34.0	−7.28
$Cu(NH_3)_3^{2+}$	aq	−58.7	−17.48
$Cu(NH_3)_4^{2+}$	aq	−83.3	−26.60
$Cu(OH)_2$	c	−106.1	−85.3
CuS	c	−11.6	−11.7
Cu_2S	c	−19.0	−20.6
Cu_2SO_4	c	−179.2	−156
$CuSO_4$	c	−184.00	−158.2
$CuCO_3$	c	−142.2	−123.8
	aq		−119.9
$Cu(CO_3)_2^{2-}$	aq		−250.5

Table of ΔH_f° and ΔG_f° values at 25 °C (Cont'd).

Formula	State	ΔH_f°	ΔG_f°
Hydrogen			
H_2	g	0	0
H^+	aq	0	0
OH^-	aq	−54.97	−37.59
H_2O	l	−68.31	−56.68
Iron			
Fe^{2+}	aq	−21.0	−20.30
Fe^{3+}	aq	−11.6	−1.1
$FeCl_3$	c	−95.48	−79.84
$Fe(OH)_2$	c	−135.8	−115.57
$FeOH^+$	aq	−77.6	−66.3
FeS	c	−23.9	−24.0
$FeSO_4$	c	−221.9	−196.2
$FeCO_3$	c	−178.70	−161.06
$Fe(OH)_3$	c	−196.7	−166.5
$FeOH^{2+}$	aq	−67.4	−55.91
$Fe(OH)_2^+$	aq		−106.2
$FePO_4$	c	−299.6	−272
$Fe_2(SO_4)_3$	c	−617	
Lead			
Pb^{2+}	aq	0.39	−5.81
Pb^{4+}	aq		72.3
$Pb(OH)_2$	c	−123.0	−100.6
$Pb(OH)_3^-$	aq		−137.6
$PbCO_3$	c	−167.3	−149.7
$Pb_3(OH)_2(CO_3)_2$	c		−406.0
PbS	c	−22.54	−22.15
$PbSO_4$	c	−219.50	−193.89
$PbHPO_3$	c	−234.5	−208.3
$Pb_3(PO_4)_2$	c	−620.3	−581.4
Magnesium			
Mg^{2+}	aq	−110.41	−108.99
$MgCO_3$	c	−266.0	−264.0
	aq		−239.85
$Mg(OH)_2$	c	−221.0	−199.27
$MgOH^+$	aq		−150.10
$MgHCO_3^+$	aq		−250.88
$MgSO_4$	c	−350.5	−280.5
	aq		−289.55
$MgNH_4PO_4$	c		−390.0
Manganese			
Mn^{2+}	aq	−53.3	−54.4
MnO_2	c	−124.2	−111.1
MnO_4^{2-}	aq		−120.4
$Mn(OH)_2$	c	−166.2	−147.0
$MnCO_3$	c	−212.0	−194.3
	aq	−213.9	−179.6

Formula	State	ΔH_f°	ΔG_f°
Nickel			
Ni^{2+}	aq	−12.9	−10.9
$Ni(OH)_2$	c	−128.6	−108.3
NiS	c		−17.7
$NiCO_3$	c	−158.7	−147.0
Nitrogen			
NO_2^-	aq	−25.4	−8.25
NO_3^-	aq	−49.37	−26.43
HNO_2	aq	−28.5	−13.3
HNO_3	aq	−49.56	−26.61
Oxygen			
O_2	g	0	0
Phosphorus			
PO_4^{3-}	aq	−306.9	−245.1
HPO_4^{2-}	aq	−310.4	−261.5
$H_2PO_4^-$	aq	−311.3	−271.3
H_3PO_4	aq	−308.2	−274.2
$P_2O_7^{4-}$	aq	−542.8	−458.7
$HP_2O_7^{3-}$	aq	−543.7	−471.6
$H_2P_2O_7^{2-}$	aq	−544.6	−480.5
$H_3P_2O_7^-$	aq	−544.1	−483.6
$H_4P_2O_7$	aq	−542.2	−485.7
Silicon			
SiO_2	c	−205.4	−192.4
H_4SiO_4	aq		−300.3
$H_3SiO_4^-$	aq		−286.8
Sodium			
Na^+	aq	−57.28	−62.59
$NaCl$	c	−98.23	−91.78
Na_2CO_3	c	−270.3	−250.4
	aq		−251.4
$NaHCO_3$	c	−226.5	−203.6
	aq		−202.56
$NaOH$	aq		−99.23
$NaCO_3^-$	aq		−190.54
$NaSO_4^-$	aq		−240.91
Sulfur			
S^{2-}	aq	7.9	20.5
HS^-	aq	−4.2	2.88
H_2S	g	−4.93	−8.02
SO_4^{2-}	aq	−216.90	−177.34
HSO_4^-	aq	−211.70	−179.94
H_2SO_4	aq	−216.90	−177.34
SO_2	g	−70.96	−71.79
SO_3^{2-}	aq	−151.9	−116.1

Table of ΔH_f° and ΔG_f° values at 25 °C (Cont'd).

Formula	State	ΔH_f°	ΔG_f°
Zinc			
Zn^{2+}	aq	−36.43	−35.18
$Zn(NH_3)_4^{2+}$	aq		−73.5
$Zn(OH)_2$	c	−153.42	−132.31
$ZnOH_3^-$	aq		−78.9
$ZnOH^+$	aq		−165.95
$Zn(OH)_4^-$	aq		−205.23
$ZnCO_3$	c	−94.26	−174.85
ZnS	c	−48.5	−47.4

Sources:

Garrels, R.M. and Christ, C.L., *Solutions, Minerals, and Equilibria*, Freeman, Cooper and Company, San Francisco, California (1965).

Lange's Handbook of Chemistry, Edited by John A. Dean, McGraw-Hill Book Company, New York, N.Y. (1973).

Rossini, F.D., Wagmen, D.D., Evans, W.H., Levine, S. and Irving, J., "Selected Values of Chemical Thermodynamic Properties", *National Bureau of Standards Circular* **500**, U.S. Department of Commerce (1952).

APPENDIX II

Table of ionization constants of acids at 25°C.

Acid	Equilibrium Reaction	K
Acetic	$CH_3\overset{O}{\overset{\|}{C}}\text{-OH} + H_2O = H_3O^+ + CH_3\overset{O}{\overset{\|}{C}}\text{-O}^-$	1.8×10^{-5} (K_a)
Aluminum ion	$Al^{3+} + 2H_2O = H_3O^+ + AlOH^{2+}$	1.4×10^{-5} (K_{h1})
Ammonium ion	$NH_4^+ + H_2O = H_3O^+ + NH_3$	5.6×10^{-10} (K_h)
Carbonic	$H_2CO_3^* + H_2O = H_3O^+ + HCO_3^-$	4.2×10^{-7} (K_{a1})
	$HCO_3^- + H_2O = H_3O^+ + CO_3^{2-}$	4.8×10^{-11} (K_{a2})
Chromic	$H_2CrO_4 + H_2O = H_3O^+ + HCrO_4^-$	1.8×10^{-1} (K_{a1})
	$HCrO_4^- + H_2O = H_3O^+ + CrO_4^{2-}$	3.2×10^{-7} (K_{a2})
Chromium (III) ion	$Cr^{3+} + 2H_2O = H_3O^+ + CrOH^{2+}$	1×10^{-4} (K_{a1})
Hydrochloric	$HCl + H_2O = H_3O^+ + Cl^-$	1×10^6 (K_a)
Hydrocyanic	$HCN + H_2O = H_3O^+ + CN^-$	4.8×10^{-10} (K_a)
Hydrofluoric	$HF + H_2O = H_3O^+ + F^-$	6.9×10^{-4} (K_a)
Hydrosulfuric	$H_2S + H_2O = H_3O^+ + HS^-$	1×10^{-7} (K_{a1})
	$HS^- + H_2O = H_3O^+ + S^{2-}$	1.3×10^{-13} (K_{a2})
Hypochlorous	$HOCl + H_2O = H_3O^+ + OCl^-$	3.2×10^{-8} (K_a)
Iron (III) ion	$Fe^{3+} + 2H_2O = H_3O^+ + FeOH^{2+}$	4.0×10^{-3} (K_{h1})
Iron (II) ion	$Fe^{2+} + 2H_2O = H_3O^+ + FeOH^+$	1.2×10^{-6} (K_{h1})
Magnesium ion	$Mg^{2+} + 2H_2O = H_3O^+ + MgOH^+$	2×10^{-12} (K_{h1})
Nitric	$HNO_3 + H_2O = H_3O^+ + NO_3^-$	1×10^2 (K_a)
Nitrous	$HNO_2 + H_2O = H_3O + NO_2^-$	4.5×10^{-4} (K_a)
Phosphoric	$H_3PO_4 + H_2O = H_3O^+ + H_2PO_4^-$	7.5×10^{-3} (K_{a1})
	$H_2PO_4^- + H_2O = H_3O^+ + HPO_4^{2-}$	6.2×10^{-8} (K_{a2})
	$HPO_4^{2-} + H_2O = H_3O^+ + PO_4^{3-}$	2.0×10^{-13} (K_{a3})
Sulfuric	$H_2SO_4 + H_2O = H_3O^+ + HSO_4^-$	large (K_{a1})
	$HSO_4^- + H_2O = H_3O^+ + SO_4^{2-}$	1.26×10^{-2} (K_{a2})
Zinc ion	$Zn^{2+} + H_2O = H_3O^+ + ZnOH^+$	2.5×10^{-10} (K_{a1})

Constants for anionic acids and molecular acids are denoted by K_a while the constants for cationic acids are given as K_h. The K_h notation emphasizes the common use of the term hydrolysis to describe the general reaction

$$Me^{m+} + 2H_2O = H_3O^+ + Me(OH)^{(m-1)+}$$

Source:
  Moeller, T. and O'Connor, R., *Ions in Aqueous Systems*, McGraw-Hill Book Company, New York, N.Y. (1972).

APPENDIX III

Table of ionization constants of bases at 25 °C.

Base	Equilibrium Reaction	K
Ammonia	$NH_3 + H_2O = NH_4^+ + OH^-$	1.8×10^{-5} (K_b)
Carbonate ion	$CO_3^{2-} + H_2O = HCO_3^- + OH^-$	2.1×10^{-4} (K_{h1})
	$HCO_3^- + H_2O = H_2CO_3^* + OH^-$	2.4×10^{-8} (K_{h2})
Chromate ion	$CrO_4^{2-} + H_2O = HCrO_4^- + OH^-$	3×10^{-8} (K_{h1})
Cyanide ion	$CN^- + H_2O = HCN + OH^-$	2.1×10^{-5} (K_h)
Fluoride ion	$F^- + H_2O = HF + OH^-$	1.5×10^{-11} (K_h)
Nitrate ion	$NO_3^- + H_2O = HNO_3 + OH^-$	4.0×10^{-16} (K_h)
Nitrite ion	$NO_2^- + H_2O = HNO_2 + OH^-$	2.2×10^{-11} (K_h)
Phosphate ion	$PO_4^{3-} + H_2O = HPO_4^{2-} + OH^-$	5.0×10^{-2} (K_{h1})
	$HPO_4^{2-} + H_2O = H_2PO_4^- + OH^-$	1.6×10^{-7} (K_{h2})
	$H_2PO_4^- + H_2O = H_3PO_4 + OH^-$	1.3×10^{-12} (K_{h3})
Sulfate ion	$SO_4^{2-} + H_2O = HSO_4^- + OH^-$	8.0×10^{-13} (K_{h1})
Sulfide ion	$S^{2-} + H_2O = HS^- + OH^-$	7.7×10^{-2} (K_{h1})
	$HS^- + H_2O = H_2S + OH^-$	1×10^{-7} (K_{h2})
Calcium hydroxide	$CaOH^+ = Ca^{2+} + OH^-$	3.5×10^{-2} (K_{b2})
Magnesium hydroxide	$MgOH^+ = Mg^{2+} + OH^-$	2.6×10^{-3} (K_{b2})

Source:
Moeller, T. and O'Connor, R., *Ions in Aqueous Systems*, McGraw-Hill Book Company, New York, N.Y. (1972).

APPENDIX IV

Table of solubility product constants at 25 °C.

Anion	Equilibrium Reaction	K_{sp}
Carbonates		
$MgCO_3$	$MgCO_{3(s)} = Mg^{2+} + CO_3^{2-}$	4.0×10^{-5}
$NiCO_3$	$NiCO_{3(s)} = Ni^{2+} + CO_3^{2-}$	1.4×10^{-7}
$CaCO_3$	$CaCO_{3(s)} = Ca^{2+} + CO_3^{2-}$	4.7×10^{-9}
$MnCO_3$	$MnCO_{3(s)} = Mn^{2+} + CO_3^{2-}$	4.0×10^{-10}
$CuCO_3$	$CuCO_{3(s)} = Cu^{2+} + CO_3^{2-}$	2.5×10^{-10}
$FeCO_3$	$FeCO_{3(s)} = Fe^{2+} + CO_3^{2-}$	2.0×10^{-11}
$ZnCO_3$	$ZnCO_{3(s)} = Zn^{2+} + CO_3^{2-}$	3.0×10^{-11}
$CdCO_3$	$CdCO_{3(s)} = Cd^{2+} + CO_3^{2-}$	5.2×10^{-12}
$PbCO_3$	$PbCO_{3(s)} = Pb^{2+} + CO_3^{2-}$	1.5×10^{-13}
Chromate		
$CaCrO_4$	$CaCrO_{4(s)} = Ca^{2+} + CrO_4^{2-}$	7.1×10^{-4}
$PbCrO_4$	$PbCrO_{4(s)} = Pb^{2+} + CrO_4^{2-}$	1.8×10^{-14}
Fluoride		
MgF_2	$MgF_{2(s)} = Mg^{2+} + 2F^-$	8×10^{-8}
CaF_2	$CaF_{2(s)} = Ca^{2+} + 2F^-$	1.7×10^{-10}
Hydroxide		
$Mg(OH)_2$	$Mg(OH)_{2(s)} = Mg^{2+} + 2OH^-$	8.9×10^{-12}
$Mn(OH)_2$	$Mn(OH)_{2(s)} = Mn^{2+} + 2OH^-$	2.0×10^{-13}
$Cd(OH)_2$	$Cd(OH)_{2(s)} = Cd^{2+} + 2OH^-$	2.0×10^{-14}
$Pb(OH)_2$	$Pb(OH)_{2(s)} = Pb^{2+} + 2OH^-$	4.2×10^{-15}
$Fe(OH)_2$	$Fe(OH)_{2(s)} = Fe^{2+} + 2OH^-$	1.8×10^{-15}
$Ni(OH)_2$	$Ni(OH)_{2(s)} = Ni^{2+} + 2OH^-$	1.6×10^{-16}
$Zn(OH)_2$	$Zn(OH)_{2(s)} = Zn^{2+} + 2OH^-$	4.5×10^{-17}
$Cu(OH)_2$	$Cu(OH)_{2(s)} = Cu^{2+} + 2OH^-$	1.6×10^{-19}
$Cr(OH)_3$	$Cr(OH)_{3(s)} = Cr^{3+} + 3OH^-$	6.7×10^{-31}
$Al(OH)_3$	$Al(OH)_{3(s)} = Al^{3+} + 3OH^-$	5.0×10^{-33}
$Fe(OH)_3$	$Fe(OH)_{3(s)} = Fe^{3+} + 3OH^-$	6.0×10^{-38}
Phosphate		
$MgNH_4PO_4$	$MgNH_4PO_{4(s)} = Mg^{2+} + NH_4^+ + PO_4^{3-}$	2.5×10^{-13}
$AlPO_4$	$AlPO_{4(s)} = Al^{3+} + PO_4^{3-}$	6.3×10^{-19}
$Mn_3(PO_4)_2$	$Mn_3(PO_4)_{2(s)} = 3Mn^{2+} + 2PO_4^{3-}$	1.0×10^{-22}
$Ca_3(PO_4)_2$	$Ca_3(PO_4)_{2(s)} = 3Ca^{2+} + 2PO_4^{3-}$	1.3×10^{-32}
$Mg_3(PO_4)_2$	$Mg_3(PO_4)_{2(s)} = 3Mg^{2+} + 2PO_4^{3-}$	10^{-32}
$Pb_3(PO_4)_2$	$Pb_3(PO_4)_{2(s)} = 3Pb^{2+} + 2PO_4^{3-}$	1.0×10^{-32}
Sulfate		
$CaSO_4$	$CaSO_{4(s)} = Ca^{2+} + SO_4^{2-}$	2.5×10^{-5}
$PbSO_4$	$PbSO_{4(s)} = Pb^{2+} + SO_4^{2-}$	1.3×10^{-8}

Anion	Equilibrium Reaction	K_{sp}
Sulfide		
MnS	$MnS_{(s)} = Mn^{2+} + S^{2-}$	7.0×10^{-16}
FeS	$FeS_{(s)} = Fe^{2+} + S^{2-}$	4.0×10^{-19}
NiS	$NiS_{(s)} = Ni^{2+} + S^{2-}$	3.0×10^{-21}
ZnS	$ZnS_{(s)} = Zn^{2+} + S^{2-}$	1.6×10^{-23}
CdS	$CdS_{(s)} = Cd^{2+} + S^{2-}$	1.0×10^{-28}
PbS	$PbS_{(s)} = Pb^{2+} + S^{2-}$	7.0×10^{-29}
CuS	$CuS_{(s)} = Cu^{2+} + S^{2-}$	8.0×10^{-37}
Cu_2S	$Cu_2S_{(s)} = 2Cu^{+} + S^{2-}$	1.2×10^{-49}
Fe_2S_3	$Fe_2S_{3(s)} = 2Fe^{3+} + 3S^{2-}$	1×10^{-88}

Source:
Moeller, T. and O'Connor, R., *Ions in Aqueous Systems*, McGraw-Hill Book Company, New York, N. Y. (1972).

APPENDIX V

Table of cumulative formation constants
for metal complexes with inorganic ligands.

Formula	K_1	K_2	K_3	K_4	K_5
Ammonia					
Cadmium	$10^{2.65}$	$10^{4.75}$	$10^{6.19}$	$10^{7.12}$	$10^{6.80}$
Copper (I)	$10^{5.93}$	$10^{10.86}$			
Iron (II)	$10^{1.4}$	$10^{2.2}$			
Nickel	$10^{2.80}$	$10^{5.04}$	$10^{6.77}$	$10^{7.96}$	$10^{8.71}$
Zinc	$10^{2.37}$	$10^{4.81}$	$10^{7.31}$	$10^{9.46}$	
Chloride					
Cadmium	$10^{1.95}$	$10^{2.50}$	$10^{2.60}$	$10^{2.80}$	
Iron (II)	$10^{0.36}$				
Iron (III)	$10^{1.48}$	$10^{2.13}$	$10^{1.99}$	$10^{0.01}$	
Lead	$10^{1.62}$	$10^{2.44}$	$10^{1.70}$	$10^{1.60}$	
Zinc	$10^{0.43}$	$10^{0.61}$	$10^{0.53}$	$10^{0.20}$	
Cyanide					
Cadmium	$10^{5.48}$	$10^{10.60}$	$10^{15.23}$	$10^{18.78}$	
Copper (I)		$10^{24.0}$	$10^{28.59}$	$10^{30.30}$	
Fluoride					
Iron (III)	$10^{5.28}$	$10^{9.30}$	$10^{12.06}$		
Hydroxide					
Aluminum	$10^{9.27}$			$10^{33.03}$	
Cadmium	$10^{4.17}$	$10^{8.33}$	$10^{9.02}$	$10^{8.62}$	
Chromium (III)	$10^{10.1}$	$10^{17.8}$		$10^{29.9}$	
Iron (II)	$10^{5.56}$	$10^{9.77}$	$10^{9.67}$	$10^{8.58}$	
Iron (III)	$10^{11.87}$	$10^{21.17}$	$10^{29.67}$		
Lead (II)	$10^{7.82}$	$10^{10.85}$	$10^{14.58}$		
Magnesium	$10^{2.58}$				
Nickel	$10^{4.97}$	$10^{8.55}$	$10^{11.33}$		
Zinc	$10^{14.3}$	$10^{28.3}$	$10^{41.9}$	$10^{55.3}$	

Source:
Lange's Handbook of Chemistry, Edited by John A. Dean, McGraw-Hill Book
Company, New York, N. Y., (1973).

APPENDIX VI

Table of standard reduction potentials at 25 °C.

Electrode Half-reaction	E°, Volts
$Ag^+ + e^- = Ag$	+0.799
$Ag^{2+} + 2e^- = Ag^+$	+1.98
$AgCl(s) + e^- = Ag + Cl^-$	+0.22
$Al^{3+} + 3e^- = Al$	−1.66
$AlO_2^- + 2H_2O + 3e^- = Al + 4OH^-$	−2.35
$As + 3H^+ + 3e^- = AsH_3$	−0.60
$H_3AsO_4 + 2H^+ + 2e^- = H_3AsO_3 + H_2O$	+0.56
$Au^{3+} + 3e^- = Au$	+1.50
$Ba^{2+} + 2e^- = Ba$	−2.90
$Be^{2+} + 2e^- = Be$	−1.85
$BiO^+ + 2H^+ + 3e^- = Bi + H_2O$	+0.28
$Br_2(aq) + 2e^- = 2Br^-$	+1.09
$2HBrO(aq) + 2H^+ + 2e^- = Br_2(aq) + 2H_2O$	+1.57
$2BrO^- + 2H_2O + 2e^- = Br_2(aq) + 4OH^-$	+0.43
$BrO_3^- + 6H^+ + 5e^- = \frac{1}{2}Br_2(aq) + 3H_2O$	+1.50
$BrO_4^- + 2H^+ + 2e^- = BrO_3^- + H_2O$	+1.76
$CO_2(g) + 2H^+ + 2e^- = CO(g) + H_2O$	−0.10
$C + 4H^+ + 4e^- = CH_4(g)$	−0.13
$C_2H_4(g) + 2H^+ + 2e^- = C_2H_6(g)$	+0.52
$CH_3OH(aq) + 2H^+ + 2e^- = CH_4(g) + H_2O$	−0.59
$HCHO(aq) + 2H^+ + 2e^- = CH_3OH(aq)$	−0.19
$HCOOH(aq) + 2H^+ + 2e^- = HCHO(aq) + H_2O$	−0.06
$CO_2(g) + 2H^+ + 2e^- = HCOOH(aq)$	−0.20
$2CO_2(g) + 2H^+ + 2e^- = H_2C_2O_4(aq)$	−0.49
$CNO^- + H_2O + 2e^- = CN^- + 2OH^-$	−0.97
$(CNS)_2 + 2e^- = 2CNS^-$	+0.77
$C_6H_4O_2(aq) + 2H^+ + 2e^- = C_6H_4(OH)_2(aq)$	+0.70
$Ca^{2+} + 2e^- = Ca$	−2.87
$Ca(OH)_2(s) + 2e^- = Ca + 2OH^-$	−3.03
$Cd^{2+} + 2e^- = Cd$	−0.40
$Ce^{3+} + 3e^- = Ce$	−2.33
$Ce^{4+} + e^- = Ce^{3+}$	+1.49
$Cl_2(g) + 2e^- = 2Cl^-$	+1.360
$HClO + H^+ + e^- = \frac{1}{2}Cl_2 + H_2O$	+1.64
$ClO_3^- + 6H^+ + 5e^- = \frac{1}{2}Cl_2 + 3H_2O$	+1.47
$ClO_4^- + 2H^+ + 2e^- = ClO_3^- + H_2O$	+1.19
$Co^{2+} + 2e^- = Co$	−0.277
$Co^{3+} + e^- = Co^{2+}$	+1.82
$Co(OH)_3 + e^- = Co(OH)_2 + OH^-$	+0.17
$Cr^{2+} + 2e^- = Cr$	−0.91
$Cr^{3+} + 3e^- = Cr$	−0.74
$Cr^{3+} + e^- = Cr^{2+}$	−0.41

Elements are in the standard state unless otherwise indicated while ions are at unit activity in aqueous solution.

Electrode Half-reaction	$E°$, *Volts*
$\frac{1}{2}Cr_2O_7^{2-} + 7H^+ + 3e^- = Cr^{3+} + 7/2H_2O$	$+1.33$
$Cs^+ + e^- = Cs$	-2.92
$Cu^+ + e^- = Cu$	$+0.521$
$\frac{1}{2}Cu_2O + \frac{1}{2}H_2O + e^- = Cu + OH^-$	-0.358
$Cu^{2+} + 2e^- = Cu$	$+0.337$
$Cu^{2+} + e^- = Cu^+$	$+0.15$
$Cu(OH)_2 + e^- = \frac{1}{2}Cu_2O + OH^- + \frac{1}{2}H_2O$	-0.080
$Cu^{2+} + I^- + e^- = CuI$	$+0.86$
$2D^+ + 2e^- = D_2$	-0.003
$F_2 + 2e^- = 2F^-$	$+2.87$
$F_2 + 2H^+ + 2e^- = 2HF(aq)$	$+3.06$
$F_2O + 2H^+ + 4e^- = H_2O + F^-$	$+2.15$
$Fe^{2+} + 2e^- = Fe$	-0.440
$Fe^{3+} + e^- = Fe^{2+}$	$+0.771$
$Fe(CN)_6^{3-} + e^- = Fe(CN)_6^{4-}$	$+0.36$
$FeO_4^{2-} + 8H^+ + 3e^- = Fe^{3+} + 4H_2O$	$+2.2$
$FeO_4^{2-} + 2H_2O + 3e^- = FeO_2^- + 4OH^-$	$+0.9$
$2H^+ + 2e^- = H_2$	0.000
$2H_2O + 2e^- = H_2 + 2OH^-$	-0.828
$H_2 + 2e^- = 2H^-$	-2.25
$\frac{1}{2}Hg_2^{2+} + e^- = Hg$	$+0.789$
$Hg^{2+} + 2e^- = Hg$	$+0.854$
$Hg^{2+} + e^- = \frac{1}{2}Hg_2^{2+}$	$+0.920$
$\frac{1}{2}Hg_2Cl_2(s) + e^- = Hg + Cl^-$	$+0.27$
$I_2(aq \text{ or as } I_3^-) + 2e^- = 2I^-$	$+0.54$
$IO_3^- + 6H^+ + 5e^- = \frac{1}{2}I_2 + 3H_2O$	$+1.20$
$K^+ + e^- = K$	-2.92
$Li^+ + e^- = Li$	-3.03
$Mg^{2+} + 2e^- = Mg$	-2.37
$Mg(OH)_2 + 2e^- = Mg + 2OH^-$	-2.69
$Mn^{2+} + 2e^- = Mn$	-1.18
$MnO_4^- + 8H^+ + 5e^- = Mn^{2+} + 4H_2O$	$+1.51$
$N_2H_5^+ + 3H^+ + 2e^- = 2NH_4^+$	$+1.27$
$N_2H_4 + 2H_2O + 2e^- = 2NH_3(aq) + 2OH^-$	$+0.1$
$N_2 + 5H^+ + 4e^- = N_2H_5^+$	-0.23
$HN_3(aq) + 3H^+ + 2e^- = NH_4^+ + N_2$	$+1.96$
$HN_3(aq) + 11H^+ + 8e^- = 3NH_4^+$	$+0.69$
$\frac{1}{2}N_2 + H_2O + 2H^+ + e^- = NH_3OH^+$	-1.89
$\frac{3}{2}N_2 + H^+ + e^- = HN_3$	-3.40
$\frac{1}{2}N_2 + 4H^+ + 3e^- = NH_4^+$	$+0.27$
$2HNO_2 + 4H^+ + 4e^- = N_2O = 3H_2O$	$+1.29$
$HNO_2 + H^+ + e^- = NO + H_2O$	$+1.00$
$NO_2^- + H^+ + e^- = HNO_2$	$+1.07$
$NO_3^- + 2H^+ + e^- = NO_2^- + H_2O$	$+0.81$
$NO_3^- + 3H^+ + 2e^- = HNO_2 + H_2O$	$+0.94$
$NO_3^- + H_2O + 2e^- = NO_2^- + 2OH^-$	$+0.01$
$NO_3^- + 4H^+ + 3e^- = NO + 2H_2O$	$+0.96$
$NO_3^- + 6H^+ + 5e^- = \frac{1}{2}N_2 + 3H_2O$	$+1.24$
$NO_3^- + 6H_2O + 8e^- = NH_3(aq) + 90H^-$	-0.13
$Na^+ + e^- = Na$	-2.71

Electrode Half-reaction	E°, Volts
$Ni^{2+} + 2e^- = Ni$	-0.250
$NiO_2 + 2H_2O + 2e^- = Ni(OH)_2 + 2OH^-$	$+0.49$
$H_2O_2(aq) + 2H^+ + 2e^- = 2H_2O$	$+1.77$
$O_2 + 4H^+ + 4e^- = 2H_2O$	$+1.229$
$O_2 + 2H_2O + 4e^- = 4OH^-$	$+0.401$
$O_2 + 2H^+ + 2e^- = H_2O_2$	$+0.68$
$O_3 + 2H^+ + 2e^- = O_2 + H_2O$	$+2.07$
$P + 3H_2O + 3e^- = PH_3 + 3OH^-$	-0.89
$H_2PO_2^- + e^- = P + 2OH^-$	-2.05
$PO_4^{3-} + 2H_2O + 2e^- = HPO_3^{2-} + 3OH^-$	-1.12
$H_3PO_4 + 2H^+ + 2e^- = H_3PO_3 + H_2O$	-0.28
$Pb^{2+} + 2e^- = Pb$	-0.126
$Pb^{4+} + 2e^- = Pb^{2+}$	$+1.69$
$PbO_2 + 4H^+ + 2e^- = Pb^{2+} + 2H_2O$	$+1.46$
$PbO_2 + H_2O + 2e^- = PbO + 2OH^-$	$+0.28$
$PbO_2 + 4H^+ + SO_4^{2-} + 2e^- = PbSO_4(s) + 2H_2O$	$+1.685$
$Ra^{2+} + 2e^- = Ra$	-2.92
$Rb^+ + e^- = Rb$	-2.92
$S + 2e^- = S^{2-}$	-0.48
$S + 2H^+ + 2e^- = H_2S$	$+0.14$
$2H_2SO_3 + 2H^+ + 4e^- = S_2O_3^{2-} + 3H_2O$	$+0.40$
$2SO_3^{2-} + 3H_2O + 4e^- = S_2O_3^{2-} + 6OH^-$	-0.58
$H_2SO_3 + 4H^+ + 4e^- = S + 3H_2O$	$+0.47$
$SO_4^{2-} + 4H^+ + 2e^- = H_2SO_3 + H_2O$	$+0.17$
$SO_4^{2-} + H_2O + 2e^- = SO_3^{2-} + 2OH^-$	-0.93
$S_4O_6^{2-} + 2e^- = 2S_2O_3^{2-}$	$+0.09$
$S_2O_8^{2-} + 2e^- = 2SO_4^{2-}$	$+2.01$
$Sb + 3H^+ + 3e^- = SbH_3(g)$	-0.51
$SbO^+ + 2H^+ + 3e^- = Sb + H_2O$	$+0.21$
$Sb_2O_5 + 6H^+ + 4e^- = 2SbO^+ + 3H_2O$	$+0.58$
$Se + 2H^+ + 2e^- = H_2Se(g)$	-0.40
$Se + 2e^- = Se^{2-}$	-0.92
$SiO_3^{2-} + 3H_2O + 4e^- = Si + 6OH^-$	-1.70
$Sn^{2+} + 2e^- = Sn$	-0.14
$Sn^{4+} + 2e^- = Sn^{2+}$	$+0.15$
$Sn(OH)_6^{2-} + 2e^- = SnO(OH)^- + H_2O + 3OH^-$	-0.90
$Sr^{2+} + 2e^- = Sr$	-2.89
$Te + 2H^+ + 2e^- = H_2Te(g)$	-0.72
$Te + 2e^- = Te^{2-}$	-1.14
$TeO_2 + 4H^+ + 4e^- = Te + 2H_2O$	$+0.59$
$Th^{4+} + 4e^- = Th$	-1.90
$Ti^{2+} + 2e^- = Ti$	-1.63
$TiO^{2+} + 2H^+ + e^- = Ti^{3+} + H_2O$	$+0.1$
$TiO^{2+} + 2H^+ + 4e^- = Ti + H_2O$	-0.89
$Tl^{3+} + 2e^- = Tl^+$	$+1.25$
$U^{3+} + 3e^- = U$	-1.80
$U^{4+} + e^- = U^{3+}$	-0.61
$UO_2^{2+} + 4H^+ + 2e^- = U^{4+} + 2H_2O$	$+0.62$
$V^{2+} + 2e^- = V$	-1.2
$V^{3+} + e^- = V^{2+}$	-0.26

Electrode Half-reaction	$E°$, Volts
$VO^{2+} + 2H^+ + e^- = V^{3+} + H_2O$	$+0.34$
$VO_2^+ + 2H^+ + e^- = VO^{2+} + H_2O$	$+1.00$
$XeO_3(g) + 6H^+ + 6e^- = Xe + 3H_2O$	$+1.8$
$H_4XeO_6 + 2H^+ + 2e^- = XeO_3 = 3H_2O$	$+2.3$
$Zn^{2+} + 2e^- = Zn$	-0.763

Source:
Selley, N. J., *Experimental Approach to Electrochemistry*, John Wiley and Sons, New York, N.Y. (1977).

APPENDIX VII

Table of common chemicals used to treat water.

Chemical	Common Name	Typical Specs	Equiv. Weight	Bulk Density lb/cu ft or lb/gal	Approx. pH 1% Solution	Solubility
Aluminum sulfate $Al_2(SO_4)_3 \cdot 14H_2O$	Alum	Lump—17% Al_2O_3 Liquid—8.5% Al_2O_3	100[1]	60 11	3.4	4.2 lb/gal @ 60°F
Bentonitic clay	Bentonite	—	—	60	—	Insoluble
Calcium carbonate $CaCO_3$	Limestone	96% $CaCO_3$	50	80	9	Insoluble
Calcium hydroxide $Ca(OH)_2$	Hydrated lime Slaked lime	96% $Ca(OH)_2$	40[1]	40	12	Insoluble
Calcium hypochlorite $Ca(OCl)_2 \cdot 4H_2O$	HTH	70% Cl_2	103	55	6–8	3% @ 60°F
Calcium oxide CaO	Burned lime, Quicklime	96% CaO	30[1]	60	12	Slake @ 10–20%
Calcium sulfate $CaSO_4 \cdot 2H_2O$	Gypsum	98% Gypsum	86[1]	55	5–6	Insoluble
Chlorine (Cl_2)	Chlorine	Gas—99.8% Cl_2	35.5	gas	—	0.07 lb/gal @ 60°F
Copper sulfate $CuSO_4 \cdot 5H_2O$	Blue vitriol	98% Pure	121[1]	75	5–6	2 lb/gal @ 60°F
Dolomitic lime $Ca(OH)_2 \cdot MgO$	Dolomitic lime	36–40% MgO	67[2]	40	12.4	Insoluble
Ferric chloride $FeCl_3 \cdot 6H_2O$	Iron chloride	Lump—20% Fe Liquid—20% Fe	91[1]	70 13	3–4	45% @ 60°F
Ferric sulfate $Fe_2(SO_4)_3 \cdot 3H_2O$	Iron sulfate	18.5% Fe	51.5[1]	70	3–4	30% @ 60°F

Table of common chemicals used to treat water (Cont'd)

Chemical	Common Name	Typical Specs	Equiv. Weight	Bulk Density lb/cu ft or lb/gal	Approx. pH 1% Solution	Solubility
Ferrous sulfate $FeSO_4 \cdot 7H_2O$	Copperas	20% Fe	139[1]	70	3–4	1 lb/gal @ 60°F
Hydrochloric acid HCl	Muriatic acid	30% HCl 20° Baume	120[1]	9.6	1–2	35% @ 60°F
Sodium aluminate $NaAlO_2$	Aluminate	Flake—46% Al_2O_3 Liquid—26% Al_2O_3	100[1]	50 13	11–12	40% @ 60°F
Sodium chloride $NaCl$	Rock salt, Salt	98% Pure	58.5	60	6–8	2.6 lb/gal @ 60°F
Sodium carbonate Na_2CO_3	Soda ash	98% Pure 58% Na_2O	53	60	11	1.5 lb/gal @ 60°F
Sodium hydroxide $NaOH$	Caustic, Lye	Flake—99% NaOH Liquid—50–70%	40	65 12	12.8	70% @ 60°F
Sodium phosphate Na_2HPO_4	Disodium phosphate	49% P_2O_5	47.3	55	9	20% @ 60°F
Sodium metaphosphate $NaPO_3$	Hexameta-phosphate	66% P_2O_5	34	47	5–6	1 lb/gal @ 60°F
Sulfuric acid	Oil of vitriol	94–96% 66° Baume	50[1]	15	1–2	Infinite

(1) Effective equivalent weight of commercial product.
(2) Effective equivalent weight based on $Ca(OH)_2$ content.

Source:
Water: The Universal Solvent, Edited by Frank N. Kemmer, Nalco Chemical Company, Oak Brook, IL (1977).

APPENDIX VIII

International atomic weights.

Name	Symbol	Atomic number	Atomic weight	Name	Symbol	Atomic number	Atomic weight
Actinium	Ac	89	—	Mercury	Hg	80	200.59
Aluminum	Al	13	26.9815	Molybdenum	Mo	42	95.94
Americium	Am	95	—	Neodynium	Nd	60	144.24
Antimony	Sb	51	121.75	Neon	Ne	10	20.183
Argon	Ar	18	39.948	Neptunium	Np	93	—
Arsenic	As	33	74.9216	Nickel	Ni	28	58.71
Astatine	At	85	—	Niobium	Nb	41	92.906
Barium	Ba	56	137.34	Nitrogen	N	7	14.0067
Berkelium	Bk	97	—	Nobelium	No	102	—
Berylium	Be	4	9.0122	Osmium	Os	76	190.2
Bismuth	Bi	83	208.980	Oxygen	O	8	15.9994
Boron	B	5	10.811	Palladium	Pd	46	106.4
Bromine	Br	35	79.904	Phosphorus	P	15	30.9738
Cadmium	Cd	48	112.40	Platinum	Pt	78	195.09
Calcium	Ca	20	40.08	Plutonium	Pu	94	—
Californium	Cf	98	—	Polonium	Po	84	—
Carbon	C	6	12.01115	Potassium	K	19	39.102
Cerium	Ce	58	140.12	Praseodymium	Pr	59	140.907
Cesium	Cs	55	132.905	Promethium	Pm	61	—
Chlorine	Cl	17	35.453	Protactinium	Pa	91	—
Chromium	Cr	24	51.996	Radium	Ra	88	—
Cobalt	Co	27	58.9332	Radon	Rn	86	—
Copper	Cu	29	63.546	Rhenium	Re	75	186.2
Curium	Cm	96	—	Rhodium	Rh	45	102.905
Dysprosium	Dy	66	162.50	Rubidium	Rb	37	85.47
Einsteinium	Es	99	—	Ruthenium	Ru	44	101.07
Erbium	Er	68	167.26	Samarium	Sm	62	150.35
Europium	Eu	63	151.96	Scandium	Sc	21	44.956
Fermium	Fm	100	—	Selenium	Se	34	78.96
Flourine	F	9	18.9984	Silicon	Si	14	28.086
Francium	Fr	87	—	Silver	Ag	47	107.868
Gadolinium	Gd	64	157.25	Sodium	Na	11	22.9898
Gallium	Ga	31	69.72	Strontium	Sr	38	87.62
Germanium	Ge	32	72.59	Sulfur	S	16	32.064
Gold	Au	79	196.967	Tantalum	Ta	73	189.948
Hafnium	Hf	72	178.49	Technetium	Tc	43	—
Helium	He	2	4.0026	Tellurium	Te	52	127.60
Holmium	Ho	67	164.930	Terbium	Tb	65	158.924
Hydrogen	H	1	1.00797	Thallium	Tl	81	204.37
Indium	In	49	114.82	Thorium	Th	90	232.038
Iodine	I	53	126.9044	Thulium	Tm	69	168.934
Iridium	Ir	77	192.2	Tin	Sn	50	118.69
Iron	Fe	26	55.847	Titanium	Ti	22	47.90
Krypton	Kr	36	83.80	Tungsten	W	74	183.85
Lanthanum	La	57	138.91	Uranium	U	92	238.03
Lead	Pb	82	207.19	Vanadium	V	23	50.942
Lithium	Li	3	6.939	Xenon	Xe	54	131.30
Lutetium	Lu	71	174.97	Ytterbium	Yb	70	173.04
Magnesium	Mg	12	24.312	Yttrium	Y	39	88.905
Manganese	Mn	25	54.9380	Zinc	Zn	30	65.37
Mendelevium	Md	101	—	Zirconium	Zr	40	91.22

APPENDIX IX

Useful regeneration data.

A. SULFURIC ACID

% H_2SO_4	Grams H_2SO_4/Liter	Normality	Specific Gravity	Pounds per U.S. Gallon
1	10.05	0.205	1.0051	0.08388
2	20.24	0.413	1.0118	0.1689
3	30.55	0.623	1.0184	0.2550
4	41.00	0.836	1.0250	0.3422
5	51.59	1.05	1.0317	0.4305
6	62.31	1.27	1.0385	0.5200
8	84.18	1.72	1.0522	0.7025
10	106.6	2.17	1.0661	0.8897
12	129.6	2.64	1.0802	1.082
15	165.3	3.37	1.1020	1.379
20	227.9	4.65	1.1394	1.902
50	697.6	14.2	1.3951	5.821
96	1762.0	35.9	1.8355	14.71
100	1831.0	37.3	1.8305	15.28

B. HYDROCHLORIC ACID

% HCl	Grams HCl/Liter	Normality	Specific Gravity	Pounds per U.S. Gallon
1	10.03	0.275	1.0032	0.08372
2	20.16	0.553	1.0082	0.1683
4	40.72	1.12	1.0181	0.3399
6	61.67	1.69	1.0279	0.5147
8	83.01	2.28	1.0376	0.6927
10	104.7	2.87	1.0474	0.8741
12	126.9	3.48	1.0574	1.059
16	172.4	4.72	1.0776	1.439
20	219.6	6.02	1.0980	1.833
30	344.8	9.46	1.1492	2.877
40	479.2	13.1	1.1980	3.999

C. SODIUM HYDROXIDE

% NaOH	Grams NaOH/Liter	Normality	Specific Gravity	Pounds per U.S. Gallon
1	10.10	0.262	1.0095	0.08425
2	20.41	0.511	1.0207	0.1704
3	30.95	0.774	1.0318	0.2583
4	41.71	1.04	1.0428	0.3481
5	52.69	1.32	1.0538	0.4397
6	63.89	1.60	1.0648	0.5332
8	86.95	2.17	1.0869	0.7256
10	110.9	2.77	1.1089	0.9254
12	135.7	3.39	1.1309	1.333
16	188.0	4.70	1.1751	1.569
20	243.8	6.10	1.2191	2.035
50	762.7	19.1	1.5253	6.365

D. AMMONIA

% NH$_3$	Grams NH$_3$/Liter	Normality	Specific Gravity	Pounds per U.S. Gallon
1	9.939	0.583	0.9939	0.08294
2	19.79	1.16	0.9895	0.1652
4	39.24	2.31	0.9811	0.3275
6	58.38	3.43	0.9730	0.4872
8	77.21	4.53	0.9651	0.6443
10	95.75	5.62	0.9575	0.7991
12	114.0	6.70	0.9501	0.9515
16	149.8	8.79	0.9362	1.250
20	184.6	10.8	0.9229	1.540
30	267.6	17.0	0.8920	2.233

E. SODIUM CARBONATE

% Na$_2$CO$_3$	Grams Na$_2$CO$_3$/Liter	Normality	Specific Gravity	Pounds per U.S. Gallon
1	10.09	0.190	1.0086	0.08417
2	20.38	0.384	1.0190	0.1701
4	41.59	0.786	1.0398	0.3471
6	63.64	1.20	1.0606	0.5311
8	86.53	1.63	1.0816	0.7221
10	110.3	2.08	1.1029	0.9204
12	134.9	2.54	1.1244	1.126
14	160.5	3.03	1.1463	1.339

Useful regeneration data (Cont'd).

F. Sodium Chloride

% NaCl	Grams NaCl/*Liter*	Specific Gravity	Pounds per U.S. Gallon
1	10.05	1.0053	0.08390
2	20.25	1.0125	0.1690
4	41.07	1.0268	0.3428
6	62.48	1.0413	0.5214
8	84.47	1.0559	0.7050
10	107.1	1.0707	0.8935
12	130.3	1.0857	1.087
16	178.6	1.1162	1.490
20	229.6	1.1478	1.916
26	311.3	1.1972	2.598

Source:
  Duolite Ion Exchange Manual, Diamond Shamrock Chemical Company, Redwood City, CA (1969).

INDEX

INDEX

design criteria (*Table*), 235
mechanisms, 232
mixing:
    compressed air, 234
    mechanical, 235
power input equation, 234
theory, 232-35
velocity gradients, 233, 235
Fluoride
    fluorosis, 405
    industrial sources, 405
    maximum contaminent level (MCL), 411
    occurrence, 405
    removal (*see* Defluoridation)
    solubility, 406-8, 418-19
Free chlorine species, 450
Free energy (*see* Chemical thermodynamics)
Freudlich adsorption isotherm, 204-6, 209, 369-72

G

Gibbs free energy (*see* Chemical thermodynamics)
Güntelberg relationship, 27

H

Hardness:
    classification of waters, 267-68
    forms, 268, 269
    removal (*see* Water softening)
Hazen-Williams coefficient, 264
Heavy metal precipitation, 137-39
Henderson-Hasselbach equation, 44
Hydrogen:
    electrode, 145
    as a reducing agent, 150

Hydrolysis, 12, 119, 220
Hypochlorous acid, 30, 450

I

Ion exchange:
    applications, 307
    capacity, 315-19
    chromium removal, 345-55
    degree of column utilization, 335-37
    diffusion constants, 329-31
    equilibrium, 207-12
    exchange coefficients, 330
    fluoride removal, 409-10
    isochrones, 332
    isoplanes, 331
    nitrogen removal, 319-20, 355-62
    regeneration efficiency, 337
    resin:
        moisture retention, 322-23
        types, 312-14
    salt splitting, 312
    system operation, 323-29
    theoretical plates, 334
    theory, 329-37
Ionization:
    constants for acids and bases, 486-87
    definition, 13
    fraction, carbonic acid, 84
    of water, 43
Ionic strength:
    calculation, 23
    effect on equilibrium constant, 28
    relation to dissolved solids, 24
Ion pairs, 118
Ion ratio method, 133
Iron:
    in drinking water, 457
    hydroxo complexes, 120, 123, 459
    kinetics of oxidation, 467-69
    oxidation, 465-67
    removal, 476
    solubility of hydroxide, 123-25
    stability in water, 152-58